D0218511

Space Technology Series

Editorial Board

Understanding Space
An Introduction to Astronautics

Jerry Jon Sellers

With Contributions by
William J. Astore
K. Stephen Crumpton
Chris Elliot
Robert B. Giffen

Editor
Wiley J. Larson
United States Air Force Academy

Text design by
Anita Shute

Illustrations by
Dale Gay

McGraw-Hill, Inc.
College Custom Series

New York   St. Louis   San Francisco   Auckland   Bogotá
Caracas   Lisbon   London   Madrid   Mexico   Milan   Montreal
New Delhi   Paris   San Juan   Singapore   Sydney   Tokyo   Toronto

This book is dedicated to the thousands of people
who've devoted their lives to exploring space.

UNDERSTANDING SPACE
An Introduction to Astronautics

3 4 5 6 7 8 9 0 DOW DOW 9 0 9 8 7

ISBN 0-07-057027-5

Custom Publisher: Margaret A. Hollander
Editor: Wiley J. Larson
Cover Design: Dale Gay
Text Design: Anita Shute
Printer/Binder: R.R. Donnelly & Sons

Understanding Space
An Introduction to Astronautics

vii

Preface

Space travel and exploration are exciting topics; yet, many people shy away from them because they seem complex. The study of astronautics and space missions can be difficult at times, but our goal in this book is to *bring space down to Earth*. If we're successful—and you'll be the judge—after studying this book you should understand the concepts and principles of space flight, space vehicles, launch systems, and space operations.

We want to help you understand space missions while developing enthusiasm and curiosity about this very exciting topic. We've been inspired by the thousands of people who've explored space— from the people who've studied and documented the heavens to the people who've given their lives flying in them. We hope to inspire *you*. Whether you're interested in engineering, business, politics, or teaching—*you* can make a difference. We need talented people to lead the way to exploring space, the stars, and galaxies, and *you* are our hope for the future.

This book is intended for use in a first course in astronautics. It will also help you if you're a professional or educator interested in space. If you're a junior or senior in high school and have a strong background in physics and math, come on in— you'll do fine.

If you're scared of equations—don't worry! The book is laid out so you can learn the necessary concepts from the text *without* having to read or manipulate the equations. The equations are for those of you who want to be more fully grounded in the basics of astronautics.

We've included helpful features in this book to make it easier for you to use. The first page of each chapter contains

- ☞ An outline of the chapter, so you know what's coming

- ☞ An "In This Chapter You'll Learn To . . ." box that tells you what you should learn in the chapter

- ☞ A "You Should Already Know . . ." box, so you can review the material you'll need to understand the chapter

Each section of the chapter contains

- ☞ An "In This Section You'll Learn To . . ." box that gives you learning outcomes
- ☞ A detailed section review which summarizes key concepts and lists key terms and equations

Within each chapter you'll find

- ☞ Key terms *italicized* and defined
- ☞ Diagrams and pictures "worth a thousand words"
- ☞ Tables summarizing important information and concepts
- ☞ Key equations boxed
- ☞ Detailed, step-by-step solutions to real-world example problems
- ☞ "Astro Fun Facts" to provide interesting insights and space trivia

At the end of each chapter you'll find

- ☞ A list of references for further study
- ☞ Problems and discussion questions, so you can practice what you've read. Astronautics is not a spectator sport—the real learning happens when you actually *do* what you've studied.
- ☞ Mission profiles designed to give you insight on specific programs and a starting point for discussion

We hope these features help you learn how exciting astronautics can be!

Leadership, funding, and support essential to developing this book were provided by the Space and Missile System Center, Air Force Space Command, Naval Research Laboratory, Office of Naval Research, Army Laboratory Command, Space Defense Initiative Office, Goddard Space Flight Center, and the Lewis Research Center. The Air Force Phillips Laboratory played a key role in making this book a success. Getting money to develop much needed reference material is exceptionally difficult in the aerospace community. We are deeply indebted to the sponsoring organizations for their support and their recognition of the importance of projects such as this one.

This book is the result of several years of effort by an international team
government, industry, and academic professionals. The Department
Astronautics, United States Air Force Academy, provided unwavering support f
the project. Robert B. Giffen, the Department Head, furnished the tim
encouragement, and resources necessary to complete this draft. David J. Clou
Jackson Ferguson, and Peter D. McQuade did detailed and useful technic
reviews of the material and provided many key insights.

Many faculty members, staff, external reviewers, and especially studen
graciously sacrificed their time to review and comment on this book. T
Academy's graphics division was helpful and responsive to our deadlines. We
especially like to thank our illustrator, Dale Gay, for his creative ideas and Debl
Porter for her help and support. We'd also like to thank the NASA Public Affa
Offices at Johnson Space Center in Houston, Texas, and at the Jet Propulsi
Laboratory in Pasadena, California, for their help with photographs.

The contributing authors—Bill Astore, Steve Crumpton, Chris Elliot, and E
Giffen—were very easy to work with. Their names are on the chapters th
contributed. Special thanks go to Steve Crumpton who devoted many hours in
early stages of the project helping to establish the outline and initial chapte
McGraw-Hill was exceptionally helpful during the development. We'd especia
like to thank our publisher, Margaret Hollander, and Rose Arlia for their patier
and guidance.

We also owe thanks to Perry Luckett for his pain-staking review of all drafts. I
attention to detail improved our organization, kept us out of grammatical pit-fa
and sharpened the clarity and consistency of the entire book. The O.
Corporation in Colorado Springs, Colorado, provided exceptional conti
support for the project. Ed Warrell and Sandy Welsh were particularly help
throughout the development period. James Qualey did a thorough techn
review of the entire book.

We owe special thanks to Anita Shute for literally making this book happen.
took our crude, often illegible drafts and sketches and created the product yo
be reading. Her creative ideas and talent are surpassed only by her hard wor

We sincerely hope this book will be useful to you in your study of astronautics. We've made every effort to eliminate mathematical and factual errors, but some may have slipped by us. Please send any errors, omissions, corrections, or comments to us, so we can incorporate them in the next edition of the book. Good luck and aim for the stars!

December, 1993

Wiley J. Larson
Managing Editor

Jerry Jon Sellers
Author and Editor

Department of Astronautics
United States Air Force Academy
Colorado Springs, Colorado 80840
Phone: 719-472-4110   FAX: 719-472-3723

From a cosmic perspective, the Earth's a very small place. This view of another galaxy gives some idea where Earth would be if we could look at the Milky Way from the outside. *(Courtesy of NASA)*

Why Space?

1

■ In This Chapter You'll Learn To...

- ☞ Describe the unique advantages of space and some of the missions that capitalize on them
- ☞ Identify the elements that make up a space mission

■ You Should Already Know...

- ❏ Nothing about space yet. That's why we're here!

Space. The Final Frontier. These are the voyages of the Starship Enterprise. Its continuing mission—to explore strange new worlds, to seek out new life and new civilizations, to boldly go where no one has gone before!

Star Trek—The Next Generation

Why study space? Why should you invest the time and effort needed to understand the basics of planetary and satellite motion, rocket propulsion, and spacecraft design—this vast area of knowledge we call astronautics? The reasons are both poetic and practical.

The poetic reasons are embodied in the quotation at the beginning of this chapter. Trying to understand the mysterious beauty of the universe, "to boldly go where no one has gone before," has always been a fundamental human urge. Gazing into the sky on a starry night, you can share an experience common to the entire history of humankind. As you ponder the fuzzy swirl of the Milky Way and the brighter shine and odd motion of the planets, you can almost feel a bond with ancient shepherds who looked at the same sky and pondered the same questions thousands of years ago.

The changing yet predictable face of the night sky has always inspired our imagination and caused us to ask questions about something greater than ourselves. This quest for an understanding of space has ultimately given us greater control over our destiny on Earth. Early star gazers learned to construct calendars, enabling them to predict spring flooding and decide when to plant and harvest. Today, using space enables us to communicate and manage resources on a global scale. To study space, then, is to grapple with questions as old as humanity. Understanding generally how the complex mechanisms of the universe work gives us a greater appreciation for the graceful and poetic beauty of the cosmos.

The practical reasons for studying space are much more down to Earth, but you can see them when you gaze at the night sky on a clear night. The intent sky watcher can witness a sight that only the last generation of humans has been able to see—tiny points of light streaking across the background of stars. They move too fast to be stars or planets. They don't brighten and die out like meteors or "falling stars." This now common sight would have startled and terrified ancient star gazers, for they're not the work of gods but the work of people. They are satellites. We see sunlight glinting off their shiny surfaces as they patiently circle the Earth.

Since the dawn of the Space Age only a few decades ago, we have come to rely more and more on satellites for a variety of needs. Daily weather forecasts, instantaneous world-wide communication, and a constant ability to keep an eye on not-so-friendly neighbors are all examples of space technology that we've come to take for granted. Studying space offers us a chance to understand and appreciate the complex requirements of this technology.

Throughout this book, we'll focus primarily on the practical aspects of space—What's it like? And how do we get there? But the poetic aspects of space exploration should remind you of our links with the past and our journey to the future. In this chapter we'll lay the groundwork for our study of space. We'll look at the "big picture" to see how space affects our lives and what goes into mounting a mission into the final frontier.

1.1 Space In Our Lives

▬ In This Section You'll Learn To...

- ☞ List and discuss the advantages offered by space and the unique space environment
- ☞ Describe current space missions

The Space Imperative

Getting into space is dangerous and expensive. So why bother? Space offers several compelling advantages for modern society

- A global perspective—the ultimate high ground
- A clear view of the heavens—unobscured by the atmosphere
- A free-fall environment—enabling the development of advanced materials impossible to make on Earth
- Abundant resources—such as solar energy and extraterrestrial materials
- A unique challenge as the final frontier

Let's explore each of these advantages in turn to see their potential benefit to Earth.

Space offers a global perspective. As you can see in Figure 1-1, the higher you are, the more you can see. For thousands of years, kings and rulers took advantage of this fact by putting lookout posts atop the tallest mountains to survey more of their realm and fend off would-be attackers. Throughout history, many battles have been fought to "take the high ground." Space takes this quest for greater perspective to its ultimate end. From the vantage point of space, we can view large parts of the Earth's surface. Orbiting satellites can thus serve as "eyes in the sky" to provide a variety of useful services.

Space offers a clear view of the heavens. When we look at stars in the night sky, we see their characteristic twinkle. This twinkle, caused by the blurring of "starlight" as it passes through the atmosphere, is known as *scintillation.* Not only is the light blurred, but some of it is blocked or attenuated altogether. This attenuation is frustrating for astronomers who need access to all the regions of the spectrum to fully explore the universe. By placing observatories in space, we can get above the atmosphere and gain an unobscured view of the universe, as depicted in Figure 1-2. The Hubble Space Telescope and the Gamma Ray Observatory are armed with sensors operating far beyond the range of human senses. Already, results from these instruments are revolution-izing our understanding of the cosmos.

Figure 1-1. A Global Perspective. Space is the ultimate high ground; it allows us to view large parts of the Earth at once for various applications.

Figure 1-2. Space Astronomy. The Earth's atmosphere obscures our view of space, so we put satellites above the atmosphere to see better.

Earth's gravity makes some manufacturing processes difficult if not impossible. To form certain new metal alloys, for example, we must blend two or more metals in just the right proportion. Unfortunately, gravity tends to pull the heavier metal to the bottom, making a uniform mixture difficult to obtain. But space offers the solution. A manufacturing plant in orbit is literally falling toward Earth but never hitting it. This is a condition known as free-fall (NOT zero gravity, as we'll see later). In *free-fall* there are no contact forces on an object, so it is said to be weightless. We'll explore this concept in greater detail in Chapter 3. Unencumbered by the weight felt on the Earth's surface, factories in orbit can create exotic new metals for computers or other advanced technologies, as well as for promising new pharmaceutical products to battle disease on Earth.

Space offers abundant resources. While some on Earth argue about how to carve the pie of Earth's resources into smaller and smaller pieces, others have argued that we need only bake a bigger pie. The bounty of the solar system offers an untapped reserve of minerals and energy to sustain the spread of mankind beyond the cradle of Earth. Spacecraft now use only one of these abundant resources—limitless solar energy. But scientists have speculated that lunar resources, or even those from the asteroids, could be used to fuel a growing space-based economy. Lunar soil, for example, is known to be rich in oxygen and aluminum. This oxygen could be used as rocket propellant and for humans to breathe. Aluminum is an important metal for various industrial uses. These resources, coupled with the human drive to explore, mean the sky is truly the limit!

Finally, space offers an advantage simply as a frontier. The human condition has always improved as new frontiers were challenged. As a stimulus for increased technological advance and a crucible for creating

Astro Fun Fact
Shot Towers

In the mid sixteenth century, Italian weapon makers developed a secret method to manufacture lead shot for use in muskets. Finding that gravity tended to misshape the shots when traditionally cast, the Italians devised a system that employed principles of free-fall. In this process, molten lead was dropped through a tiny opening at a height of about 100 m (300 ft.) from a "shot tower." As the molten lead plummeted, it cooled into a near perfect sphere. At journey's end, the lead fell into a pool of cold water where it quickly hardened. As time passed, shot towers became common throughout Europe and the United States. More cost-effective and advanced methods have now replaced them.

Burrard, Sir Major Gerald. The Modern Shotgun Volume II: The Cartridge. London: Herbert Jenkins Ltd., 1955.

Deane. Deane's Manual of Fire Arms. London: Longman, Brown, Green, Longmans & Robers, 1858.

Contributed by Troy Kitch, United States Air Force Academy

greater economic expansion, space offers a limitless challenge that compels our attention. Many have compared the challenges of space to those faced by the first explorers to the New World. As humans crossed the narrow bridge between what is now Alaska and Siberia, they moved onto a vast new continent. Later, European settlers began to explore the apparently limitless resources of the New World.

We're still a long way from placing colonies on the Moon or Mars. But already the lure of this final frontier has affected us. Audiences spend millions of dollars each year on inspiring movies like *Star Wars* and *Star Trek*. The Apollo Moon landings and scores of Space Shuttle flights have captured the wonder and imagination of people across the planet. Future missions promise to be even more captivating as a greater number of humans join in the quest for space, as illustrated in Figure 1-3.

Figure 1-3. Space. Space is many things to many people. It's the wonder of the stars, rockets, spacecraft, and all the other aspects of the final frontier.

Using Space

Although the full potential of space is yet to be realized, over the years we've learned to take advantage of several of its unique attributes in ways that affect all of us. The most common space missions fall into four general areas

- Communications
- Remote sensing
- Navigation
- Science and exploration

Let's briefly look at each of these missions to see how they are changing the way we live in and understand our world.

Communications

In October 1945, scientist and science-fiction writer Arthur C. Clarke (author of classics such as *2001: A Space Odyssey*) proposed an idea that would change the course of civilization.

> One orbit, with a radius of 42,000 km, has a period (the time it takes to go once around the Earth) of exactly 24 hours. A body in such an orbit, if its plane coincided with that of the earth's equator, would revolve with the earth and would thus be stationary above the same spot . . . [a satellite] in this orbit could be provided with receiving and transmitting equipment and could act as a repeater to relay transmissions between any two points on the hemisphere beneath . . . . A transmission received from any point on the hemisphere could be broadcast to the whole visible face of the globe. (From *Wireless World* [Chagas and Canuto, 1978].)

The information age was born. Clarke proposed a unique application of the global perspective space offers. Although two people on Earth may be too far apart to see each other directly, they can both "see" a satellite in high orbit, as shown in Figure 1-4, and the satellite can relay messages from one point to another.

Few ideas have had a greater impact in shrinking the apparent size of the world. With the launch of the first experimental communications satellite, Echo I, into Earth orbit in 1960, Clarke's fanciful idea showed promise of becoming reality. Although Echo was little more than a reflective balloon in low-Earth orbit, radio signals were bounced off it, demonstrating that space could be used to broaden our horizons of communication. An explosion of technology to exploit this idea quickly followed. Live television broadcasts by satellite from remote regions of the globe are now common on the nightly news.

Satellites are now used for a large percentage of commercial and government communications and for most domestic cable television. Relief workers can now stay in constant contact with their organizations, enabling them to better distribute aid to refugees hungry for food. News reporters on remote assignment can report live from the scene to a world hungry for information. Figure 1-5 shows reporters in Baghdad during the Gulf War sending back updates by satellite. Our modern military now relies almost totally on satellites to communicate with forces deployed world-wide. Without satellites, global communication as we know it today would not be possible.

We're only now appreciating the impact of communication satellites on world development. Many credit the world-wide marketplace of ideas ushered in by satellites with the collapse of the Soviet Union and the rejection of closed, authoritarian regimes. Communication satellites have also been a boon to developing nations. Chagas and Canuto [1978] showed

Figure 1-4. Communication Through Satellite. A satellite's global perspective allows users in remote parts of the world to talk to each other.

how the launch of the Palapa A and B satellites, for example, allowed the island country of Indonesia to expand telephone service from a mere 625 phones in 1969, to more than 233,000 only five years later. This veritable explosion in the ability to communicate has been credited with greatly improving the nation's economy and expanding its gross national product, thus benefiting all citizens. All this from only two satellites! Other developing nations have also realized the benefits of space.

Figure 1-5. Satellite Communications. The explosion in satellite communication technology has shrunk the world and linked us more tightly together in a global community. Here we see reporters sending back news during the Gulf War through a portable satellite ground station. *(Courtesy of Mobile Telesystems, Inc.)*

Eventually, a large collection of satellites in low-Earth orbit will form a world-wide cellular telephone network. With this network in place, anyone with a small portable phone will be able to call anyone else on the planet at the touch of a button. Suddenly, no matter where you go on Earth, you'll always be able to phone home. We can only imagine how this expanded ability to communicate will further shrink the global village.

Remote Sensing

Satellites operating from the global perspective of space have also made possible the science of remote sensing. *Remote sensing* is the act of observing Earth and other objects from space. For decades, military "spy satellites" have kept tabs on the activities of potential adversaries using remote-sensing technology. This data has been essential in determining troop movements and violations of international treaties. During the Gulf War, for example, remote-sensing satellites gave the United Nations alliance a decisive edge. United Nations forces knew nearly all Iraqi troop deployments, whereas the Iraqis, lacking these sensors, didn't know where allied troops were positioned. Furthermore, early-warning satellites, originally orbited to detect enemy missile launches against the United States, proved equally effective in detecting the launch of SCUD missiles against allied targets. This early warning gave the Patriot antimissile batteries time to prepare to engage the missiles.

This same technology has been adapted for civilian use. The United States Landsat and the French SPOT (Satellite Pour l'Observation de la Terre) systems are good examples. Figure 1-6 shows a Landsat image of Washington, D.C., and Figure 1-7 shows a SPOT image of Kansas. These satellites "spy" on crops, ocean currents, and natural resources to aid farmers, resource managers, and planners on Earth. In countries where the failure of a harvest may mean the difference between bounty and starvation, spacecraft have helped planners manage scarce resources and head off potential disasters before insects or other blights could wipe out an entire crop. For example, in agricultural regions near the fringes of the Sahara desert in Africa, Landsat images have been used to predict where locust swarms were breeding. Scientists were then able to control the locusts before they could swarm, saving large areas of crop land.

Figure 1-6. City Planning from Space. Remote-sensing images, like this one from the Landsat spacecraft, can be used for urban planning. In this photo, we see the Washington D.C. area in the infrared band clearly showing the difference between concrete and vegetation. The runways of Washington National Airport can be seen as crossed lines on the left bank of the Potomac River. *(Reproduced by permission of Earth Observation Satellite Company, Lanham, MD, U.S.A.)*

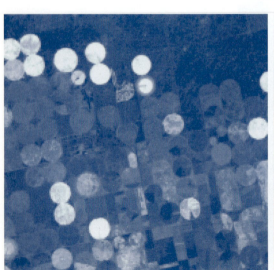

Figure 1-7. Two Views of Kansas. A remote-sensing image from the French SPOT satellite shows irrigated fields (the circular areas) in Kansas. On the right, astronaut Joe Engle, a native of Kansas, gives his opinion of being in space. *(Courtesy of EoSAT and NASA)*

Remote-sensing data can also help in managing other scarce resources. Careful interpretation of the data can indicate the best places to drill for water or oil. From space, astronauts can easily see fires burning in the rain forests of South America as trees are cleared for farms and roads. Remote-sensing spacecraft have become one of the most formidable weapons against the destruction of the environment because they can systematically monitor large areas to assess the spread of pollution and other damage.

Remote-sensing technology has been a boon to planners and map makers as well. With the advent of satellite imagery, maps can be made in a fraction of the time it would take for a ground survey. This has allowed city planners to keep up better with urban sprawl and given troops deployed in new areas the latest maps of unfamiliar terrain.

Weather forecasting is a further application of remote-sensing technology—one we've all come to rely on. National weather forecasts usually begin with a current satellite view of the Earth. At a glance, any of us can tell which parts of the country are clear or cloudy. When the satellite map is put in motion, we can easily see the direction of clouds and storms. An untold number of lives are saved every year by this simple ability to track the paths of hurricanes and other deadly storms, such as the one shown in Figure 1-8. By providing farmers valuable climatic data and agricultural planners information about flood control and other weather-related disasters, this technology has markedly improved food production and crop management.

Overall, we've come to rely more and more on the ability to monitor and map our entire planet. As the pressure builds to better manage scarce resources and assess environmental damage, we'll call upon remote-sensing spacecraft to do even more.

Figure 1-8. Viewing Hurricanes From Space. It's hard to imagine a world without weather satellites. Images of hurricanes and other severe storms provide timely warning to those in their path and save countless lives every year. *(Courtesy of NASA)*

Navigation

Satellites have revolutionized *navigation*—the process of determining where you are and where you're going. If you have a low-cost receiver, the Global Positioning System (GPS) can give you instant information on where you are—with mind boggling accuracy. With a GPS receiver no one need ever get lost again! GPS was developed by the Department of Defense to help airplanes, ships, and ground troops navigate around the globe, but it also offers incredible civilian applications. With four GPS satellites in view, as shown in Figure 1-9, the system can provide you with a "fix" of your position accurate to within a few meters. In fact, the biggest problem you may face is that the fix from GPS is actually more accurate than many maps! With a GPS receiver installed in your car, along with some additional computing power, you'll be able to find your way across a strange city without ever consulting a map. You can simply put in the location you're trying to reach, and the GPS receiver will tell you how to get there. No more stops at gas stations to ask directions!

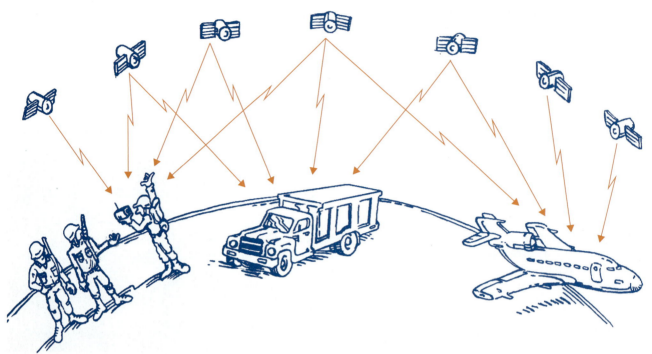

Figure 1-9. The Global Positioning System (GPS). The GPS space segment consists of a constellation of satellites deployed and operated by the U.S. Air Force. GPS has literally revolutionized navigation by providing highly accurate position, velocity, and time information to users on Earth.

Science and Exploration

Since the dawn of the space age, dozens of satellites have been launched for purely scientific purposes. These mechanical explorers have helped to answer (and raise) basic questions about the nature of Earth, the solar system, and the universe.

In the 1960s and 1970s, the United States launched the Pioneer series of spacecraft, which told us about the nature of Venus, Mercury, and the Sun. The Mariner spacecraft flew by Mars to give us the first close-up view of the Red Planet. In 1976, two Viking spacecraft landed on Mars to perform experiments designed to search for life on the one planet in the solar system whose environment most closely resembles Earth's. In the 1970s and 1980s, the Voyager spacecraft took us on a grand tour of the outer planets, beginning with Jupiter and followed by Saturn, Uranus, and Neptune. The Magellan spacecraft, launched in 1989, has mapped the surface of Venus beneath its dense layer of clouds, as shown in Figure 1-10. While all of these missions have answered many questions about the composition of the solar system, they have also raised many other questions which await future generations of robotic and manned explorers.

Figure 1-10. Images of Venus. Magellan, another interplanetary spacecraft, has provided a wealth of scientific data. Using its powerful synthetic aperture radar to pierce the dense clouds of Venus, it has mapped Venusian surface in detail. *(Courtesy of NASA)*

Since the launch of cosmonaut Yuri Gagarin on April 12, 1961, space has been home to humans as well as machines. In less than ten years, we went from day-long missions in low-Earth orbit to setting foot on the Moon. The motivation for sending humans into space was at first purely political, as we'll explore in Chapter 2. But scientific advances in exploration, physiology, material processing, and environmental observation have proved that, for widely varying missions, humans' unique ability to adapt under stress to changing conditions can make them essential to mission success.

Future Missions

What does the future hold? In these times of changing world order and constant budget fluctuation, it's impossible to predict. An international space station is envisioned. Discussion is underway to send humans back to the Moon, this time to stay, and then on to Mars, as shown in Figure 1-11.

As we become more concerned about damage to the Earth's environment, we look to space for answers. The proposed "Mission to Planet Earth" would use spacecraft from several different countries designed to monitor the health of Earth from space. Data collected by these spacecraft can help us assess the extent of environmental damage. In this way, scientists will be better able to suggest programs that can help clean up the environment and prevent future damage. One example of monitoring the environment from space is shown in Figure 1-12. These images, taken by the upper atmospheric research satellite (UARS), are used to track the concentration of ozone. The ozone layer protects us from harmful ultra-violet radiation. Scientists are concerned that depletion of this protective layer may cause health hazards, such as skin cancer, on Earth.

Figure 1-11. Going to Mars. Future human missions may explore the canyons of Mars for signs that life may have once flourished there, only to be extinguished as the planet's atmosphere diminished. *(Courtesy of NASA)*

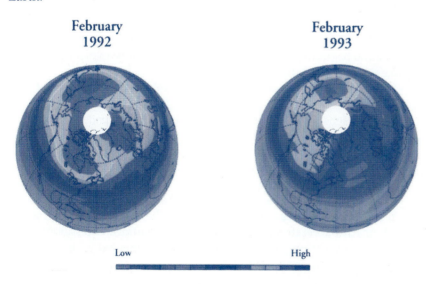

February 1992   **February 1993**

Low   High

Figure 1-12. Monitoring Ozone. Data from the upper atmospheric research satellite (UARS) is used to track changes to the Earth's protective ozone layer. *(Courtesy of Jet Propulsion Laboratory)*

Increasing spending on space research has its critics. Many argue that the money spent on such programs could be better spent on improving the human condition here on Earth. But others point out that voyages of discovery and exploration have always helped to improve the lot of humans. The human condition, they say, can't be improved by turning our backs to the potential wonders and benefits of an entire solar system. A society that fails to invest in its future may have no future in the end.

The eventual course of the space program is very much up to you. Whether we continue to push out and test the boundary of human experience or retreat from it depends on the level of interest and technical competence of the general public. By reading this book, you've already accepted the challenge to study the stars. By studying astronautics, you too can explore the final frontier.

▰ Section Review

Key Terms

free-fall
navigation
remote sensing
scintillation

Key Concepts

➤ Space offers several unique advantages which make its exploration essential for modern society
  • Global perspective
  • A clear view of the universe without the adverse affects of the atmosphere
  • A free-fall environment
  • Abundant resources
  • A final frontier

➤ Since the beginning of the space age, missions have evolved to take advantage of space
  • Communications satellites have tied together remote regions of the globe
  • Remote-sensing satellites observe the Earth from space, providing weather forecasts, essential military information, and valuable data to help us better manage Earth's resources
  • Navigation satellites have revolutionized how we travel on Earth
  • Scientific spacecraft have explored the Earth and the outer reaches of the solar system and peered to the edge of the universe
  • Manned spacecraft have provided valuable information about living and working in space and have experimented with processing important materials

1.2 Elements of a Space Mission

▬ In This Section You'll Learn to...

☞ Identify the elements common to all space missions

Now that you understand a little more about why we go to space, let's begin our exploration of the wonderful world of astronautics by seeing exactly what we need to get there.

As you watch the weather map on the nightly news or pick up the phone to make an overseas phone call, your mind is probably not focused on the complex space infrastructure necessary to make these things possible. At most, you might have visualized a small electronic box encumbered with solar panels and antennae somewhere out in space—a satellite. However, while a satellite may represent the end result of decades of planning, designing, and testing by a veritable army of engineers, managers, operators, and technicians, it is only one small piece of a vast array of technology needed to do the job.

We'll define the *space mission architecture*, shown in Figure 1-13, to be the collection of orbits, boosters, operations, and all other things that make a space mission possible. Let's briefly look at each of these elements to see how they fit together.

The Mission

All space missions begin with a need, such as the need to communicate between different parts of the world or to monitor pollution in the upper atmosphere. This need creates the mission itself, which is central to understanding the entire space mission architecture. Simply stated, the *mission* is why we're going to space. For any mission, no matter how complex, we should be able to write a succinct *mission statement* that tells us three things:

- The mission *objective—why* we're doing the mission
- The mission *users—who* will benefit
- The mission *operations concept—how* all the mission elements work together

We'll explore the why, who, and how of a space mission in much greater detail in Chapter 11. For now, simply realize that we must answer each of these important questions before we can envision the entire mission architecture.

Before we step too far off the deep end in our understanding of space hardware, you should realize that people are the most important element of any space mission. Without people handling various jobs and services, all the expensive hardware is useless. Hollywood shows us only

14

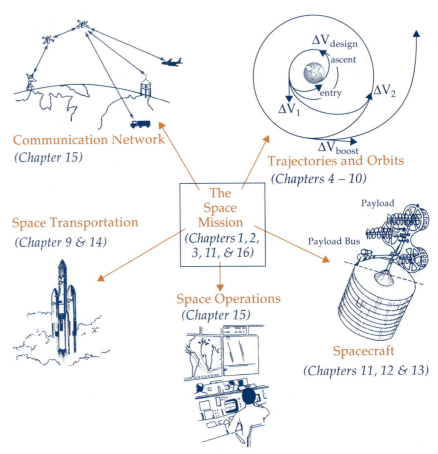

Figure 1-13. Space Mission Architecture. A space mission required a variety of interrelated elements. Collectively these elements define the space mission architecture.

astronauts doing "glamorous" jobs during a space walk or "nerdish" looking engineers hunched over a computer. But you don't have to be a rocket scientist or even an astronaut to work with space. Thousands of jobs in the aerospace industry require only a desire to work hard and get the job done. Support people are needed at every level—from food services to legal services—to get any space mission off the ground. Sure, an astronaut turning a bolt to fix a satellite will get his or her picture on the evening news, but someone had to make the wrench, and someone else had to place it in the toolbox.

Let's begin our investigation of the elements of a space mission architecture by looking at one that depends on people to make it happen—space operations.

Space Operations

The first word spoken by humans from the surface of the Moon was "Houston." Neil Armstrong was calling back to the Mission Control Center at Johnson Space Center in Houston, Texas, to let them know the

Eagle had successfully landed. To design an operations concept to support our mission statement, we have to consider how the mission data will be collected, stored, and delivered to users on Earth. Furthermore, we have to factor in how the flight control team will receive and monitor data on the spacecraft's health and build in ground control for commanding some of the spacecraft's functions.

It would be nice if, once a spacecraft has been deployed to its final orbit, it would work day after day on its own. Then users on Earth could go about their business without concern for the satellite's care and feeding. Unfortunately, this is not the case. Modern spacecraft, despite their sophistication, require almost constant attention from a team of flight controllers on the ground.

Figure 1-14. The Space Shuttle Mission Control Center in Houston, Texas. Space operations involves monitoring and controlling spacecraft from the ground. Here, we see attentive flight controllers at the Guidance (GuiDO) and Flight Dynamics (FIDO) consoles. *(Courtesy of NASA)*

The *flight control team* monitors the spacecraft's health and status to ensure its correct operation. Should trouble arise, flight controllers have an arsenal of procedures they can use to nurse the spacecraft back to health. The flight control team operates out of a Mission Control Center (MCC) such as the one in Houston, Texas, used for U.S. manned missions and shown in Figure 1-14. Some unmanned Air Force satellites are controlled from a similar MCC at Falcon Air Force Base in Colorado Springs, Colorado. Department of Defense satellites and some commercial ones are controlled from the Satellite Control Facility in Sunnyvale, California.

Within the MCC, the team members hold positions that follow the spacecraft's functional lines. For example, one person may monitor the spacecraft's path through space while another keeps an eye on the electrical-power system. The lead flight controller, called a *flight director* for manned operations, orchestrates the inputs from each of the flight-control disciplines. Flight directors make decisions based on recommendations and their own experience and judgment.

The Spacecraft

The word "spacecraft" tends to conjure up images of the starship Enterprise or sleek flying saucers from all those 1950s SciFi movies. In reality, however, spacecraft tend to be more squat and ungainly than sleek and streamlined. The reasons for this are purely practical—spacecraft are built to perform a particular mission in an efficient, cost-effective manner. In the vacuum of space, there's no need to be streamlined.

For discussion, we'll divide the spacecraft into two basic parts—the payload and the spacecraft bus. The *payload* is that part of the spacecraft which actually performs the mission. Thus, the type of payload a spacecraft has depends directly on the type of mission it's performing. For example, the payload for a mission to monitor the Earth's ozone layer could be a series of scientific sensors, each designed to measure one aspect of the environment. A payload is designed to interact with the primary focus of interest for the mission, called the *subject*. So, in this example, the subject is the ozone layer.

The spacecraft bus does not arrive every morning at 7:16 to deliver the payload to school. But we'll see that the functions performed by a spacecraft bus aren't that different from a common school bus. Without the spacecraft bus, the payload couldn't do its job. The *spacecraft bus* provides all the "housekeeping" functions necessary to make the payload work. It provides electrical power, maintains the right temperature, processes information, communicates with Earth and other spacecraft, controls the spacecraft's orientation, and holds everything together.

Trajectories and Orbits

Now let's look at where the spacecraft goes. A *trajectory* is the path an object follows through space. In getting from the launch pad and into space, a rocket follows a particular ascent trajectory designed to lift it efficiently out of Earth's atmosphere. Once in space, the spacecraft resides in an orbit. We'll look at orbits in great detail in later chapters, but for now realize an *orbit* is essentially a fixed "racetrack" on which the spacecraft travels around whatever planet or body you're interested in. Like racetracks, orbits usually have an oval shape as shown in Figure 1-15. Just as planets orbit the Sun, we can place satellites into orbit around the Earth.

When selecting an orbit for a particular satellite mission, we need to know where on Earth we need to look. As it turns out, we can put a spacecraft into an almost limitless number of different orbits, but we must choose the orbit which best fulfills the mission. For instance, suppose our mission is to provide continuous communication between New York and Los Angeles. Our subjects—the primary focus for the mission—are these two cities, so we would want to position our spacecraft on a "racetrack" that would allow it to always have both cities in sight. The orbit's size, shape, and orientation will determine whether or not a given orbit will allow the payload to observe its subjects.

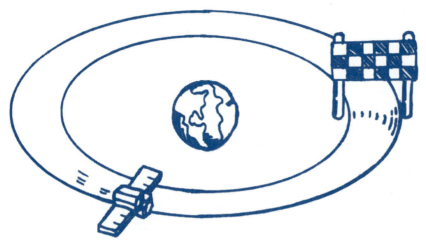

Figure 1-15. The Orbit. An orbit can be thought of as a fixed racetrack in space that the spacecraft drives on. Depending on the mission, this racetrack's size, shape, and orientation will vary.

Unfortunately, the larger an orbit is, the harder it is to attain. Just as climbing ten flights of stairs takes more energy than climbing only one, putting a spacecraft into a higher (larger) orbit requires more energy, meaning a bigger rocket and greater expense.

The size of orbit will also dictate the spacecraft's field of view as shown in Figure 1-16. The *field-of-view (FOV)* is the angle which describes the amount of the Earth's surface the spacecraft can see at any one time. In other words, the higher you are the more you can see. The total amount of ground seen at any one time is the *swath width*.

Some missions require continuous coverage of a point on Earth or the ability to communicate from one side of the planet to the other. When this happens, a single spacecraft usually can't satisfy mission requirements. Instead, we build several identical spacecraft and place them in different orbits to cover the subject. A *constellation* is a collection of spacecraft deployed in several orbits to accomplish a single mission. With higher mission orbits, fewer satellites are needed to provide world-wide coverage.

The Global Positioning System (GPS) is a good example of a mission requirement that dictated a constellation of satellites to do the job. Designers mandated that every point on Earth always be in view of at least four GPS satellites. This was impossible to do with just four satellites. Instead, the GPS constellation was designed to contain 21 spacecraft to continuously cover the world. A system to provide world-wide coverage for cellular telephones is envisioned to require more than 60 satellites!

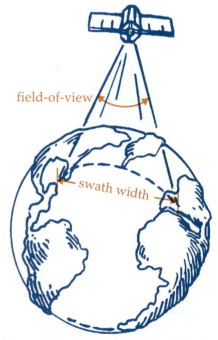

field-of-view

swath width

Figure 1-16. Field-of-View (FOV). The field-of-view of a spacecraft defines the area of coverage on the Earth's surface, called the swath width.

Space Transportation

Now that we know where we're going, we can determine *how* we're going to get there. As we said, it takes energy to get into orbit—the higher the orbit, the greater the energy. Because the size of a satellite's orbit determines its energy, we need something to deliver the spacecraft to the right mission orbit—a rocket. The thunderous energies released in a rocket's fiery blast-off provide the velocity for our spacecraft to "slip the surly bonds of Earth" and enter the realm of space, as the Shuttle demonstrates in Figure 1-17. To get the spacecraft into its final orbit and keep it there, we need three general categories of rockets—boosters, upperstages, and thrusters.

The *booster* is the rocket we see sitting on the pad during countdown. It provides the necessary change in velocity to get the spacecraft into space. The booster blasts almost straight up to gain altitude rapidly and get out of the dense atmosphere which slows it down through drag. Then it begins to pitch over to start gaining horizontal velocity. As we'll see later, this horizontal velocity keeps the spacecraft in orbit.

Because of the limits of rocket technology, we can't construct a single rocket which can deliver a spacecraft efficiently into orbit. Instead, a large booster is actually a series of smaller rockets which light and then burn out in succession, each one handing off to the other like runners in a relay race. These smaller rockets are called *stages*. In most cases, a booster will require at least three stages before reaching orbit. For example, the Ariane booster is a three-stage rocket used by the European Space Agency (ESA), as shown in Figure 1-18.

Normally, the booster can't deliver the spacecraft to its final orbit by itself. Instead, when the booster finishes its job and burns out, the spacecraft remains in a parking orbit. A *parking orbit* is a temporary orbit where the spacecraft will stay until transferring to its final mission orbit. Once a spacecraft is in its parking orbit, a final "kick" must send it onto a transfer orbit. The transfer orbit is an intermediate orbit that takes the spacecraft from its parking orbit to the final mission orbit. Once in the final mission orbit, the spacecraft starts imaging the Earth, as shown in Figure 1-19.

Figure 1-17. Lift Off! The Space Shuttle acts as a booster to lift satellites into low-Earth orbit. From there, an upperstage is used to move into a higher orbit. *(Courtesy of NASA)*

Figure 1-18. Ariane. The European Space Agency's Ariane booster lifts commercial satellites into orbit. Here we see it lifting off from its pad in Kourou, French Guiana, South America. *(Courtesy of NASA)*

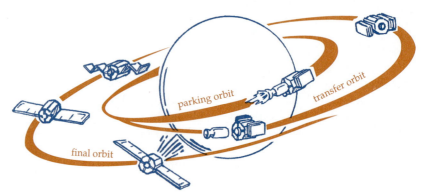

Figure 1-19. Space Mission Orbits. The booster is used primarily to deliver a spacecraft into a low-altitude parking orbit. From this point an upperstage is used to move the spacecraft onto a transfer orbit then out to the final mission orbit.

The extra kick of energy needed to transfer from the parking orbit to the final orbit comes from the *upperstage*. In some cases, the upperstage is actually integral to the spacecraft itself, sharing the plumbing and propellant which the spacecraft will later use to orient itself and maintain its orbit. In other cases, the upperstage is an autonomous spacecraft with the one-shot mission of delivering the spacecraft to its final orbit. In the latter case, the upperstage breaks off once its job is done. Regardless of how it is configured, the upperstage consists mainly of a rocket engine (or engines) and the propellent needed to change the spacecraft's energy enough to enter the desired final orbit. The upperstage used to send the Magellan spacecraft to Venus is shown in Figure 1-20.

Once the spacecraft reaches its final orbit, it may still need thrusters to keep it in place or maneuver around. *Thrusters* are relatively small rockets used to adjust the spacecraft's orientation and maintain the orbit's size and shape, both of which can change over time due to external forces.

Figure 1-20. The Inertial Upperstage (IUS). The IUS is shown here attached to the Magellan spacecraft bound for Venus. *(Courtesy of NASA)*

Communication Network

The communication network is the "glue" that holds the mission together. An entire communication network keeps the flight-control team in nearly continuous contact with the spacecraft. For the Space Shuttle this network is the Spaceflight Tracking and Data Network (STDN), consisting of both ground-based and space-based elements. As the name implies, this network does two things. First, it uses radar and other sensors to keep track of where spacecraft are at all times. The United States Air Force Space Command uses a similar world-wide tracking network to catalog the whereabouts of literally thousands of pieces of space junk. Second, the communication network relays data between the spacecraft and the flight control team to support mission operation. Figure 1-21 shows the elements of a typical communication network. One of the critical aspects of linking all these far-flung elements together is the communication process itself. Whether you're talking to your friend across a noisy room or to a spacecraft on the edge of the solar system, the basic problems are the same. We'll see how to deal with these problems in Chapter 13.

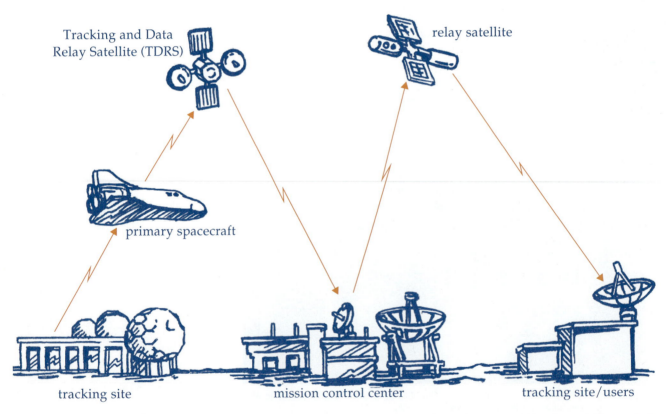

Figure 1-21. Communication Network. The flight control team relies on complex assembly of control centers, tracking sites, satellites, and relay satellites to keep them in contact with spacecraft and users. In this example, data is sent to the Space Shuttle from a tracking site and relayed through another satellite, such as the Tracking and Data Relay Satellite (TDRS), back to the control center. Data is then passed to users through a third relay satellite.

Using the Space Mission Architecture

Now that we've defined all these separate mission elements, let's look at an actual space mission to give you an example of how they all fit together in practice. Space Shuttle mission STS-54 was launched from the Kennedy Space Center (KSC) in Florida on January 13, 1993. The primary objective of this mission was to deploy a Tracking and Data Relay Satellite (TDRS) to be used by ground controllers to track and monitor NASA and DoD spacecraft, including the Space Shuttle itself. In Figure 1-22, you can see how all the elements for this particular mission tie together.

Throughout the rest of this text, we'll focus our attention on individual elements that make up a space mission. We'll begin putting missions into perspective by reviewing the history of space flight in Chapter 2. Next, we'll set the stage for our understanding of space by exploring space itself to learn about the unique demands of this hostile environment. Then we'll begin our discussion of the individual mission elements with orbits and trajectories. In Chapters 4 – 10, we'll consider orbits and how their behavior affects mission planning. In Chapters 11 – 13, we'll turn our attention to the spacecraft to learn how all the subsystems onboard tie together to make an effective mission. In Chapter 14 we'll focus on rockets to see how they provide the transportation we need to get into space. Chapter 15 will look at the remaining two elements of a space mission—operation and communication. There we'll see how ground controllers use a complex communication network to talk to satellites all over the solar system. Finally, in Chapter 16, we'll again look at missions as a whole but from a different perspective—money. We'll learn how economic laws can be as important as physical laws in determining what gets into space.

Communication Network

NASA spacecraft tracking and data network

STS-54 Crew Members

Clockwise from top: Susan J. Helms, Mario Runco Jr., John H. Casper, Donald R. McMonagle, and Gregory J. Harbaugh

Trajectory and Orbit

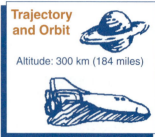

Altitude: 300 km (184 miles)

Space Transportation

Space Shuttle for delivery to parking orbit. Inertial upperstage (IUS) to deliver TDRS to its final orbit.

The STS-54 Mission
CASPER McMONAGLE RUNCO HARBAUGH HELMS
Objective: Deploy TDRS Satellite

Spacecraft

Tracking and Data Relay Satellite (TDRS)

Space Operations

Operation Concept: Ground controllers in Houston, TX, and White Sands, NM, will work to deploy and operate the spacecraft to relay data to users on the ground.

Figure 1-22. STS-54 Space Mission Architecture. *(All photos courtesy of NASA)*

▰ Section Review

Key Terms

booster
constellation
field-of-view (FOV)
flight control team
flight director
mission
mission statement
objective
operations concept
orbit
parking orbit
payload
space mission architecture
spacecraft bus
stages
subject
swath width
thrusters
trajectory
upperstage
users

Key Concepts

➤ Central to understanding any space mission is the mission itself

- The mission statement clearly identifies the major objectives of the mission (why it is being done), the users (who it is being done for), and the operations concept (how all the pieces will fit together)

➤ A space mission architecture includes the following elements:

- Space operations—the network of ground- and space-based resources that allow the flight control team to monitor the spacecraft and deliver mission data to users
- The spacecraft—composed of the bus, which does essential housekeeping, and the payload, that performs the mission
- The trajectory and orbits—the path the spacecraft will follow through space. This includes the orbit (or racetrack) the spacecraft follows around the Earth.
- Space transportation—the rockets which propel the spacecraft into space and maneuver it along its mission trajectory
- The communication network is the "glue" that holds the mission together. It consists of tracking sites, flight controllers, satellites, and users all tied together by communication links.

References

Canuto, Vittorio and Carlos Chagas. *The Impact of Space Exploration on Mankind*. Pontificaia Academia Scientiarum, proceedings of a study week held October 1–5, 1984, Ex Aedibus Academicis In Civitate Vaticana, 1986.

Wilson, Andrew (ed.), *Space Directory 1990–91*. Jane's information group. Alexandria, Virginia, 1990.

Mission Problems

1.1   Space In Our Lives

1   What five unique advantages of space make its exploitation imperative for modern society?

2   What are the four primary space missions in use today? Give an example of how each has affected, or could affect, your life.

1.2   Elements of a Space Mission

3   The mission statement tells us what three things?

4   What are the elements of a space mission?

5   What is the mission operations concept and how does it apply to space operations?

6   List the two basic parts of a spacecraft and discuss what they do for the mission.

7   What is an orbit? How does changing its size affect the energy required to get into the orbit and the swath width available to any payload in this orbit?

8   Describe what the booster, upperstage, and thrusters do.

9   What is a transfer orbit?

10   Why do we say that the communication network is the "glue" that holds the other elements together?

11   Use this mission scenario to answer the following questions. The space shuttle is launched from Kennedy Space Center to deploy a satellite that will monitor Earth's upper atmosphere. Once deployed from the low shuttle orbit, an inertial upperstage (IUS) will boost the spacecraft onto a transfer orbit and these out to its mission orbit. Once in place, it will monitor Earth's atmosphere and relay the data back to scientists on Earth through the Tracking and Data Relay Satellite (TDRS).

a)   What is the mission of this shuttle launch?

b)   What is the mission of the spacecraft?

c)   Discuss the trajectory, spacecraft, operations, booster, and upperstage for the scenario.

d)   Who are the mission users?

e)   What is the subject of the mission?

f)   What part does TDRS play in this mission?

g)   Briefly discuss your ideas for the operations concept of this mission.

For Discussion

12   What future missions which could exploit the free-fall environment of space?

13   What future space missions which could exploit lunar-based resources?

14   You hear a television commentator say the Space Shuttle's missions are a waste of money. How would you respond to this charge?

Projects

15   Moderate a debate between sides for and against space exploration. Outline what points you'd expect each side to make.

16   Given the following mission statement, select appropriate elements to accomplish the task.

Mission Statement: To monitor iceflows in the Arctic Ocean and warn ships in the area.

17   Obtain information from NASA on an upcoming space mission and prepare a short briefing on it to present to your class.

18   Write a justification for a manned mission to Mars. List and explain each element of the mission. Compile a list of skills needed by each member of the astronaut crew and the mission team.

Mission Profile—Voyager

The Voyager program consisted of two spacecraft launched by NASA in late 1977. Voyager 2 was actually launched a month prior to Voyager 1, which would fly on a faster, shorter path. This would enable Voyager 1 to arrive at the first planet, Jupiter, four months before Voyager 2. The timing of the operation was critical. Jupiter, Saturn, Uranus, and Neptune align themselves for such a mission only once every 175 years. The results from the Voyager program have answered and raised many basic questions about the origin of the solar system.

Mission Overview

NASA designed the Voyager spacecraft with two objectives in mind. First, they would build two identical spacecrafts for redundancy. They feared that the available technology meant at least one of the spacecrafts would fail. Second, they only planned to visit Jupiter and Saturn, with a possibility of visiting Neptune and Uranus if the spacecraft lasted that long. It was generally agreed that five years was the limit on spacecraft lifetime. In the end, both spacecraft performed far better than anyone's wildest imagination. Today they continue their voyage through empty space, their mission complete.

Mission Data

✓ The Voyager spacecraft used the gravity of the planets they visited to slingshot them to their next target. This gravity assist (described in Chapter 7) shortened the spacecraft's voyage by many years.

✓ Voyager 1 headed into deep space after probing the space behind Saturn's rings. Voyager 2, however, successfully probed Neptune and Uranus as well.

✓ Voyager 1 discovered that one of Jupiter's moons, Io, has an active volcano spouting lava 160 km (100 miles) into space.

✓ Miranda, one of Uranus' 15 known moons, has been called the "strangest body in the solar system." Discovered by Voyager 2, it's only 480 km (300 miles) across, which means that it should be nearly spherical in shape. Instead, it's constantly churning itself inside-out. Scientists believe this is caused by the strong gravity from Uranus reacting with a process called differentiation (where the densest material on the moon migrates to the core). The result is a moon which looks like "scoops of marble-fudge ice cream"—the dense and light materials mixed randomly in jigsaw fashion.

Mission Impact

The overwhelming success of the Voyager mission has prompted a new surge of planetary exploration by NASA. In 1996, NASA hopes to launch the Cassini mission to explore Saturn. In November 1995, the Galileo spacecraft will reach Jupiter to further study its system. These are but two of many new missions NASA has planned and begun, trying to answer the new questions the Voyager missions have uncovered.

Images from Voyager. This composite view taken by the Voyager spacecraft shows the planet Jupiter and its four largest moons. *(Courtesy of NASA)*

For Discussion

• The major problem with space exploration is exorbitant cost. Do you think the United States should sink more money into future exploratory missions? What about teaming up with other advanced countries?

• What is the benefit for humans to uncover the mysteries and perplexities of our solar system?

Contributor

Troy Kitch, United States Air Force Academy

References

Davis, Joel. *FLYBY: The Interplanetary Odyssey of Voyager 2*. New York: Atheneum, 1987.

Evans, Barry. *The Wrong Way Comet and Other Mysteries of Our Solar System*. Blue Ridge Summit: Tab Books, 1992.

Vogt, Gregory. *Voyager*. Brookfield: The Millbrook Press, 1991.

Buzz Aldrin poses against the stark lunar landscape. Neil Armstrong can be seen reflected in his helmet. *(Courtesy of NASA)*

Exploring Space

2

William J. Astore
United States Air Force Academy

In This Chapter You'll Learn To...

- ☛ Explain how early space explorers used their eyes and minds to explore space and contribute to our understanding of it
- ☛ Explain the beginnings of the Space Age and the significant events that have led to our current capabilities in space

You Should Already Know...

- ❏ Nothing yet; we'll explore space together

It is difficult to say what is impossible, for the dream of yesterday is the hope of today and the reality of tomorrow.

Robert H. Goddard

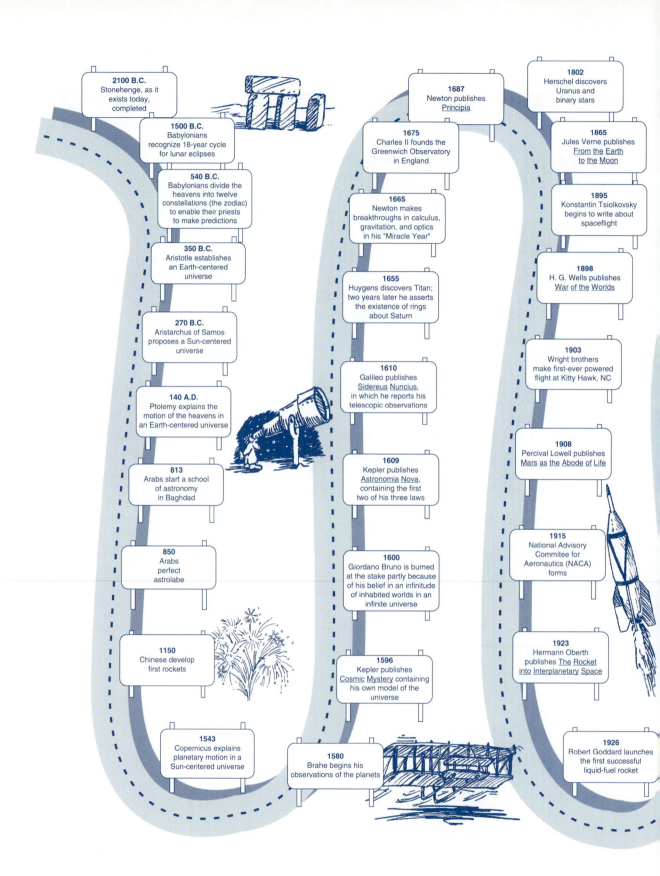

2100 B.C.
Stonehenge, as it exists today, completed

1500 B.C.
Babylonians recognize 18-year cycle for lunar eclipses

540 B.C.
Babylonians divide the heavens into twelve constellations (the zodiac) to enable their priests to make predictions

350 B.C.
Aristotle establishes an Earth-centered universe

270 B.C.
Aristarchus of Samos proposes a Sun-centered universe

140 A.D.
Ptolemy explains the motion of the heavens in an Earth-centered universe

813
Arabs start a school of astronomy in Baghdad

850
Arabs perfect astrolabe

1150
Chinese develop first rockets

1543
Copernicus explains planetary motion in a Sun-centered universe

1580
Brahe begins his observations of the planets

1596
Kepler publishes Cosmic Mystery containing his own model of the universe

1600
Giordano Bruno is burned at the stake partly because of his belief in an infinitude of inhabited worlds in an infinite universe

1609
Kepler publishes Astronomia Nova, containing the first two of his three laws

1610
Galileo publishes Sidereus Nuncius, in which he reports his telescopic observations

1655
Huygens discovers Titan; two years later he asserts the existence of rings about Saturn

1665
Newton makes breakthroughs in calculus, gravitation, and optics in his "Miracle Year"

1675
Charles II founds the Greenwich Observatory in England

1687
Newton publishes Principia

1802
Herschel discovers Uranus and binary stars

1865
Jules Verne publishes From the Earth to the Moon

1895
Konstantin Tsiolkovsky begins to write about spaceflight

1898
H. G. Wells publishes War of the Worlds

1903
Wright brothers make first-ever powered flight at Kitty Hawk, NC

1908
Percival Lowell publishes Mars as the Abode of Life

1915
National Advisory Committee for Aeronautics (NACA) forms

1923
Hermann Oberth publishes The Rocket into Interplanetary Space

1926
Robert Goddard launches the first successful liquid-fuel rocket

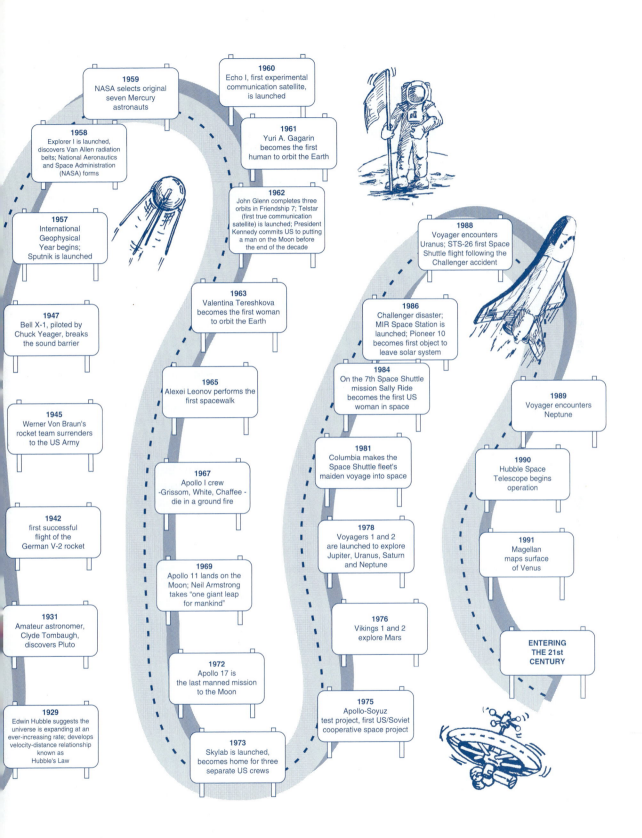

1959
NASA selects original seven Mercury astronauts

1960
Echo I, first experimental communication satellite, is launched

1958
Explorer I is launched, discovers Van Allen radiation belts; National Aeronautics and Space Administration (NASA) forms

1961
Yuri A. Gagarin becomes the first human to orbit the Earth

1962
John Glenn completes three orbits in Friendship 7; Telstar (first true communication satellite) is launched; President Kennedy commits US to putting a man on the Moon before the end of the decade

1957
International Geophysical Year begins; Sputnik is launched

1988
Voyager encounters Uranus; STS-26 first Space Shuttle flight following the Challenger accident

1963
Valentina Tereshkova becomes the first woman to orbit the Earth

1986
Challenger disaster; MIR Space Station is launched; Pioneer 10 becomes first object to leave solar system

1947
Bell X-1, piloted by Chuck Yeager, breaks the sound barrier

1984
On the 7th Space Shuttle mission Sally Ride becomes the first US woman in space

1965
Alexei Leonov performs the first spacewalk

1989
Voyager encounters Neptune

1945
Werner Von Braun's rocket team surrenders to the US Army

1981
Columbia makes the Space Shuttle fleet's maiden voyage into space

1967
Apollo I crew -Grissom, White, Chaffee - die in a ground fire

1990
Hubble Space Telescope begins operation

1942
first successful flight of the German V-2 rocket

1978
Voyagers 1 and 2 are launched to explore Jupiter, Uranus, Saturn and Neptune

1969
Apollo 11 lands on the Moon; Neil Armstrong takes "one giant leap for mankind"

1991
Magellan maps surface of Venus

1931
Amateur astronomer, Clyde Tombaugh, discovers Pluto

1976
Vikings 1 and 2 explore Mars

1972
Apollo 17 is the last manned mission to the Moon

ENTERING THE 21st CENTURY

1929
Edwin Hubble suggests the universe is expanding at an ever-increasing rate; develops velocity-distance relationship known as Hubble's Law

1975
Apollo-Soyuz test project, first US/Soviet cooperative space project

1973
Skylab is launched, becomes home for three separate US crews

You don't have to leave Earth to explore space. Long before rockets and interplanetary probes escaped the Earth's atmosphere, people explored the heavens with just their eyes and imagination. Later, with the aid of telescopes and other instruments, humans continued their struggle to bring order to the heavens. With order came some understanding and a concept of our place in the universe.

Thousands of years ago, the priestly classes of ancient Egypt and Babylon carefully observed the heavens to plan religious festivals, to control the planting and harvesting of various crops, and to understand at least partially the realm in which they believed many of their gods lived. Later, philosophers such as Aristotle and Ptolemy developed complex theories to explain and predict the motions of the Sun, Moon, planets, and stars.

The theories of Aristotle and Ptolemy dominated the world of astronomy and our understanding of the heavens well into the 1600s. Combining ancient traditions with new observations and insights, natural philosophers such as Copernicus and Kepler offered rival explanations from the 1500s onward. Using their models and Isaac Newton's new tools of physics, astronomers in the 1700s and 1800s made several startling discoveries, including two new planets—Uranus and Neptune. As we moved into the 20th century, physical exploration of space became possible. Rapid advances in technology, accelerated by World War II, made missiles and eventually large rockets available, allowing us to escape Earth entirely. In this chapter, we'll follow trailblazers who have led us from our earliest attempts to explore space to our explorations of the Moon and beyond. You will blaze our future trail in the heavens!

2.1 Early Space Explorers

▬ In This Section You'll Learn To...

- ☞ Explain the two traditions of thought established by Aristotle and Ptolemy, which dominated astronomy well into the 1600s
- ☞ Discuss the contributions to astronomy made by prominent philosophers and scientists from 1500 to the 20th Century

Astronomy Begins

More than 4000 years ago, the Egyptians and Babylonians were, for the most part, content with practical and religious applications of their heavenly observations. They developed calendars to control agriculture and star charts both to predict eclipses and to show how the movements of the Sun and planets influenced human lives (astrology). But the ancient Greeks took a more contemplative approach to studying space. They held that *astronomy*—the science of the heavens—was a divine practice best pursued through physical theories. Based on observations, aesthetic arguments, and common sense, the Greek philosopher Aristotle (384–322 B.C.) developed a complex, mechanical model of the universe. He also developed a complex physics to explain change in the universe.

Explaining how and why objects change their position can be difficult, and Aristotle often made mistakes. For example, he reasoned that if two balls, one heavy and one light, were dropped at precisely the same time, the heavier ball would fall faster to hit the ground first, as shown in Figure 2-1. Galileo would later prove Aristotle wrong. But with his rigorous logic and his respect for experimental evidence, Aristotle set a good example for future natural philosophers.

Looking to the heavens, Aristotle saw perfection. Because the circle was the only perfect shape, he surmised that the paths of the planets and stars must be circular as well. Furthermore, because the gods must consider Earth to be of central importance in the universe, it must occupy the center of creation with everything else revolving around it.

In this *geostatic* (Earth not moving) and *geocentric* (Earth-centered) universe, Aristotle believed solid crystalline spheres carried the five known planets, as well as the Moon and Sun, in circular orbits about the Earth. An outermost crystalline sphere held the stars and bounded the universe. In Aristotle's model, all of these spheres were inspired to circle the Earth by some sort of "unmoved mover" or god.

Aristotle further divided his universe into two sections—a *sublunar realm* (everything beneath the Moon's sphere) and a *superlunar realm* (everything from the Moon on up to the sphere of the fixed stars), as seen in Figure 2-2. Humans lived in the imperfect sublunar realm, which

Figure 2-1. Aristotle's Rules of Motion. Aristotle determined that heavy objects fall faster than light objects.

Figure 2-2. Aristotle's Model. The universe divided into two sections—a sublunar and a superlunar realm—each having its own distinct elements and physical laws.

consisted of four elements—earth, water, air, and fire. Earth and water naturally moved *down*—air and fire moved *up*. The perfect superlunar realm, in contrast, was made up of a fifth element (Aether) whose natural motion was circular. In separating the Earth from the heavens and using different laws of physics for each, Aristotle complicated the efforts of future astronomers.

Although Aristotle's model of the universe may seem strange to the modern student, it was developed from extensive observations combined with a heavy dose of common sense. What should concern us most is not the accuracy but the audacity of Aristotle's vision of the universe. With the power of his mind alone, Aristotle explored and ordered the heavens. His geocentric model of the universe dominated astronomy for 2000 years.

Astronomy in the ancient world reached its peak of refinement in about 140 A.D. with Ptolemy's *Almagest*. Following Greek tradition, Ptolemy calculated the orbits of the Sun, Moon, and planets using combinations of circles. These combinations, known as eccentrics, epicycles, and equants, were not meant to represent physical reality—they were merely devices for calculating and predicting orbits. Like Aristotle, Ptolemy held that solid crystalline spheres drove heavenly bodies about the Earth in perfect circular orbits. In the eyes of the ancients, describing orbits (kinematics) and explaining the causes of orbits (dynamics) were two unrelated problems. It would take almost 1500 years before Johannes Kepler healed this split.

Astronomers made further strides during the Middle Ages, with the Arabs' contributions being especially noteworthy. While the West struggled through the Dark Ages, Arab astronomers translated the *Almagest* and other ancient texts. They built up a learned tradition of commentary about and criticism of these texts, which Copernicus later found invaluable for his reform of Ptolemy. Arabs also perfected the astrolabe, a sophisticated observational instrument that could be used to chart the courses of the stars and aid travellers in navigation, as shown in Figure 2-3. Their observations, collected in the Toledan tables, formed the basis of the Alfonsine Tables used for astronomical calculations in the West from 1272 to the mid-16th century. Moreover, Arabic numerals, combined with the Hindu concept of zero, replaced the far clumsier Roman numerals. Along with Arab advances in trigonometry, this new numbering system greatly enhanced computational astronomy. Our language today bears continuing witness to Arabs' contributions—we adopted algebra, nadir, zenith, and other words from them.

With the fall of Toledo, Spain, in 1085, Arabic translations of and commentaries about ancient works became available to the West, touching off a renaissance in twelfth-century Europe. Once again, western Europeans turned their attention to the heavens, but because medieval scholasticism made Aristotle into dogma, hundreds of years would pass before fundamental breakthroughs occurred in astronomy.

Figure 2-3. The Astrolabe. Developed by Arab Scholars in the Middle Ages, it revolutionized astronomy and navigation. *(Courtesy of Greenwich Royal Observatory)*

Reordering The Universe

With the Renaissance and humanism came a new emphasis on the accessibility of the heavens to human thought. Nicolaus Copernicus (1473–1543), a Renaissance humanist and Catholic clergyman, reordered the universe and enlarged man's horizons. He placed the Sun at the center of the solar system, as shown in Figure 2-4, and had the Earth rotate on its axis once a day while revolving about the Sun once a year. Copernicus promoted his *heliocentric* (sun-centered) vision of the universe in his *On the Revolutions of the Celestial Spheres*, which he dedicated to Pope Paul III in 1543. A heliocentric universe, he explained, is more symmetrical, simpler, and matches observations better than Aristotle's and Ptolemy's geocentric model. For example, Copernicus explained it was simpler to attribute the observed rotation of the sphere of the fixed stars (he didn't abandon Aristotle's notion of solid crystalline spheres) to the Earth's own daily rotation than to imagine the immense sphere of the fixed stars rotating at near infinite speed about a fixed Earth.

Copernicus further observed that, with respect to a viewer located on the Earth, the planets occasionally appear to back up in their orbits as they move against the background of the fixed stars. Ptolemy and others resorted to complex combinations of circles to explain this backward motion of the planets, but Copernicus cleverly explained that this motion was simply the effect of the Earth overtaking, and being overtaken by, the planets as they all revolved about the Sun.

However, Copernicus' heliocentric system had its drawbacks. He couldn't prove the Earth moved, and he couldn't explain why the Earth rotated on its axis while revolving about the Sun. He also adhered to the Greek tradition that orbits follow uniform circles, so his geometry was nearly as complex and physically erroneous as Ptolemy's. In addition, Copernicus wrestled with the problem of *parallax*——the apparent shift in the position of bodies when viewed from different distances. If the Earth truly revolved about the Sun, critics observed, a viewer stationed on the Earth should see an apparent shift in position of a closer star with respect to its more distant neighbors. Because no one saw this shift, Copernicus' sun-centered system was suspect. In response, Copernicus speculated that the stars must be at vast distances from the Earth, but such distances were far too great for most people to contemplate at the time, so this idea was also widely rejected.

Copernicus saw himself more as a reformer than as a revolutionary. Nevertheless, he did revolutionize astronomy and challenge humanity's view of itself and the world. The reality of his system was quickly denied by Catholics and Protestants alike, with Martin Luther bluntly asserting: "This fool [Copernicus] wishes us to reverse the entire science of astronomy. . . sacred Scripture tells us that Joshua commanded the Sun to stand still [Joshua 10:12–13], and not the Earth." Still, Catholics and Protestants could accept the Copernican hypothesis as a useful tool for astronomical calculations and calendar reform as long as it wasn't used to represent reality.

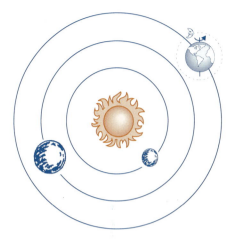

Figure 2-4. Copernican Model of the Solar System. Copernicus placed the Sun near the center of the universe with the planets moving around it in circular orbits. [Not to scale]

Because of these physical and religious problems, only a few scholars dared to embrace Copernicanism. They were staggered by the implications of this system. If the Earth were just another planet, and the heavens were far more vast than they'd believed, then perhaps an infinite number of inhabited planets were going about an infinite number of suns, and perhaps the heavens themselves were infinite. Giordano Bruno (1548–1600) embraced and advanced these implications in the 1580s and 1590s, but for his radical views on the universe and religious doctrine he burned at the stake in 1600.

Ironically, Bruno's vision of an infinite number of inhabited worlds occupying an infinite universe derived from his belief that an omnipotent God could create nothing less. Eventually, other intrepid explorers seeking to plumb the depths of space would share his vision. But his mystical insights were ultimately less productive than more traditional observational astronomy, especially as practiced at this time by Tycho Brahe (1546–1601).

Brahe rebelled against his parents, who wanted him to study law and serve the Danish king at court in typical Renaissance fashion. Instead, he studied astronomy and built, on the island of Hven in the Danish Sound, a castle-observatory known as Uraniborg, or "heavenly castle." Brahe's castle preserved his status as a knight, and a knight he was, both by position and temperament. Never one to duck a challenge, Brahe once dueled with another Danish nobleman and lost part of his nose, which he ingeniously reconstructed out of gold, silver, and wax.

Brahe brought the same tenacity to observational astronomy. He obtained the best observing instruments of his time and pushed them to the limits of their accuracy to achieve observations precise to approximately four to five minutes of arc. If you were to draw a circle and divide it into 360 equal parts, the angle described would be a *degree*. If you then divide each degree into 60 equal parts, you would get one *minute of arc*. Figure 2-5 gives you some feel for how small one minute of arc really is.

Brahe observed the supernova of 1572 and the comet of 1577. He calculated that the nova was far beyond the sphere of the Moon and that the comet's orbit intersected those of the planets. Thus, he concluded that change does occur in the superlunar realm and that solid crystalline spheres don't exist in space. In a sense, he shattered Aristotle's solid spheres theory, concluding that space was imperfect and empty except for the Sun, Moon, planets, and stars.

Although Brahe's ideas were revolutionary, he couldn't bring himself to embrace the Copernican concept of the solar system. Instead, he kept the Earth at the center of everything in a complex, geo-heliocentric model of the universe. In this model, the Moon and Sun revolved about the Earth, with everything else revolving about the Sun. It was a compromise which preserved many of the merits of the Copernican system while keeping Christians safely at the center of everything. Many who could not accept Copernicanism, such as the Jesuits, adopted Brahe's system.

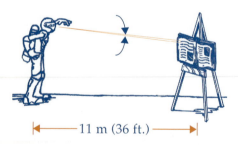

|← ——— 11 m (36 ft.) ——— →|

Figure 2-5. What is One Minute of Arc? This represents the angle subtended by a single letter on this page when viewed from a distance of 11 m (36 ft.).

Brahe himself didn't take full advantage of his new, more precise observations, partly because he wasn't a skilled mathematician. But Johannes Kepler (1571–1630), shown in Figure 2-6, was. Astronomers, Kepler held, were priests of nature who were called by God to interpret His creation. Because God plainly chose to manifest Himself in nature, the study of the heavens would undoubtedly be pleasing to God and as holy as the study of Scripture.

Inspired by this perceived holy decree, Kepler explored the universe, trying to redraw in his own mind God's harmonious blueprint for it. By the age of twenty-five, Kepler published the *Cosmic Mystery*, revealing God's model of the universe. Although his model attracted few supporters in 1596 and seems delightfully screwy to students today, Kepler insisted throughout his life that this was his monumental achievement. He even tried to sell his duke on the idea of creating, out of gold and silver, a mechanical miniature of this model which would double as an elaborate alcoholic drink dispenser! Having failed with this clever appeal for princely support, Kepler sought out and eventually found himself working for Brahe in 1600.

The Brahe-Kepler collaboration would be short-lived, for Brahe died in 1601. Before his death, Brahe challenged Kepler to calculate the orbit of Mars. Brahe's choice of planets was fortunate, for of the six planets then known, Mars had the second most eccentric orbit. Eccentric means "off center," and *eccentricity* describes the deviation of a shape from a perfect circle. A circle is said to have an eccentricity of zero, and an ellipse has eccentricity between zero and one. But as Kepler began to pore through Brahe's observations of Mars, he found a slight problem. Mars' orbit wasn't circular. He consistently came up with a difference of eight minutes of arc between what would be expected for a circular orbit and Brahe's observations.

Figure 2-6. Johannes Kepler. He struggled to find harmony in the motion of the planets. *(Courtesy of NASA)*

Astro Fun Fact

"There is Nothing New Under the Sun"

Was Copernicus the first scientist to place the Sun, not the Earth, in the center of the solar system? No! In approximately 320 B.C., Aristarchus was born in Samos of the Ancient Greek Empire. The profession he grew into was astronomy, so he moved to Alexandria, then the cultural hub for all kinds of scientists and natural philosophers. His work centered on determining the distance from the Earth to the Sun and the Moon. He did this through geometric measurements of the Moon's phases and the size of the Earth's shadow during lunar eclipses. He eventually showed that the Sun was enormously larger than the Earth. Therefore, he believed that the Sun and not the Earth occupied the center of the known universe. His findings were too revolutionary for his day, and if not for the fact that Archimedes mentioned Aristarchus' work in some of his writings, the resounding rejection of his work might have caused it to be lost in obscurity.

Asimov, Isaac. <u>Asimov's Biographical Encyclopedia of Science and Technology</u>. Garden City, NJ: Doubleday & Co. Inc., 1972.

Contributed by Thomas L. Yoder, United States Air Force Academy

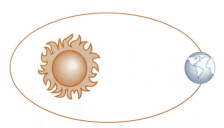

Figure 2-7. Kepler's First Law. Kepler's First Law states that the orbits of the planets are ellipses with the Sun at one of the foci as shown here in this greatly exaggerated view of Earth's orbit around the Sun.

Most astronomers would have simply ignored this small discrepancy and proclaimed it "close enough for government work." But it wasn't close enough for Kepler. He simply couldn't disregard all of Brahe's data. Instead, he began to look for other shapes that would match the observations. After wrestling with ovals for a short time he finally arrived at the idea that the planets moved around the Sun in elliptical orbits, with the Sun not at the center but at a *focus*. (Focus comes from a Latin term meaning hearth or fireplace, which is a fairly good description of the Sun.) So confident was Kepler in his own mathematical abilities and in Brahe's data that he codified this discovery into a Law of Motion. Although this was actually the second law he discovered, we call this Kepler's First Law, shown in Figure 2-7.

Kepler's First Law. The orbits of the planets are ellipses with the Sun at one focus.

With Kepler's Second Law, he began to hint at the Law of Universal Gravitation which Newton would discover decades later. By studying individual "slices" of the orbit of Mars versus the time between observations, Kepler noticed that a line between the Sun and Mars swept out equal areas in equal times. For instance, a planet moving through two arcs of its orbit in 30 days would have the same area contained in each "piece of pie." To account for this, Kepler reasoned that as a planet draws closer to the Sun it must move faster (to sweep out the same area), and when it is farther from the Sun, it must slow down. Figure 2-8 shows this varying motion.

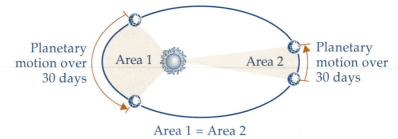

Planetary motion over 30 days | Area 1 | Area 2 | Planetary motion over 30 days

Area 1 = Area 2

Figure 2-8. Kepler's Second Law. Kepler's Second Law states that planets (or anything else in orbit) sweep out equal areas in equal time.

Kepler's Second Law. The line joining a planet to the Sun sweeps out equal area in equal times.

Kepler developed his first two laws between 1600 and 1606 and published them in 1609. Ten years later, Kepler discovered his third law while searching for the notes he believed the planets sang as they orbited the Sun! Again, with much trial and error, Kepler stumbled onto a relationship which he called his third law.

Kepler's Third Law. The square of the orbital *period*—how long it takes to go once around the orbit—is directly proportional to the cube of the mean or average distance between the Sun and the planet.

Figure 2-9 illustrates these parameters. As we'll see in greater detail in Chapter 4, when we look at orbital motion in earnest, Kepler's Third Law allows us to predict the orbits not only of planets but also of moons, satellites, and space shuttles.

Besides his new laws for describing the motion of the planets, Kepler's astronomy brought a new emphasis on finding and quantifying the physical causes of these motions. Kepler fervently believed that God had drawn His plan of the universe along mathematical lines and implemented it using only physical causes. Thus, he united the geometrical or kinematic description of orbits with their physical or dynamic cause.

Kepler continued to use his imagination to explore space, "travelling" to the constellation of Orion in an attempt to prove the universe was finite. (Harmony and proportion were everything to Kepler, and an infinite universe seemed to lack both.) In 1608 he wrote a fictional account of a Moon voyage (the *Somnium*) which was published posthumously in 1634 [Kepler, 1634]. In the *Somnium*, Kepler "mind-trips" to the Moon with the help of magic, where he discovers the Moon is an inhospitable place inhabited by specially-adapted Moon creatures. Kepler's *Somnium* would eventually inspire other authors to explore space through fiction, including Jules Verne in *From the Earth to the Moon* (1865) and H.G. Wells in *The First Men in the Moon* (1900).

Up to Kepler's time, man's efforts to explore the universe had been remarkably successful but constrained by the limits of human eyesight. But this was to change. In 1609 an innovative mathematician, Galileo Galilei (1564–1642), shown in Figure 2-10, heard of a new optical device which could magnify objects so they would appear to be closer and brighter than when seen with the naked eye. Building a telescope that could magnify an image 20 times, Galileo ushered in a new era of space exploration. He made some startling telescopic observations of the Moon, the planets, and the stars, thereby attaining stardom in the eyes of his peers and potential patrons. Observing the Moon, Galileo noticed it looked remarkably like the Earth's surface, with mountains, valleys, and even seas. This disproved Aristotle's claim that the Moon was perfect and wholly different from the Earth. Observing the planets, Galileo noticed that Jupiter had four moons or satellites (a word coined by Kepler in 1611) that moved about it. This disproved Aristotle's claim that everything revolved about the Earth.

Observing the stars, Galileo solved the mystery of the "milkiness" of our Milky Way galaxy. He explained it was due to the radiance of countless faint stars which the naked eye couldn't resolve. Galileo also noticed that his telescope didn't magnify the stars, which seemed to confirm Copernicus' guess about their vast distance from the Earth.

Galileo quickly published his telescopic discoveries in the *Starry Messenger* in 1610. This book, written in a popular, non-technical style, presented a formidable array of observational evidence against Aristotle's and Ptolemy's geocentric universe. Galileo at first had to overcome people's suspicions, especially as to the trustworthiness of telescopes,

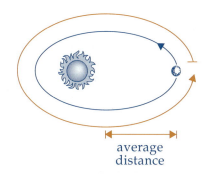

average
distance

Figure 2-9. Kepler's Third Law. Kepler's Third Law states that square of an orbit's period is proportional to the cube of the average distance between the planet and the Sun.

Figure 2-10. Galileo Galilei. Galileo used the telescope to revolutionize our understanding of the universe. *(Courtesy of Western Civilization Collection, United States Air Force Academy)*

which used glass of poor quality. As he did so, he used his telescope to subvert Aristotle's distinction between the sub- and superlunar realms and to unify terrestrial and heavenly phenomena.

Almost immediately after he published the *Starry Messenger*, people began to argue by analogy that, if the Moon looks like the Earth, perhaps it too was inhabited. The search for extraterrestrial life encouraged experts and laymen to explore the heavens. As early as 1638, in his book *The Discovery of a World in the Moone*, John Wilkins encouraged men to colonize the Moon by voyaging out into space in "flying chariots." Wilkins was strongly influenced by England's colonizing of the Western Hemisphere.

Galileo also took on Aristotle's physics. He rolled a sphere down a grooved ramp and used a water clock to measure the time it took to reach the bottom. He repeated the experiment with heavier and lighter spheres, as well as steeper and shallower ramps, and cleverly extended his results to objects in free-fall. Through these experiments, Galileo discovered, contrary to Aristotle, that all objects fall at the same rate regardless of their weight, as shown in Figure 2-11.

Figure 2-11. Everything Falls at the Same Rate. Galileo was the first to demonstrate through experiment that all masses, regardless of size, fall at the same rate when dropped from the same height.

Galileo further contradicted Aristotle as to why objects, once in motion, tend to keep going. Aristotle held that objects in "violent" motion, such as arrows shot from bows, keep going only as long as something is physically in touch with them, pushing them onward. Once this push died out, they resumed their natural motion and dropped straight to Earth. Galileo showed that objects in uniform motion keep going unless disturbed by some outside influence. He wrongly held that this uniform motion was circular, and he never used the term "inertia." Nevertheless, we applaud Galileo today for greatly refining the concept of inertia which we'll explore in Chapter 4.

Another concept Galileo refined was *relativity* (often termed Galilean relativity to separate it from Einstein's theory of relativity). Galileo wrote

> Imagine two observers, one standing on a ship's deck at sea, the other standing still on shore. A sailor near the top of the ship's mast drops an object to the observer on deck. To this observer, the object falls straight down. To the observer on shore, however, who does not share the horizontal motion of the ship, the object follows a parabolic course as it falls. Both observers are correct! [Galileo, 1632]

In other words, motion depends on the perspective or frame of reference of the observer.

To complete the astronomical revolution, which Copernicus had almost unwittingly started and which Brahe, Kepler, and Galileo had advanced, the terrestrial and heavenly realms had to be united under one set of natural laws. Isaac Newton (1642–1727), shown in Figure 2-12, answered this challenge. Newton was a mercurial person, a brilliant natural philosopher and mathematician, who provided a majestic vision of nature's unity and simplicity. 1665 proved to be Newton's "miracle year," in which he significantly advanced the study of calculus, gravitation, and

Figure 2-12. Isaac Newton. Newton was perhaps the greatest scientist who ever lived. He developed calculus (independent from the German Leibniz), worked out laws of motion and gravity, and experimented with optics. Yet, he spent most of his time studying alchemy and biblical chronology! *(Courtesy of Western Civilization Collection, United States Air Force Academy)*

optics. Extending the groundbreaking work of Galileo in dynamics, Newton published his three laws of motion and the law of universal gravitation in the *Principia* in 1687. With these laws one could explain and predict motion not only on Earth but also in tides, comets, moons, planets—in other words, motion everywhere, as we'll see in Chapter 4.

Newton's crowning achievement helped inspire the enlightenment of the eighteenth century, an age when philosophers asserted the universe was thoroughly rational and understandable. Motivated by this belief, and Newton's shining example, astronomers in the eighteenth century confidently explored the universe. Some worked in state-supported observatories to determine longitude at sea by using celestial observations. Others, like William Herschel (1738–1822), tried to find evidence of extraterrestrial life. Herschel never found his moon-dwellers, but with help from his sister Caroline, he shocked the world in 1781 when he accidently discovered Uranus. As astronomers studied Uranus, they noticed its orbit wobbled slightly. John Couch Adams (1819–1892) and Urbain Leverrier (1811–1877) used this wobble, known as an orbit *perturbation*, to calculate the location of a new planet which, obeying Newton's Law of Gravity, would cause the wobble. Observing the specified coordinates, astronomers at the Berlin observatory quickly located Neptune in 1846.

The 20th century witnessed even more remarkable discoveries by astronomers. Up to about 1918, astronomers believed that our solar system was near the center of the Milky Way, that the Milky Way was the only galaxy in the universe, and that its approximate size was a few thousand light-years across. (A *light year* is the distance light can travel at a speed of 300,000 km/s [186,000 miles/s] in one year or about 9,460,000,000,000 km [5.88×10^{12} miles].) By 1930 astronomers realized that our solar system was about half way out from the center of our galaxy, that there were other galaxies external to our Milky Way, that the universe was expanding, and that previous estimates of our galaxy's size had been ten times too small.

Two American astronomers, Harlow Shapley (1885–1972) and Edwin Hubble (1889–1953), were most responsible for this radical shift. Shapley determined in 1918 that our solar system was near the fringes of the Milky Way. Hubble, using the 250-cm reflecting telescope at Mount Wilson observatory, roughly determined the size and structure of the universe through a velocity-distance relationship now known as Hubble's Law. By examining the Doppler or red-shift of stars, he also determined the universe was expanding at an increasing rate. At this point, astronomers began to speculate that a huge explosion, or "Big Bang," had created the universe.

While Hubble and others contemplated an expanding universe, Albert Einstein (1879–1955), shown in Figure 2-13, revolutionized physics with his concepts of relativity, the space-time continuum, and his equation, $E = mc^2$. This equation showed for the first time, the equivalence of mass and energy related by a constant, the speed of light—300,000 km/s. Combined with the discovery of radioactivity in 1896 by Henri Becquerel

Figure 2-13. Albert Einstein. Einstein revolutionized physics with his concepts of relativity, space-time, and the now famous equation, $E = mc^2$. *(Courtesy of Western Civilization Collection, United States Air Force Academy)*

(1852–1908), Einstein's equation, $E = mc^2$, explained in broad terms the inner workings of the Sun.

Neptune's discovery in the previous century had been a triumph for Newton's laws. But those laws didn't explain why, among other things, Mercury's orbit changes slightly over time. Einstein's general theory of relativity accurately predicted this motion. Einstein explained that any amount of mass curves the space surrounding it, and that gravity is a manifestation of this "warped" space. Furthermore, Einstein showed that the passage of time is not constant, but relative to the observer. This means two objects travelling at different speeds will observe time passing at different rates. This concept has profound implications for satellites like GPS which have highly accurate atomic clocks.

By the dawn of the Space Age, scientists had constructed a view of the universe radically different from that of 1900. We continue to explore the universe with our minds and Earth-based instruments. But since 1957, we've been able to launch probes into space to explore the universe directly. Thus, advances in space science increasingly depends on efforts to send these probes, and people, into space.

Section Review

Key Terms

astronomy
degree
eccentricity
focus
geocentric
geostatic
heliocentric
light year
minute of arc
parallax
period
perturbation
relativity
sublunar realm
superlunar realm

Key Concepts

➤ Two distinct traditions existed in astronomy through the early 1600s
  • Aristotle's geocentric universe of concentric spheres
  • Ptolemy's complex combinations of circles used to calculate orbits for the Sun, Moon, and planets

➤ Several natural philosophers and scientists completely reformed our concept of space from 1500 to the 20th Century
  • Copernicus defined a heliocentric (sun-centered) universe
  • Brahe vastly improved the precision of astronomical observations
  • Kepler developed his three laws of motion
    - The orbits of the planets are ellipses with the Sun at one focus
    - Orbits sweep out equal areas in equal times
    - The square of the orbit period is proportional to the cube of the mean distance from the Sun
  • Galileo developed dynamics and made key telescopic discoveries
  • Newton developed his three laws of motion and the law of universal gravitation
  • Shapley proved our solar system was near the fringe, not the center, of our galaxy
  • Hubble helped show that our galaxy was only one of billions of galaxies, and that the universe was expanding at an ever-increasing rate, perhaps due to a "Big Bang" at the beginning of time
  • Einstein developed the theory of relativity and the relationship between mass and energy described by $E = mc^2$

2.2 Entering Space

▤ In This Section You'll Learn To...

☞ Describe the rapid changes in space exploration in the 20th century from the first crude rockets to space shuttles

We've shown that we need not actually "slip the surly bonds of Earth" to explore space. With our senses, imagination, and instruments such as the telescope, we can discover new features of and raise new questions about the universe. But somehow, that's not enough. People long to go there. Even before the myth of Daedalus and Icarus, we dreamed of flying. On December 17, 1903, Orville and Wilbur Wright made this dream come true and challenged the skies above us. From then on, advances in flight and rocketry have led us right to the edge of space and beyond.

The Age of Rockets Begins

Kepler journeyed to the Moon by magic, and Jules Verne's hero in *From the Earth to the Moon* was fired out of an immensely powerful cannon. But most scientists have agreed that rockets offer the best promise for space flight. Rocket development was driven largely by military requirements. Incendiary devices such as fire arrows have appeared in war for thousands of years, but the first recorded military use of rockets came in 1232 A.D., when the Chin Tartars defended Kai-feng-fu in China by firing rockets at the attacking Mongols.

In the early nineteenth century Sir William Congreve (1772–1828), a British colonel and artillery expert, developed incendiary rockets based on models captured in India. Congreve's rockets, powered by black powder, ranged in size from 3 to 23 kg (6.6 to 50 lbs.). During the Napoleonic wars, the British fired two hundred of these rockets in thirty minutes against the French at Boulogne in 1806, setting the town on fire. They also fired approximately three hundred against Copenhagen in 1807 to prevent the French from seizing the Danish fleet.

Although their inaccuracy limited their use on the battlefield, rockets demonstrated their potential at the Battle of Leipzig in 1813. The Russian general, Ludwig Wittgenstein (1769–1843), remarked that the rockets "look as if they were made in hell, and surely are the devil's own artillery." [Haythornthwaite, 1990] The British created a rocket corps as an adjunct to the Royal Artillery, but after 1815 conventional artillery rapidly improved, so the British Army became less interested in researching more powerful rockets.

The waning military interest in rockets in Britain didn't deter all theoretical studies, however. The first person to research rocket-powered spaceflight was Konstantin Tsiolkovsky (1857–1935), the father of Russian cosmonautics. In the 1880s he calculated the velocity required for a journey beyond the Earth's atmosphere. He also suggested that burning a combination of liquid hydrogen and liquid oxygen could improve rocket efficiency. (The Space Shuttle's main engines run on these propellants!) Inspired by Tsiolkovsky's brilliance, the Soviet Union was the first country to endorse and support the goal of spaceflight, creating in 1924 the Bureau for the Study of the Problems of Rockets.

The United States, in contrast, lagged far behind, except for a single visionary, Robert H. Goddard (1882–1945), shown with one of his first rockets in Figure 2-14. He experimented with liquid-fuel rockets, successfully launching the first in history on March 16, 1926. A skilled engineer and brilliant theorist, Goddard believed that a powerful-enough rocket could reach the Moon or Mars, but he had no support from the United States government.

A far different state of affairs existed in Germany. Hermann Oberth's work on the mathematical theory of space flight fostered the growth of rocketry in Germany and led to the founding of the Society for Space Travel in July, 1927. Several German rocket societies flourished in the 1920s and 1930s, composed mainly of students and their professors. Strong government backing of these organizations began in the mid-1930s, resulting directly from the Treaty of Versailles, which ended World War I. The treaty severely limited Germany's development and production of heavy artillery. After Adolf Hitler assumed power in 1933, the German military saw rockets as a way to deliver warheads over long distances without violating the treaty. They thus supported several of the rocket societies. Wernher Von Braun (1912–1977), a young member of one of these societies, progressed rapidly and finally led the development of the V-2 rocket (the world's first ballistic missile). Von Braun's true desire was to develop launch vehicles for interplanetary flight. His life-long rocket research culminated in the Saturn V moon rocket.

The Germans launched more than two thousand V-2s, armed with one-ton warheads, before the end of World War II. The rockets exploded in Antwerp, London, and other parts of England. Towards the end of the war, all the Allies frantically sought to recruit German rocket scientists. The United States hit paydirt with Project Paperclip when Von Braun and his research team, carrying all their records with them, surrendered to the Americans in 1945. The Soviets also recruited heavily, signing on German scientists as the technical nucleus of the effort that produced Sputnik.

Besides recruiting many German rocket scientists, the United States captured the enormous Mittelwerke underground rocket factory in the Harz Mountains of central Germany, as well as enough components to assemble 68 V-2 rockets. Using these captured V-2s as sounding rockets, with scientific payloads in place of warheads, the V-2 Upper Atmosphere Research Panel studied the Earth's atmosphere between 1946 and 1952 and inaugurated X-ray astronomy. The V-2s launched at White Sands, New Mexico, yielded "a rich harvest of information on atmospheric temperatures, pressures, densities, composition, ionization, winds, and atmospheric and solar radiations, the Earth's magnetic field at high

Figure 2-14. Robert Goddard. Goddard pioneered the field of liquid-fueled rocketry. He's shown here with one of his early models. *(Courtesy of NASA)*

altitudes, and cosmic rays." [Newell, 1980] Until 1957, American astronomers studied space at a leisurely pace. Cold War tensions made national security, not astronomy, a priority. Scientists developed rockets such as the Thor Intermediate Range Ballistic Missile (IRBM) and Atlas Intercontinental Ballistic Missile (ICBM), not to explore space, but to deliver nuclear warheads.

Sputnik: The Russian Moon

By the end of the 1950s, the distinction between airplane and rocket research began to blur. Pilots assigned to the National Advisory Committee for Aeronautics (NACA) in the United States, were flying experimental aircraft, such as the Bell X-1A and X-2, to the edge of the Earth's atmosphere. In this era, many aerospace experts believed that space would first be explored by pilots flying "spaceplanes." A strong candidate was North American Aviation's X-15, shown in Figure 2-15, a

Figure 2-15. X-15 Rocket Plane. The X-15, shown hanging from the wing of a B-52, was piloted to the edge of space and helped develop modern aeronautics and astronautics. *(Courtesy of NASA)*

rocket-propelled spaceplane able to exceed Mach 8 (eight times the speed of sound) and climb more than 112 km (70 miles) above the Earth. Confident in its technological supremacy, the United States was shocked when the Soviets launched the unmanned satellite Sputnik, shown in Figure 2-16, into orbit on October 4, 1957. Sputnik changed everything—space became the new high ground in the Cold War, and, in the crisis atmosphere following Sputnik, Americans believed they must occupy it first.

At least initially, however, the Soviets surged ahead in what the media quickly labeled the Space Race. Less than one month after Sputnik, the Soviets launched Sputnik II on November 3, 1957. Sputnik II carried a dog named Laika, the first living creature to orbit the Earth. In the meantime,

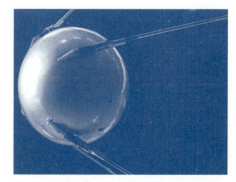

Figure 2-16. Sputnik. Sputnik shook the world when it became the first-ever man-made satellite in 1957. *(Courtesy of NASA)*

the United States was desperate to orbit its own satellite, but the first attempt, a Navy Vanguard rocket, exploded on the launch pad on December 6, 1957, in front of a national television audience. Fortunately, Von Braun and his team had been working since 1950 for the Army Ballistic Missile Agency on the Redstone rocket, a further development of the V-2. Using a modified Redstone, the United States successfully launched its first satellite, Explorer 1, on January 31, 1958.

After the launch of Explorer 1, shown in Figure 2-17, Americans clamored for more space feats to match and surpass the Communists. Most shocking to American scientists was the ability of their Soviet counterparts to orbit heavy payloads. Explorer 1 weighed a feather-light 14 kg (30 lbs), whereas the first Sputnik weighed 84 kg (185 lbs). Sputnik III was a geophysical laboratory orbited on May 15, 1958, and weighed a whopping 1350 kg (2970 lbs). The Soviets also enthralled the world in October, 1959, when their space probe Luna 3 took the first photographs of the dark side of the Moon.

Without question the Soviet Premier, Nikita Khrushchev (1894–1971), was mounting a deliberate and effective propaganda effort. Khrushchev exploited the space feats of Sergei Korolev, the "Grand Designer" of Soviet Rocketry from the 1930s until his death in 1966, to prove that Communism was superior to Democracy.

Meanwhile, calls went out across the Western world to improve science education in schools and to mobilize the best and brightest scientists to counter the presumed Soviet technical superiority. The capstone for this effort was the creation of the National Aeronautics and Space Administration (NASA) on October 1, 1958.

NASA was created to put a man in space before the Soviets. It also managed the rapidly mushrooming budget devoted to the American space effort, which rocketed from $90 million for the *entire* effort in 1958 to $3.7 billion for NASA *alone* in 1963. These were heady days for NASA, a time of great successes and equally spectacular failures. In this scientific dimension of the Cold War, American astronauts became more like swashbucklers defending the ship of state than sage scientists seeking knowledge. President Dwight D. Eisenhower (1890–1969) reinforced this nationalistic mood when he recruited the first Mercury astronauts exclusively from the ranks of military test pilots. (M. Scott Carpenter, Jr., L. Gordon Cooper, Jr., John H. Glenn, Jr., Virgil I. "Gus" Grissom, Walter M. Schirra, Jr., Alan B. Shepard, Jr., and Donald K. "Deke" Slayton were the original seven.) The Mercury program in particular suffered through serious initial teething pains. Mercury-Atlas 1, launched on July 29, 1960, exploded one minute after lift-off. On November 21, 1960, Mercury-Redstone 1 ignited momentarily, climbed about 10 cm (4 inches) off the launch pad before its engine cutoff, then settled back down on the launch pad as the escape tower blew. Despite these failures, NASA persisted. On January 21, 1961, Mercury-Redstone 2 successfully launched a chimpanzee called Ham on a suborbital flight.

Once again, however, the Soviets caused the world to hold its collective breath when on April 12, 1961, Major Yuri A. Gagarin, shown in Figure 2-

Figure 2-17. Explorer I. Explorer 1 was the first United States satellite launched after many frustrating failures. *(Courtesy of NASA)*

18, completed an orbit about the Earth in Vostok I. Here was proof positive, in Khrushchev's eyes, of the superiority of Communism. After all, the United States could only launch astronauts Alan Shepard and Gus Grissom on suborbital flights in May and July, 1961. The Soviets astounded the world one more time on August 7, 1961, by launching cosmonaut Gherman S. Titov on a day-long, seventeen-orbit flight about the Earth. After what some Americans no doubt thought was an eternity, the United States finally orbited its first astronaut, John Glenn, on February 20, 1962. Glenn's achievement revived America's sagging morale, although he completed only three orbits due to a problem with his capsule's heat shield.

From 1961–1965, the Soviets orbited more astronauts (including one woman, Valentina Tereshkova, on June 16, 1963), for longer periods, and always sooner than their American rivals. The Soviets always seemed to be a few weeks or months ahead. For example, cosmonaut Alexei Leonov beat astronaut Edward H. White II by eleven weeks for the honor of the first spacewalk, which took place on March 18, 1965.

Figure 2-18. Yuri Gagarin. The Soviet Union scored another first in the space race when Yuri Gagarin became the first human to orbit the Earth. *(Courtesy of NASA)*

Americans were able to record some space firsts in communication, however. Echo 1, launched on August 12, 1960, was an aluminum-coated, 30.4 meter-diameter plastic sphere which passively reflected voice and picture signals, demonstrating the feasibility of satellite communications. Based on Echo's experimental success, the first active communications satellite was Telstar, launched on July 10, 1962, which relayed communication between far-flung points on the Earth. Echo and Telstar were the forerunners of the incredible variety of communication satellites now circling the Earth.

Astro Fun Fact

Disney and the Soviet Rocket Program

The Soviet rocket program was unwittingly boosted by the Walt Disney movie, <u>Man In Space</u>, shown in August, 1955, at an international space conference in Copenhagen which two Soviet rocket scientists attended. The Soviet newspaper, Pravda, reported on the conference and increased the Soviet feeling that the United States was serious about space. The Pravda article also emphasized the need for more Soviet public awareness of spaceflight. Professor Leonid I. Sedov, Chairman of the U.S.S.R. Academy of Science's Interdepartmental Commission on Interplanetary Communications, addressed this issue in his Pravda article of September 26, 1955: "A popular-science artistic cinema film entitled <u>Man In Space</u>, released by the American director Walt Disney and the German rocket-missile expert Von Braun, chief designer of the V-2, was shown at the Congress. . . it would be desirable that new popular-science films devoted to the problems of interplanetary travels be shown in our country in the near future. It is also extremely important to increase the interest of the general public in the problem of astronautics. Here is a worthwhile field of activity for scientists, writers, artists, and for many works of Soviet culture." As a result, the Soviets became more concerned with the United States' space effort and focused more on their own efforts to launch satellites.

Krieger, F.J. <u>Behind the Sputniks—A Survey of Soviet Space Science</u>. Washington, D.C.: Public Affairs Press, 1958.

Contributed by Dr. Jackson R. Ferguson, Jr., United States Air Force Academy

Although Echo, Telstar, and interplanetary probes such as the Mariner series helped reinflate sagging American prestige in the early 1960s, President John F. Kennedy (1917–1963) was not satisfied. In a speech before a joint session of Congress in 1961, he launched the nation in a bold, new direction. "I believe this nation should commit itself to achieving the goal, before this decade is out, of landing a man on the Moon and returning him safely to the Earth." For Kennedy, the clearest path to American pre-eminence in space led directly to the Moon.

Armstrong's Small Step

The Apollo program to put an American on the Moon by 1970 was meant to accept the challenge of the space race with the Soviets and prove which country was technically superior. By 1963, Apollo already consumed two-thirds of America's total space budget. Support for Apollo was strong among Americans, but the program was not without its critics. Former President Eisenhower remarked that "Anybody who would spend 40 billion dollars in a race to the Moon for national prestige is nuts."

Critics notwithstanding, Neil Armstrong and Buzz Aldrin's footprints in the lunar soil, as shown in Figure 2-19, on July 20, 1969, demonstrated an amazing technical achievement. Most of the world watched the dramatic exploits of Apollo 11 with great excitement and a profound sense of awe. President Richard M. Nixon was perhaps only slightly exaggerating when he described Apollo 11's mission as the "greatest week in the history of the world since the Creation."

Begun as a nationalistic exercise in technological chest-thumping, Apollo became a spiritual adventure, a wonderous experience that made people stop and think about life and its true meaning. The plaque left behind on the Apollo 11 lunar module perhaps best summed up the mission. "**Here men from the planet Earth first set foot upon the Moon July 1969, A.D. We came in peace for all mankind.**" A small gold olive branch remains today at the Sea of Tranquility, left behind by the Apollo 11 astronauts as a symbol of peace.

Apollo was NASA's greatest triumph. Even when disaster struck 330,000 km (205,000 miles) from the Earth during the Apollo 13 mission, NASA quickly rallied and brought the astronauts home safely. After Apollo 11, there would be five additional Moon landings between November, 1969, and December, 1972. The six Apollo Moon missions combined brought back about 382 kg (840 lbs) of Moon rocks and soil, which scientists hoped would be a "Rosetta stone," helping them date the very beginning of the solar system. Regrettably, the rocks turned out to be millions of years younger than expected. Still, the Apollo missions revealed much about the nature of the Moon, although many scientists remained critical of Apollo's ultimate scientific worth. This criticism came partly because only on the last mission (Apollo 17) did NASA send a scientist, Dr. Harrison H. (Jack) Schmitt, shown in Figure 2-20, to explore the Moon.

Figure 2-19. First on the Moon. Neil Armstrong and Buzz Aldrin left their footprints on the surface of the Moon. *(Courtesy of NASA)*

Figure 2-20. Lunar Goal Achieved. Astronaut Harrison Schmitt poses with the American flag and Earth in the background. *(Courtesy of NASA)*

After the clear-cut triumph of Apollo, NASA's achievements were considerable if somewhat anticlimactic. For example, in a display of brilliant improvisation, NASA converted a Saturn V third-stage into a laboratory module known as Skylab, shown in Figure 2-21. After suffering external damage during launch on May 14, 1973, Skylab was successfully repaired by space-walking astronauts and served three separate crews as a laboratory in space for 28, 59, and 84 days, respectively.

In July, 1975, space also became a forum for international cooperation between strong and sometimes bitter rivals when Apollo 18 docked with a Soviet Soyuz spacecraft. A triumph for detente, Apollo-Soyuz was perhaps a harbinger of the future of space exploration—progress through cooperation and friendly competition.

With the end of the Apollo program, NASA began work on a new challenge—the Space Shuttle. On April 12, 1981, twenty years to the day after Gagarin's historic flight, astronauts John Young and Robert Crippen piloted the reusable spacecraft Columbia on its maiden flight. Columbia, joined by her sister ships Challenger, Discovery, and later Endeavour, proved the versatility of the shuttle fleet. Shuttle missions have been used to deploy satellites, launch interplanetary probes, rescue and repair satellites, conduct experiments, as shown in Figure 2-22, and monitor the Earth.

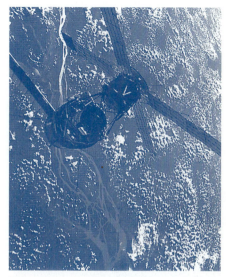

Figure 2-21. Skylab. Launched on May 14, 1973, Skylab was America's first space station. During nine months, three different crews called it home. Pulled back to Earth because of atmospheric drag, Skylab burned up in 1979. *(Courtesy of NASA)*

Figure 2-22. Experiments on the Space Shuttle. Guion Bluford works on a Space Shuttle mid-deck experiment. A variety of experiments in biology, life sciences, and materials processing have been conducted by astronauts aboard the Shuttle. *(Courtesy of NASA)*

While the U.S. focused on the Shuttle, the Soviet space program took a different path. Relying on the expendable Soyuz booster, the Soviets concentrated their efforts on a series of space stations designed to extend the length of human presence in space. From 1971–1982, seven increasingly capable variations of Salyut space stations were launched, home to dozens of crews. In 1986, the expandable Mir Space Station marked a new generation of space habitat. In 1988, cosmonauts Vladimir Titov and Musa Manarov became the first humans to spend more than one year in space aboard Mir.

Satellites and Interplanetary Probes

Today, space is a part of all our lives. Satellites serve as navigation beacons, relay stations for radio and television signals, and other forms of communication. They also watch our weather and help us verify international treaties. Satellites are able to discover phenomena that cannot be observed or measured from the Earth's surface because of atmospheric interference. While the race to the Moon was on in the 1960s, scientists used satellites to explore the atmosphere. They studied gravitational, electrical, and magnetic fields, energetic particles, and other space phenomena. In addition, projects such as the United States Land Remote Sensing Program (LandSat) have proved invaluable for tracking and managing Earth's resources.

We've also achieved great results with interplanetary probes. In the 1960s the Mariner probes took close-up photographs of Mercury, Venus, and Mars. On February 12, 1961, the Soviet Union launched Venera 1, which passed within 60,000 km (37,200 miles) of Venus and analyzed its atmosphere. The United States explored Jupiter and Saturn with the Pioneer probes. Some of the most recent successes include the exploration of Mars in 1976 by Vikings 1 and 2 and the exploration of Jupiter and Saturn in 1980–1981 by Voyagers 1 and 2. Voyager 2 went on to explore Uranus in 1986 and Neptune in 1989. The Magellan probe in 1991 completed a radar mapping of more than 80% of Venus's surface, revealing extraordinary geological diversity and activity.

We're also moving some of our most complex and expensive scientific instruments into space to escape the interference of the Earth's atmosphere. Perhaps the most famous of these space observatories was the Hubble Space Telescope, named in honor of astronomer Edwin Hubble and launched in April, 1990. Although initially hampered by an optical flaw, it has enabled us to peer back nearly to the dawn of time. The Shuttle has also carried into space other astronautical observatories with specialized instruments to observe ultraviolet and X-ray phenomena, bands of the electromagnetic spectrum that we simply can't perceive with our own eyes. Other projects with great potential include the Ulysses solar-polar observatory, the Galileo mission to Jupiter, and the proposed "Mission to Planet Earth."

The Future

What does the future of space exploration hold? First, we must recognize that *space exploration is inherently dangerous*. During ground testing of the Apollo 1 command module on January 17, 1967, a fire killed

Virgil I. (Gus) Grissom, Edward H. White II, and Roger B. Chaffee. The Soviets have also had their tragedies, with cosmonaut Vladimir Komarov dying in April 1967, when the parachute lines of his capsule became entangled during his return to Earth. The Challenger explosion on January 28, 1986, killed the seven crew members, shown in Figure 2-23, including Teacher in Space participant Christa McAuliffe. It also sobered Americans to the dangers of space exploration. Goals worth achieving, however, usually carry with them a measure of risk, and human beings have shown themselves willing to take great risks for even greater rewards.

Figure 2-23. The Crew of the Space Shuttle Challenger. These seven astronauts paid the ultimate price of exploration on January 28, 1986. Front row: Mike Smith, Dick Scobee, and Ron McNair. Back row: El Onizuka, Christa McAuliffe, Greg Jarvis, and Judy Resnik. *(Courtesy of NASA)*

Interplanetary probes will no doubt continue to amaze us with new information about our solar system. Many practical lessons have come from studying Earth's neighbors—such as the large dust storms Mariner 9 observed on Mars in 1971. Scientists now believe these storms cooled Mars' surface. Some have since speculated about potentially disastrous cooling of the Earth's surface from similar large dust clouds that would be raised by a nuclear war. Similarly, further studies of Venus's atmosphere may shed some light on the greenhouse effect and global warming from carbon dioxide emissions on our own planet.

Many, many scientific questions remain unanswered. No doubt, some also remain unasked. Scientists are searching for some form of "Dark Matter," which they believe constitutes at least 90% of the universe's mass. Ironically, some scientists today are contemplating reviving Aristotle's distinction between terrestrial and heavenly physics, although in a vastly different form and for very different reasons. We need a special physics—quantum mechanics—to describe the motions and attributes of subatomic particles or the "very small." Perhaps we need a different physics—cosmological mechanics—to describe the motions and attributes of galaxies or the "very big." Black holes, naked singularities, pulsars, quasars, the Big Bang theory, the expanding universe—scientists have much left to explore.

Interstellar travel over immense distances will most likely remain impractical for some time. But interplanetary travel, though expensive, is quite possible. For example, we can travel to Mars by the year 2019, although an international mission may be necessary to meet the enormous expense and risk. Meanwhile, as of 1991, the United States has already spent billions of dollars planning space station Freedom. The Soviets, and later the Russians, have kept their smaller space station, Mir or Peace, almost continually manned since 1986. Also under development in the United States is the X-30 National Aerospace Plane, shown in Figure 2-24, which will take-off from a runway and fly directly into low-Earth orbit.

Figure 2-24. The National Aerospace Plane (NASP). NASP promises to offer cheap access to low-Earth orbit. (Courtesy of NASA)

The exact steps of humankind's future explorations and adventures in space are not yet known. But these future steps, and astonishingly giant leaps, are certain. Men and women will always have more to explore in space, as long as we remain creative, imaginative, and strong-minded.

Section Review

Key Concepts

➤ Rockets were first developed by the Chinese, and later were used to explore space after the Second World War

➤ Sputnik, launched by the Soviet Union on October 4, 1957, was the first artificial satellite to orbit the Earth

➤ Yuri Gagarin was the first human to orbit the Earth on April 12, 1961

➤ The space race between the United States and Soviet Union culminated in Apollo 11, when Armstrong and Aldrin became the first humans to walk on the Moon

➤ Space exploration thrives today through manned and unmanned probes. New ideas, new discoveries, and new worlds await future explorers—perhaps you?!

References

Asimov, Isaac. *Asimov's Biographical Encyclopedia of Science and Technology.* Garden City, New York: Doubleday and Company, Inc., 1972.

Kepler, Johannes. The complete title of the *Somnium* is *IOH. Keppleri Mathematitici Olim Imperatorii Somnium, Sev Opus posthumum De Astronomia Lunari. Divulgatum M. Ludovico Kepplero Filio, Medicinae Candidato.* 1634.

Edinburgh Evening Courant, 20 January 1814, quoted in Philip J. Haythornthwaite, *The Napoleonic Source Book.* New York: Facts on File, 1990. p. 92.

Galilei, Galileo. *Dialogue Concerning the Two Chief World Systems.* (orig. 1632) Translated by Stillman Drake. Berkeley, California: University of California Press, 1967.

Newell, Homer E. *Beyond the Atmosphere: Early Years of Space Science.* Washington, D.C.: Scientific and Technical Information Branch, NASA, 1980. p. 37.

Rycroft, Michael. *The Cambridge Encyclopedia of Space.* France: Cambridge University Press, 1990.

Wilson, Andrew. *Space Directory.* Alexandria, Virginia: Jane's Information Group Inc., 1990.

For Further Reading

Baker, David. *The History of Manned Space Flight.* New York: Crown Publishers, 1981.

Compton, William D. *Where No Man Has Gone Before: A History of Apollo Lunar Exploration Missions.* Washington, D.C.: National Aeronautics and Space Administration, 1989.

Crowe, Michael J. *Extraterrestrial Life Debate, 1750 - 1900: The Idea of a Plurality of Worlds.* Cambridge and New York: Cambridge University Press, 1986.

Quoted in Allen G. Debus, *Man and Nature in the Renaissance.* Cambridge: Cambridge University Press, 1978. p. 96.

Dick, Steven J. *Plurality of Worlds: The Origins of the Extraterrestrial Life Debate from Democritus to Kant.* Cambridge and New York: Cambridge University Press, 1982.

Galilei, Galileo. *Sidereus Nuncius.* 1610. Translated with introduction, conclusion, and notes by Albert Van Helden. Chicago: University of Chicago Press, 1989.

Gingerich, Owen. *Islamic Astronomy.* Scientific American, April 1986. Pages 74-83.

Hearnshaw, J.B. *The Analysis of Starlight: One Hundred and Fifty Years of Astronomical Spectroscopy.* Cambridge: Cambridge University Press, 1986.

Hetherington, Norriss S. *Science and Objectivity: Episodes in the History of Astronomy.* Ames: Iowa University Press, 1988.

Krieger, Firmin J. *Behind the Sputniks: A Survey of Soviet Space Science.* The Rand Corporation: Washington, D.C. Public Affairs Press, 1958.

Lewis, Richard. *Space in the 21st Century.* New York: Columbia University Press, 1990.

McDougall, Walter A.*...the Heavens and the Earth: A Political History of the Space Age.* New York: Basic Books, 1985.

Michener, James. *Space.* New York: Ballantine Books, 1982.

Sheehan, William. *Planets and Perception: Telescopic Views and Interpretations, 1609–1909.* Tucson: University of Arizona Press, 1988.

Smith, Robert W. *The Expanding Universe: Astronomy's 'Great Debate' 1900–1931.* Cambridge: Cambridge University Press, 1982.

Van Helden, Albert. *Measuring the Universe: Cosmic Dimensions from Aristarchus to Halley.* Chicago: University of Chicago Press, 1985.

Winter, Frank H. *Rockets into Space.* Cambridge, Mass: Harvard University Press, 1989.

Winter, Frank H. *The First Golden Age of Rocketry; Congreve and Hale Rockets of the Nineteenth Century.* Washington, D.C.: Smithsonian Institution Press, 1989.

Wolfe, Tom. *The Right Stuff.* New York: Farrar, Straus & Giroux, 1979.

Mission Problems

2.1 Early Space Explorers

1 Why did astronomers continue to believe the planets were in circular orbits from Aristotle's time until Kepler's discovery?

2 Why were people reluctant to adopt Copernicus's heliocentric system of the universe?

3 Why might we call Brahe and Kepler a perfect team?

4 What role did Galileo's belief in God play in his view of astronomy?

5 How did Newton complete the Astronomical Revolution of the seventeenth century?

6 Why are instruments like the telescope such powerful tools for learning and discovery?

7 How did our concept of the universe change in the first few decades of the 20th century?

2.2 Entering Space

8 What are the various ways that man can explore space?

9 Why might we call 1957–1965 the years of Soviet supremacy in space?

10 What was the chief legacy of the Apollo manned missions to the Moon?

For Discussion

11 Do we need to launch men and women into space, or should we rely exclusively on probes and earth-based instruments to explore space?

12 In exploring space, can we learn anything that can help us solve problems here on Earth?

13 How much do movies and television shows such as *2001: A Space Odyssey*, *Alien*, *Star Wars*, and *Star Trek* point the way to humanity's future in space?

14 Has the search for extraterrestrials been important to the development of astronomy? Should we continue searching for other life forms and inhabited planets?

Mission Profile—V-2

The V-2 rocket was first developed by the German Army in 1944. V-2 stood for "Vengeance Weapon 2," indicative of Hitler's wish for a weapon which could conquer the world. Yet the scientists who developed the missile worked independently of Hitler's influence until the missile was fully developed. They called their experimental rocket the A-4. Hitler actually had little interest in the rocket's development and failed to adequately fund the project until it was too late to decisively employ it in the Second World War. The rocket itself had no single inventor. Rather, it was the result of a team effort of individual Germans who envisioned both the good and ill of applied modern rocketry.

Mission Overview

The V-2 (or A-4) program had obscure beginnings in the late 1920s. Many of the members of the initial V-2 team wished to create a long range rocket which might serve as a stepping stone for future spaceflight applications. Yet the mission remained open-ended: the small team of scientists at Peenemuende (located on the Baltic coast) did not focus on the future. Because Germans were not allowed to produce mass artillery, as a result of the Treaty of Versailles ending World War I, the project's initial drive was to devise a new powerful weapon not outlawed by the treaty. The missile was officially constructed and funded as a tactical weapon with improved capabilities and a longer range than the existing long-range artillery.

Mission Data

✓ The V-2 was a single-stage rocket which burned a liquid oxygen and kerosene mixture for a thrust of 244,600 N (55,000 lbs.)

✓ The maximum design range of the missile was 275 km (171 miles), enabling Hitler to bomb London as well as continental allied countries.

✓ The actual maximum altitude reached by the V-2 was 83 km (52 miles), for a total trajectory distance of 190 km (118 miles). This was extraordinarily better than any previous missile.

✓ The overall missile length was 14 m (46 ft.), with a maximum width of 3.57 m (11 ft. 8.4 in.)

✓ The V-2 was never mass-produced at Peenemuende. Once production was moved to mainland Germany in 1944, 3745 were produced and launched for the Axis war effort.

✓ The V-2 design team produced 60,000 necessary changes after it was considered ready for mass production.

✓ The primary advantage of the V-2 was its cost of $38,000. This compared favorably to the $1,250,000 cost of a manned German bomber.

Mission Impact

While many of the original Peenemuende team dreamed of possible implications for space travel, the V-2 was first and foremost a machine of war. Yet, it was the first supersonic rocket and is generally regarded as a monumental step in the history of modern rocket technology. After the war was over, the technology of the V-2 and many German scientists came to the United States, forming the foundation of the future U.S. space program.

For Discussion

• Can you think of any other technological advances that were initiated and developed through military channels?

• Was the V-2 truly the beginning of the drive towards space travel? What important developments led to the U.S. drive to place a man on the moon?

• What happened to many of the top scientists from Peenemuende after World War II? Do you think that scientific interest should supersede political agendas?

Contributor

Troy Kitch, United States Air Force Academy

References

Emme, Eugene M., (ed.) *The History of Rocket Technology, "The German V-2"* by Walter R. Dornberger. Detroit: Wayne State University Press, 1964.

Gatland, Kenneth. *Missiles and Rockets.* New York: Macmillan Publishing Co., 1975.

An astronaut's eye view of our blue planet. *(Courtesy of NASA)*

The Space Environment

3

In This Chapter You'll Learn to...

☛ Explain where space begins and describe our place in the universe

☛ List the major hazards of the space environment and describe their effects on spacecraft

☛ List and describe the major hazards of the space environment that pose a problem for humans living and working in space

You Should Already Know...

❏ The elements of a space mission (Chapter 1)

In space, no one can hear you whine.

Anonymous

Space is a place. Some people think of space as a nebulous region far above their heads—extending out to infinity. But for us, space is a place where things happen: satellites orbit the Earth, planets orbit the Sun, and the Sun revolves around the center of our galaxy.

In this chapter we'll look at this place we call space. We'll explore where it begins and how far it extends. You'll see that space is actually very close. Then, starting with our "local neighborhood," we'll take a mind-expanding tour beyond the galaxy to see what's in space. Next we'll see what space is like. Before taking any trip, we usually check the weather, so we'll know whether to pack a swim suit or a parka. In the same way, we'll look at the space environment to see how we must prepare ourselves and our machines to handle this hostile environment.

3.1 Cosmic Perspective

▬ In This Section You'll Learn to...

- ☞ Explain where space is and how it's defined
- ☞ Describe the primary outputs from the Sun that dominate the space environment
- ☞ Provide some perspective on the size of space

Where is Space?

If space is a place, where is it? Safe within the cocoon of Earth's atmosphere, we can stare into the night sky at thousands of stars spanning millions of light years. We know space begins somewhere above our heads, but how far? If you get into a powerful jet fighter plane and "push the envelope" of its ability, you can barely make it to a height where the sky takes on a purplish color and stars become visible in the light of day. But even then, you're not quite in space. Only by climbing aboard a rocket can you escape entirely above Earth's atmosphere into the realm we normally think of as space.

But the line between where the atmosphere ends and space begins is by no means clear. In fact, there is no universally accepted definition of precisely where space begins. If you ask NASA or the U.S. Air Force, you'll find their definition of space is somewhat arbitrary. To earn the right to wear astronaut wings, for example, you must reach an altitude of more than 50 nautical miles (92.6 km) but don't actually have to go into orbit, as illustrated in Figure 3-1. (That's why X-15 pilots and the first United States astronauts to fly suborbital flights in the Mercury program were able to wear these much-coveted wings.)

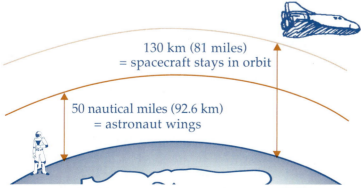

Figure 3-1. Where is Space? For awarding astronaut wings, NASA defines space at an altitude of 50 nautical miles (92.6 km). For our purposes, space begins where satellites can maintain orbit—about 130 km (81 miles).

Figure 3-2. Shuttle Orbit Drawn Closer to Scale. (If drawn exactly to scale, you wouldn't be able to see it!) As you can see, space is very close. Space Shuttle orbits are just barely above the atmosphere.

Although this definition works, it's not very meaningful. For our purposes, as seen in Figure 3-1, space begins at the altitude where an object in orbit will remain in orbit briefly (only a day or two in some cases) before the wispy air molecules in the upper atmosphere drag it back to Earth. This occurs above an altitude of about 130 km (about 80 miles). That's about the distance you can drive in your car in just over an hour! So the next time someone asks you, "how do I get to space?" just tell them to "turn straight up and go about 130 km (80 miles) until the stars come out."

As you can see, space is very close. Normally, when you see drawings of orbits around the Earth (as you'll see in later chapters), they look far, far away. But these diagrams are seldom drawn to scale. To put low-Earth orbits (LEO), like the ones flown by the Space Shuttle, into perspective, imagine the Earth were the size of a peach—then a typical Shuttle orbit would be just above the fuzz. A diagram closer to scale (but not exactly) is shown in Figure 3-2.

Now that we have some idea of where space is, let's take a grand tour of our "local neighborhood" to see what's out there. We'll begin by looking at the solar system and expand our view to cover the galaxy.

The Solar System

At the center of the solar system is the star closest to the Earth—the Sun. As we'll see, the Sun has the biggest effect on the space environment. As stars go, our Sun is quite ordinary. The Sun is just one small, yellow star out of billions in the galaxy.   Fueled by nuclear fusion, the Sun combines or "fuses" 600 million tons of hydrogen each second. (Don't worry, at that rate the Sun won't run out of hydrogen for about 5,000,000,000 years!). We're most interested in two by-products of the fusion process:

- Electromagnetic radiation
- Charged particles

The energy released by nuclear fusion is governed by Einstein's famous $E = mc^2$ formula. This energy, of course, makes life on Earth possible. And the Sun puts out lots of energy, enough each second to supply all the energy the United States needs for about 624 million years! This energy is primarily in the form of electromagnetic radiation. In a clear, blue sky, the Sun appears as an intensely bright circle of light. With your eyes closed on a summer day, you can feel the Sun's heat beating down on you. But light and heat are only part of the Sun's *electromagnetic (EM) radiation*. The term "radiation" often conjures up visions of nuclear wars and mutant space creatures, but EM radiation is something we live with every day. EM radiation is a way for energy to get from one place to another. You can think of the Sun's intense energy as radiating out from its surface in all directions in waves. We classify these waves of radiant energy in terms of the distance between wave crests, or *wavelength, λ*, as in Figure 3-3.

What difference does changing the wavelength make? If you've ever seen a rainbow on a sunny spring day, you've seen the awesome beauty of changing the wavelength of EM radiation by only 0.0000003 meters (9.8 ×

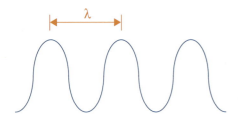

Figure 3-3. Electromagnetic (EM) Radiation. We classify EM radiation in terms of the wavelength, λ, (or frequency) of the energy.

10^{-7} ft.)! The colors of the rainbow, from violet to red, represent only a very small fraction of the entire electromagnetic spectrum. This spectrum spans from high energy X-rays (like you get in the dentist's office) at one end to long-wavelength radio waves (like your favorite FM station) at the other. Light and all radiation move at the speed of light—300,000 km/s or more than 671 million miles per hour! As we'll see, solar radiation can be both helpful and harmful to spacecraft and humans in space. We'll learn more about the uses for EM radiation in Chapter 11.

The other fusion by-product we're concerned with is charged particles. Scientists model atoms with three building-block particles—protons, electrons, and neutrons, as illustrated in Figure 3-4. Protons and electrons are known as *charged particles*. Protons have a positive charge, and electrons have a negative charge. The neutron, because it doesn't have a charge, is said to be neutral. Protons and neutrons make up the nucleus or center of an atom. Electrons swirl around this dense nucleus.

During the fusion process, intense heat is generated in the Sun's interior (more than 1,000,000° C). At these temperatures, a fourth state of matter exists. We're all familiar with the other three states of matter—solid, liquid, and gas. If we take a block of ice (a solid) and heat it up, we'll get water (a liquid). If we continue to heat the water, it will begin to boil, and we'll get steam (a gas). However, if we continue to heat the steam, we'd eventually get to a point where the water molecules begin to break down. Eventually, the atoms themselves would break into their basic particles and form a hot *plasma*. Thus, inside the Sun, we have a swirling hot soup of charged particles—free electrons and protons. (A neutron quickly decays into a proton plus an electron.)

These charged particles in the Sun don't stay put. All charged particles respond to electric and magnetic fields. Your television set, for example, takes advantage of this by using a magnet to focus a beam of electrons at the screen to make it glow. Similarly, the Sun has an intense magnetic field, so electrons and protons shoot away from the Sun at speeds of 300 to 700 km/s. This stream of charged particles flying off the Sun is called the *solar wind*.

Occasionally, areas of the Sun's surface erupt in gigantic bursts of charged particles called *solar particle events* or *solar flares*, which make all of the nuclear weapons on Earth look like pop guns. Lasting only a few days or less, these flares are sometimes so violent they extend out to Earth's orbit (148 million km or 93 million miles)! Fortunately, such large flares are infrequent (every few years or so) and concentrated in specific regions of space, so they usually miss the Earth entirely. Later, we'll see what kinds of problems these charged particles from the solar wind and solar flares pose to humans and machines in space.

Besides the star of the show, the Sun, nine planets, dozens of moons, and thousands of asteroids are in our solar system. The planets range from the small terrestrial-class ones—Mercury, Venus, Earth, and Mars—to the mighty gas giants—Jupiter, Saturn, Uranus, and Neptune. Tiny Pluto is all alone at the edge of the solar system and may be a lost Moon of Neptune. Figure 3-5 tries to give you some perspective on the size of the solar

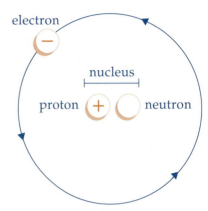

Figure 3-4. The Atom. The nucleus of an atom contains positively charged protons and neutral neutrons. Around the nucleus are negatively charged electrons.

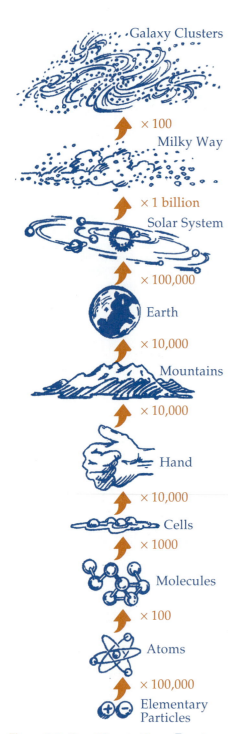

Figure 3-5. The Solar System in Perspective. If the Earth were the size of a baseball, about 10 cm (~4 in.) in diameter, the Moon would be only 2.54 cm (1 in.) in diameter and about 5.6 m (18 ft.) away. At the same scale the Sun would be a ball 10 m (33 ft.) in diameter (about the size and volume of a small two-bedroom house); it would be more than 2 km (nearly 1.3 miles) away. Again, keeping the same scale, the smallest planet Pluto would be about the same size as Earth's Moon, 2.54 cm (1 in.), and 86.1 km (53.5 miles) away from the house-sized Sun.

The Cosmos

Space is big. Really BIG. Besides our Sun, more than 300 billion other stars are in our neighborhood—the Milky Way galaxy. Because the distances involved are so vast, normal human reckoning (miles or kilometers) loses meaning. When trying to understand the importance of charged particles in the grand scheme of the universe, for example, the mind boggles. Figure 3-6 tries to put human references on a scale with the other micro and macro dimensions of the universe.

One convenient yardstick we use to discuss stellar distances is the light year. One light year is the distance light can travel in one year. At 300,000 km/s, this is about 9.46×10^{12} km (about 5.88 trillion miles). Using this measure, we can begin to describe our location with respect to everything else in the universe. The Milky Way galaxy is spiral shaped and is about 100,000 light years across. Earth is located about half way out from the center (about 25,000 light years) on one of the spiral arms. The Milky Way (and we along with it) slowly revolves around the galactic center, completing one revolution every 240 million years or so. The time it takes to revolve once around the center of the galaxy is sometimes called a *cosmic year*. In these terms, our solar system is thought to be about 20 cosmic years old (about 4.8 billion Earth years).

Stars in our galaxy are very spread out. The closest star to our solar system is Alpha Centauri at 4.3 light years or 4.1×10^{13} km away. The Space Shuttle, moving at its orbital velocity of 7.7 km/s (about 17,500 m.p.h.), would take more than 180 million years to get there! Trying to imagine

Figure 3-6. From Micro to Macro. To get an idea about the relative size of things in the universe, start with elementary particles—protons and electrons. You can magnify them 100,000 times to reach the size of an atom, etc.

system, and Appendix D gives some physical data on all the planets. However, because we tend to spend most of our time near Earth, we'll focus our discussion of the space environment on spacecraft and astronauts in Earth orbits.

these kinds of distances gives most of us a headache. The nearest galaxy to our own is Andromeda, which is about 2 million light years away. Beyond Andromeda are billions and billions of other galaxies, all arranged in strange configurations which astronomers are only now beginning to catalog. Figure 3-7 puts the distance between us and the next closest star into understandable terms. Figure 3-8 tries to do the same thing with the size of our galaxy. In the next section we'll beam back closer to home to understand the practical effects of sending men and machines to explore the vast reaches of the cosmos.

Figure 3-7. Stellar Distances. Let our Sun (1.4×10^6 km or 8.6×10^6 miles in diameter) be the size of a large marble, roughly 2.54 cm (1 in.) in diameter. At this scale, the nearest star to our solar system, Alpha Centauri, would be more than 1500 km (932 miles) away. So, if the Sun is the size of a large marble (2.54 cm or 1 in. in diameter) in Denver, Colorado, the nearest star would be in Chicago, Illinois. At this stellar scale, the diameter of the Milky Way galaxy would then be 33.8 million km (21 million miles) across! Still too big for us to visualize!

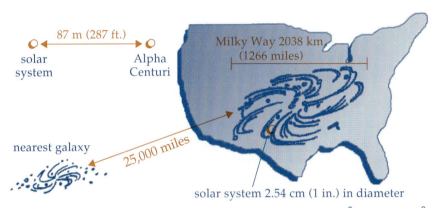

Figure 3-8. Galactic Distances. Imagine the entire solar system (11.8×10^9 km or 7.3×10^9 miles across) were just the size of a large marble 2.54 cm (1 in.) in diameter. At this scale, the nearest star is now 87 m (287 ft.) away. The diameter of the Milky Way galaxy is then 2038 km (1266 miles). So, if the solar system were the size of a marble in Denver, Colorado, the Milky Way galaxy would cover most of the western United States. At this scale, the nearest galaxy is 40,000 km (25,000 miles) away.

▰ Section Review

Key Terms

charged particles
cosmic year
electromagnetic (EM) radiation
plasma
solar flares
solar particle events
solar wind
wavelength

Key Concepts

➤ For our purposes, space begins at an altitude where a satellite can briefly maintain an orbit. Thus, space is close. It's only about 130 km (80 miles) straight up.

➤ The Sun is a fairly average yellow star which burns by the heat of nuclear fusion. Its surface temperature is more than 6000° K and its output includes

• Electromagnetic radiation that we see and feel here on Earth as light and heat

• Streams of charged particles that sweep out from the Sun as part of the solar wind

• Solar particle events or solar flares, which are brief but intense periods of charged-particle emissions

➤ Our solar system is about half way out on one of the Milky Way galaxy's spiral arms. Our galaxy is just one of billions and billions of galaxies in the universe.

3.2 The Space Environment and Spacecraft

▀▀ In This Section You'll Learn to...

☛ List and describe major hazards of the space environment and their effect on spacecraft

To build spacecraft that will survive the harsh space environment, we must first understand what hazards they may face. The Earth, the Sun, and the cosmos combined offer six different challenges to spacecraft designers, as shown in Figure 3-9. We'll begin by looking at the gravitational environment a spacecraft faces. Then we'll

- See how the Earth's atmosphere can affect a spacecraft even in orbit
- Investigate how the vacuum in space above the atmosphere gives us another challenge to deal with
- See collision hazards posed by natural and man-made objects in space
- See how radiation and charged particles from the Sun and the rest of the universe can severely damage unprotected spacecraft

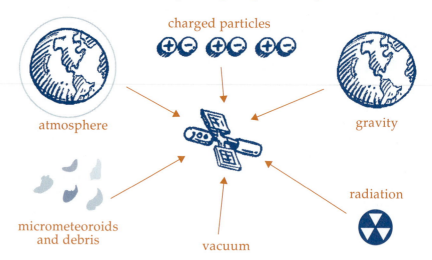

Figure 3-9. Factors Affecting Spacecraft in the Space Environment. There are six challenges unique to the space environment we deal with—gravity, the atmosphere, vacuum, micrometeoroids and debris, radiation, and charged particles.

Gravity

When we see astronauts on television floating around the Space Shuttle mid-deck, as in Figure 3-10, we often hear they are in "zero gravity." But this is not true! As we'll see in Chapter 4, all objects attract each other with a gravitational force which depends on their mass (how much "stuff" they have).   This force tends to decrease as objects get farther away from each other. So that means gravity doesn't just disappear when you get to space. In a low-Earth orbit, for example, say at an altitude of 300 km, the pull of gravity is still 91% of what it is on the Earth's surface.

So why do astronauts float around in their spacecraft? A spacecraft and everything in it are in free-fall. As the term implies, an object in *free-fall* is falling under the influence of gravity, free from any other forces. Free-fall is that momentary feeling you get when you jump off a diving board. It's what skydivers feel before their parachutes open. In free-fall you don't feel the force of gravity even though gravity is present. As you sit there in your chair, you don't *feel* gravity on your behind. You feel the *chair* pushing up at you with a force equal to the force of gravity. Forces which act only on the surface of an object are called *contact forces*. Astronauts in orbit experience no contact forces because they and their spacecraft are in free-fall, not in contact with Earth's surface. But if everything in orbit is falling, why doesn't it hit the Earth? As we'll see in Chapter 4, an object in orbit must have enough horizontal velocity so that, as it falls, it keeps missing the Earth.

Earth's gravitational pull dominates objects close to it. But as spacecraft move into higher and higher orbits, the gravitational pull of the Moon and Sun begin to exert their influence. As we'll see in Chapter 4, for Earth-orbiting applications, we can assume the Moon and Sun have no effect. However, as we'll see in Chapter 7, for interplanetary spacecraft, this isn't true.

Gravity dictates the size and shape of a spacecraft's orbit. Booster rockets must first overcome gravity to fling spacecraft into space. Once a satellite is in orbit, gravity determines the amount of propellant its engines must use to move between orbits or link up with other spacecraft. Beyond Earth, the gravitational pull of the Moon, the Sun, and other planets similarly shape the spacecraft's path. Gravity is so important to the space environment that an entire branch of astronautics, called *astrodynamics* deals with quantifying its effects on spacecraft and planetary motion. Chapters 4 through 9 will focus on understanding spacecraft trajectories and the exciting field of astrodynamics.

As we mentioned in Chapter 1, the free-fall environment of space offers many potential opportunities for space manufacturing. On Earth, if two materials are mixed, such as rocks and water, the heavier rocks will sink to the bottom. In free-fall, we can mix materials in proportions impossible on Earth. Thus, we can make exotic and useful metal alloys for electronics and other applications, or biological materials for new types of drugs and testing methods.

Figure 3-10. Astronauts in Free-Fall. In space there is no such thing as "zero gravity." Rather, astronauts and satellites are in free-fall—falling but never hitting the Earth. *(Courtesy of NASA)*

However, free-fall has its drawbacks. One area of frustration for engineers is handling fluids in space. Think about the gas gauge in a car. By measuring the height of a floating bulb, you can constantly track the amount of fuel in the tank. But in orbit nothing "floats" in the tank because the liquid and everything else is sloshing around in free-fall. Thus, fluids are much harder to pump and measure in free-fall. But these problems are relatively minor compared to the profound physiological problems humans experience when exposed to free-fall for long periods. We'll look at these problems separately in the next section.

Atmosphere

The Earth's atmosphere affects a spacecraft in low-Earth orbit (below about 966 km [about 600 miles] altitude), in two ways

- Drag—shortens orbit lifetimes
- Atomic oxygen—degrades spacecraft surfaces

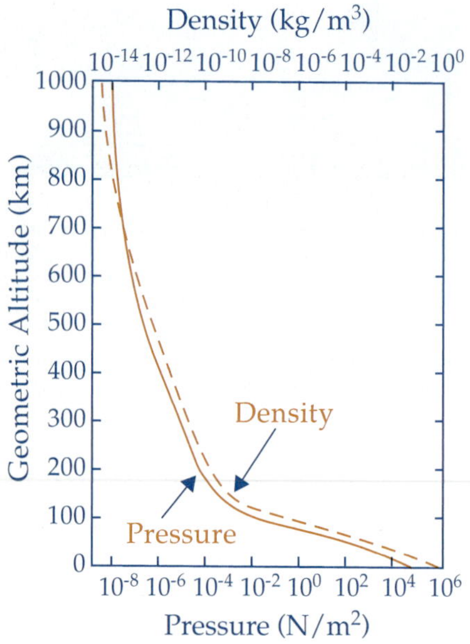

Take a deep breath. The air you breathe makes up the Earth's atmosphere. Without it, of course, we'd all die in a few minutes. While this atmosphere forms only a thin layer around the Earth less than a hundred kilometers thick, satellites in low-Earth orbit can still feel its effects. Over time, it can work to drag a satellite back to Earth, and the oxygen in the atmosphere can wreak havoc on many spacecraft materials.

Two terms are important to understanding the atmosphere—pressure and density. *Atmospheric pressure* represents the amount of force per unit area exerted by the weight of the atmosphere pushing down on us. *Atmospheric density* tells us how much air is packed into a given volume. As you go higher and higher into the atmosphere, both the pressure and density begin to decrease in a nearly exponential way, as shown in Figure 3-11. Visualize a column of air extending above you out into space. As you go higher, there is less volume of air above you, so the pressure (and thus the density) goes down. If we were to go up in an airplane with a pressure and density meter, we would see that as we go higher and higher, the density begins to drop off more and more rapidly.

Earth's atmosphere doesn't just end abruptly. Even at fairly high altitudes, up to 480 km (300 miles), the atmosphere continues to create a drag on orbiting spacecraft. *Drag* is the force you feel pushing your hand backward when you stick it out the window of a car rushing down the freeway. The amount of drag you feel on your hand depends on the air's density, your speed, the shape and size of your hand, and the orientation of your hand with respect to the airflow.

Drag immediately affects spacecraft returning to Earth. For example, as the Space Shuttle enters the atmosphere enroute to a landing at Edwards AFB in California, the force of drag is used to slow the Shuttle from an orbital velocity of over 25 times the speed of sound (7.75 km/s or 17,300 m.p.h.) to a runway landing at about 360 km/hr (225 m.p.h.). Drag quickly affects spacecraft in very low orbits (less than 130 km or 80 miles altitude), causing them to be pulled back to a fiery encounter with the atmosphere.

Figure 3-11. Structure of Earth's Atmosphere. The density of Earth's atmosphere decreases exponentially as you go higher. Even in low-Earth orbit, however, you can still feel the effects of the atmosphere in the form of drag.

The effect of drag on spacecraft in higher orbits is much more subtle. Between 130 km and 600 km, it will vary greatly depending on how the atmosphere changes (expands or contracts) due to variations in solar activity. Acting over months or years, drag can cause spacecraft in these orbits to gradually lose altitude until they enter the atmosphere to burn up. In 1979, the Skylab space station succumbed to the long-term effects of drag and plunged back to Earth. Above 600 km (375 miles), the atmosphere is so thin the drag effect is almost insignificant. Thus, spacecraft in orbits above 600 km are fairly safe from drag.

Besides drag, we must also consider the nature of air. At sea level, air is composed of about 21% oxygen, 78% nitrogen, and 1% miscellaneous other gasses, such as argon and carbon dioxide. Normally, oxygen atoms like to hang out in groups of two called molecules. The oxygen molecule is abbreviated O_2. Under normal conditions, when an oxygen molecule splits apart for some reason, the atoms quickly reform into a new molecule. In the upper parts of the atmosphere, oxygen molecules are few and far between. When radiation and charged particles cause them to split apart, they're left by themselves as *atomic oxygen*, abbreviated just O.

So what's the problem with O? We've all seen the results of exposing a piece of steel to water for a few days—it starts to rust. Chemically speaking, rust is *oxidation*. It occurs when oxygen molecules in the air combine with the metal. This oxidation problem is bad enough with O_2, but when O by itself is present, the reaction is much, much worse. Spacecraft exposed to atomic oxygen experience breakdown or "rusting" of their surfaces, which weakens components, changes their thermal characteristics, and can degrade sensors. One of the goals of NASA's Long Duration Exposure Facility (LDEF), shown in Figure 3-12, was to determine the extent of atomic oxygen damage over time.

On the good side, most atomic oxygen floating around in the upper atmosphere combines with oxygen molecules to form a special molecule, O_3, which we call *ozone*. Ozone acts like a window shade to block harmful radiation, especially the ultraviolet radiation that causes sunburn and skin cancer. In Chapter 11 we'll learn more about how the atmosphere blocks various types of radiation.

Figure 3-12. Long Duration Exposure Facility (LDEF). The mission of LDEF, deployed and retrieved by the Space Shuttle in the 1980s was to determine the extent of space environment hazards such as atomic oxygen and micrometeoroids. *(Courtesy of NASA)*

Vacuum

Beyond the thin skin of Earth's atmosphere we enter the cold vacuum of space. This vacuum environment creates three potential problems for spacecraft:

- Outgassing—release of gasses from spacecraft materials that can cloud sensors
- Cold welding—fuses metal parts together
- Heat transfer—only effective through radiation

As we've seen, atmospheric density decreases exponentially with altitude. At a height of about 80 km (50 miles), particle density is 10,000 times less than what it is at sea level.   If we go to 960 km (596 miles), we would find a given volume of space to contain one trillion times less air than at the surface. A pure vacuum, by the strictest definition of the word, would be a volume of space completely devoid of all material. In practice, however, a pure vacuum is nearly unattainable. Even at an altitude of 960 km (596 miles), you'll still find about 1,000,000 particles per cubic centimeter. So when we talk about the vacuum of space, we're talking about a "near" or "hard" vacuum.

Under standard atmospheric pressure at sea level, more than 101,325 N/m^2 (14.7 lb/in^2) of force is exerted on all material. The soda inside a soda can is under slightly higher pressure, forcing carbon dioxide (CO_2) into the solution. When you open the can, you release the pressure, causing some of the CO_2 to come out of the solution and making it foam. Spacecraft face a similar, but less tasty, problem. Some materials used in their construction, especially composites, contain tiny bubbles of gas which are trapped while under atmospheric pressure. When this pressure is released in the vacuum of space, the gasses begin to escape. This release of trapped gasses in a vacuum is called *outgassing*. Usually, outgassing is not a big problem; however, in some cases, the gasses can actually coat delicate sensors, such as lenses. When this happens, outgassing can be destructive. For this reason, we must carefully select and test materials used on spacecraft. We often "bake" a spacecraft in a vacuum oven prior to flight, as shown in Figure 3-13, to ensure it won't outgas in space.

Another problem created by vacuum is cold welding. *Cold welding* occurs between mechanical parts which have very little separation between them. When the moving part is tested on Earth, a tiny air space may allow the parts to move freely. After launch, the hard vacuum in space will eliminate this tiny air space, causing the two parts to "weld" together. When this happens, ground controllers must try various techniques to "unstick" the two parts. For example, they may expose one part to the Sun and the other to shade so that differential heating will cause the parts to expand and contract, respectively, and thus free them.

In many cases, just as with moving parts in the engine of your car, lubricants between mechanisms allow them to move easily and prevent this welding problem. However, because of the surrounding vacuum, we must select these lubricants carefully, so they don't evaporate or outgas. Dry graphite (the so-called "lead" in your pencil) is common because it lubricates well and won't evaporate into the vacuum as a common oil would.

Finally, the vacuum environment creates a problem with heat transfer. As we'll see in greater detail in Chapter 13, heat gets from one place to another in three ways. *Conduction* is heat flow directly from one point to another through some solid medium. If you hold a piece of metal in a fire, you'll quickly discover how conduction works. The second method of heat transfer is convection. *Convection* takes place when gravity, wind, or some other force moves a fluid, air, or water over a hot surface. Heat transfers from the surface to the fluid. Convection takes place whenever we feel chilled by a breeze or boil water on the stove. Both of these methods can

Figure 3-13. Spacecraft in a Vacuum Chamber. Prior to flight, spacecraft go through a rigorous testing procedure, including exposure to a hard vacuum in vacuum chambers like the one shown here. In this way we can test for problems with outgassing, cold welding, or heat transfer. *(Courtesy of Ball Aerospace)*

move heat around inside a spacecraft but they can't remove heat from it in the free-fall, vacuum environment of space. So we're left with the third method—radiation. We've already discussed EM radiation. *Radiation* is a way to transfer energy from one point to another. The heat you feel coming from the glowing coils of a space heater is radiated heat. Because radiation doesn't need a solid or fluid medium, it's the primary method of moving heat into and out of a spacecraft. We'll explore ways to do this in Chapter 13.

Micrometeoroids and Space Junk

Space is not empty. It's full of debris, most of which we're used to. If you've seen a falling star, you've witnessed just one part of the more than 20,000 tons of natural materials—dust, meteoroids, chunks of asteroids, and comets—which hit the Earth every year. For spacecraft and astronauts in orbit, the risk of getting hit by a meteoroid, as such naturally occurring objects are called, is remote.

However, since the beginning of the space age, debris has begun to accumulate from another source—human beings. With nearly every space mission, broken satellites, pieces of old booster segments or satellites, and even an astronaut's glove have remained in space. The environment near Earth is getting full of this junk (about 2200 tons of it), which poses an increasing risk to satellites and astronauts in orbit. A spacecraft in low orbit is now more likely to hit a piece of junk than a piece of natural material.

Keeping track of all this junk is the job of the North American Aerospace Defense Command (NORAD) in Colorado Springs, Colorado. NORAD tracks more than 7000 objects, baseball sized and larger, in Earth's orbit. Some estimates say at least 40,000 golf-ball-sized pieces (too small for NORAD to track) are also in orbit [Larson, Wertz, 1992]. To make matters worse, there are also billions of much smaller pieces—paint flakes, slivers of metal, etc.

If you get hit by a paint flake no big deal, right? Wrong! In low-Earth orbit, this small material is moving at fantastic speeds—7000 m/s or greater when it hits. This gives it a great amount of energy—much more than a rifle bullet! The potential danger of all this space junk was brought home during a Space Shuttle mission in 1983. During the mission, a paint flake only 0.2 mm (0.008 in.) in diameter hit the Challenger, making a crater 4 mm (0.16 in.) wide in its window. The crater, shown in Figure 3-14, cost more than $50,000 to repair. Analysis of other spacecraft shows collision with very small objects is common. Russian engineers believe a piece of space debris may have incapacitated one of their spacecraft in a transfer orbit.

Because there are billions of very small objects and only thousands of very large objects, you have a greater chance of getting hit by a very small object. For a spacecraft with a cross-sectional area of 50–200 m² at an altitude of 300 km (186 miles) (typical for Space Shuttle missions), the chance of getting hit by an object larger than a baseball during one year in orbit is about one in 100,000 or less [Larson, Wertz, 1992]. The chance of

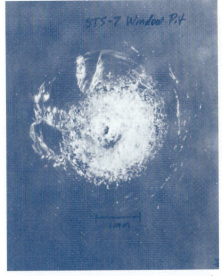

Figure 3-14. Shuttle Hit by Space Junk. At orbital speeds, even a paint flake can cause significant damage. The Space Shuttle was hit by a tiny paint flake, causing this crater in the front windshield. *(Courtesy of NASA)*

getting hit by something only 1 mm in diameter, however, is about one thousand times more likely, or about one in a thousand during one year in orbit.

Right now, there are no plans to clean up this space junk. Some international agreements aim at decreasing the rate at which the junk accumulates—for instance, by requiring worn-out satellites to be boosted to "graveyard" orbits. Who knows? Maybe a lucrative 21st century job will be "removing trash from orbit."

The Radiation Environment

As we saw in the previous section, one of the Sun's main outputs is electromagnetic (EM) radiation. Most of this radiation is in the visible and near-infrared parts of the EM spectrum. Of course, we see the light and feel the heat every day. However, a smaller but significant part of the Sun's output is at other wavelengths of radiation, such as X-rays and gamma rays.

Spacecraft and astronauts are well above the atmosphere, so they bear the full brunt of the Sun's output. The effect on a spacecraft depends on the wavelength of the radiation. In many cases, visible light hitting the spacecraft produces electrical energy through *solar cells* (sometimes called photovoltaic (PV) cells). This is a cheap, abundant, and reliable source of power for a spacecraft; we'll explore it in greater detail in Chapter 12. This radiation can also lead to several problems for spacecraft:

- Heating on exposed surfaces
- Degradation or damage to surfaces and electronic components
- Solar pressure

The infrared or thermal radiation a spacecraft endures leads to heating on exposed surfaces which can be either helpful or harmful to the spacecraft, depending on the overall thermal characteristics of its surfaces. Electronics in a spacecraft need to operate at about normal room temperatures (20° C or 68° F). The vacuum of space is normally very cold (–200° C or more). In some cases, the Sun's thermal energy can help to warm electronic components. In other cases, this solar input—in addition to the heat generated onboard from operation of the electronic components—can make the spacecraft too hot. As we'll see in Chapter 13, we must design the spacecraft's thermal control system to moderate its temperature.

Normally, the EM radiation in the other regions of the spectrum have little effect on a spacecraft. However, prolonged exposure to ultraviolet radiation can begin to degrade spacecraft coatings. This radiation is especially harmful to solar cells, but it can also harm electronic components, requiring them to be shielded or *hardened* to handle the environment. In addition, during intense solar flares, bursts of radiation in the radio region of the spectrum can interfere with communications equipment onboard.

When you hold your hand up to the Sun, all you feel is heat. However, all that light hitting your hand is also exerting a very small amount of pressure. Earlier, we said EM radiation could be thought of as waves, like ripples on a pond. Another way to look at it is as tiny bundles of energy called photons. *Photons* are massless bundles of energy that move at the speed of light. These photons strike your hand, exerting pressure similar in effect to atmospheric drag. But this *solar pressure* is much, much smaller than drag. In fact, it's only about 5 N of force (about one pound) for a square kilometer of surface (one-third square mile). While that may not sound like much, over time this solar pressure can disturb the orientation of spacecraft and cause them to point in the wrong direction. We'll learn more about solar pressure effects in Chapter 12. In Chapter 14 we'll see how we may use them to sail around the solar system.

Charged Particles

Perhaps the most dangerous aspect of the space environment is the pervasive influence of charged particles. Three primary sources for these particles are

- The solar wind and flares
- Galactic cosmic rays (GCRs)
- The Van Allen radiation belts

As we've already seen, the Sun puts out a stream of charged particles (protons and electrons) as part of the solar wind—at a rate of 1×10^9 kg/s (2.2×10^9 lb/s). During intense solar flares the number of particles ejected can increase dramatically.

As if this source of charged particles wasn't enough, we must also be concerned with high-energy particles from *galactic cosmic rays (GCRs)*. GCRs are particles similar to those found in the solar wind or in solar flares, but they originate outside of the solar system. GCRs represent the solar wind from distant stars, the remnants of exploded stars, or, perhaps, shrapnel from the "Big Bang" explosion which may have created the Universe. In many cases, however, GCRs are much more massive and energetic than particles of solar origin. Ironically, the very thing that protects us on Earth from these sources of charged particles creates a third source potentially harmful to orbiting spacecraft and astronauts—the Van Allen radiation belts.

To understand the Van Allen belts, we must begin by remembering that the Earth has a strong magnetic field as a result of its liquid iron core. This magnetic field behaves in much the same way as those toy magnets you used to play with as a kid, but it's vastly more powerful. Although we can't *feel* this field around us, it's always there. Pick up a compass and you'll see how the field moves the needle to point north. Magnets always come with a North Pole at one end and a South Pole at the other. If you've ever played with magnets, you've discovered that the North Pole attracts the South Pole (and vice versa), whereas two north poles (or south poles)

north compass pole

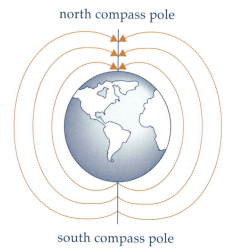

south compass pole

Figure 3-15. Earth's Magnetosphere. The Earth's liquid iron core creates a strong magnetic field. This field is represented by field lines extending from the south pole to the north pole. The volume this field encloses is the magnetosphere.

repel each other. These magnetic field lines wrap around the Earth to form the *magnetosphere*, as shown in Figure 3-15.

Remember, magnetic fields affect charged particles. This basic principle allows us to "steer" electron beams with magnets inside television sets. Similarly, the solar wind's charged particles and the GCRs form streams which hit the Earth's magnetic field like a hard rain hitting an umbrella deflects the raindrops over its curved surface, the Earth's magnetic field wards off the charged particles, keeping us safe. (For SciFi buffs, perhaps a more appropriate analogy is the fictional force field or "shields" from *Star Trek*, used to divert Romulan disruptor beams and protect the ship.)

The point of contact between the solar wind and the Earth's magnetic field is the *shock front* or *bow shock*. As the solar wind bends around the Earth, it stretches out the Earth's magnetic field lines along with it, as you can see in Figure 3-16. In this way, the Earth resembles a boat traveling through the water with a wake out behind it. Inside the shock front, the point of contact between the charged particles of the solar wind and the magnetic field lines is called the *magnetopause*, and the area directly behind the Earth is called the *magnetotail*. As we'll see, charged particles can affect spacecraft orbiting well within the Earth's protective magnetosphere.

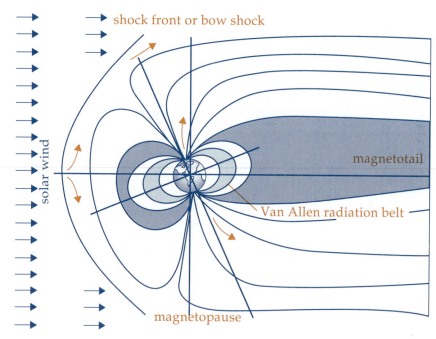

Figure 3-16. Interaction Between Solar Wind and Earth's Magnetic Field. As the solar wind and GCRs hit the Earth's magnetosphere they are deflected, keeping us safe.

As the solar wind interacts with the Earth's magnetic field, some high-energy particles get trapped and concentrated between field lines. These areas of concentration are called the *Van Allen radiation belts*, which are named after Professor James Van Allen of the University of Iowa. Van

Allen discovered them based on data collected by Explorer I, America's first satellite, launched in 1958.

Although they are called radiation belts, space is not really radioactive. Scientists often lump charged particles in with EM radiation and call both radiation because their effects are similar. Realize, however, that we're really dealing with charged particles in this case. (Perhaps the radiation belts should more accurately be called "charged-particle suspenders" because they're really full of charged particles and occupy a region from shoulder to waist [pole to pole] around the Earth!)

Whether charged particles come directly from the solar wind, indirectly from the Van Allen belts, or from the other side of the galaxy, they can harm spacecraft in three ways:

- Charging

- Sputtering

- Single-event phenomenon

Spacecraft charging isn't something the government does to buy a spacecraft. As the charged particles from the solar wind get trapped in the magnetotail, the effect is similar to walking across a carpeted floor wearing socks. You eventually build up a static charge which discharges when you touch something metal—resulting in a nasty shock. *Spacecraft charging* results when charges build up on different parts of the spacecraft as it moves through areas of concentrated charged particles. Once this charge builds up, discharge can occur with disastrous effects—damage to surface coatings, degrading of solar panels, loss of power, or switching off or permanently damaging electronics.

Sometimes, these charged particles trapped by the magnetosphere interact with the Earth's atmosphere in a dazzling display called the Northern Lights or Aurora Borealis, as shown in Figure 3-17. This light show is caused by charged particles streaming toward the Earth along magnetic field lines converging at the poles. As the particles interact with the atmosphere, the result is similar to what happens in a neon light— charged particles interact with a gas, exciting it and causing it to glow. On Earth we see an eerie curtain of light in the sky.

These particles can also damage a spacecraft's surface because of their high speed. It's as if the spacecraft were being "sand blasted" by atomic-sized particles—sometimes referred to as *sputtering*. Over time, sputtering can damage a spacecraft's thermal coatings and sensors.

Finally, a single charged particle can penetrate deep into the guts of the spacecraft to disrupt electronics. Each disruption is known as a *single event phenomenon (SEP)*. Both solar flares and GCR can cause SEP.

One type of SEP is a *single event upset (SEU)* or "bitflip." This occurs when the impact of a high-energy particle resets one part of a computer's memory from 1 to 0, or vice versa. This can cause subtle but significant changes to spacecraft functions. For example, setting a bit from 1 to 0 may cause the spacecraft to turn off or forget which direction to point its antenna. Some scientists believe an SEU was the cause of problems with

Figure 3-17. Lights in the Sky. As charged particles from the solar wind interact with the Earth's upper atmosphere, they create a spectacular sight known as the Northern (or Southern) Lights. This light show can be seen by people living in high latitudes and was captured on film by Shuttle astronauts from space. *(Courtesy of NASA)*

the Magellan spacecraft when it first went into orbit around Venus and started acting erratically.

It's difficult for us to prevent these random impacts. Spacecraft shielding offers some protection, but spacecraft operators must be aware of the possibility of these events and know how to save the spacecraft should they occur.

Astro Fun Fact
Message In A Bottle

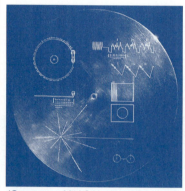

(Courtesy of NASA)

(Courtesy of NASA)

If you were going to send a "message in a bottle" to another planet, what would you say? Dr. Carl Sagan and a committee of scientists, artists, and musicians tried to answer this question before the Voyager launched in 1977. They developed a multi-media program containing two hours of pictures, greetings, sounds, and music they felt represented Earth's variety of culture. The collection contains such items as Chuck Berry's "Johnny B. Goode," people laughing, a Pakistan street scene, and a map to find your way to the Earth from other galaxies. A record company manufactured a 12-inch copper disc to hold the information, and the record was sealed in a container along with a specially designed phonograph. This package, which scientists estimate could last more than a billion years, was placed onboard each of the Voyager spacecraft. We hope another life form will encounter one of the Voyager probes, construct the phonograph, and play back the sounds of the Earth. But don't count on hearing from any space beings—the Voyagers won't visit another star until 40,000 years from now!

Eberhard, Jonathan. <u>The World on a Record</u>. Science News. Aug. 20, 1977, p. 124–125.

Wilford, John Noble. <u>Some Beings Out There Just May Be Listening</u>. New York Times Magazine. Sept. 4, 1977, p. 12–13.

Contributed by Donald Bridges, United States Air Force Academy

■ Section Review

Key Terms

astrodynamics
atmospheric density
atmospheric pressure
atomic oxygen
bow shock
cold welding
conduction
contact forces
convection
drag
free-fall
galactic cosmic rays (GCRs)
magnetopause
magnetosphere
magnetotail
outgassing
oxidation
ozone
photons
shock front
single event phenomena (SEP)
single event upset (SEU)
solar cells
solar pressure
spacecraft charging
sputtering
Van Allen radiation belts

Key Concepts

➤ Six major environmental factors affect spacecraft in Earth orbit.
  • Gravity
  • Atmosphere
  • Vacuum
  • Micrometeoroids and space junk
  • Radiation
  • Charged particles

➤ The Earth exerts a gravitational pull which keeps spacecraft in orbit. The condition of a spacecraft or astronaut in orbit is best described as free-fall because they're both falling around the Earth.

➤ The Earth's atmosphere isn't completely absent in low-Earth orbit. It can cause
  • Drag—which shortens orbit lifetimes
  • Atomic oxygen—which can damage exposed surfaces

➤ In the vacuum of space, spacecraft can experience
  • Outgassing—a condition in which trapped gas particles in a material are released when atmospheric pressure is removed
  • Cold welding—a condition that can cause metal parts to fuse together
  • Heat transfer problems—a spacecraft can rid itself of heat only through radiation

➤ Micrometeoroids and space junk can damage spacecraft from their speed of impact

➤ Radiation, primarily from the Sun, can cause
  • Heating on exposed surfaces
  • Damage to electronic components and disruption in communication
  • Solar pressure, which can change a spacecraft's orientation

➤ Charged particles come from three sources:
  • Solar wind and flares
  • Galactic Cosmic Rays (GCRs)
  • Van Allen radiation belts

➤ The Earth is protected from charged particles by its magnetic field or magnetosphere. The Van Allen radiation belts were formed by charged particles trapped and concentrated by this magnetosphere.

➤ Charged particles from all sources can cause
  • Charging
  • Sputtering
  • Single-Event Phenomena (SEP)

3.3 Living and Working in Space

▬ In This Section You'll Learn to...

- ☞ Describe the free-fall environment's three effects on the human body
- ☞ Discuss the hazards posed to humans from radiation and charged particles
- ☞ Discuss the potential psychological challenges of space flight

Humans and other living things on Earth have evolved to deal with the Earth's unique environment. We developed a strong backbone, along with muscle and connective tissue, to support ourselves against the pull of gravity. On Earth, we're protected from radiation and charged particles by the ozone layer and the magnetosphere, so we don't have any natural, biological defenses against them. When we leave Earth to travel into space, however, we must learn to adapt in an entirely different environment. In this section, we'll discover how free-fall, radiation, and charged particles can harm humans in space. Then we'll see some of the psychological challenges for astronauts venturing into the final frontier.

Free-Fall

Earlier, we learned that in space there is no such thing as "zero gravity"; you're actually in a free-fall environment. While free-fall can benefit engineering and materials processing, it poses a significant hazard to humans. Free-fall causes three potentially harmful physiological changes to the human body, as summarized in Figure 3-18:

- Decreased hydrostatic gradient—fluid shift
- Altered vestibular functions—motion sickness
- Reduced load on weight-bearing tissues

Hydrostatic gradient refers to the distribution of fluids in our body. On Earth's surface, gravity acts on this fluid and pulls it into our legs. So, normally, blood pressure is higher in our legs than in our arms or shoulders. Under free-fall conditions, the fluid no longer pools in our legs but distributes equally. As a result, fluid pressure in the legs decreases while pressure in the higher parts of the body increases. The shift of fluid from our legs to our upper body is called a *decreased hydrostatic gradient* or *fluid shift*. This effect leads to several changes. To begin with, the kidneys detect this fluid shift as "extra" fluid which is eliminated through increased urination. This leads to dehydration and other associated problems which can be especially pronounced during entry as astronauts again experience g-forces. To alleviate these problems, Shuttle astronauts are required to re-hydrate by drinking several quarts of liquid prior to entry.

Figure 3-18. The Free-Fall Environment and Humans. The free-fall environment offers many hazards to humans living and working in space. These include fluid shift, motion sickness, and reduced load on weight-bearing tissue.

The fluid shift also causes *edema* of the face (a red "puffiness"), so astronauts in space appear to be blushing. In addition, the heart begins to beat faster with greater irregularity and it loses mass because it doesn't have to work as hard in free-fall. Finally, astronauts experience a minor "head rush" on return to Earth. We call this condition *orthostatic intolerance*—that feeling you sometimes get when you stand up too fast after sitting or lying down for a long time. For astronauts returning from space, this condition is sometimes very pronounced and can cause blackouts.

Vestibular functions have to do with a human's built-in ability to sense movement. Close your eyes and move your head around. Tiny sensors in your inner ear can detect this movement. Together, your eyes and inner ears are able to determine your body's orientation and sense acceleration. Our vestibular system allows us to walk without falling over. Sometimes, what you *feel* with your inner ear and what you *see* with your eyes gets out of synch (such as on a high-speed roller coaster). When this happens, you can get disoriented or sick. That also explains why you tend to experience more motion sickness riding in the back seat of a car than while driving— you can feel the motion, but your eyes don't see it.

Because our vestibular system was calibrated to work under a constant 1 g on the Earth's surface, this calibration is thrown off when you go into orbit and enter a free-fall environment. As a result, nearly all astronauts experience motion sickness during the first few days in space until they can re-calibrate. Veteran astronauts report that over repeated space flights this calibration time decreases.

If you're bedridden for a long time, your muscles will grow weak from lack of use and begin to atrophy. Astronauts in free-fall experience a similar *reduced load on weight bearing tissue* such as on muscles (including the heart) and bones. Muscles lose mass and weaken. Bones lose calcium and weaken. Bone marrow, which produces blood, is also affected, leading to a reduced number of red blood cells.

Scientists are still working on ways to alleviate all these problems of free-fall. Vigorous exercise offers some promise in preventing long-term atrophy of muscles, but no one has found a way to prevent changes within the bones. Some scientists suggest astronauts should have "artificial gravity" for very long missions to Mars. Spinning the spacecraft would produce this force, which would feel like gravity pinning them to the wall. This is the same force you feel when you take a corner very fast in a car and you're pushed to the outside of the curve. This artificial gravity could maintain the load on all weight-bearing tissue and alleviate some of the other detrimental effects of free-fall. However, building and operating such a system is an engineering challenge.

Radiation and Charged Particles

As we've seen, the ozone layer and magnetosphere protect us from charged particles and EM radiation down here on Earth. In space, however, we're well above the ozone layer and may enter the Van Allen radiation belts or even leave the Earth altogether, thus exposing ourselves to the full force of galactic cosmic rays (GCRs).

Astro Fun Fact
The "Vomit Comet"

(Courtesy of NASA)

How do astronauts train for the free-fall environment of space? They take a ride in modified Air Force KC-135 aircraft owned and operated by NASA. Affectionately called the "Vomit Comet" by those who've experienced the fun as well as the not-so fun aspects of free-fall, this plane flies a series of parabolas, alternately climbing and diving to achieve almost a minute of free-fall. This is like the momentary lightness you feel when you go over a hill at a high speed in a car. During these precious few seconds of free-fall, astronauts can practice getting into space suits or experiment with other equipment specifically designed to function in space.

(Courtesy of NASA)

Until now, we've been careful to delineate the differences between the effects of EM radiation and charged particles. However, from the standpoint of biological damage, we can treat exposure to EM radiation and charged particles in much the same way. The overall severity of this damage depends on the total dosage. *Dosage* is a measure of accumulated radiation or charged particle exposure.

Quantifying the dosage depends on the energy contained in the radiation or particles and the *relative biological effectiveness (RBE)* rating of the exposure. Dosage energy is measured in terms of *RAD*, with one RAD representing 100 erg of energy per gram of target material (1.08×10^{-3} cal/lb). (This is about as much energy as it takes to lift a paper clip 1 mm [3.9×10^{-2} in.] off your desk). The RBE represents the destructive power of the dosage on living tissue. This depends on whether the exposure is EM radiation (photons) with an RBE of one or charged particles with an RBE of as much as ten or more. An RBE of ten is much more destructive to tissue than an RBE of one. The total dosage is then quantified as the product of RAD and RBE to get a dosage measurement in *roentgen equivalent man (REM)*. The REM dosage is cumulative over a person's entire lifetime.

The potential effects on humans exposed to radiation and charged particles depend to some extent on the time over which a dosage occurs. For example, a 50-REM dosage accumulated in one day will be much more harmful than the same dosage spread over one year. Such short-term dosages are called *acute dosages*. They tend to be more damaging, primarily because of their effect on fast reproducing cells within our bodies, specifically in the gastrointestinal tract, bone marrow, and testes. Table 3-1 gives the effects of acute dosages on humans, including blood count changes, vomiting, diarrhea, and death. The cumulative effects of dosage spread over much longer periods include cancer, cataracts, and leukemia.

Table 3-1. Effects of Acute Radiation and Charged Particle Dosages on Humans (from Nicogossian, et al).

Effect	Dosage (REM)
Blood count changes in the population	15–50
Vomiting "effective threshold"*	100
Mortality "effective threshold"*	150
LD_{50}** with minimal supportive care	320–360
LD_{50}** with full supportive medical treatment required	480–540

   * Effective threshold is the lowest dosage causing these effects in at least one member of the exposed population

  ** LD_{50} is the lethal dosage in 50% of the exposed population

Just living on Earth, we all accumulate dosage. For example, living one year in Houston, Texas, (at sea level) gives you a dosage of 0.1 REM. As you get closer to space there is less atmosphere protecting you, so living in Denver, Colorado, (the "Mile-High City") gives you a dosage of twice that amount. Certain medical procedures also contribute to our lifetime dosage. One chest X-ray, for example, gives you 0.01 REM exposure. Table 3-2 shows some typical dosages for various events.

Table 3-2. Dosages for Some Common Events (from SICSA Outreach and Nicogossian, et al).

Event	Dosage (REM)
Transcontinental round trip in a jet	0.004
Chest X-ray (lung dose)	0.01
Living one year in Houston, Texas (sea level)	0.1
Living one year in Denver, Colorado (elev. 1600 m)	0.2
Skylab 3 for 84 days (skin)	17.85
Space Shuttle Mission (STS-41D)	0.65

Except for solar flares, which can cause very high short-term dosages with the associated effects, astronauts are most concerned with dosage spread over an entire mission or career. NASA sets dosage limits for astronauts at 50 REM per year. Few astronauts will be in space for a full year, so their dosages will be much less than 50 REM. By comparison, workers in the nuclear industry are limited to one tenth that, or five REM per year. A typical Shuttle mission exposes the crew to a dosage of less than one REM. The main concern is for very long missions, such as in the space station or on a trip to Mars.

For the most part, it is relatively easy to build shielding made of aluminum or other light metals to protect astronauts from the solar EM radiation and the protons from solar wind. In the case of solar flares, long missions may require "storm shelters"—small areas deep within the ship that would protect astronauts for a few days until the flare has subsided. GCRs cause our greatest concern. Because these particles are so massive, it's impractical to provide enough shielding. To make matters worse, as the GCRs interact with the shield material itself, they produce secondary radiation (sometimes called "Bremsstrahlung" radiation after a German word for braking), which is also harmful.

Space-mission planners try to avoid areas of concentrated charged particles such as those in the Van Allen belts. For example, because astronauts are much more lightly shielded in just their space suits, extra vehicular activities (EVA—or space walks) are not planned while they pass through the "South Atlantic Anomaly." This area between South America and Africa, shown in Figure 3-19, is where the Van Allen belts "dip" toward the Earth. Long missions, such as those to Mars, will require "storm shelters" and a constant watch to warn crews when solar flares erupt. As for GCRs, there's really nothing more we can do except conduct more research to better quantify this hazard and try to decrease trip times.

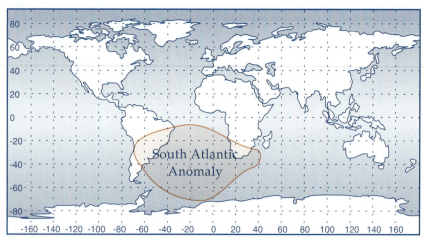

Figure 3-19. The South Atlantic Anomaly. The South Atlantic Anomaly is an area over the Earth where the Van Allen belts "dip" closer to the surface. Astronauts should avoid space walks in this region because of the high concentration of charged particles.

Psychological Effects

Because sending humans to space costs so much, we typically try to get our money's worth by scheduling grueling days of activities for the crew. This excessive workload can begin to exhaust even the best crews, seriously degrade their performance, and even endanger the mission. It can also lead to morale problems. During one United States Skylab mission, the crew actually went on strike for a day to protest the excessive demands on their time. Similar problems have been reported aboard the Russian MIR space station.

The crew's extreme isolation also adds to their stress and may cause loneliness and depression on long missions. Being forced into tight living conditions with the same people day-after-day can also take its toll. Tempers can flare, and team performance may suffer. This problem is not unique to missions in space. Scientists at remote Antarctic stations during the long, lonely winters have reported similar episodes of extreme depression and friction between team members.

We must take these human factors into account when planning and designing missions. Crew members must be able to take a break or "mini-vacation" once in awhile. On long missions, they'll need frequent contact with loved ones at home to alleviate their isolation. Planners must also select crew members who can work closely, in tight confines, with other group members for long periods. Psychological diversions such as music, video games, and movies will be required on very long missions to relieve boredom.

Section Review

Key Terms

acute dosages
decreased hydrostatic gradient
dosage
edema
fluid shift
hydrostatic gradient
orthostatic intolerance
RAD
relative biological effectiveness (RBE)
roentgen equivalent man (REM)
vestibular functions

Key Concepts

➤ Effects of the space environment on humans come from
  • Free-fall
  • Radiation and charged particles
  • Psychological effects

➤ The free-fall environment can cause
  • Decreased hydrostatic gradient—a condition where fluid in the body shifts to the head
  • Altered vestibular functions—motion sickness
  • Decreased load on weight bearing tissue—causing weakness in bones and muscles

➤ Depending on the dosage, the radiation and charged particle environment can cause short-term and long-term damage to the human body, or even death

➤ Psychological stresses on astronauts include
  • Excessive workload
  • Isolation, loneliness, and depression

References

Air University Press. *Space Handbook*. AV-18. Maxwell AFB, Alabama: 1985.

"Astronomy", August 1987.

Bate, Roger R., Donald D. Mueller, and Jerry E. White. *Fundamentals of Astrodynamics*. New York, NY: Dover Publications, Inc., 1971.

Bueche, Frederick J. *Introduction to Physics for Scientists and Engineers*. New York, NY: McGraw-Hill, Inc., 1980.

Chang, Prof. I. Dee (Stanford University), Dr. John Billingham (NASA Ames), and Dr. Alan Hargen (NASA Ames), Spring, 1990. "Colloquium on Life in Space."

Concepts in Physics. Del Mar, CA: Communications Research Machines, Inc., 1973.

Concise Science Dictionary. Oxford, U.K.: Oxford University Press, 1984.

Glover, Thomas J. *Pocket REF*. Morrison, CO: Sequoia Publishing, Inc., 1989.

Goldsmith, Donald. *The Astronomers*. New York, NY: Community Television of Southern California, Inc., 1991.

Gonick, Larry and Art Huffman. *The Cartoon Guide to Physics*. New York, NY: Harpee Perennial, 1990.

Hartman, William K. *Moon and Planets*. Belmont, California: Wadsworth, Inc., 1983.

Hewitt, Paul G. *Conceptual Physics. . .A New Introduction to Your Environment*. Boston, MA: Little, Brown and Company, 1981.

Jursa, Adolph S. (ed.). *Handbook of Geophysics and the Space Environment*. Air Force Geophysics Laboratory, Air Force Systems Command USAF, 1985.

King-Hele, Desmond. *Observing Earth Satellites*. New York, NY: Van Nostrand Reinhold Company, Inc., 1983.

Nicogossian, Arnauld E., Carolyn Leach Huntoon, Sam L. Pool. *Space Physiology and Medicine*, 2nd Ed. Philadelphia, PA: Lea & Febiger, 1989.

Rycroft, Michael (ed.), *The Cambridge Encyclopedia of Space*. New York, NY: Press Syndicate of the University of Cambridge, 1990.

Sasakawa International Center for Space Architecture (SICSA) Outreach. July-September 1989. Special Information Topic Issue, "Space Radiation Health Hazards: Assessing and Mitigating the Risks," Vol. 2, No. 3.

Tascione, Maj. T.F., Maj. R.H. Bloomer, Jr., and Lt. Col. D.J. Evans. *SRII, Introduction to Space Science: Short Course*. U.S.A.F. Academy, Department of Physics.

Wertz, James R. and Wiley J. Larson. *Space Mission Analysis and Design*. Netherlands: Kluwer Academic Publishers, 1991.

Woodcock, Gordon, *Space Stations and Platforms*, Malabar, Florida: Orbit Book Company, 1986.

The World Almanac and Book of Facts. 1991. New York, NY: Pharos Books, 1990.

Mission Problems

3.1 Cosmic Perspective

1 Where does Space begin?

2 What object most strongly affects the space environment?

3 What is the star closest to the Earth? The second-closest star?

4 List and describe the Sun's two forms of energy output.

5 What are solar flares? How do they differ from the solar wind?

3.2 The Space Environment and Spacecraft

6 List the six major hazards to spacecraft in the space environment.

7 Why are astronauts in space not in a "zero gravity" environment? Why is free-fall a better description of the gravity environment?

8 How does the density and pressure of Earth's atmosphere change with altitude?

9 What is atmospheric drag?

10 What is atomic oxygen?

11 What are the major problems in the vacuum environment of space?

12 Describe the potential hazards to spacecraft from micrometeoroids and space junk.

13 Describe the mechanism which protects the Earth from the effects of solar and cosmic charged particles.

14 What are Galactic Cosmic Rays?

15 What are the Van Allen radiation belts and what are they filled with?

16 Describe the potential harmful effects on spacecraft from charged particles.

3.3   Living and Working in Space

17   List and describe the three physiological changes to the human body during free-fall.

18   How are dosages of radiation and charged particles quantified?

19   What are the potential short-term and long-term effects of exposure to radiation and charged particles?

20   How do long space flights affect astronauts psychologically?

For Discussion

21   Using a basketball to represent the size of the Sun, lay out a scale model of the solar system. How far away would the nearest star have to be?

22   As a spacecraft designer for a manned mission to Mars, you must protect the crew from the space environment. Compile a list of all the potential hazards they may face during this multi-year mission and discuss how you plan to deal with them.

Mission Profile—SETI

In the fall of 1992, 500 years after Columbus discovered America, NASA officially began an exciting ten-year mission to search for extraterrestrial intelligence (SETI). Because we can't send a spacecraft over interstellar distances, the NASA mission focuses on radio astronomy as the most probable way to contact extraterrestrial life. As Seth Shostak of the SETI Institute says, "our generation is the first with the capability to address one of mankind's most fundamental questions."—is there other intelligent life in the universe?

Mission Overview

The mission, formally called the NASA SETI Microwave Observing Project (MOP), is designed to intercept a radio signal from other intelligent beings. Through a computer-linked network, NASA has combined the efforts of many radio telescopes around the world. These telescopic dishes will survey the stars and select certain areas for a more sensitive search. The operation is scheduled to continue until the year 2001.

Mission Data

✓ The NASA radio telescope network scans 15 million frequencies every second with 300 times the sensitivity of any previous system.

✓ In the first minutes of the mission, NASA searched more comprehensively for extraterrestrial intelligence than in all previous attempts combined.

✓ NASA listens for frequencies between 1 and 10 GHz. Below 1 GHz, natural radiation in our galaxy makes communication difficult to discern. Above 10 Ghz, radio noise from the atmosphere makes it impossible to hear a transmission. The area between these frequencies offers the strongest possible radio transmission.

✓ NASA will scan the sky using mobile radio antennae. Using stationary dishes, they will target only stars similar to our own Sun (about 10% of all known stars).

Mission Impact

With the onset of the ten year NASA mission, SETI was given a substantial credibility boost. Whether or not intelligent life is found, we will undoubtedly accrue new knowledge of our universe as well as upgrade our technological ability. This alone will prove the mission worthy. Yet, if SETI succeeds to find other life in the universe, it will truly be a turning point in the history of man.

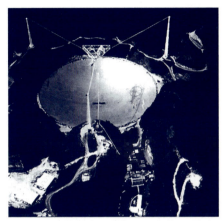

The SETI project will employ the 305-m (1000 ft.) Arecibo dish in Puerto Rico shown here, as well as the 34-m (111.5 ft.) Deep Space Network antennae in California and numerous other antennae in the Southern Hemisphere. *(Courtesy of NASA)*

For Discussion

• What is the next step once intelligent life is found?

• Given that radio astronomy is only a best guess in contacting extraterrestrial intelligence, how can NASA be sure it is worth the effort and money? Should this mission be continued?

• Is it simply an Earth bias to specifically look at stars similar to our Sun? Could life exist in other scenarios?

Contributor

Troy E. Kitch, United States Air Force Academy

References

Blum, Howard. *Out There*. New York: Simon and Schuster, 1990.

White, Frank. *The SETI Factor*. New York: Walker and Company, 1990.

The lunar module orbits above the Moon's stark landscape with the blue Earth rising above the horizon. (Courtesy of NASA)

Understanding Orbits

4

In This Chapter You'll Learn to...

☛ Explain the basic concepts of orbit motion and describe how to analyze it

☛ Explain and use the basic laws of motion Isaac Newton developed

☛ Use laws of motion to develop a mathematical and geometric representation of orbits

☛ Use two constants of orbit motion—specific mechanical energy and specific angular momentum—to determine important orbit variables

You Should Already Know...

❏ Elements of a space mission (Chapter 1)

❏ Orbit concepts (Chapter 1)

❏ Concepts of vector mathematics (Appendix A)

❏ Calculus concepts (Appendix B)

❏ Kepler's Laws of Planetary Motion (Chapter 2)

Outline

Space is for everybody. It's not just for a few people in science or math, or a select group of astronauts. That's our new frontier out there and it's everybody's business to know about space.

Christa McAuliffe
teacher and astronaut on the
ill-fated Challenger Space Shuttle

Spacecraft live in orbits. In Chapter 1, we described an orbit as a "racetrack" which a satellite drives around, as seen in Figure 4-1. Orbits and trajectories are two of the basic elements of any space mission. Rockets must follow specific paths to get into space. Spacecraft must deploy at just the right altitude and velocity to carry out their mission. Upperstages are used to maneuver from one orbit to another. Understanding this motion may at first seem rather intimidating. After all, to fully describe orbital motion we need some basic physics along with a healthy dose of calculus and geometry. However, as we'll see, the complex trajectories of rockets flying into space aren't all that different from the paths of baseballs pitched across home plate. In fact, in most cases, both can be described in terms of the single force pinning you to your chair right now—gravity.

Armed only with an understanding of this single pervasive force, we can predict, explain, and understand the motion of nearly all objects in space, from baseballs to entire galaxies. Chapter 4 is just the beginning. Here we'll explore the basic tools for analyzing orbits. In the next several chapters we'll see that, in a way, understanding orbits gives us a crystal ball to see into the future. Once we know an object's position and velocity, as well as the nature of the local gravitational field, we can gaze into this crystal ball to predict exactly where the object will be minutes, hours, or even years from now.

We'll begin by taking a conceptual approach to understanding orbits. Once we have a basic feel for how they work, we can take a more rigorous approach to understanding and describing satellite motion. We'll use tools provided by Isaac Newton, who developed some fundamental laws more than 200 years ago that we can use to explain orbits today. Finally, we'll look at some interesting implications of orbit motion that allow us to describe their shape and determine which aspects remain constant—when left undisturbed by outside non-gravitational forces.

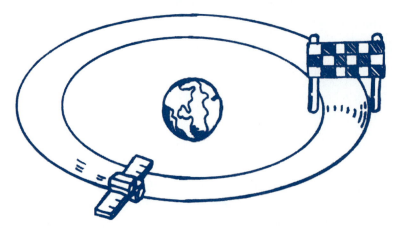

Figure 4-1. Orbits as Racetracks. Orbits are like giant racetracks on which spacecraft "drive" around the Earth.

4.1 Orbital Motion

In This Section You'll Learn to...

- Explain, conceptually, how an object is put into orbit
- Describe how to analyze the motion of any object

Baseballs in Orbit

What is an orbit? Sure, we said it was a type of "racetrack" in space that an object drives around, but what makes these racetracks? Throughout the rest of this chapter we'll explore the physical and mathematical principles that allow orbits to exist. But before diving into a complicated explanation, let's begin with a simple experiment that illustrates, conceptually, how orbits work. To do this, we'll arm ourselves with a bunch of baseballs and travel to the top of a tall mountain.

Imagine you were standing on top of this mountain prepared to pitch baseballs to the east. As the balls sail off the summit, what would you see? Besides unsuspecting hikers panting up the trail and running for cover, you could see that the balls follow a curved path. Why is this? The force of your throw is causing them to go outward, but the force of gravity is pulling them down. Therefore, the "compromise" shape of the baseball's path is a curve.

The faster you throw the balls, the farther they go before hitting the ground, as you can see in Figure 4-2. This could lead you to conclude that the faster you throw them the longer it takes before they hit the ground. But is this really the case? Let's try another experiment to see. As you watch, two astronauts, standing on flat ground, will release baseballs. The first one will simply drop a ball from a fixed height. At exactly the same time, a second astronaut will throw an identical ball horizontally as hard as possible. What will you see? If the second astronaut throws a fast ball, it'll travel out about 20 m (60 ft.) or so before it hits the ground. But, the ball dropped by the first astronaut will hit the ground at exactly the same time as the pitched ball, as Figure 4-3 shows!

How can this be? To understand this seeming paradox, we must recognize that, in this case, the motion in one direction is *independent* of motion in another. Thus, while the second astronaut's ball is moving horizontally at 30 km/hr (20 m.p.h.) or so, it's still falling at the same rate as the first ball. This rate is the constant gravitational acceleration of all objects near the Earth's surface, 9.798 m/s^2. Thus, they hit the ground at the same time. The only difference is that the pitched ball, because it has horizontal velocity, manages to travel some distance before intercepting the ground.

Figure 4-2. Throwing Baseballs Off the Top of a Mountain. At faster and faster velocities, the balls travel further before hitting the ground.

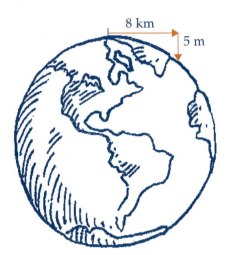

Figure 4-4. Earth's Curvature. The Earth's curvature means the surface curves down about 5 m for every 8 km.

Figure 4-3. Both Balls Hit at the Same Time. A dropped ball and a ball thrown horizontally from the same height will both hit the ground at the same time. This is because horizontal and vertical motion are independent. Gravity is acting on both balls equally, pulling them to the ground at exactly the same rate of 9.798 m/s².

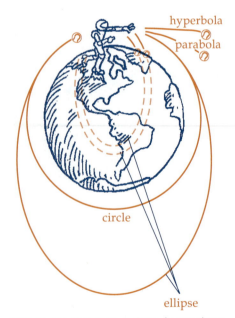

Figure 4-5. Baseballs in Orbit. As you throw baseballs faster and faster, eventually you can reach a speed at which the Earth curves away as fast as the baseball falls, placing the ball in orbit. At exactly the right speed it will be in a circular orbit. A little faster and it's in an elliptical orbit. Even faster and it can escape the Earth altogether on a parabolic or hyperbolic trajectory

Now let's return to the top of our mountain and start throwing our baseballs faster and faster to see what happens. No matter how fast we throw them, the balls still fall at the same rate. However, as we increase their horizontal velocity, they're able to travel farther and farther before they hit the ground. Because the Earth is basically spherical in shape, something interesting happens. The Earth's spherical shape causes the surface to drop approximately five meters for every eight km we travel horizontally across it, as shown in Figure 4-4. So, if we were able to throw a baseball at eight km/s (assuming no air resistance), its path would exactly match the rate of curvature of the Earth. That is, gravity would pull it down about five meters for every eight kilometers it travels, and it would continue around the Earth at a constant height. If we don't remember to duck, it will hit us in the back of the head about 85 minutes later. (Actually, because of the rotation of the Earth, it would miss your head.) A ball thrown at a speed slower than eight km/s will fall faster than the Earth will curve away beneath it. Thus, it will end up striking the surface. The results of our baseball throwing experiment are shown in Figure 4-5.

If we analyze our various baseball trajectories, we see a whole range of different shapes. Only at exactly one particular velocity do we get a circular trajectory. Any slower than that and our trajectory hits the Earth at some point. If we were to project this shape into the Earth, we'd find the trajectory we see is really a piece of an ellipse (contrary to popular opinion which says they're in a parabolic trajectory—it only looks parabolic, it's actually elliptical.) If we throw the ball a bit harder than the circular velocity, we also obtain an ellipse. If we eat our Wheaties and throw the ball too hard, it leaves the Earth altogether on a parabolic or hyperbolic trajectory, never to return. No matter how hard we throw, our trajectory resembles either a circle, ellipse, parabola, or hyperbola. As we'll see in Section 4.4, these four shapes are called *conic sections*.

So an object in orbit is literally falling around the Earth, but because of its horizontal velocity it never quite hits the ground. Throughout this book we'll see how important having the right velocity at the right place is to determining the kind of orbit we're in.

Analyzing Motion

Now that we've looked at orbits conceptually, let's see how we could analyze this motion in a more rigorous way. Chances are, when you first learned to play catch with a baseball, you had problems. Your poor partner had to run all over the field trying to catch your first tentative throws, which never seemed to go where you wanted. But gradually, after a little bit of practice (and several exhausted partners), you began to get the hang of it. Eventually, you could place the ball right into your partner's glove, almost without conscious thought.

In fact, if you talk to expert pitchers, you find they don't think about *how* to throw; they simply concentrate on *where* to throw. Somehow, their brain calculates the precise path needed to deliver the ball to the desired location. Then it commands the arm to a predetermined release point and time with exactly the right amount of force. All this happens in a matter of seconds, without a thought given to the likes of Isaac Newton and the equations that describe the baseball's motion. "So what?" you may wonder. After all, you know *how* to throw a baseball over home plate. Why bother with all the equations which describe *why* it travels the way it does?

Unfortunately, to build a pitching machine for a batting cage or to launch a satellite into orbit, we can't simply tell the machine or rocket to "take aim and throw." In the case of the rocket, we must carefully study its motion between the launch pad and space.

We'll define a systematic approach that we can use to analyze all types of motion. It's called the Motion Analysis Process (MAP) checklist and is shown in Figure 4-6. To see the MAP in action, pretend that, as a school project, you're asked to describe the motion of a baseball thrown by our two astronauts in Figure 4-7. How would you go about it?

> **MOTION ANALYSIS PROCESS (MAP)**
> **CHECKLIST**
>
> ❏ Coordinate System
>
> ❏ Equation of Motion
>
> ❏ Simplifying Assumptions
>
> ❏ Initial Conditions
>
> ❏ Error Analysis
>
> ❏ Testing the Model

Figure 4-6. Motion Analysis Process (MAP) Checklist.

Figure 4-7. Baseball Motion. To analyze the motion of a baseball, or a spacecraft, we must step through the Motion Analysis Process (MAP) checklist.

First of all, you would need to define some frame of reference or *coordinate system*. For example, do you want to describe the motion with respect to a nearby building or to the center of the Earth? In either case, you must define a reference point and a coordinate frame for the motion you're describing. Next you need some short-hand way of describing this motion and its relation to the forces involved—a short-hand way we'll call an *equation of motion*.

Once you've determined which equation best describes the baseball's motion, you need to decide how to make your job easier. After all, you don't want to try to deal with how the motion of the baseball changes due to the gravitational pull of Venus or every little gust of wind in the park. So, to simplify the problem, you must make some reasonable *simplifying assumptions*. For instance, you could easily assume that the gravitational attraction on the baseball from Venus, for example, is too small to worry about and the drag on the baseball due to air resistance is insignificant. And, in fact, as a good approximation, you could assume that the only force on the baseball comes from the Earth's gravitational pull.

With these assumptions made, you could then turn your attention to the finer details of the baseball problem. For example, you would want to carefully define where and how the motion of the baseball begins. We call these the *initial conditions* of the problem. If these initial conditions are varied somehow (e.g., you throw the baseball a little harder or with a slightly different angle), the motion of the baseball will change. By assessing how these variations in initial conditions affect where the baseball goes, you can find out how sensitive the trajectory is to small changes or errors in the initial conditions.

Finally, once you've completed all of these steps, you'd want to verify the entire process by *testing the model* of baseball motion you've developed. You'd actually throw some baseballs, measure their deviation, and perform *error analysis* between the motion you predict for the baseball and what you find from your tests. If you find significant differences, you may have to change your coordinate system, equation of motion, assumptions, initial conditions, or all of these. With the MAP in mind, let's begin our investigation of orbit motion by considering some fundamental laws of motion Isaac Newton developed.

▦ Section Review

Key Terms

conic sections
coordinate system
equation of motion
error analysis
initial conditions
simplifying assumptions

Key Concepts

➤ From a conceptual standpoint, orbit motion involves giving something enough horizontal velocity so that, by the time gravity pulls it down, it has traveled far enough to have the Earth's surface curve away from it. As a result, it stays above the surface. An object in orbit is essentially falling around the Earth but going so fast it never hits it.

➤ The Motion Analysis Process is a general approach for understanding the motion of any object through space. It consists of
  • Coordinate system
  • Equation of motion
  • Simplifying assumptions
  • Initial conditions
  • Error analysis
  • Testing the model

4.2 Newton's Laws

▬ In This Section You'll Learn to...

- ☞ Explain the concepts of weight, mass, and inertia
- ☞ Explain Newton's laws of motion
- ☞ Use Newton's laws to analyze the simple motion of objects

Since the first caveman threw a rock at a sabre-toothed tiger, we have been intrigued by the study of motion. In our quest to understand nature, we've looked for simple, fundamental laws which all objects must obey. These Laws of Motion would apply universally for everything from gumdrops to galaxies. They would be unbreakable and would empower us to explain the motion of the heavens, understand the paths of the stars, and predict the destiny of our Earth. In Chapter 2, we saw how the Greek philosopher Aristotle defined concepts of orbit motion which held favor until finally challenged by such thinkers as Galileo and Kepler. Recall that Kepler gave us three laws of planetary motion. These laws, while useful for describing the planets' motion, didn't tell us *why* planets moved that way. That's where Isaac Newton comes in.

Reflecting back on his lifetime of scientific accomplishments, Newton rightly observed that he was able to do so much because he "stood on the shoulders of giants." Armed with Galileo's two basic principles of motion—inertia and relativity—and Kepler's laws of planetary motion, Isaac Newton was poised to determine the basic laws of motion which have revolutionized our understanding of the world.

No single person has had as great an impact on science as Isaac Newton. His numerous discoveries and fundamental breakthroughs would easily fill this volume. Inventing calculus (math students still haven't forgiven him for that one), inventing the reflecting telescope, and discovering gravity are just some of his many accomplishments. For our purposes, we'll see that all of astrodynamics builds on four of Newton's laws: three of motion and one describing gravity.

Weight, Mass, and Inertia

Before plunging into a discussion of Newton's many laws, let's take a moment to complicate a topic that, until now, you probably thought you understood very well—weight. When you order a "quarter-pounder with cheese," you're describing the weight of the hamburger (before cooking). To measure this weight (say, to determine what it weighs after cooking), we would slap the burger on a scale and read the results. If our scale gave weight in metric units, we'd see our quarter-pounder weighs about one Newton. This property we call weight is really the result of another, more basic property of the hamburger and an outside influence we'll later learn is gravity. This more basic property of the hamburger is its mass. A hamburger that weighs one Newton (on Earth) has a mass of 1/9.798 kg or about 0.1 kg. Mass is a very useful property. In fact, when we know the mass of our hamburger, we automatically know three different things about it, as illustrated in Figure 4-8.

First, *mass* is a measure of how much matter or "stuff" you have. The more mass, the more stuff. If we take the quarter-pounder and start dividing it into smaller and smaller pieces, we'll eventually get down to the atomic level. These atoms are made up of the basic building blocks of all matter: protons, neutrons, and electrons. Therefore, if you tell me the mass of the quarter-pounder is 0.1 kg, I know basically how much stuff I have.

But that's not all. Knowing the mass of an object also tells us how much inertia it has. As you may remember, Galileo first put forth the principle of *inertia* in terms of an object's tendency to stay at rest or in motion unless acted on by an outside influence. Your own personal analogy of inertia perhaps comes when you're in "couch potato" mode in front of the TV, with your homework sitting on the desk at the other side of the room, calling for your attention. Somehow, you just can't seem to motivate yourself to get up from the couch and start those math problems. You have too much "inertia," so it takes an outside influence (your mother or a deep-rooted fear of failing the next test) to overcome that "inertia." For a given quantity of mass, inertia works in much the same way. An object at rest has a certain quantity of inertia, represented by its mass, which must be overcome to get it into motion. Thus, to get the quarter-pounder from its package and into your mouth, you must overcome the inertia inherent in the mass of the hamburger.

Finally, if we know an object's mass, we also know how it affects other objects merely by its presence. There's an old, corny riddle which asks "Which weighs more—a pound of feathers or a pound of lead?" Of course, they both weigh the same—one pound. Why is that? Weight is a result of two things—the amount of mass, or "stuff," and gravity. So, assuming we take the weight measurements for the feathers and lead at a point where the gravity is the same, we'll see that we can conclude their masses are the same. *Gravity*, as we'll see in greater detail later, is the tendency for two (or more) chunks of stuff to attract each other. The more stuff (or mass) they have, the more they will attract. This natural attraction between chunks of stuff is always there. Thus, your quarter-pounder lying there in its package will cause a very slight gravitational pull on your fries, milk shake, and all other mass in the universe. (You'd better eat fast!)

Now that you'll never be able to look at a quarter-pounder the same way again, let's see how Isaac Newton was able to use these concepts of mass to develop some basic laws of motion and gravity.

Figure 4-8. What is Mass? The amount of mass an object has tells us 3 things about it: (1) how much "stuff" it contains, (2) how much it resists changes in motion—its inertia, and (3) how much gravitational force it exerts and is exerted on it by other masses in the universe.

Momentum

In 1655 a great plague ravaged England. The plague was so severe that universities actually shut down. At the time, Isaac Newton was a 23-year-old student at Cambridge. Instead of hitting the beach for an extended "spring break," the more scholarly Newton hit the apple orchard, or so legend has it. Although his findings would not be published for twenty years, Newton did his greatest work on the nature of motion and gravity during this period. The famous story of Newton being hit on the head with an apple and discovering gravity is the most well known example.

Newton's First Law of Motion was actually a variation on Galileo's concept of inertia. Published in 1687, his monumental work *The Mathematical Principles of Natural Philosophy* (actually, Latin was the vogue among scholars of his day, so the real title read: *Philosophiae Naturalis Principia Mathematica*) stated:

Newton's First Law of Motion. A body continues in its state of rest, or of uniform motion in a straight line, unless compelled to change that state by forces impressed upon it.

Newton's First Law says that any object (or chunk of mass) which is at rest will *stay* at rest forever, unless some force *makes* it move. Similarly, any object in motion will *stay* in motion forever, with a constant speed in exactly the same straight-line direction, until some force *makes* it change either its speed or direction of motion. Try to stop a speeding bullet like the one in Figure 4-9 and you get a good idea of how profound Newton's First Law is.

One very important aspect of the first law to keep in mind, especially when we start looking at satellite motion, is that motion tends to *stay in a straight line*. Therefore, if you ever see something *not* moving in a straight line, like something moving around in a circle, then some force must be acting on it.

We know that an object at rest is lazy; it doesn't want to start moving and will resist movement to the fullest extent of its mass. We've also discovered that, once it's in motion, it resists any change in its speed or direction. But the amount of resistance for an object at rest and one in motion are not the same! This seeming paradox is due to the concept of momentum. *Momentum* is the amount of resistance an object in motion has to changes in its speed or direction of motion. This momentum is the result of combining an object's mass with its velocity. Because an object's velocity can be either linear or angular, there are two types of momentum: linear and angular.

We'll start with linear momentum. To see how it works, consider the difference between a bulldozer and a baby carriage moving down the street, as shown in Figure 4-10. Bulldozers are massively constructed machines designed to savagely rip tons of dirt from the Earth. Baby carriages are delicately constructed, four-wheeled carts designed to carry cute little babies around the neighborhood. Obviously, a bulldozer has much more mass than a baby carriage, but how does their momentum

Figure 4-9. Newton's First Law. Any object in motion, such as a speeding bullet, will tend to stay in motion, in a straight line, unless acted on by some outside force (like gravity or hitting a brick wall.)

Figure 4-10. Bulldozer, Baby Carriage, and Momentum Inertia. The momentum of any object is the product of its mass and velocity.

compare? Unlike inertia, which is a function only of an object's mass, *linear momentum, p̄*, is the product of an object's mass, m, and its velocity, V̄. [Note: because velocity and momentum must be described in terms of magnitude and direction, we treat them and other important concepts as *vector* quantities. Appendix A reviews vector notation and concepts.]

Note: We'll consistently refer to the *velocity vector* as V̄, which denotes both speed *and* direction. The magnitude of the velocity vector we'll call *speed* or simply *velocity, V*.

$$\vec{p} = m\vec{V} \qquad (4\text{-}1)$$

where
$\vec{p}$ = linear momentum vector (kg · m/s)
m = mass (kg)
$\vec{V}$ = velocity vector (m/s)

To compare the linear momentum of the bulldozer and the baby carriage, we'd have to know how fast both were moving. For the two to have the same linear momentum, the baby carriage, being much less massive, would have to be going much, much faster! Example 4-1, at the end of this section, shows this relationship.

Linear momentum is fairly basic because it involves motion in a straight line. Angular momentum, on the other hand, is slightly harder to understand. Let's look at a top. If you set a top on a table it will simply fall over, but if you spin it, the top will seem to defy gravity.

A spinning object tends to resist changes in the direction and rate of spin, like the toy top shown in Figure 4-11, just as an object moving in a straight line resists changes to its speed and direction of motion. *Angular momentum, H̄*, is the amount of resistance of a spinning object to change in spin rate or direction of spin. We found linear momentum to be the product of the object's mass, m, (which represents its inertia, or resistance to change in speed and direction), and its velocity, V̄. Similarly, angular momentum is the product of an object's resistance to change in spin rate or direction, and its rate of spin. An object's resistance to spin is its *moment of inertia, I*. The *angular velocity*, which is a vector, is represented by Ω̄. So the angular momentum vector, H̄, can be found using Equation (4-2).

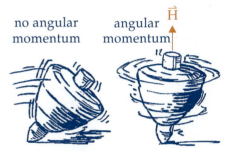

Figure 4-11. Angular Momentum. A non-spinning top falls right over, but a spinning top, because of its angular momentum, resists the torque applied by gravity and stays upright.

$$\vec{H} = I\,\vec{\Omega} \qquad (4\text{-}2)$$

where
$\vec{H}$ = angular momentum vector (kg · m²/s)
I = moment of inertia (kg · m²)
$\vec{\Omega}$ = angular velocity vector (rad/s)

To characterize the direction of angular momentum, we need to examine the angular velocity, Ω̄. Look at the spinning wheel in Figure 4-12 and apply the right-hand rule. With your fingers curled in the direction of the spin, the angular velocity vector, Ω̄, and the angular momentum vector, H̄, point in the direction of your thumb.

As Equation (4-2) implies, H̄ will always be in the same direction as the angular velocity vector, Ω̄. In the next section we'll see that, because of angular momentum, a spinning object will resist change to both its spin direction and speed.

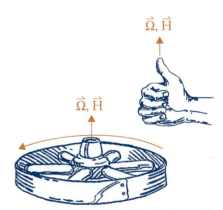

Figure 4-12. The Right-Hand Rule. The direction of the angular velocity vector, Ω̄, and the angular momentum vector, H̄, are found using the right-hand rule.

Astro Fun Fact

Spiralling Football

Why does a spiralling football go farther than a tumbling one? The spiralling football will resist change to the spin (and therefore the angular momentum) direction, so it will present its streamlined profile to the wind throughout its flight. This allows the ball to go farther than a tumbling ball, which will encounter greater wind resistance. This same principle is applied in a rifle barrel. "Rifling" refers to the spiral grooves cut inside the barrel. This causes the bullet to spin as it's fired, giving it angular momentum. The spin keeps the bullet from tumbling, thus making it less susceptible to wind resistance, so it can go further and straighter.

Figure 4-13. Describing Angular Momentum. The direction of the angular momentum vector, $\vec{H}$, is perpendicular to both $\vec{R}$ and $\vec{V}$ and is found using the right-hand rule.

We can describe angular momentum in another way. A mass spinning on the end of a string also has angular momentum. In this case, the angular momentum is found by using the instantaneous tangential velocity of the spinning mass, $\vec{V}$, and the length of the string, $\vec{R}$, also called the *moment arm*. These two are combined with the mass, m, using a cross product relationship to get $\vec{H}$.

$$\vec{H} = \vec{R} \times m\vec{V}$$

(4-3)

where
$\vec{H}$ = angular momentum vector (kg · m²/s)
$\vec{R}$ = position (m)
m = mass (kg)
$\vec{V}$ = velocity vector (m/s)

By the nature of the cross product, you can tell that $\vec{H}$ must be perpendicular to both $\vec{R}$ and $\vec{V}$. Once again, we can use the right hand rule to find $\vec{H}$, as shown in Figure 4-13. Example 4-2 analyzes the mass on the end of the string in more detail.

Later in this chapter, we'll see that angular momentum is a very important property of spacecraft orbits. In Chapter 12, we'll find angular momentum is also a useful property for gyroscopes and spacecraft.

Changing Momentum

Now that we understand a little bit about momentum, let's go back to Newton's Laws of Motion. As we saw, whether we're dealing with linear or angular momentum, both represent the amount a moving object resists change in its direction or speed. Now we can determine what it will take to overcome this resistance. That's where Newton's Second Law comes in.

Newton's Second Law of Motion. The time rate of change of an object's momentum is equal to the applied force.

In other words, to change an object's momentum very quickly, such as when we hit a fast ball with a bat, the force applied must be relatively high. On the other hand, if we're in no hurry to change the momentum, we can apply a much lower force over much more time.

Let's pretend you see a bulldozer creeping down the street at 1 m/s (3.28 ft./s), as in Figure 4-14. To stop the bulldozer dead in its tracks, you must apply some force, usually by pressing on the brakes. How much force you need depends on how fast you want to stop the bulldozer. If, for instance, you wanted to bring the "dozer" to a complete halt in one second, you'd have to overcome all of its momentum quickly. In this case, you'd need to apply a tremendous force. On the other hand, if you want to bring the "dozer" to a halt over one hour, you could apply a much smaller force.

Now let's look at how we can summarize the relationship implied by Newton's Second Law. The short-hand symbol we'll use to represent a force is $\vec{F}$. We've already seen that momentum is represented by the symbol $\vec{p}$. To represent how fast a quantity is changing, we must introduce some notation from calculus. (See Appendix B for a complete review of these concepts.) We'll use the Greek symbol "delta," Δ, to represent a change in any quantity. Thus, we can represent the rate of change of a quantity, such as momentum, $\vec{p}$, over some length of time, t, as

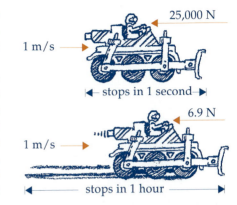

Figure 4-14. Newton's Second Law. The force we must apply to stop a moving object depends on how fast we want to change its momentum. If the two bulldozers are both moving at 1 m/s (about the speed of a brisk walk), you must apply a much, much larger force to stop a bulldozer in one second than to stop it in one hour.

$$\frac{\Delta \vec{p}}{\Delta t} = \frac{\text{change in momentum}}{\text{change in time}} \qquad (4\text{-}4)$$

This shows how fast momentum is changing. We can now express Newton's Second Law in symbolic shorthand as

$$\vec{F} = \frac{\Delta \vec{p}}{\Delta t} = \frac{\Delta (m\vec{V})}{\Delta t} \qquad (4\text{-}5)$$

We can expand this equation using another concept from calculus, to get

$$\vec{F} = m\frac{\Delta \vec{V}}{\Delta t} + \frac{\Delta m}{\Delta t}\vec{V} \qquad (4\text{-}6)$$

So what can we do with this relationship? Let's begin with $\Delta m / \Delta t$ in the second term. This ratio represents how fast the mass of the object is changing. For many cases, the mass of the object won't be changing, so this term goes to zero. (In Chapter 14, we'll see this isn't the case for rockets.) Now we have only the first term in the relationship. $\Delta \vec{V}/\Delta t$ represents how fast velocity is changing over time. But this is just the definition of acceleration, $\vec{a}$. If we substitute $\vec{a}$ for $\Delta \vec{V}/\Delta t$ into Equation (4-6), we get the more familiar

$$\boxed{\vec{F} = m\vec{a}} \qquad (4\text{-}7)$$

where
$\vec{F}$ = force vector (kg $\cdot$ m/s^2 = N)
m = mass (kg)
$\vec{a}$ = acceleration (m/s^2)

Equation (4-7) is arguably one of the most useful equations in all of physics. It allows us to understand how forces will affect the motion of objects. Armed only with this simple relationship, we can determine everything from how hard we must kick a football to put it between the goal posts to how long we must light our rocket engines to blast a spacecraft to Mars. Example 4-3 shows this equation in action.

Action and Reaction

These two laws alone would have made Newton famous. But he went on to discover a very important relationship between action and reaction. If you turn to the students next to you and punch them in the arm, chances are, you'll get punched back. This is an example of your taking an action (punching the other students in the arm) and then experiencing a reaction (getting punched back). Although there are no firm reports of Newton getting into lots of fights as a kid, he was somehow able to determine that this principle of action and reaction applied universally for all mechanical systems.

A less facetious and more precise example of Newton's Third Law applies in roller skating. Imagine that two astronauts are on roller skates standing in the middle of the rink, as shown in Figure 4-15. If one gives the other a push, what happens? They both move backward! The first astronaut exerted a force on the second, but in turn an equal but opposite force is exerted on him, thus sending him backwards! In fact, Newton found that the reaction is exactly equal in magnitude but opposite in direction to the original action.

Newton's Third Law of Motion. When body A exerts a force on body B, body B will exert an equal, but opposite, force on body A.

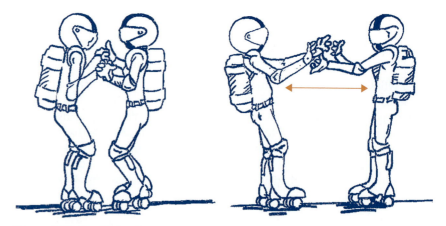

Figure 4-15. Two Roller Skaters Demonstrate Newton's Third Law. If they initially start at rest and the first one pushes against the second, they'll both go backwards. The first astronaut applied an action—pushing—and received an equal but opposite reaction.

In the free-fall environment of space an astronaut must be very conscious of this fact. Suppose an astronaut tries to use a wrench to turn a simple bolt without the force of gravity to anchor her in place. Unless she braces herself somehow, *she'll* start to spin instead of the bolt!

Gravity

The image most people have of Newton is of a curly-haired man clad in the tights and lace common to the 17th century, seated under an apple tree with a steady stream of apples landing on his head. Finally, after being hit by one too many apples, he suddenly jumped up and shouted "Eureka! (borrowing a phrase from Archimedes) I've invented gravity!" While this image is more the stuff of Hollywood than historical fact, it contains some truth. Newton did observe falling objects, such as apples, and read extensively the work done by Galileo on the investigation of falling objects.

The breakthrough came when Newton reasoned the force due to gravity must decrease with the square of the distance from the attracting body. In other words, an object twice as far away from the Earth will be attracted only one fourth as much. Newton excitedly took observations of the Moon to verify this model of gravity. Unfortunately, his measurements consistently disagreed with his model by one-sixth. Finally, in frustration, Newton abandoned his work on gravity. Years later, however, he found that the size of the Earth he had been using in his calculations was off by exactly one-sixth. Thus, his model of gravity had been correct all along!

We call Newton's Law of Gravitation "Universal" because we believe the same principle must apply anywhere in the universe. In fact, much of modern cosmology—all we know about the structure of the universe—depends on applying this simple law. We can see it applied most simply in Figure 4-16.

Newton's Universal Law of Gravitation. The force of gravity between two bodies is directly proportional to the product of their two masses and inversely proportional to the square of the distance between them.

We can express this in symbolic shorthand as

$$F_g = \frac{G\, m_1 m_2}{R^2}$$

(4-8)

where

F_g	= force due to gravity (N)
G	= universal gravitational constant = 6.67×10^{-11} N·m²/kg²
m_1, m_2	= mass of two bodies (kg)
R	= distance between the two bodies (m)

So what does this tell us? If we have two bodies, say Earth and the Moon, the force of attraction between the two is equal to the product of

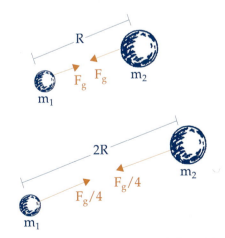

Figure 4-16. Newton's Universal Law of Gravitation. The force of attraction between any two masses is directly proportional to the product of their masses and inversely proportional to the square of the distance between them. Thus, if we double the distance between two objects, the gravitation force decreases to 1/4 the original amount.

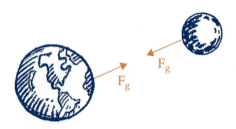

Figure 4-17. Earth and Moon in Tug-of-War. Because of gravity, the Earth and Moon pull on each other with incredible force, which causes tides on Earth.

their two masses times a constant divided by the square of the distance between them. Let's look at some real numbers to see just how hard the Earth tugs on the Moon and vice versa, as shown in Figure 4-17. The mass of the Earth, m_{Earth}, is 5.98×10^{24} kg (give or take a couple of mountains!), and the mass of the Moon, m_{Moon}, is 7.35×10^{22} kg. The average distance between the Earth and Moon is about 3.84×10^8 m. We already know the gravitational constant, G. Using the relationship for gravitational force we just discovered, we find

$$F_g = \frac{Gm_{Earth}m_{Moon}}{R^2}$$

$$F_g = \frac{\left(6.67 \times 10^{-11} \frac{Nm^2}{kg^2}\right)(5.98 \times 10^{24} \text{ kg})(7.35 \times 10^{22} \text{ kg})}{(3.84 \times 10^8 \text{ m})^2}$$

$$F_g = 1.98 \times 10^{20} \text{ N (or about } 4.46 \times 10^{19} \text{ lb}_f)$$

In other words, there's a heck of a lot of force pulling the Earth and Moon together. But do we here on Earth experience the result of this age-old tug-of-war between the Earth and its largest satellite? You bet we do! The biggest result we see is in ocean tides. The side of the Earth closest to the Moon is attracted more than the side away from the Moon (gravity decreases as the square of the distance). Thus, all the ocean water on the side closest to the Moon swells toward the Moon; on the other side, the water swells away from the Moon. Depending on the height and shape of the ocean floor, tides can raise and lower the sea level in some places more than 5 m (16 ft.). If you think about how much force it would take you to lift half the ocean this much, the incredibly large force we computed above makes sense.

It's important to remember that the force of gravity decreases as the square of the distance between masses increases. This means that if you want to lose weight you should take a trip to the mountains! If you normally live in Houston, Texas, (elevation ~0 ft.) and you take a trip to Leadville, Colorado, (elevation 3048 m or 10,000 ft.), you will weigh less. That's because you're somewhat farther away from the attracting body (the center of the Earth). But before you start packing your bags, look closely at what is happening. Your *weight* will change because the force of gravity will be slightly less, but your *mass* won't change. Remember, weight measures how much gravity is pulling you down. Mass measures how much stuff you have. So even though the force pulling down on the scale will be slightly less, you'll still have those unwanted bulges.

The acceleration due to gravity also changes with the square of the distance. We can compute the acceleration due to gravity by combining the relationships expressed in Newton's Second Law of Motion and Newton's Universal Law of Gravitation. We know from Newton's Second Law (dropping vector notation because we're interested only in magnitudes) that

$$F = ma \qquad (4\text{-}9)$$

We can substitute this expression into Newton's relationship for gravity to get an expression for the acceleration of any mass due to Earth's gravity.

$$ma_g = \frac{mG \; m_{Earth}}{R^2}$$

which simplifies to

$$a_g = \frac{G \; m_{Earth}}{R^2}$$

For convenience, we typically combine G and the mass of the central body to get a new value we call the *gravitational parameter, μ* (mu), where $\mu \equiv G\,m$. For Earth, this is denoted with a subscript, μ_{Earth}.

$$\boxed{a_g = \frac{\mu_{Earth}}{R^2}} \tag{4-10}$$

where
a_g = acceleration due to gravity (m/s^2)
μ_{Earth} $\equiv G\,m_{Earth} = 3.986005 \times 10^{14} \, m^3/s^2$
R   = distance to center of Earth (m)

If we substitute the values for μ_{Earth} and use the mean Earth radius (6378.137 km) we get $a_g = 9.798$ m/s^2 at the Earth's surface.

Astro Fun Fact

Galileo Was Correct

Nearly 400 years later and more than 400,000 km away, one of Galileo's ideas was finally put to the test. On the Moon during the Apollo 15 mission, in the summer of 1971, astronaut Dave Scott performed a simple experiment: "In my left hand I have a feather. In my right hand, a hammer. I guess one of the reasons we got here today was because of the gentleman named Galileo a long time ago who made a rather significant discovery about falling objects in gravity fields, and we thought, 'where would be a better place to confirm his findings than on the Moon?' And so we'll try it here for you. The feather happens to be appropriately a falcon feather for our Falcon [the name of the lunar lander] and I'll drop the two of them here and, hopefully, they'll hit the ground at the same time." With that, Scott dropped the two objects which impacted the lunar surface simultaneously in the absence of any air resistance. "How about that," Scott exclaimed, "this proves that Mr. Galileo was correct!"

David Baker, PhD, The History of Manned Space Flight. New York, NY: Crown Publishers Inc., 1981.

Section Review

Key Terms

angular momentum, $\vec{H}$
angular velocity, $\vec{\Omega}$
gravitational parameter, μ
gravity
inertia
linear momentum, $\vec{p}$
mass
moment arm
moment of inertia, I
momentum

Key Equations

$$\vec{p} = m\vec{V}$$

$$\vec{H} = I\,\vec{\Omega}$$

$$\vec{H} = \vec{R} \times m\vec{V}$$

$$\vec{F} = m\vec{a}$$

$$F_g = \frac{G\,m_1 m_2}{R^2}$$

$$a_g = \frac{\mu_{Earth}}{R^2}$$

Key Concepts

➤ The mass of an object denotes three things about it:
  • How much "stuff" it has
  • How much it resists motion—its inertia
  • How much gravitational attraction it has

➤ Newton's three laws of motion are
  • First Law. A body continues in its state of rest, or in uniform motion in a straight line, unless compelled to change that state by forces impressed upon it.
    - The first law refers to an object's linear or angular momentum
    - Linear momentum, $\vec{p}$, is related to an object's mass, m, and velocity, $\vec{V}$
    - Angular momentum, $\vec{H}$, is the product of an object's moment of inertia, I, (the amount it resists angular motion) and its angular velocity, $\vec{\Omega}$
    - Angular momentum can also be expressed as a vector cross product of an object's position from the center of rotation, $\vec{R}$ (called its moment arm), and the product of its mass and instantaneous tangential velocity, $\vec{V}$
  • Second Law. The time rate of change of an object's momentum is equal to the applied force.
  • Third law. When body A exerts a force on body B, body B will exert an equal, but opposite, force on body A.

➤ Newton's Universal Law of Gravitation. The force of gravity between two bodies (m_1 and m_2) is directly proportional to the product of the two masses and inversely proportional to the square of the distance between them (R).
  • G = universal gravitational constant = 6.67×10^{-11} Nm^2/kg^2
  • We often use the gravitational parameter, μ, to replace G and m. $\mu \equiv G\,m$
    - The gravitational parameters of Earth, μ_{Earth}, is

$$\mu_{Earth} \equiv G\,m_{Earth} = 3.986005 \times 10^{14}\ m^3/s^2$$

$$= 3.986005 \times 10^5\ km^3/s^2$$

Example 4-1

Problem Statement

How fast would a 25 kg baby carriage have to be going to have the same linear momentum as a 25,000 kg bulldozer moving at 1 m/s?

Problem Summary

Given: $m_{bulldozer} = 25,000$ kg

$V_{bulldozer} = 1$ m/s

$m_{baby\ carriage} = 25$ kg

Find: $V_{baby\ carriage}$ to equal momentum of the bulldozer

Problem Diagram

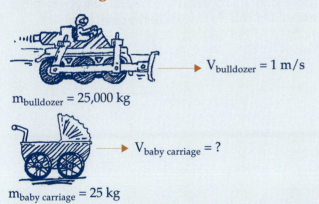

$V_{bulldozer} = 1$ m/s

$m_{bulldozer} = 25,000$ kg

$V_{baby\ carriage} = ?$

$m_{baby\ carriage} = 25$ kg

Conceptual Solution

1) Determine the magnitude of the linear momentum of the bulldozer

$$p_{bulldozer} = m_{bulldozer} V_{bulldozer}$$

2) Using the momentum of the bulldozer and the mass of the baby carriage, solve for the required velocity of the baby carriage

$$V_{baby\ carriage} = \frac{p_{bulldozer}}{m_{baby\ carriage}}$$

Analytical Solution

1) Determine linear momentum of bulldozer

$$p_{bulldozer} = m_{bulldozer} V_{bulldozer}$$
$$= (25,000\ kg)\ (1\ m/s)$$
$$= 25,000\ kg \cdot m/s$$

2) Solve for required baby carriage velocity

$$V_{baby\ carriage} = \frac{p_{bulldozer}}{m_{baby\ carriage}}$$
$$= \frac{25,000\ kg \cdot m/s}{25\ kg}$$
$$= 1000\ m/s$$

Interpreting the Results

For a baby carriage to have the same linear momentum as a massive bulldozer moving at only 1 m/s (about the speed of a brisk walk), it would have to go 1000 m/s—almost three times the speed of sound!

Example 4-2

Problem Statement

Astroboy is spinning a 0.1 kg ball at the end of a 1.0 m string. The angular momentum of the spinning system is known to be 10 kg m²/s. If Astroboy lets go of the string, how fast and in what direction will the ball go?

Problem Summary

Given: $m_{ball} = 0.1$ kg
$H = 10$ kg m²/s
$R = 1.0$ m
Find: V_{ball} direction when released

Problem Diagram

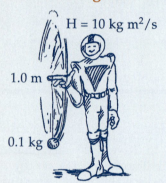

$H = 10$ kg m²/s

1.0 m

0.1 kg

Analytical Solution

1) Solve for tangential velocity of the ball

$$V_{ball} = \frac{H_{ball}}{R_{ball} m_{ball}}$$

$$= \frac{10 \text{ kg} \cdot \text{m}^2/\text{s}}{(1.0 \text{ m}) (0.1 \text{ kg})}$$

$$= 100 \text{ m/s}$$

2) By inspection—when the ball is released, the force of the string is no longer forcing it to go in a circular path, so it will move off on a straight line tangent at the point of release.

Interpreting the Results

A 0.1 kg ball on a circular path with a radius of 1.0 m and an angular momentum of 10 kg · m²/s must be moving at a tangential velocity of 100 m/s. When released, it will fly tangent to the point of release.

Conceptual Solution

1) Solve the angular momentum equation for the tangential velocity of the ball

$$\vec{H}_{ball} = \vec{R}_{ball} \times m_{ball} \vec{V}_{ball}$$

$$H_{ball} = R_{ball} m_{ball} V_{ball}$$

$$V_{ball} = \frac{H_{ball}}{R_{ball} m_{ball}}$$

2) By inspection—determine which direction the ball will travel when released.

Example 4-3

Problem Statement

A placekicker is able to apply a 100 N force to a 1 kg football for a total of 0.1 seconds. Ignoring gravity, how fast will the football be going?

Problem Summary

Given: $m_{football} = 1$ kg
$F_{kicker} = 100$ N
$\Delta t = 0.1$ s

Find: $\Delta V_{football}$

Problem Diagram

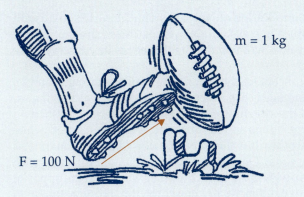

m = 1 kg

F = 100 N

Conceptual Solution

1) Use Newton's Second Law of Motion to solve for the change in velocity of an object in terms of a force applied over some length of time

$$\vec{F} = m\vec{a}$$

$$F = ma$$

$$F = m\frac{\Delta V}{\Delta t}$$

$$\Delta V = \frac{F\Delta t}{m}$$

Analytical Solution

1) Solve for ΔV

$$\Delta V = \frac{F\,\Delta t}{m}$$

$$\Delta V = \frac{(100\ N)\,(0.1s)}{1kg} = 10\ m/s$$

Interpreting the Results

You can use Newton's Second Law of Motion to analyze the results of applying a given force to an object for some length of time. In this case, a kicker applying 100 N of force will kick a football to a speed of 10 m/s (22 m.p.h.).

4.3   Laws of Conservation

▰ In This Section You'll Learn to...

☞ Describe the basic laws of conservation of momentum and energy and apply them to simple problems

As we'll see in this section, nature invented the concept of recycling long before we came up with it. Nature hates to create or destroy anything it doesn't have to. For any given system, many basic properties, such as momentum and energy, remain fixed. In physics we say that if a certain property or quantity remains unchanged for a given system, that property or quantity is *conserved*. So let's take a look at two basic properties of systems—momentum and energy.

Momentum

One very important implication of Newton's Third Law has to do with the amount of momentum in a system. Newton's Third Law implies the total momentum in a system will remain unchanged, or be conserved. We call this *conservation of momentum*.

To understand this concept let's go back to our roller skating example. When the two astronauts faced each other, neither of them was moving, so the total momentum of the system was zero. Then the first one pushed on the second, and he went rolling in one direction with some speed while she went rolling in the other. But their speeds won't be the same unless their masses are exactly equal. One astronaut will roll in one direction with a speed which depends on his mass while the other will roll in the opposite direction with a speed depending on her mass. Yet, the second astronaut's momentum (the product of her mass and velocity) will be exactly equal in magnitude, but opposite in direction, to his. Depending on how we define our frame of reference, the first astronaut's momentum could be negative while the other's positive. If we add them, they'd cancel each other out and we'd get zero! The original momentum of the system hasn't changed. Thus, as Figure 4-18 shows, we say that the total momentum of the system is conserved. Example 4-4 shows this principle in action.

This principle works equally well for angular momentum. You've probably seen a good example of this with figure skaters. At one point in their routine they always go into a spin. Remember, once an object (or skater) begins to spin, it has angular momentum. If you watch these skaters closely, you see that they can vary the rate of their spin. How do they do this? By moving their arms outward or inward. But how does this change the spin rate?

We know from Equation (4-2) that angular momentum, $\vec{H}$, is equal to the product of the moment of inertia, I, and the spin rate. Moment of

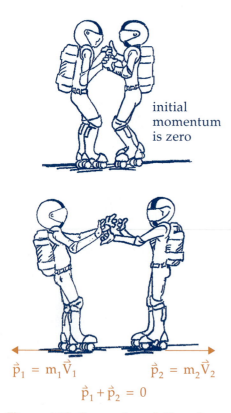

initial momentum is zero

$$\vec{p}_1 = m_1\vec{V}_1 \qquad \vec{p}_2 = m_2\vec{V}_2$$

$$\vec{p}_1 + \vec{p}_2 = 0$$

Figure 4-18. Conservation of Momentum. Two astronauts on roller skates demonstrate the concept of conservation of linear momentum. Initially the two are at rest; thus, the momentum of the system is zero. But as one astronaut pushes on the other, they both start rolling in opposite directions. Adding their two momentum vectors together still gives us zero; thus, momentum of the system is conserved.

inertia is the product of the object's mass and the square of its radius. To change their moment of inertia, skaters move their arms outward or inward, which increases or decreases the radius, changing I. Because momentum is conserved, it must stay constant as moment of inertia changes. But the only way this can happen is for the angular velocity, $\vec{\Omega}$, to change. Thus, if skaters put their arms out, as in Figure 4-19, they increase their moment of inertia and spin slower to maintain the same angular momentum. If they bring their arms in, as in Figure 4-20, they decrease their moment of inertia and increase the spin rate to maintain the same momentum.

Energy

We've all had those days when somehow we just don't seem to have any energy. But what exactly is energy? Energy can take electrical, chemical, and mechanical forms. For now, let's deal only with mechanical energy because it's the most important for understanding motion. Any time you've jumped off a chair or played with a spring, you've experienced mechanical energy. *Total mechanical energy, E,* is due to an object's position and motion. It's composed of *potential energy, PE,* which is due entirely to an object's position and *kinetic energy, KE,* which is due entirely to the object's motion. Total mechanical energy can be only potential, kinetic, or some combination of both

$$\boxed{E = KE + PE} \tag{4-11}$$

where
E = total mechanical energy (kg m^2/s^2)
PE = potential energy (kg m^2/s^2)
KE = kinetic energy (kg m^2/s^2)

To better understand what this trade-off between potential and kinetic energy means, we need to understand where it takes place. Gravity is said to be a *conservative field*—a field in which total energy is conserved. Thus, the sum of PE and KE, or the total E, in a conservative field is constant. Potential energy is the energy an object in a conservative field has entirely because of its position. For example, if you pick up a 1 kg (2.2 lb.) mass and raise it above your head, you give it a "potential" energy because of its higher position. To quantify this form of energy, we need to know three things: the amount of mass, m; its position above some reference point, h; and the acceleration due to gravity, a_g.

$$PE = m\ a_g h \tag{4-12}$$

where
m = mass (kg)
a_g = acceleration due to gravity (m/s^2)
h = height above reference point (m)

As we know from the last section, the gravitational acceleration varies depending on your distance from the center of the Earth, R. In Equation (4-12), we describe the position of the mass in terms of h, its height above

Figure 4-19. Spinning Slowly. Skaters extend their arms to increase moment of inertia—spinning more slowly.

Figure 4-20. Spinning Quickly. Skaters bring in their arms to decrease moment of inertia—spinning more quickly. Total angular momentum is the same in both cases.

some reference point (say the ground). We can just as easily describe this position in terms of R (the distance from the object to the center of the Earth) so we get a more useful relation for potential energy. Substituting a_g from Equation (4-10) into Equation (4-12) we get

$$PE = -\frac{m\mu}{R}$$

(4-13)

where
PE = potential energy (kg km^2/s^2)
m = mass (kg)
μ = gravitational parameter (km^3/s^2)
R = position measured from the center of the Earth (km)

Notice the negative sign in Equation (4-13). This is due to the convention we're using, which defines R to be positive outward from the center of the Earth. We know potential energy should increase as we raise an object up above the Earth, so is this still consistent? Yes! As something moves up, R gets bigger. As R gets bigger, PE gets less negative—which means it gets bigger too. Remember, –3 is a bigger number than –4 because it's less negative. As R gets bigger and bigger, PE gets less and less negative until, when R reaches infinity (or close enough), PE approaches zero.

One way to visualize this strange situation is to think about the center of the Earth being at the bottom of a deep, deep well like in that Figure 4-21. At the bottom of the well R is zero, so PE is at a minimum (its largest negative value). As we begin to climb out of the well, our PE begins to increase (gets less and less negative) until we reach the lip of the well at R near infinity. At this point, our PE is effectively zero and, for all practical purposes, we have left Earth's gravitational influence entirely. Of course, you never really reach an "infinite" distance from Earth, but as we'll see when we discuss interplanetary travel in Chapter 7, you would essentially leave Earth's "gravity well" at a distance of about one million km (621,000 miles).

If you have a 1 kg mass suspended above your head, how do you realize the "potential" of its energy? You let go! Gravity will then cause the mass to accelerate downward so, when it hits the ground (and hopefully not your foot), it's moving at considerable speed and thus has energy of a different kind—energy of motion which we call kinetic energy. Similar to linear momentum, kinetic energy is solely a function of an object's mass and its velocity.

$$KE = \frac{1}{2}mV^2$$

(4-14)

where
KE = kinetic energy (kg km^2/s^2)
m = mass (kg)
V = velocity (km/s)

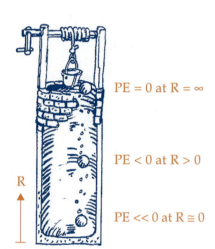

PE = 0 at R = ∞

PE < 0 at R > 0

PE << 0 at R ≅ 0

R

Figure 4-21. Potential Energy (PE). PE increases as you get farther from the center of the Earth by becoming less negative. It's as if we're climbing out of a deep well.

As we said, total mechanical energy in a conservative field stays constant. The endless trade-off between PE and KE to make this happen goes on all around us—but you may never have been aware of it. We've all played on a simple playground swing like the one in Figure 4-22. As you swing up and down, you are constantly trading back and forth between KE and PE. At the bottom of the arc, you are moving the fastest, so your KE is at a maximum and PE is at a minimum. As you swing up, your speed begins to diminish as you get higher and higher until, at the top of the arc, you actually stop briefly. At this point, your KE is zero because you're not moving, but your PE is at a maximum. The reverse happens as you swing back, this time turning your PE back into energy of motion. If it weren't for the friction and wind resistance, once you got started on a swing, you'd go on forever. Another way to understand this trade-off between KE and PE is to take a ride on a roller coaster, such as the one illustrated in Example 4-5 at the end of this section.

We can now combine KE and PE to get a new expression for total mechanical energy

$$E = \frac{1}{2}m\ V^2 - \frac{m\mu}{R} \qquad (4\text{-}15)$$

where
E = total mechanical energy (kg km^2/s^2)
m = mass (kg)
V = velocity (km/s)
μ = gravitational parameter (km^3/s^2)
R = position (km)

Later we'll use this expression to develop some useful tools for analyzing orbital motion.

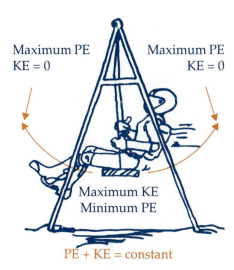

Maximum PE
KE = 0

Maximum PE
KE = 0

Maximum KE
Minimum PE

PE + KE = constant

Figure 4-22. Mechanical Energy is Conserved. The total mechanical energy, the sum of kinetic and potential energy, is constant in a conservative field. This can be seen with a simple swing. At the bottom of the arc, speed is greatest and height is lowest; hence, KE is at a maximum and PE is at a minimum. As the swing rises to the top of the arc, KE is traded for PE until you stop momentarily at the top where PE is maximum and KE is zero.

▰ Section Review

Key Terms

conservation of momentum
conservative field
conserved
kinetic energy, KE
potential energy, PE
total mechanical energy, E

Key Equations

$E = KE + PE$

$PE = -\dfrac{m\mu}{R}$

$KE = \dfrac{1}{2}mV^2$

$E = \dfrac{1}{2}m\ V^2 - \dfrac{m\mu}{R}$

Key Concepts

➤ A property is said to be conserved if it stays constant in a system

➤ In the absence of outside forces, linear and angular momentum are conserved

➤ A conservative field, such as gravity, is one in which total mechanical energy is conserved

➤ Total mechanical energy, E, is the sum of potential and kinetic energy:
  • Kinetic energy, KE, is energy of motion
  • Potential energy, PE, is energy of position

Example 4-4

Problem Statement

A 50 kg roller skater is motionless holding a 0.5 kg baseball. If the skater throws the baseball eastward at a velocity of 10 m/s what happens to the skater?

Problem Summary

Given: $m_{skater} = 50 \text{ kg}$

$V_{skater\ initial} = 0 \text{ m/s}$

$V_{mass\ initial} = 0 \text{ m/s}$

$m_{baseball} = 0.5 \text{ kg}$

$V_{baseball\ final} = 10 \text{ m/s}$

Find: $V_{skater\ final}$ and direction of motion

Problem Diagram

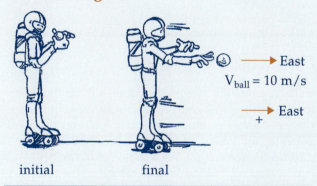

initial   final

Conceptual Solution

1) Apply Newton's First Law of Motion and the concept of conservation of momentum. The total momentum of the roller skater plus the baseball must be the same before and after the baseball is thrown.

$$\vec{P}_{initial} = \vec{P}_{final}$$

$$m_{skater}\vec{V}_{skater\ initial} + m_{baseball}\vec{V}_{baseball\ initial}$$

$$= m_{skater}\vec{V}_{skater\ final} + m_{baseball}\vec{V}_{baseball\ final}$$

2) Solve for the unknown, $\vec{V}_{skater\ final}$. The direction is found from Newton's Third Law. If the baseball is thrown eastward (action) the skater must go westward (reaction).

Analytical Solution

1) $(m_{skater}\vec{V}_{skater\ initial}) + (m_{baseball}\vec{V}_{baseball\ initial})$

$= (m_{skater}\vec{V}_{skater\ final}) + (m_{baseball}\vec{V}_{baseball\ final})$

$[(50 \text{ kg})\ (0 \text{ m/s}) + [0.5 \text{ kg})\ (0 \text{ m/s})]$

$= [(50 \text{ kg})\ (\vec{V}_{skater\ final})] + [(0.5 \text{ kg})\ (10 \text{ m/s})]$

$0 = (50 \text{ kg})\ \vec{V}_{skater\ final} + 5\dfrac{\text{kg} \cdot \text{m}}{\text{s}}$

2) Solve for $\vec{V}_{skater\ final}$

$$\vec{V}_{skater\ final} = \dfrac{-5 \text{kg} \cdot \text{m}}{50 \text{kg} \cdot \text{s}}$$

$$= -0.1 \text{ m/s}$$

Negative sign indicates westward travel because we choose positive for eastward.

Interpreting the Results

When a skater throws a baseball in one direction, he or she will go in the opposite direction according to Newton's Third Law. The velocity in the opposite direction is found by using the principle of conservation of momentum. This same basic idea is used to propel rockets. They eject mass at some high velocity in one direction and therefore move in the opposite direction.

Example 4-5

Problem Statement

A roller coaster car (on a frictionless track) begins from a dead stop at the top of the first hill at a height of 50 m. How fast will it be going at the top of the second hill at a height of 40 m?

Problem Summary

Given: $V_{initial} = 0 \text{ m/s}$
$\quad\quad\quad h_{final} = 40 \text{ m}$
$\quad\quad\quad h_{initial} = 50 \text{ m}$

Find: V_{final}

Problem Diagram

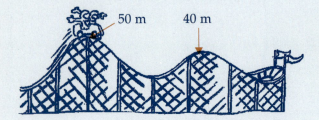

50 m   40 m

Conceptual Solution

1) Find total mechanical energy at the beginning of the problem [hint: use PE convention from Equation (4-12)]

$E_{initial} = KE + PE$

$PE_{initial} = m \, a_g \, h_{initial}$

$KE_{initial} = 1/2 \, mV^2$

2) By conservation of energy, set this total equal to the total mechanical energy at the end of the problem and solve for V_{final}

$E_{initial} = E_{final}$

$m \, a_g \, h_{initial} + 1/2 \, m \, V_{initial}^2$

$\quad = m \, a_g \, h_{final} + 1/2 \, m \, V_{final}^2$

[Assume a_g = constant]

Analytical Solution

1) Find $E_{initial}$

$E_{initial} = m \, a_g \, h_{initial} + 1/2 \, m \, V_{initial}^2$

$\dfrac{E_{initial}}{m} = (9.798 \text{ m/s}^2)(50 \text{ m}) + 1/2 \, (0 \text{ m/s})^2$

$\quad\quad = 489.9 \text{ m}^2/\text{s}^2$

2) Set $E_{initial} = E_{final}$

$\dfrac{E_{initial}}{m} = \dfrac{E_{final}}{m}$

$489.9 \text{ m}^2/\text{s}^2 = a_g \, h_{final} + 1/2 \, V_{final}^2$

$\quad\quad\quad = (9.798 \text{ m/s}^2)(40 \text{ m}) + 1/2 \, V_{final}^2$

$489.9 \text{ m/s} - 391.9 \text{ m/s} = 1/2 \, V_{final}^2$

$V_{final}^2 \cong 196 \text{ m}^2/\text{s}^2$

$V_{final} \cong 14 \text{ m/s}$

Interpreting the Results

Starting from the top of the 50 m hill on the roller coaster, you are at the point of maximum potential energy. As you go down that first hill, you trade-off potential for kinetic energy and gain speed. As you start up the second hill, the trade-off turns around and you lose speed, but you're still going 14 m/s at the top of that second hill (over 30 m.p.h.). Notice we didn't given the mass of the car in this problem and didn't need it to find the velocity. This implies the car would reach the same velocity no matter what its mass is. We'll use this same concept when we analyze satellite motion and introduce *specific* mechanical energy.

4.4 The Restricted Two-Body Problem

In This Section You'll Learn to...

- Explain the approach used to develop the restricted two-body equation of motion, including coordinate systems and assumptions

- Explain how the solution to the two-body equation of motion dictates orbit geometry

- Define and use the terms to describe orbit geometry

Earlier, we outlined a general approach to analyzing the motion of an object called the MAP, shown again in Figure 4-23. There we described the motion of a baseball. Now we can use the first three steps of this same method to understand the motion of any object in orbit. A special application of the MAP is the *restricted two-body problem*. Why restricted? As we'll see later in this section, we must restrict our analysis with assumptions we need to make our lives easier. Why two bodies? That's one of the assumptions. Why a problem? Finding an equation to represent this motion has been a classic problem solved and refined by students and mathematicians since Isaac Newton. In this section, we'll rely on the work of the mathematicians who have come before us. So at the end of this section you'll say, "The motion of two bodies? Hey, no problem!"

❏ Coordinate System

❏ Equation of Motion

❏ Simplifying Assumptions

❏ Initial Conditions

❏ Error Analysis

❏ Testing the Model

Figure 4-23. Motion Analysis Process (MAP) Checklist.

Coordinate Systems

To be valid, Newton's Laws must be expressed in an inertial frame of reference, meaning a frame that is not accelerating. To illustrate this, think about trying to describe the acceleration of a fly buzzing around in your car. You see the fly accelerating with respect to you. But that's not the whole story. Your car may be accelerating with respect to a police car behind you. Both your car and the police car may be accelerating with respect to the surface of the Earth. And of course we must consider the motion of the Earth spinning on its axis, the Earth's motion around the Sun, the Sun's motion in the Galaxy, the Galaxy's motion through the universe, and the expansion of the universe!

So you can see how this reference frame stuff can get complicated very quickly. Indeed, from astronomical observations, it looks like everything in the universe is accelerating. So how can we find any purely non-accelerating reference? We can't. To apply Newton's Laws to our buzzing fly, we must select a reference frame that's close enough to, or "sufficiently," inertial for our problem.

Any reference frame is just a collection of unit vectors at right angles to each other which allow us to specify the magnitude and direction of other vectors. This collection of unit vectors allows us to establish the components of vectors in 3-D space. By rigidly defining these unit vectors, we define a coordinate system.

To create a coordinate system we need to specify four pieces of information—an origin, a fundamental plane, a principal direction, and a third axis, as shown in Figure 4-24. The *origin* defines a physically identifiable starting point for the coordinate system. The other two parameters fix the orientation of the frame. The *fundamental plane* will contain two axes of the system. Once we know the plane, we can define a direction perpendicular to that plane. The unit vector in this direction at the origin is one axis. Next, we need a *principal direction* within the plane. Again, we pick something that is physically significant, like a star. Now that we have two directions, the principal direction and an axis perpendicular to the fundamental plane, we can find the third axis using the right-hand rule.

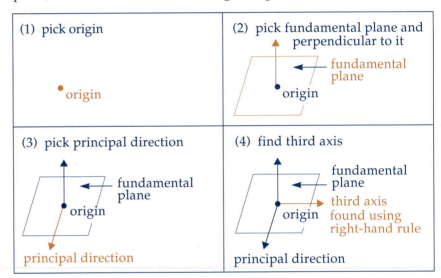

Figure 4-24. Defining a Coordinate System. Coordinate systems are found by (1) selecting a convenient origin; (2) selecting a convenient fundamental plane containing the origin and then selecting an axis perpendicular to the plane; (3) selecting a convenient principal direction within the plane; and (4) completing the 3-axis system using the right-hand rule.

Remember—coordinate systems should make our lives easier. If we choose the correct coordinate system, developing the equations of motion can be simple. If we choose the wrong system, it can be nearly impossible.

For Earth-orbiting spacecraft, we'll choose a tried-and-true system that we know makes solving the equations of motion relatively easy. We call this system the *geocentric-equatorial coordinate system* and here's how it's defined:

- Origin—center of the Earth (hence the name *geo*centric)
- Fundamental plane—Earth's equator (hence geocentric-*equatorial*). Perpendicular to plane—North Pole direction
- Principal direction—vernal equinox direction, ♈, or the vector pointing to the first point of Aries. This points at the zodiac constellation Aries and is found by drawing a line from the Earth to the Sun on the first day of Spring, as shown in Figure 4-25. While this

first day of Spring

Figure 4-25. Vernal Equinox Direction. The vernal equinox direction is used as the principal direction for the geocentric equatorial coordinate system. It points at the zodiac constellation Aries (hence, we use the zodiac symbol for rams horns ♈). It's found by drawing a line from the Earth through the Sun on the first day of spring, March 21. On this day, the Sun appears to cross the equator from south to north and the length of daylight equals the length of night. That's why it's called an equinox.

direction may not seem "convenient" to you, it's significant to astronomers who originally defined the system.

- Third axis found using right-hand rule

The entire coordinate system is shown in Figure 4-26.

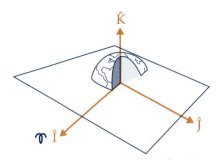

Figure 4-26. Geocentric-Equatorial Coordinate System. This system is defined by
- Origin—center of Earth
- Fundamental plane—equatorial plane
- Perpendicular to plane—north pole
- Principal direction—vernal equinox direction (Υ)

It's the primary coordinate system we'll use for Earth-orbiting spacecraft.

Equation of Motion

Now that we've fixed our reference, we can safely apply Newton's Second Law to examine the external forces affecting the system, or in this case, a satellite. So let's place ourselves on an imaginary spaceship in orbit around the Earth and see if we can list the forces on our ship.

- Earth's gravity (Newton wouldn't let us forget this one)
- Drag—if we're a little too close to the atmosphere
- Thrust—if we fire rockets
- 3rd body—gravity from the Sun or Moon
- Other—just in case we miss something

Summing all these forces, shown in Figure 4-27, we end up with the following equation of motion

$$\sum \vec{F}_{external} = \vec{F}_{gravity} + \vec{F}_{drag} + \vec{F}_{thrust} + \vec{F}_{3rd\ body} + \vec{F}_{other} \quad (4\text{-}16)$$

If we tried to put in mathematical expressions for the various forces and come up with a solution to the equation, we would create a difficult problem—not to mention an enormous headache. So let's examine some reasonable assumptions we can make to simplify the problem.

Simplifying Assumptions

Luckily, we can assume some things about orbit motion that will simplify the problem. However, they will "restrict" our solution to cases in which these assumptions apply. Fortunately, this includes most of the situations we'll use. We'll assume

- Satellites travel high enough above the Earth's atmosphere to say that the drag force is small, $\vec{F}_{drag} \cong 0$
- The satellite will not be maneuvering or changing its path, so we can ignore the thrust force, $\vec{F}_{thrust} \cong 0$
- We are considering the motion of the satellite close to the Earth, so we can ignore the gravitational attraction of the Sun, the Moon, or any third body, $\vec{F}_{3rd\ body} \cong 0$. (That's why we call this the two-body problem.)
- Compared to Earth's gravity, other forces such as those due to solar radiation, electromagnetic fields, etc. are negligible, $\vec{F}_{other} \cong 0$
- The mass of the Earth is much, much larger than the mass of any spacecraft, $m_{Earth} \gg m_{satellite}$

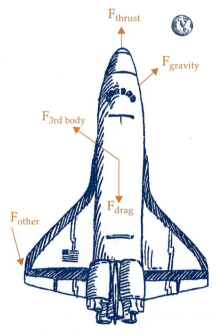

Figure 4-27. Possible Forces on a Spacecraft. We can brainstorm all the possible forces on a spacecraft to include gravity, drag, thrust, third bodies, and other forces.

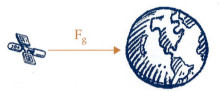

Figure 4-28. The Force of Gravity. In the restricted two-body problem, we reduce the forces acting on a spacecraft to a single force—the force of gravity due to the Earth.

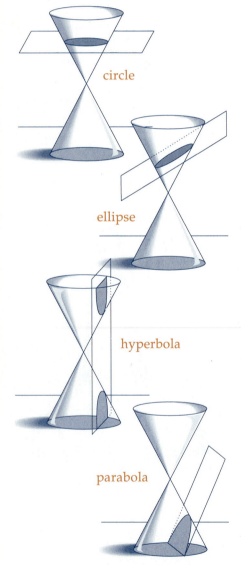

Figure 4-29. Conic Sections. The solution to the restricted two-body equation of motion gives the polar equation for a conic section. Conic sections are found by slicing right cones at various angles.

- The Earth is spherically symmetrical with uniform density and can thus be treated as a point mass

After all these assumptions, we're left with gravity as the only force, so $\sum \vec{F}_{external} = \vec{F}_{gravity} = m\vec{a}$, as shown in Figure 4-28. Now we can apply Newton's Universal Law of Gravitation in vector form

$$\vec{F}_{gravity} = -\frac{\mu m}{R^2}\hat{R} \tag{4-17}$$

Substituting into the equation of motion

$$\vec{F}_{gravity} = -\frac{\mu m}{R^2}\hat{R} = m\vec{a} = m\ddot{\vec{R}}$$

we arrive at the *restricted two-body equation of motion*

$$\boxed{\ddot{\vec{R}} + \frac{\mu}{R^2}\hat{R} = 0} \tag{4-18}$$

where
R = distance to center of Earth (km)
$\ddot{\vec{R}}$ = acceleration (km/s²)
μ = gravitational parameter (km³/s²)
$\hat{R}$ = unit vector in direction of $\vec{R}$

[Note, we use the engineering convention for the second derivative of $\vec{R}$ with respect to time, which is $\ddot{\vec{R}}$, better known as acceleration, $\vec{a}$.]

What can the two-body equation of motion tell us about the movement of a satellite around the Earth? Unfortunately, in its present form—a second-order, non-linear, vector differential equation—it doesn't help us visualize anything about this movement. So what good is it? To understand the significance of the two-body equation of motion, we must first "solve" it using rather complex mathematical slight-of-hand (see Appendix C). When the smoke clears, we're left with an expression for the position of an object in space in terms of some odd variables.

$$R = \frac{k_1}{1 + k_2 \cos \nu}$$

where
R = magnitude of $\vec{R}$
k_1 = constant that depends on μ, $\vec{R}$, and $\vec{V}$
k_2 = constant that depends on μ, $\vec{R}$, and $\vec{V}$
ν = (Greek letter "nu") polar angle measured from a principle axis to $\vec{R}$

This equation represents the solution to the restricted two-body equation of motion and describes the location, $\vec{R}$, of a satellite in terms of two constants and a polar angle, v. Students of geometry may recognize that this equation also represents a general relationship for any conic section. The four conic sections shown in Figure 4-29 are the circle, ellipse, parabola, and hyperbola. Now, here's the really significant part of all this—we just proved Kepler's Laws of Planetary Motion! Based on Brahe's data, Kepler was able to show the orbits of the planets were ellipses. He could describe *how* the orbits looked but not *why*. We've just shown *why*. Any object moving in a gravitational field must follow only one of the conic sections. In the case of planets or satellites in orbit, this path is an ellipse or a circle (which is just a special case of an ellipse).

Now that we know orbits must follow conic section paths, we can look at some ways to describe the size and shape of an orbit.

Orbit Geometry

We've just shown the two-body equation of motion can be represented as an equation of a conic. Let's examine what the conic sections are and how the parts are labeled. Several parameters define each of the conic sections. We'll begin with the most common of the orbit shapes—the ellipse.

$\vec{R}$ = satellite position vector, measured from the center of the Earth.

$\vec{V}$ = satellite velocity vector

F and F' = primary and vacant foci of ellipse

R_p = Radius of perigee (closest approach)

R_a = Radius of apogee (farthest approach)

2a = major axis

a = semi-major axis

v = true anomaly

ϕ = flight-path angle

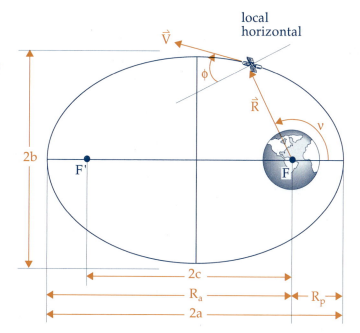

Figure 4-30. Geometry of an Elliptical Orbit.

With Figure 4-30 as a reference, let's now define some important parameters in the ellipse

- R is the radius from the focus of the ellipse (in this case, the center of the Earth) to the satellite
- F and F' are the primary (occupied) and vacant (unoccupied) foci. The center of the Earth is at the occupied focus.
- R_p is the *radius of periapsis* (radius of closest approach of the satellite to the occupied focus); it's called the radius of *perigee* when the orbit is around the Earth
- R_a is the *radius of apoapsis* (radius of farthest approach of the satellite to the occupied focus); it's called the radius of *apogee* when the orbit is around the Earth
- 2a is the major axis or the length of the ellipse. One-half of this is called "a," or the *semi-major axis* (semi means one half).

$$a = \frac{R_a + R_p}{2} \qquad (4\text{-}19)$$

- 2b is the minor axis or width of the ellipse. One-half of this is called "b," or the *semi-minor axis*
- 2c is the distance between the foci, $R_a - R_p$
- ν is the *true anomaly* or polar angle measured from perigee to the satellite position vector $\vec{R}$, in the direction of satellite motion. It locates the satellite in the orbit. For example, if $\nu = 180°$ the satellite would be 180° from perigee, putting it at apogee. The range for true anomaly is 0° to 360°.
- ϕ is the *flight-path angle*, measured from the local horizontal to the velocity vector, $\vec{V}$. The local horizontal is a line perpendicular to the position vector, $\vec{R}$. When traveling from perigee to apogee (outbound), the velocity vector will always be above the local horizon (gaining altitude), so $\phi > 0°$. When traveling from apogee to perigee (in-bound), the velocity vector will always be below the local horizon (losing altitude), so $\phi < 0°$. At exactly perigee and apogee on an ellipse, the velocity vector is parallel to the local horizon, so $\phi = 0$. The maximum value of flight-path angle is 90°.
- e is the *eccentricity*, which is the ratio of the distance between the foci (2c) to the length of the ellipse (2a)

$$e = \frac{2c}{2a}$$

Eccentricity defines the shape or type of conic section. Eccentricity is a medieval term representing the degree of noncircularity (meaning "out of center") of a conic. Because circular motion was once considered perfect, any deviation was abnormal, or eccentric (maybe you know someone like that). Because the distance between the foci in an ellipse is always less than the length of the ellipse, its eccentricity is between 0 and 1. A circle has e = 0. A very long, narrow ellipse has e approaching 1. A parabola has e = 1 and a hyperbola has e > 1.

With all these things defined we can now return to our polar equation of a conic and substitute for the constants we used. $k_1 = a(1 - e^2)$ and $k_2 = e$. Thus, we have

$$R = \frac{a(1 - e^2)}{1 + e\cos\upsilon} \qquad (4\text{-}20)$$

where

R = position (km)
a = semi-major axis (km)
e = eccentricity (dimensionless)
υ = true anomaly (deg or rad)

$$e = 2c/2a = \frac{R_a - R_p}{R_a + R_p} \qquad (4\text{-}21)$$

To determine the distances at closest approach, R_p, and farthest approach, R_a, we can use this equation.

$$\text{At } \upsilon = 0°, R = R_p = \frac{a(1 - e^2)}{(1 + e\cos(0))} = a(1 - e)$$

$$\text{At } \upsilon = 180°, R = R_a = \frac{a(1 - e^2)}{(1 + e\cos(180))} = a(1 + e)$$

Looking at the figure of the ellipse, we can see that the length of the ellipse, 2a, is equal to ($R_a + R_p$), and the distance between the foci, 2c, is equal to ($R_a - R_p$). Now, if we want to compute the eccentricity of the orbit based on the perigee and apogee radii, we can use the second part of Equation (4-21). See Example 4-6.

The same parameters we have used for the ellipse also apply to the other conic sections. Figure 4-31 shows a circular orbit. The parabolic trajectory in Figure 4-32 represents an escape trajectory or a path that takes a satellite away from the pull of the Earth's gravity. The hyperbola in Figure 4-33 also represents an escape trajectory with respect to the Earth. It's an unusual shape with a different sign convention. Because the length of the hyperbola bends back on itself, or is measured outside the conic, we define this distance, 2a, as being negative. The same convention also applies for the distance between the foci, 2c. But the magnitude of 2c will always be larger than the magnitude of 2a, so we will have an eccentricity greater than 1. Table 4-1 summarizes all of these parameters.

R = a = b

e = 0 (the foci are collocated so 2c, the distance between the foci, equals 0)

Figure 4-31. Circle. A circle is just a special case of an ellipse.

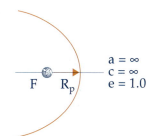

$a = \infty$
$c = \infty$
$e = 1.0$

Figure 4-32. Parabola. A parabolic trajectory is a special case which leaves the Earth altogether.

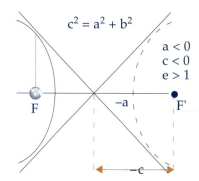

$c^2 = a^2 + b^2$

$a < 0$
$c < 0$
$e > 1$

Figure 4-33. Hyperbola. A hyperbolic trajectory is used for interplanetary missions. Notice a real trajectory is around the occupied focus and an imaginary, mirror-image trajectory is around the vacant focus.

Table 4-1. A Summary of Parameters for Conic Sections.

Conic Section	a = Semi-Major Axis	c = One-Half the Distance Between Foci	e = Eccentricity
circle	$a > 0$	$c = 0$	$e = 0$
ellipse	$a > 0$	$0 < c < a$	$0 < e < 1$
parabola	$a = \infty$	$c = \infty$	$e = 1$
hyperbola	$a < 0$	$a < c < 0$	$e > 1$

▬ Section Review

Key Terms

apogee
conic sections
eccentricity
flight-path angle
fundamental plane
geocentric-equatorial coordinate system
origin
perigee
principal direction
radius of apoapsis
radius of periapsis
restricted two-body problem
semi-major axis
semi-minor axis
true anomaly

Key Equations

$$\ddot{\vec{R}} + \frac{\mu}{R^2}\hat{R} = 0$$

$$R = \frac{a(1 - e^2)}{1 + e\ \cos v}$$

Key Concepts

➤ An object in orbit can be thought of as falling around the Earth, but because of its horizontal velocity, it never quite hits

➤ Combining Newton's Second Law and his Universal Law of Gravitation, we developed the restricted two-body equation of motion:
  • The coordinate system used to derive the two-body equation of motion is the geocentric-equatorial system:
    - Origin—center of Earth
    - Fundamental plane—equatorial plane
    - Direction perpendicular to plane—North Pole direction
    - Principal direction—vernal equinox direction, ♈
  • In deriving this equation, we assume
    - Drag force is negligible
    - Satellite is not thrusting
    - Gravitational pull of third bodies and all other forces are negligible
    - $m_{Earth} \gg m_{satellite}$
    - Earth is spherically symmetrical and of uniform density and can thus be treated as a point mass

➤ Solving the restricted two-body equation of motion results in the polar equation for a conic section

➤ Figure 4-30 shows parameters for orbit geometry, and Table 4-1 summarizes parameters for conic sections

Example 4-6

Problem Statement

A new class of remote-sensing satellites is known to be in an orbit with a radius of perigee of 7000 km and a radius of apogee of 10,000 km. What is its altitude above the Earth when the true anomaly is 90°?

Problem Summary

Given: $R_p = 7000$ km
         $R_a = 10,000$ km
Find: Altitude, when $v = 90°$

Problem Diagram

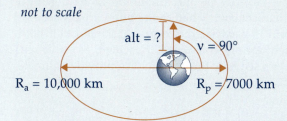

not to scale

alt = ?

$v = 90°$

$R_a = 10,000$ km $R_p = 7000$ km

Conceptual Solution

1) Find the eccentricity for the orbit using

$$e = \frac{R_a - R_p}{R_a + R_p}$$

2) Find the semi-major axis for the orbit using

$$a = \frac{R_a + R_p}{2}$$

3) Solve for the radius when $v = 90°$ using the polar equation of a conic

$$R = \frac{a(1 - e^2)}{1 + e \cos v}$$

4) The altitude when $v = 90°$ is the radius minus the radius of the Earth, R_{Earth}

$$Alt_{v = 90°} = R_{v = 90°} - R_{Earth}$$

Analytical Solution

1) Find e

$$e = \frac{R_a - R_p}{R_a + R_p} = \frac{10,000 \text{ km} - 7000 \text{ km}}{10,000 \text{ km} + 7000 \text{ km}} = 0.1765$$

2) Find a

$$a = \frac{R_a + R_p}{2} = \frac{10,000 \text{ km} + 7000 \text{ km}}{2} = 8500 \text{ km}$$

3) Find R

$$R = \frac{a(1 - e^2)}{1 + e \cos v} = \frac{(8500 \text{ km})(1 - (0.1765)^2)}{1 + 0.1765 \cos 90°}$$

$$R = 8235.3 \text{ km}$$

4) Find $Alt_{v = 90°}$

$$Alt_{v = 90°} = 8235.3 \text{ km} - 6378.14 \text{ km}$$

$$= 1857.2 \text{ km}$$

Interpreting the Results

When this new remote-sensing spacecraft has reached a point 90° past perigee, it's at an altitude of 1857.3 km.

4.5 Constants of Orbit Motion

▬ In This Section You'll Learn to...

- ☞ Define the two constants of orbit motion—specific mechanical energy and specific angular momentum
- ☞ Apply specific mechanical energy to determine orbit velocity and period
- ☞ Apply the concept of conservation of specific angular momentum to show an orbit plane is fixed in space

By now you're probably convinced that, with all these flight-path angles, true anomalies, and ellipses flying around, there is nothing consistent about orbits. Well, take heart. Life has some constants, we even have constants in astrodynamics. We saw in our discussion of motion in a conservative field that mechanical energy and momentum are conserved. Because orbit motion occurs in a conservative gravitational field, both energy and angular momentum are conserved. However, as we'll see, linear momentum is not. Let's see how these principles provide valuable tools for studying orbital motion.

Specific Mechanical Energy

In an earlier section, we referred to equations of motion being like crystal balls in that they allow us to gaze into the future to predict where an object will be. Mechanical energy provides us with one such crystal ball. Recall in our definition of mechanical energy we developed a relationship between our position, our velocity, and the local gravitational parameter. It looked like this

$$E = \frac{1}{2}mV^2 - \frac{\mu m}{R} \tag{4-22}$$

To generalize our solution so we don't have to worry about the mass of the spacecraft, we'll divide both sides of the equation by mass. We can do this by defining a new flavor of mechanical energy which we will call specific mechanical energy. *Specific mechanical energy, ε,* is simply the total mechanical energy divided by the mass of the object

$$\varepsilon \equiv \frac{E}{m} \tag{4-23}$$

or

$$\varepsilon = \frac{V^2}{2} - \frac{\mu}{R} \tag{4-24}$$

where
ε = specific mechanical energy (km^2/s^2)
V = velocity (km/s)
μ = gravitational parameter (km^3/s^2)
R = position (km)

Because the total mechanical energy, and hence the specific mechanical energy, is conserved for an orbit, at *any point along an orbit, the specific mechanical energy will be exactly the same!* Riding around the Earth in an orbit is just like swinging on a swing. As you approach apogee (the highest point in the orbit) you gain altitude, meaning your R, or distance from the center of the central attracting body, increases. This increase in R means you gain in potential energy—which actually means your PE gets less negative (because of the way we define R). At the same time, your speed is decreasing and hence you are losing kinetic energy. When you reach apogee you're at the highest point in orbit, so your PE is at its maximum value. However, because your speed is the lowest at apogee, your KE is at its lowest. But the sum of your PE and KE, the specific mechanical energy, remains constant.

Just as with a swing, at the top of the arc you start racing back toward perigee (the lowest point on the orbit) and begin to trade your PE in for KE. Consequently, your speed steadily increases until you reach perigee, at which point your KE is at its highest in the orbit, so your speed is at a maximum. To keep the specific mechanical energy constant, your PE must be at its lowest at periapsis because you're at the lowest point in the orbit. Figure 4-34 illustrates these relationships.

The fact that the specific mechanical energy is constant gives us a tremendously powerful tool for analyzing orbits. Look again at the relationship for specific mechanical energy. Notice that ε depends only on position, R, velocity, V, and the local gravitational parameter, μ. This means if we know the position and velocity of a particular satellite at any single point along its orbit, we can predict where that satellite will be at any time.

Another important concept to glean out of the constancy of our orbital energy is the relationship between R and V. Assume we know the energy for an orbit. Then, at any given position, R, on that orbit there is one and only one possible velocity, V! Thus, if we know the energy for an orbit, we can easily find the velocity at any point in the orbit if we know R. Mathematically, we can demonstrate this by simply rearranging the relationship for energy we found in Equation (4-24).

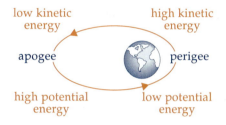

low kinetic energy   high kinetic energy

apogee   perigee

high potential energy   low potential energy

Figure 4-34. Trading Energy in an Orbit. An orbit is just like a swing. PE and KE trade off throughout the orbit.

$$V = \sqrt{2\left(\frac{\mu}{R} + \varepsilon\right)} \qquad (4\text{-}25)$$

where
V = velocity (km/s)
μ = gravitational parameter (km^3/s^2)
R = position (km)
ε = specific mechanical energy (km^2/s^2)

Note: R is defined from the center of the Earth so when you're using orbit altitude remember to add the radius of the Earth.

So, once we know the velocity we have in an orbit and the velocity we want to have in a different orbit, we can use this relationship to determine how much velocity change we need to get from our engines to "drive" over to a new orbit.

Recall from our discussion of conic-section geometry, one parameter represented the satellite's mean distance from the primary focus. This parameter was the semi-major axis, a. We can develop a new relationship for specific mechanical energy which depends only on a and μ. (See Appendix C)

$$\varepsilon = \frac{-\mu}{2a}$$

(4-26)

where
ε = specific mechanical energy (km^2/s^2)
μ = gravitational parameters (km^3/s^2)
a = semi-major axis (km)

This means simply knowing a satellite's average distance from the central body tells you all you need to know about how much specific mechanical energy the satellite has. We can also learn something about the trajectory simply from the sign of the specific mechanical energy. For a circular or elliptical orbit, specific mechanical energy is *negative* (because a is positive). For a parabola, ε = 0 (because a = ∞). For a hyperbola, specific mechanical energy is *positive* (because a is negative). These are important points to keep in mind as you work orbit problems. If the sign on your energy is wrong, you'll get the wrong answer.

Another consequence of knowing energy is that we can determine orbit period. The *orbit period, P*, is the time for a satellite to complete one revolution of its orbit. From Kepler's Third Law of Planetary Motion, which we learned in Chapter 2,

$$P^2 \text{ is proportional to } a^3$$

where "a," the semi-major axis, is used as mean, or average, distance. More specifically

$$P = 2\pi\sqrt{\frac{a^3}{\mu}}$$

(4-27)

where
P = period (seconds)
π = 3.14159. . .
a = semi-major axis (km)
μ = gravitational parameter (km^3/s^2)

The period is defined only for "closed" conics (circles or ellipses). It is infinite for a parabola and an imaginary number for a hyperbola because the values of semi-major axis for these conics are infinite and negative, respectively.

Specific mechanical energy, ε, is a very valuable constant of satellite motion. Knowing it, you don't have to painstakingly take repeated measurements of an object's position and velocity throughout its orbit to plot its path through space. With a single observation of position and velocity, we learn much about a satellite's orbit. But ε gives us only part of the story. In fact, ε, by itself, defines only the orbit's *size* and gives us V as a function of R. It doesn't tell us anything about *where* the orbit is in space. For insight into that important bit of information we must turn to the angular momentum. Example 4-7 shows one application of specific mechanical energy.

Specific Angular Momentum

Recall from our earlier discussion that we can find angular momentum from Equation (4-3).

$$\vec{H} = \vec{R} \times m\vec{V}$$

Once again, to make our life less complicated, we'd like to divide both sides of the equation by the mass of the object we're investigating. Doing this, we define the *specific angular momentum, $\vec{h}$*, as

$$\vec{h} \equiv \frac{\vec{H}}{m}$$

$$\boxed{\vec{h} = \vec{R} \times \vec{V}} \qquad (4\text{-}28)$$

where

$\vec{h}$ = specific angular momentum vector (km^2/s)

$\vec{R}$ = position vector (km)

$\vec{V}$ = velocity vector (km/s)

Notice that specific angular momentum is the result of the cross product between two vectors: position and velocity. Recall from geometry that any two lines define a plane. So in this case, $\vec{R}$ and $\vec{V}$ are two lines (as vectors that have both magnitude and direction) which define a plane. We call this plane containing $\vec{R}$ and $\vec{V}$ the *orbit plane*. Because the cross product of any two vectors results in a third vector which is perpendicular to the first two, $\vec{h}$ must be perpendicular to $\vec{R}$ and $\vec{V}$. Figure 4-35 shows $\vec{R}$, $\vec{V}$, and $\vec{h}$.

Here's where we need to apply a little deductive reasoning and consider the logical consequence of the facts we know to this point. First of all, as we saw in Section 4.3, angular momentum and, hence, specific angular momentum are a constant both in magnitude and direction. Second, $\vec{R}$ and $\vec{V}$ define the orbit plane. Next, $\vec{h}$ is perpendicular to the orbit plane. Therefore, if $\vec{h}$ is always perpendicular to the orbit plane, and $\vec{h}$ is constant, the orbit plane must also be constant. This means that in our restricted two-body problem the orbit plane is forever frozen in inertial space! However, in reality, as we'll see in Chapter 8, slight disturbances will cause the orbit plane to change gradually over time.

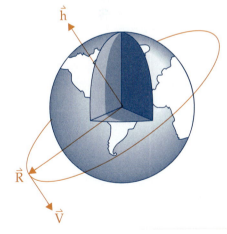

Figure 4-35. Specific Angular Momentum. The specific angular momentum vector, $\vec{h}$, is perpendicular to the orbit plane defined by $\vec{R}$ and $\vec{V}$.

Specific angular momentum is also useful in providing a key tie between the geometric solution to the two-body problem we saw in the last section and dynamics. It turns out that the numerator in the polar equation of a conic section presented in Equation (4-20), which comes from geometry, can be related to specific angular momentum by

$$\frac{h^2}{\mu} = a\,(1 - e^2) \qquad (4\text{-}29)$$

■ Section Review

Key Terms

orbit period, P
orbit plane
specific angular momentum, $\vec{h}$
specific mechanical energy, ε

Key Equations

$$\varepsilon = \frac{V^2}{2} - \frac{\mu}{R}$$

$$V = \sqrt{2\left(\frac{\mu}{R} + \varepsilon\right)}$$

$$\varepsilon = \frac{-\mu}{2a}$$

$$P = 2\pi\sqrt{\frac{a^3}{\mu}}$$

$$\vec{h} = \vec{R} \times \vec{V}$$

Key Concepts

➤ In the absence of any force other than gravity, two things remain constant for a given orbit:
  • Specific mechanical energy
  • Specific angular momentum

➤ Specific mechanical energy, ε, is defined as $\varepsilon \equiv E/m$:
  • $\varepsilon < 0$ for circular and elliptical orbits
  • $\varepsilon = 0$ for parabolic trajectories
  • $\varepsilon > 0$ for hyperbolic trajectories

➤ Specific angular momentum, $\vec{h}$ is defined as $\vec{h} \equiv \vec{H}/m$
  • It is constant for any given orbit
  • The fact that $\vec{h}$ is constant implies that orbit planes must be fixed in space

Example 4-7

Problem Statement

What is the velocity of the remote-sensing spacecraft discussed in Example 4-7 when the true anomaly is 90°? How long before the spacecraft returns to this point in the orbit?

Problem Summary

Given: $a = 8500$ km
$\quad\quad R_{\upsilon=90°} = 8235.2$ km
Find: $\quad V_{\upsilon=90°}, P$

Problem Diagram

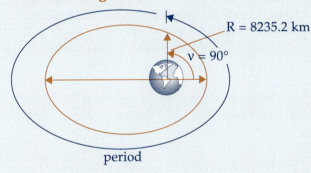

period

Conceptual Solution

1) Find the specific mechanical energy for the orbit from

$$\varepsilon = \frac{-\mu}{2a}$$

2) Find the velocity at $\upsilon = 90°$ using

$$V = \sqrt{2\left(\frac{\mu}{R} + \varepsilon\right)}$$

3) Find the orbit period from

$$P = 2\pi\sqrt{\frac{a^3}{\mu}}$$

Analytical Solution

1) Find ε

$$\varepsilon = \frac{-\mu}{2a} = \frac{-3.986005 \times 10^5 \dfrac{km^3}{s^2}}{2\,(8500\ km)} = -23.45 \frac{km^2}{s^2}$$

2) Find V

$$V = \sqrt{2\left(\frac{\mu}{R} + \varepsilon\right)}$$

$$= \sqrt{2\left(\frac{3.986005 \times 10^5 \dfrac{km^3}{s^2}}{8235.2\ km} - 23.45\frac{km^2}{s^2}\right)}$$

$$= 7.065 \frac{km}{s}$$

3) Find P

$$P = 2\pi\sqrt{\frac{a^3}{\mu}} = 2\pi\sqrt{\frac{(8500\ km)^3}{3.986005 \times 10^5 \dfrac{km^3}{s^2}}}$$

$$= 7799\ \text{seconds}$$

$$\cong 130\ \text{minutes}$$

Interpreting the Results

At a point in the orbit 90° from perigee, this spacecraft's velocity is 7.064 km/s (15,802 m.p.h.). For this orbit the period is 130 minutes, or just over two hours.

References

Bate, Robert R., Donald D. Mueller and Jerry E. White. *Fundamentals of Astrodynamics*. New York, New York: Dover Publications, Inc., 1971.

Boorstin, Daniel J. *The Discoverers*, Random House, 1983.

Concepts in Physics. Del Mar, CA: Communications Research Machines, Inc., 1973.

Feynman, Richard P., Robert B. Leighton, and Matthew Sands, *The Feynman Lectures on Physics*, Reading, MA: Addison-Wesley Publishing Co., 1963.

Gonick, Larry and Art Huffman, *The Cartoon Guide to Physics*. New York, NY: HarperCollins Publishers, 1990.

Hewitt, Paul G. *Conceptual Physics. . . A New Introduction to Your Environment*, Boston, MA: Little, Brown and Company, 1981.

King-Hele, Desmond. *Observing Earth Satellites*, New York, NY: Van Nostrand Reinhold Company, Inc., 1983.

Szebehely, Victor G. *Adventures in Celestial Mechanics*, Austin, TX: University of Texas Press, 1989.

Thiel, Rudolf. *And There Was Light*. New York: Alfred A Knopf, 1957.

Young, Louise B., Ed., *Exploring the Universe*, Oxford, MA: Oxford University Press, 1971.

Mission Problems

4.1   Orbital Motion

1   Explain how an object's horizontal velocity allows it to achieve orbit.

2   An object in a circular orbit is given a bit of extra velocity. What type of orbit will it now be in?

3   What two types of trajectories completely escape the Earth?

4   Explain how you could use the steps in the Motion Analysis Process checklist to analyze the motion of a volleyball being served.

4.2   Newton's Laws

5   What three things does an object's mass tell you about the object?

6   An astronaut on the Moon drops a hammer and a feather from the same height at the same time. Describe what happens and why. Explain the difference if this experiment occurs on Earth.

7   Describe how the recoil you feel when firing a rifle is the result of Newton's Third Law of Motion.

8   You are spinning a 0.25 kg weight over your head at the end of a 0.5 m string. If you let go of the string, the weight will sail off on a tangent at 2 m/s. What is the angular momentum of the spinning weight before release? Because angular momentum is always conserved, where does the angular momentum go after release?

9   The new quarterback throws a football with a moment of inertia of 0.001 kg · m² in a perfect spiral. If the football is spinning at 60 revolutions per minute (RPM), what is its angular momentum?

10   Two asteroids pass by each other in the void of interstellar space. Asteroid Zulu has a mass of 1×10^6 kg. Asteroid Echo has a mass of 8×10^6 kg. If the two are separated by 100 m, what is the force of gravitational attraction between them?

11 While flying in your SR-71 airplane 25,000 m above the surface of the Earth, you drop your pencil. What acceleration will the pencil have as it falls toward the cockpit floor?

12 Neglecting air resistance, how fast will a baseball (m = 0.1 kg) be travelling when it hits the ground if it's dropped from the Empire State Building (about 300 m high)? How long will it take to hit the ground?

13 Match the physical laws on the left with the best term or description on the right. [Hint: you may need to review Kepler's Laws from Chapter 2]

1) Newton's First Law

2) Newton's Second Law

3) Newton's Third Law

4) Newton's Law of Universal Gravitation

5) Kepler's First Law

6) Kepler's Second Law

7) Kepler's Third Law

a) Planetary orbits are ellipses

b) Gravity is the only force on a satellite

c) Inertia

d) Action/reaction

e) Relates orbital period to orbit size

f) A net force causes an acceleration

g) Equal areas in equal times

h) Satellites may be treated as point masses

i) Force is inversely proportional to the square of the distance

14 When does the gravitational parameter equal $3.986005 \times 10^5 \text{ km}^3/\text{s}^2$?

15 A rocket engine operates by expelling high-velocity gas from the exhaust nozzle. Discuss the physical law which explains why this occurrence produces a force on the rocket.

4.3 Laws of Conservation

16 For an isolated system (one which has no interaction with its surroundings), what quantities are constant according to the laws of physics?

17 You are floating in the cockpit of the Space Shuttle Atlantis when you decide to do somersaults, thus increasing your angular momentum. Why are you floating? What happens to the overall angular momentum of the Space Shuttle and all of its contents? Why?

18 Describe the potential, kinetic, and total energy of a baseball that's thrown into the air, reaches its highest point, then falls to the ground.

4.4 The Restricted Two-Body Problem

19 What are the origin, principle direction, and fundamental plane for the geocentric-equatorial coordinate system?

20 What simplifying assumptions do we use to "restrict" the two-body equation of motion?

21 In solving the restricted two-body equation of motion, we obtain the polar equation of a conic section. Why is this significant?

22 Match the following terms with their definitions.

1) $\vec{R}$

2) $\vec{V}$

3) F and F'

4) Radius at perigee (R_p)

5) Radius at apogee (R_a)

6) Major-axis (2a)

7) True anomaly (v)

8) Flight-path angle (ϕ)

9) Eccentricity (e)

a) Closest point in an orbit

b) Primary and vacant foci of a conic section

c) Position vector

d) Angle between perigee and the position vector

e) "Out of roundness" of a conic section

f) Distance across the long axis of an ellipse

g) Angle between local horizontal and the velocity vector

h) Velocity vector

i) Furthest point in an orbit

23 Complete the following table with the possible range of values of eccentricity, semi-major axis, and specific mechanical energy.

Conic Section	a = Semi-Major Axis	c = One-Half the Distance Between Foci	e = Eccentricity
circle			
ellipse			
parabola			
hyperbola			

24 A Soviet satellite is in Earth orbit with an altitude at perigee of 375 km and an altitude at apogee of 2000 km.

a) What is the semi-major axis of the orbit?

b) What is the eccentricity?

c) If the true anomaly is 175°, what is the satellite's altitude?

d) If the true anomaly is 290°, is the flight-path angle positive or negative? Why?

4.5   Constants of Orbit Motion

25 Where is the potential energy of a satellite greater, at perigee or apogee? Why?

26 While co-piloting a futuristic space vehicle, you receive a report of your position and velocity in the geocentric equatorial frame:

position vector $= 7000 \, \hat{I} + 0 \, \hat{J} + 0 \, \hat{K}$ km

velocity vector $= 0 \, \hat{I} - 7.063 \, \hat{J} + 0 \, \hat{K}$ km/s

a) Sketch the satellite position and velocity vectors relative to the Earth.

b) What is the specific angular momentum? Draw this vector on the sketch.

c) What does this angular momentum vector tell you about the orientation of your orbit?

d) What is your specific mechanical energy?

e) What type of orbit are you in? How can you tell?

27 You are the engineer in charge of a top-secret spy satellite. The satellite will be placed in a circular, sun-synchronous orbit (see Chapter 8), with an altitude of 759 km and a mass of 10,000 kg. A politician who dislikes the project says the satellite poses a danger to the public because of its large kinetic energy so close to the Earth. What is the kinetic energy of the satellite? Compare this to the KE of a 2000 kg truck travelling down the interstate at 65 mph. Is the comparison realistic? Why or why not? (Hint: are the two objects in the same reference frame?)

28 Calculate the altitude needed for a circular, geosynchronous orbit (an orbit whose orbital period matches the Earth's rotation rate).

29 We know that the velocity of a satellite in an elliptical orbit is greatest at perigee due to conservation of specific mechanical energy. Relate this fact to Kepler's Second Law.

30 A Mars probe is in a circular orbit around Earth with a radius of 25,000 km. The next step is to thrust so the probe can enter an escape orbit. (a) Determine the minimum change in velocity required to enter the escape orbit. (b) Determine the associated change in specific kinetic energy. (c) Now compare this result to the specific mechanical energy of the circular orbit. Are you surprised? Why or why not?

For Discussion

31 Demonstrate that two-body motion is confined to a plane fixed in space (Hint: what quantities are conserved?)

32 For the equations of motion to be correct the coordinate reference frames must be inertial. Is the geocentric equatorial frame commonly used for satellites a truly inertial reference frame? Why or why not? If not, why can we use it?

33 We based the equations of motion derived to describe satellite motion on the two-body problem. If you wish to design a trajectory for an interplanetary probe to Mars, which two bodies would you consider at the beginning of the flight? At the middle? At the end? What problems, if any, do you think this might produce?

Mission Profile—Apollo

"...I believe that this nation should commit itself to achieving the goal, before this decade is out, of landing a man on the Moon and returning him safely to Earth."

President John F. Kennedy
May 25, 1961

Less than three weeks after Alan Shepard's first suborbital flight, President Kennedy's address to Congress boldly established a Moon landing as a national goal. Over the next eleven years, project Apollo grew from a statement of national intent to a project that successfully launched 11 spacecraft and allowed 12 men to walk on the surface of the Moon.

Mission Overview

Apollo's mission was as simple as President Kennedy's quote—get a man to the Moon and back safely. Once this initial goal was accomplished, Apollo astronauts were responsible for collecting scientific data about the Moon and Earth.

Mission Data

✓ Apollo 7, October 11, 1968 (Crew: Cunningham, Eisele, Schirra): First manned Apollo flight. System checkout of command module. Earth orbit only.

✓ Apollo 8, December 21, 1968 (Crew: Anders, Borman, Lovell): First manned launch of Saturn V. First lunar orbit.

✓ Apollo 9, March 3, 1969 (Crew: McDivitt, Schweickart, Scott): First flight, test, and docking. Earth orbit only.

✓ Apollo 10, May 18, 1969 (Crew: Cernan, Stafford, Young): "Dress rehearsal" for lunar landing: included descent of lunar module to 50,000 ft. above lunar surface.

✓ Apollo 11, July 16, 1969 (Crew: Aldrin, Armstrong, Collins): First lunar landing. Armstrong first man to walk on Moon.

✓ Apollo 12, November 14, 1969 (Crew: Bean, Conrad, Gordon): Recovered parts from Surveyor 3. Conducted scientific experiments.

✓ Apollo 13, April 11, 1970 (Crew: Haise, Lovell, Swigert): Mission aborted due to explosion on command module on the way to the Moon.

✓ Apollo 14, January 31, 1971 (Crew: Mitchell, Roosa, Shepard): First golf ball hit on Moon. Conducted scientific experiments.

✓ Apollo 15, July 26, 1971 (Crew: Irwin, Scott, Worden): First use of Lunar Roving Vehicle.

✓ Apollo 16, April 16, 1972 (Crew: Duke, Mattingly, Young): Over 20 hours of extra-vehicular activity on the Moon.

✓ Apollo 17, December 7, 1972 (Crew: Cernan, Evans, Schmidt): Last Apollo Moon landing.

Mission Impact

Apollo met President Kennedy's goal and captured the imagination of the entire world. In addition, the Apollo program provided scientists with invaluable information about the Moon. Unfortunately, the public and Congress soon lost interest in the Moon. NASA shifted its focus to developing the Space Shuttle, and the technological infrastructure to take humans to the Moon and back was laid to rest in museums.

(Courtesy of NASA)

For Discussion

• Why did we go to the Moon?

• Some say we spent all that money on Apollo and all we got was "a bunch of rocks." Is this really true?

• What plans do we have to return to the Moon?

Contributor

Todd Lovell, United States Air Force Academy

References

Baker, David. *The History of Manned Spaceflight.* New York: Crown, 1981.

Yenne, Bill. *The Encyclopedia of US Spacecraft.* New York: Exeter, 1985.

Flight controllers in the Mission Control Center use this large ground track display to diligently monitor the Space Shuttle's path throughout a mission. *(Courtesy of NASA)*

Describing Orbits

5

▰ In This Chapter You'll Learn to...

☛ Define the Classical Orbital Elements (COEs) used to describe the size, shape, and orientation of an orbit and the location of a satellite in that orbit

☛ Determine the COEs given the position, $\vec{R}$, and velocity, $\vec{V}$, of a satellite at one point in its orbit

☛ Explain and use orbit ground tracks

▰ You Should Already Know...

❑ The restricted two-body equation of motion and its assumptions (Chapter 4)

❑ Orbit specific mechanical energy, ε (Chapter 4)

❑ Orbit specific angular momentum, $\vec{h}$ (Chapter 4)

❑ Definition of vectors and vector operations including dot and cross products (Appendix A)

❑ Inverse trigonometric functions $\cos^{-1}$ and $\sin^{-1}$ (Appendix A)

Space isn't remote at all. It's only an hour's drive away if your car could go straight upwards.

Sir Fred Hoyle, London Observer

I n the last chapter we looked at the restricted two-body problem and developed an equation of motion to describe in strictly mathematical terms how satellites move through space. But many times it's not enough to generate a list of numbers that give a satellite's position and velocity in inertial space. Often, we want to visualize its orbit with respect to points on Earth. For example, we may want to know when a remote-sensing spacecraft will be over a flood-damaged area.

In this chapter, we'll explore two important tools that help us "see" satellite motion—the Classical Orbital Elements (COEs) and ground tracks. Once you get the hang of it, you'll be able to use these COEs to visualize how the orbit looks in space. Ground tracks will allow you to determine when certain parts of the Earth pass into the satellite's field of view, or when an observer on the Earth can see the satellite.

5.1 Orbital Elements

▬ In This Section You'll Learn to...

☛ Define the Classical Orbital Elements (COEs)

☛ Use the COEs to describe the size, shape, and orientation of an orbit and the location of a satellite

☛ Explain when particular COEs are undefined and which alternate elements must be used

If you're flying an airplane and the ground controllers call you on the radio to ask where you are and where you're going, you'll need to tell them six things: your airplane's

- Latitude
- Longitude
- Altitude
- Horizontal velocity
- Heading (i.e. North, South, etc.)
- Vertical velocity (ascending or descending)

Knowing these things, controllers can then predict where you'll be some time in the future.

Space operators perform a similar task with spacecraft. Instead of asking where the satellite is, they use radar at tracking sites to measure a satellite's current position, $\vec{R}$, and velocity, $\vec{V}$. As we'll see in Chapter 8, this information helps them predict where the satellite will be in the future.

Notice that position, $\vec{R}$, and velocity, $\vec{V}$, are vectors with three components each. Unlike latitude and longitude used for aircraft, $\vec{R}$ and $\vec{V}$ aren't very useful in visualizing exactly what type of orbit a spacecraft is in and where it's going.

For example, suppose you're given this current position and velocity for a spacecraft:

$$\vec{R} = 10,000\ \hat{I} + 8,000\ \hat{J} - 7,500\ \hat{K}\ \ km$$

$$\vec{V} = 4.4\ \hat{I} + 3.1\ \hat{J} - 2.7\ \hat{K}\ \ km/s$$

What could you tell about the orbit's size and shape or where the satellite is?

With the tools you've learned, about the only thing you could do is plot $\vec{R}$ and $\vec{V}$ in a 3-D coordinate system and try to figure something out from there. But there's a better way. Hundreds of years ago, Johannes Kepler developed a method for describing orbits that allows us to visualize their

Recall that $\vec{V}$ is the velocity vector that describes speed *and* direction. V is used to denote speed or simply velocity.

CLASSICAL ORBITAL ELEMENTS (COEs) CHECKLIST

❏ *Size*

❏ *Shape*

❏ *Orientation*
- orbit plane in space
- orbit within plane

❏ *Location*

CLASSICAL ORBITAL ELEMENTS (COEs) CHECKLIST

☑ *Size*—semi-major axis, a

❏ *Shape*

❏ *Orientation*
- orbit plane in space
- orbit within plane

❏ *Location*

size, shape, and orientation. Because we still need six measurements to tell us all we need to know about an orbit and a satellite's place in it, Kepler came up with six orbital elements. We now call these the *Classical Orbital Elements (COEs)*, and we'll use them to tell us the four things we want to know, as summarized in the COEs checklist on the left:

- Orbit size
- Orbit shape
- Orientation
    - orbit plane in space
    - orbit within plane
- Location of satellite

Defining the Classical Orbital Elements

Let's start with orbit *size*. In Chapter 4 we were able to relate the size of an orbit to its specific mechanical energy using the relationship

$$\varepsilon = \frac{-\mu}{2a} \qquad (5\text{-}1)$$

where
ε = specific mechanical energy (km^2/s^2)
μ = gravitational parameter of central body (km^3/s^2)
a = semi-major axis (km)

We also learned that the semi-major axis describes half the distance across the major (long) axis of the orbit, as shown in Figure 5-1. Thus, we'll use the semi-major axis, a, as our first COE.

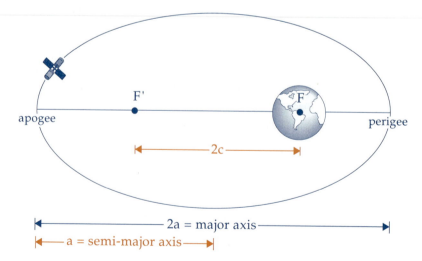

Figure 5-1. Semi-Major Axis. The semi-major axis, a, is one half the distance across the long axis of an ellipse. The distance between the foci (F and F') of the ellipse is 2c.

With the orbit size accounted for, the next thing we want to know is its shape. Once again, we gave this one away back in Chapter 4 when we discussed a way to describe the "out of roundness" of a conic section in terms of its eccentricity, e. *Eccentricity* specifies the *shape* of an orbit by looking at the ratio of the distance between the two foci and the length of the major axis.

$$e = \frac{2c}{2a} \tag{5-2}$$

The relationship between orbit shape and eccentricity is summarized in Table 5-1 and illustrated in Figure 5-2.

Table 5-1.   Relationship Between Conic Section and Eccentricity.

Conic Section	Eccentricity
circle	e = 0
ellipse	0 < e < 1
parabola	e = 1
hyperbola	e > 1

With an orbit's eccentricity in hand, we know the type of conic section we're dealing with. Now we have two pieces of our orbit puzzle: its size, a, and its shape, e. Next we can tackle its orientation in space. In Chapter 4 we saw that because specific angular momentum is constant, an orbit plane is fixed in inertial space. To describe its orientation, we refer to an inertial coordinate system used in Chapter 4—the geocentric-equatorial coordinate system shown in Figure 5-3. In the following discussion, we'll describe angles between key vectors, so make sure you know how to perform dot products and change from degrees to radians.

The first angle we'd use to describe the orientation of the orbit with respect to our coordinate system is inclination, i. *Inclination* describes the *tilt* of the orbit plane with respect to the fundamental plane. We could describe this *tilt* as simply the angle between the two planes, but this would be ambiguous (do we use the big angle or the small angle?) Instead, we'll define inclination as the angle between two vectors: one perpendicular to the orbit plane, $\hat{h}$ (the specific angular momentum vector), and one perpendicular to the fundamental plane, $\hat{K}$, as shown in Figure 5-4. Inclination can have a range of values from 0° to 180°.

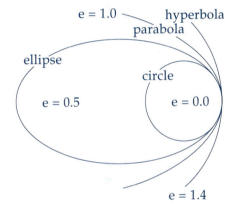

Figure 5-2. Eccentricity. Eccentricity defines the orbit shape.

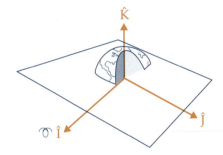

Figure 5-3. The Geocentric-Equatorial Coordinate System. We'll use the geocentric-equatorial coordinate system to reference all orbital elements. The fundamental plane is the Earth's equatorial plane, the principle direction ($\hat{I}$) points in the vernal equinox direction, ♈, the $\hat{K}$ unit vector points to the North Pole, and $\hat{J}$ completes the right-hand rule.

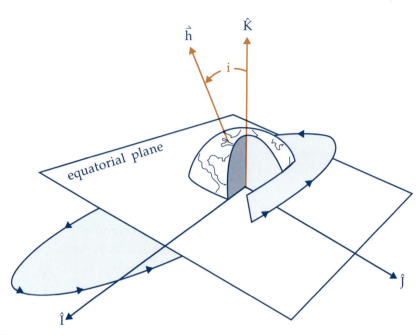

Figure 5-4. Inclination. Inclination, i, describes the *tilt* of the orbit plane with respect to the equator. The angle between the two planes is defined by looking at the angle between $\hat{K}$ (which is perpendicular to the equator) and $\vec{h}$ (which is perpendicular to the orbit plane).

We can use inclination to define several different kinds of orbits. An Earth orbit with an inclination of 0° or 180° is called an *equatorial orbit* because it goes around the equator. If the orbit has i = 90°, we call it a *polar orbit* because it travels over both the north and south poles. We also use the value of inclination to distinguish between two major classes of orbits. If 0° ≤ i < 90°, the satellite is moving with the rotation of the Earth (in an easterly direction), and the spacecraft is said to be in a *direct orbit* or *prograde orbit*. If 90° < i ≤ 180°, the spacecraft is moving against the Earth's rotation (in a westerly direction), so it's in an *indirect orbit* or *retrograde orbit*. Table 5-2 summarizes these orbits.

Thus, inclination is the third COE. It specifies the *tilt* of the orbit plane with respect to the fundamental plane and helps us understand the orbit's orientation with respect to the equator.

The fourth COE is an angle we'll use to describe orbit orientation. It's the longitude of ascending node, Ω, which describes the orientation of the orbit with respect to the principle direction, Î. Before you give up on a name like that, let's look at what each of its pieces means. First of all, what is longitude? You can describe your location on the Earth in terms of longitude, which is an angle measured around the equator. Geographers measure longitude from the Prime Meridian (Greenwich, England) either to the East or to the West. Longitude of ascending node is another angle measured around the equator, but as we'll see, it has a different starting point.

Now let's look at the other part of this new angle's name. What is an ascending node (or a node of any kind)? As we just described, the orbit plane normally has a tilt with respect to the fundamental plane (unless $i = 0°$ or $180°$). From geometry, you may remember that the intersection of two planes forms a line. In our case, the intersection of the orbit plane and the fundamental plane is called the *line of nodes*. The two points where the orbit actually crosses the equatorial plane are the nodes. The node where the satellite goes from below the equator (Southern Hemisphere) to above the equator (Northern Hemisphere) is called the *ascending node*. Similarly, when the satellite crosses the equator from the Northern Hemisphere to the Southern, it is passing through the *descending node*. See Table 5.2.

Table 5-2.   Types of Orbits and Their Inclination.

Inclination	Orbit Type	Diagram
$0°$ or $180°$	Equatorial	
$90°$	Polar	$i = 90°$
$0° \leq i < 90°$	Direct or Prograde (moves in direction of Earth's rotation)	ascending node
$90° < i \leq 180°$	Indirect or Retrograde (moves against the direction of Earth's rotation)	ascending node

Now let's put longitude and ascending node together to see what we get. We're describing the orientation of the plane with respect to the principle direction. That is, how is the orbit plane rotated in space? We use the vernal equinox direction or $\hat{I}$ (an inertial reference) as the starting point instead of Greenwich and measure the angle to the ascending node. Thus, the *longitude of the ascending node, Ω,* is the angle measured from the principle direction, $\hat{I}$, to the ascending node. By convention, this angle is always measured eastward. This new angle, Ω, then acts like a celestial map reference to give us the *swivel* of the orbit, which helps us better understand its orientation in space. The longitude of ascending node is illustrated in Figure 5-5; its range of values is $0° \leq \Omega < 360°$.

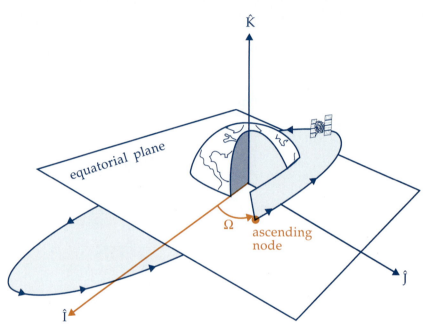

Figure 5-5. Longitude of Ascending Node, Ω. This describes the swivel of the orbit plane with respect to the principle direction. It is the angle between the principal direction and the point where the orbit plane crosses the equator from south to north (ascending node), measured eastward.

Let's recap where we are. We now know the *size* of the orbit, a, the *shape* of the orbit, e, the *tilt* of the orbit, i, and the *swivel*, Ω. But we don't know how the orbit itself is oriented within the plane. For example, for an elliptical orbit, we may want to know whether perigee (point closest to Earth) is in the Northern or Southern Hemisphere. This would be important if we wanted to take pictures of a particular point when we're closest. Picking up where we left off with the last orbital element, we can measure the angle between the ascending node and perigee and call it argument of perigee, ω. To remove any ambiguities, we will always measure the angle in the direction of satellite motion.

Why perigee? Because we're locating perigee. Why "argument"? Because we're "making clear" (from Latin) where perigee is. So our fifth COE, *argument of perigee, ω,* is the angle measured in the direction of satellite motion from the ascending node to perigee. It gives us the

---

CLASSICAL ORBITAL ELEMENTS (COEs) CHECKLIST

☑ *Size*—semi-major axis, a

☑ *Shape*—eccentricity, e

☑ *Orientation*

- orbit plane in space
  - inclination, i
  - longitude of ascending node, Ω
- orbit within plane
  - argument of perigee, ω

❑ *Location*

orientation of the orbit within the orbit plane as shown in Figure 5-6. The range on argument of perigee is $0° \leq \omega < 360°$.

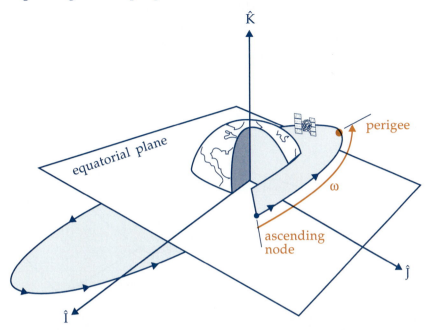

Figure 5-6. Argument of Perigee. Argument of perigee, ω, describes the *orientation* of the orbit within the orbit plane. It is the angle between the ascending node and perigee, measured in the direction of satellite motion.

After specifying the *size* and *shape* of the orbit, along with its *orientation* (tilt and swivel), we still need to find the satellite's location within the orbit. Fortunately, as we've already seen in Chapter 4, we can find it using the true anomaly. *True anomaly*, v, is the angle from perigee to the satellite's position vector, $\vec{R}$. Similar to the argument of perigee, true anomaly is measured in the orbit plane and always in the direction of satellite motion. True anomaly is shown in Figure 5-7; its range of values are $0° \leq v < 360°$.

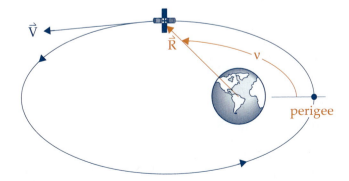

Figure 5-7. True Anomaly. True anomaly, v, specifies the *location* of the satellite within the orbit. It is the angle between perigee and the satellite's position vector measured in the direction of satellite motion. Of all the COEs, only true anomaly changes with time (as long as our two-body assumptions hold).

**CLASSICAL ORBITAL
ELEMENTS (COEs)
CHECKLIST**

☑ *Size*—semi-major axis, a

☑ *Shape*—eccentricity, e

☑ *Orientation*

- orbit plane in space
  - inclination, i
  - longitude of ascending
    node, Ω
- orbit within plane
  - argument of perigee, ω

☑ *Location*

True anomaly, ν, tells us the *location* of the satellite in the orbit. Of all the COEs, only true anomaly will change with time (as long as our two-body assumptions hold) as the satellite moves around in its orbit.

Now that you've seen all six of the classical orbital elements, we can show four of them together in Figure 5-8 (we can shown size and shape only indirectly in the way we draw the orbit). Table 5-3 summarizes all six.

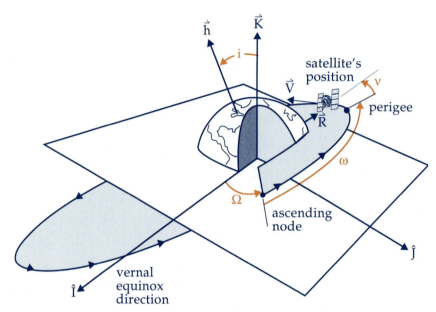

Figure 5-8. Classical Orbital Elements. Here we see four of the six COEs. We can use the COEs to find the position, $\vec{R}$, and velocity, $\vec{V}$, of the satellite. The other two COEs, semi-major axis, a, and eccentricity, e, specify the size and shape of the orbit.

Table 5-3. Summary of Classical Orbital Elements.

Element	Name	Description	Range of Values	Undefined
a	semi-major axis	*size*	Depends on conic section	never
e	eccentricity	*shape*	e = 0: circle 0 < e < 1: ellipse	never
i	inclination	*tilt*, angle from $\hat{K}$ unit vector to specific angular momentum vector h	$0 \leq i \leq 180°$	never
Ω	longitude of ascending node	*swivel*, angle from vernal equinox to ascending node	$0 \leq \Omega < 360°$	when i = 0 or 180° (equatorial orbit)
ω	argument of perigee	Angle from ascending node to perigee	$0 \leq \omega < 360°$	when i = 0 or 180° (equatorial orbit) or e = 0 (circular orbit)
ν	true anomaly	Angle from perigee to satellite position	$0 \leq \nu < 360°$	when e = 0 (circular orbit)

By now you're probably wondering what the heck these Classical Orbital Elements are good for! Let's look at an example to see how they can help us visualize a particular orbit. Suppose your communication satellite has the following COEs:

- Semi-major axis, a = 50,000 km
- Eccentricity, e = 0.4
- Inclination, i = 45°
- Longitude of ascending node, Ω = 50°
- Argument of perigee, ω = 110°
- True anomaly, ν = 170°

To begin with, as in Figure 5-9, you can sketch the size and shape of the orbit given the semi-major axis and the eccentricity. The eccentricity of 0.4 indicates an elliptical orbit (it's between 0 and 1). The semi-major axis of 50,000 km tells you how large to draw the orbit.

Now that you can picture what the orbit looks like in two dimensions, try to visualize how it's oriented in three dimensions. Because the inclination angle is 45°, we know the orbit plane tilts 45° from the equator. We can also describe inclination as the angle between the specific angular momentum vector, $\vec{h}$, of the orbit and $\hat{K}$ of our geocentric equatorial coordinate system. So we can sketch the crossing of the two planes in three dimensions as you see in Figure 5-10.

Next, to find the swivel of the orbit plane with respect to the principle direction, we use the longitude of the ascending node, Ω. After locating the principle direction in the equatorial plane, $\hat{I}$, we can swivel the orbit plane by positioning the ascending node 50° east of the $\hat{I}$ vector. What we know so far gives us the picture of the orbit in Figure 5-11.

Figure 5-9. Orbit Size and Shape. Here we see the approximate size and shape of an orbit with a semi-major axis of 50,000 km and an eccentricity of 0.4.

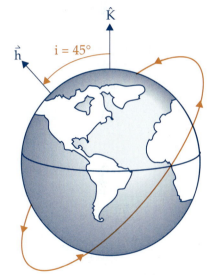

Figure 5-10. Inclination. Orientation of an orbit with an inclination of 45°.

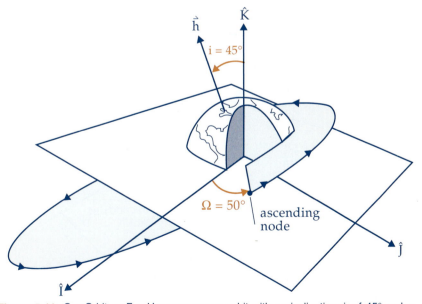

Figure 5-11. Our Orbit so Far. Here we see an orbit with an inclination, i, of 45° and a longitude of ascending node, Ω, of 50°.

Now we've completely specified the orbit's size and shape, as well as the orientation of the orbit plane in space. But we still don't know how the orbit is oriented within the plane. Luckily, argument of perigee, ω, comes to the rescue. Locating perigee within the orbit, we orient it 110° from the ascending node, measuring in the direction of satellite motion. Figure 5-12 shows us how the orbit is actually oriented in the orbit plane.

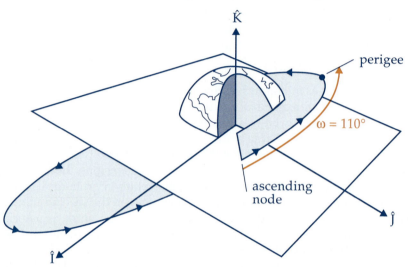

Figure 5-12. Argument of Perigee for the Example. The argument of perigee, ω, is 110°.

Finally, we can locate this satellite within the orbit. Using the value of true anomaly, ν, we measure 170° in the direction of satellite motion from perigee to locate the satellite. And there it is in Figure 5-13!

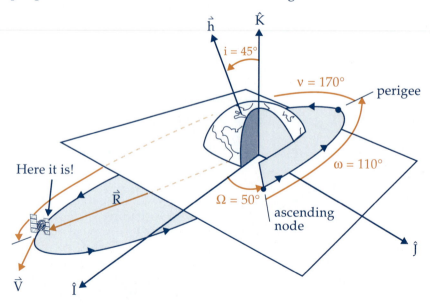

Figure 5-13. Finding the Satellite. Here we see the position of a satellite with the following COEs: a = 50,000 km, i = 45°, Ω = 50°, ω = 110°, ν = 170°.

As we already know, varying missions require different orbits, which COEs can describe. Table 5-4 shows various missions and typical orbits. A *geostationary orbit* is a circular orbit with a period of 24 hours and inclination of 0°. A satellite in a geostationary orbit appears to be stationary to an Earth-based observer. *Geosynchronous orbits* are inclined orbits with a period of 24 hours. They are typically used for communication satellites. A *semi-synchronous orbit* has a period of 12 hours. *Sun-synchronous orbits* are retrograde low-Earth orbits (LEO) inclined 95° to 105° and typically used in remote-sensing missions to observe Earth. A *Molniya orbit* is a semi-synchronous, eccentric orbit used for some communication missions. *Super-synchronous orbits* are usually circular orbits with periods longer than 24 hours.

Table 5-4. Orbital Elements for Various Missions.

Mission	Orbit Type	Semi-Major Axis (Altitude)	Period	Inclination	Other
• Communications • Early warning • Nuclear detection	Geostationary	42,158 km (35,780 km)	24 hr	~0°	$e \cong 0$
• Remote sensing	Sun-synchronous	~6500 – 7300 km (~150 – 900 km)	~90 min	~95°	$e \cong 0$
• Navigation - GPS	Semi-synchronous	26,610 km (20,232 km)	12 hr	55°	$e \cong 0$
• Space Shuttle	Low-Earth orbit	~6700 km (~300 km)	~90 min	28.5° or 57°	$e \cong 0$
• Communication/ intelligence	Molniya	26,571 km (R_p = 7971 km; R_a = 45,170 km)	12 hr	63.4°	ω = 270° e = 0.7

Alternate Orbital Elements

Now that you understand how to find all of these elements, you're ready for some bad news—they're not always defined! For example, a *circular orbit* has no perigee. In this case, we have no argument of perigee, ω, or true anomaly, v, because both use perigee as one of their references. To correct this deficiency, we bring in alternate orbital elements to replace these two missing angles. In general, whenever we face a peculiar orbit with one or more of the classical orbital elements undefined, we always work backward from the satellite's position vector (the one thing that's always defined) to the next thing that is defined. For example, a circular orbit has no perigee, so instead of using true anomaly to define position, we keep going from the position vector all the way back to the ascending node. This is how we get the first alternate element—the argument of latitude, u. *Argument of latitude, u*, is measured from the ascending node to the satellite's position. We measure this angle in the direction of satellite motion from the ascending node to the satellite's position vector.

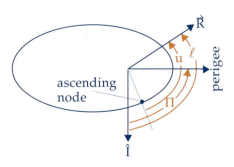

Figure 5-14. Alternate Orbital Elements. The alternate orbital elements are used when one or more of the classical orbital elements are undefined. u is the argument of latitude. Π is the longitude of perigee. ℓ is the true longitude.

Another situation that can give us grief is an equatorial orbit (i = 0° or 180°). In this case, the line of intersection between the equator and the orbit plane is missing (the line of nodes), so the ascending node doesn't exist. This time the longitude of the ascending node, Ω, and the argument of perigee, ω, are both undefined. We replace them with another alternate element, the *longitude of perigee, Π*—the angle measured from the principle direction, Î, to perigee in the direction of satellite motion.

Finally, the worst nightmare for a budding astrodynamicist is a *circular equatorial orbit*. This orbit has neither perigee *nor* ascending node, so the longitude of the ascending node, Ω, the argument of perigee, ω, and true anomaly, ν, are *all* undefined! Therefore, we must bring in a final alternate element to replace all of them—the *true longitude, ℓ.* This angle is measured from the principle direction, Î, to the satellite-position vector, R̀, in the direction of satellite motion. Figure 5-14 and Table 5-5 summarize the alternate orbital elements.

Table 5-5. Alternate Orbital Elements.

Element	Name	Description	Range of Values	When to Use
u	argument of latitude	Angle from ascending node to satellite position	$0 \leq u < 360°$	use when no perigee (e = 0)
Π	longitude of perigee	Angle from principal direction to perigee	$0 \leq \Pi < 360°$	use when equatorial (i = 0 or 180°) no ascending node
ℓ	true longitude	Angle from vernal equinox to satellite position	$0 \leq \ell < 360°$	use when no perigee and ascending node (e = 0 and i = 0 or 180°)

Astro Fun Fact
The Number 2

The number 2 plays an exceedingly critical role in our universe. As you have learned, the force of gravity is inversely proportional to the square of the distance between two bodies. But what if the distance were not squared? The answer is disturbing. If the exponent were larger than 2, the orbits of the Earth and Moon would spiral into the Sun. Yet, if the exponent were any less than 2, the orbits would expand away from the Sun into infinity. This holds true for all bodies in the known universe. Geometrically, the number 2 dictates that all orbits must be shaped as closed curves or ellipses. But don't worry about the number 2 suddenly changing! The inverse square law, as applied to the universal law of gravitation, is simply a human way of describing what universally exists in nature.

$$\ddot{R} + \frac{\mu}{R^2}\hat{R} = 0$$

Contributed by Dr. Jackson R. Ferguson and Michael Banks, United States Air Force Academy

Section Review

Key Terms

argument of latitude, u
argument of perigee, ω
ascending node
circular equatorial orbit
circular orbit
Classical Orbital Elements (COEs)
descending node
direct orbit
eccentricity
equatorial orbit
geostationary orbit
geosynchronous orbits
inclination
indirect orbit
line of nodes
longitude of perigee, Π
longitude of the ascending node, Ω
Molniya orbit
polar orbit
prograde orbit
retrograde orbit
semi-synchronous orbit
sun-synchronous orbits
super-synchronous orbits
true anomaly, ν
true longitude, ℓ

Key Equations

$$\varepsilon = \frac{-\mu}{2a}$$

Key Concepts

➤ To specify a satellite's orbit in space, you need to know four things about it:
  - Size
  - Shape
  - Orientation
  - Location

➤ The six classical orbital elements (COEs) specify these four pieces of information
  - Semi-major axis, a—one-half the distance across the long axis of an ellipse. It specifies the *size* of the orbit and is related to an orbit's energy
  - Eccentricity, e—specifies the *shape* of the orbit by telling what type of conic section we have
  - Inclination, i—specifies the *orientation* or *tilt* of the orbit plane with respect to a fundamental plane such as the equator
  - Longitude of ascending node, Ω—specifies the *orientation* or *swivel* of the orbit plane with respect to the principle direction, Î
  - Argument of perigee, ω—specifies the *orientation* of the orbit within the plane
  - True anomaly, ν—specifies the *location* of the satellite within the orbit plane

➤ Whenever one or more COEs are undefined, you must use alternate orbital elements

5.2 Computing Orbital Elements

▬ In This Section You'll Learn to...

☞ Determine all six orbital elements given only the position, $\vec{R}$, and velocity, $\vec{V}$, of a satellite at one particular time

Now let's put these classical orbital elements, COEs, to work for us. In real life, we can't measure COEs directly, but we can determine the spacecraft's inertial position and velocity, $\vec{R}$ and $\vec{V}$, using ground-tracking sites. Still we need some way to convert the $\vec{R}$ and $\vec{V}$ vectors to COEs, so we can make sense of the orbit. As we'll see, armed with just a position vector, $\vec{R}$, and a velocity vector, $\vec{V}$, at a single point in time, we can find all of the orbital elements. This shouldn't be too surprising. We already know we must have six pieces of information to define an orbit; the three components of $\vec{R}$; and the three components of $\vec{V}$ give us the six we need. Assuming we know $\vec{R}$ and $\vec{V}$, let's go through all of the COEs and see how we'd compute them, starting with the semi-major axis.

Finding Semi-Major Axis, a

Recall the semi-major axis, a, which tells us the size of the orbit, depends on the orbit's specific mechanical energy, ε. Thus, if we know the energy of the orbit, we can determine the semi-major axis. In Chapter 4, we showed specific mechanical energy could be found from

$$\varepsilon = \frac{V^2}{2} - \frac{\mu}{R} \qquad (5\text{-}3)$$

where
V = velocity (km/s)
μ = gravitational parameter (km^3/s^2)
R = position (km)

But we also know that ε is related to semi-major axis through Equation (5-1). So if we know the magnitude of $\vec{R}$ and of $\vec{V}$, we can solve for the energy and thus the semi-major axis. This is shown in Example 5-1 (Part 1).

Whenever you solve for semi-major axis (or any other parameter for that matter), it's a good idea to do a "reality check" on your result. For example, an orbiting satellite should have a semi-major axis greater than the radius of the planet it's orbiting; otherwise, it would hit the planet! Also be careful with parabolic trajectories where a = ∞ and ε = 0.

Finding Eccentricity, e

The next COE we defined was eccentricity, e. To determine the eccentricity, we need to use a little mathematical sleight-of-hand (the full derivation is in Appendix C) and define an *eccentricity vector, $\vec{e}$*, which happens to point from the center of the Earth to perigee and whose magnitude is equal to the eccentricity, e. $\vec{e}$ is related to position, $\vec{R}$, and velocity, $\vec{V}$, by

$$\vec{e} = \frac{1}{\mu}\left[\left(V^2 - \frac{\mu}{R}\right)\vec{R} - (\vec{R} \cdot \vec{V})\,\vec{V}\right] \qquad (5\text{-}4)$$

where
$\vec{e}$ = eccentricity vector (unitless, points at perigee)
μ = gravitational parameter (km^3/s^2)
V = magnitude of $\vec{V}$ (km/s)
R = magnitude of $\vec{R}$ (km)
$\vec{R}$ = position vector (km)
$\vec{V}$ = velocity vector (km/s)

Figure 5-15 shows the eccentricity vector for an orbit. Thus, all we need is $\vec{R}$ and $\vec{V}$ to solve for eccentricity, and we find out where perigee is as an added bonus. This will be useful later.

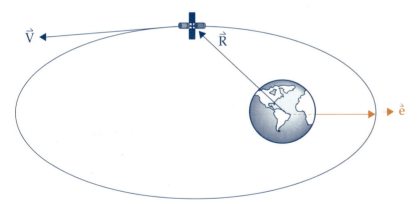

Figure 5-15. Eccentricity Vector, $\vec{e}$. The eccentricity vector, $\vec{e}$, is found from the position and velocity vectors ($\vec{R}$ and $\vec{V}$). It points at perigee and its magnitude is equal to the orbit's eccentricity.

Finding Inclination, i

The other four orbital elements are all angles. To find them, we need to use the definition of the dot product, which allows us to find an angle if we know two appropriate reference vectors. For any two vectors $\vec{A}$ and $\vec{B}$, as shown in Figure 5-16, we can say

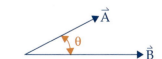

Figure 5-16. Dot Product to Find Angles. The dot product gives us the angle between two vectors.

$$\vec{A} \cdot \vec{B} = AB\cos\theta \qquad (5\text{-}5)$$

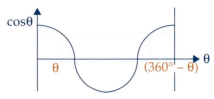

Figure 5-17. Inverse Cosine. An inverse cosine gives two possible answers: θ and (360 – θ).

where A and B are the magnitudes of the vectors, and θ is the angle between them. Solving for θ gives us

$$\theta = \cos^{-1}\frac{\vec{A} \cdot \vec{B}}{AB} \qquad (5\text{-}6)$$

But be careful! When you take an inverse cosine to find θ, as shown in Figure 5-17, two angles are possible: θ and (360 – θ).

Let's see how we can use this dot product idea to find the inclination angle, i. Recall we defined i as the angle between $\hat{K}$ and $\vec{h}$, as shown in Figure 5-18. We can then apply our dot product relationship to these two vectors to arrive at the following expression for inclination:

$$\boxed{i = \cos^{-1}\frac{\hat{K} \cdot \vec{h}}{Kh}} \qquad (5\text{-}7)$$

where
i = inclination (deg or rad)
$\hat{K}$ = unit vector through North Pole
$\vec{h}$ = specific angular momentum vector (km²/s)
K = magnitude of $\hat{K}$ = 1
h = magnitude of h (km²/s)

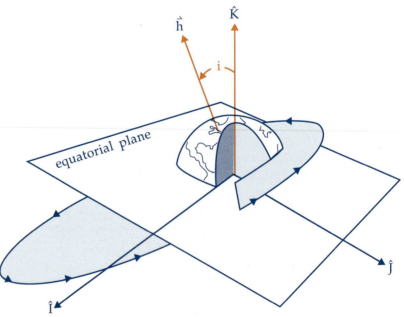

Figure 5-18. Inclination. Recall inclination, i, is defined as the angle between the $\hat{K}$ unit vector and the specific angular momentum vector, $\vec{h}$.

Because the magnitude of $\hat{K}$ is one, the denominator reduces to just the magnitude of $\vec{h}$. Recall from Figure 4-28 that $\vec{h}$ is the cross product of $\vec{R}$ and $\vec{V}$. The quantity $\hat{K} \cdot \vec{h}$ is simply the kth component of $\vec{h}$ because $\hat{K}$ is a unit vector. Do we have to worry about a quadrant check in this case?

No, because the value of inclination is always less than or equal to 180°, so the smaller number we get will always be right.

Finding Longitude of Ascending Node, Ω

We can find the longitude of the ascending node, Ω, using the same basic approach we used to find inclination. From the definition of Ω, you know it's the angle between the principle direction, $\hat{I}$, and the ascending node. Now we need some vectors. Can we define a vector that points at the ascending node? You bet! If you draw a vector from the center of the Earth pointing at the ascending node, you'll notice it lies along the intersection of two planes—the orbit plane and the equatorial plane. Thus, this new vector, which we call the *ascending node vector, $\vec{n}$*, must be perpendicular to $\hat{K}$ and $\vec{h}$, as shown in Figure 5-19. (That's because $\hat{K}$ is perpendicular to the equator, and $\vec{h}$ is perpendicular to the orbit plane.) Using the definition of cross product (and the right hand rule), we get

$$\boxed{\vec{n} = \hat{K} \times \vec{h}}$$
(5-8)

where
$\vec{n}$ = ascending node vector (unitless, points at ascending node)
$\hat{K}$ = unit vector through North Pole
$\vec{h}$ = specific angular momentum vector (km^2/s)

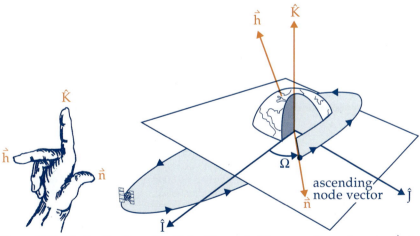

Figure 5-19. Finding the Ascending Node. We can find the ascending node vector, $\vec{n}$, by using the right-hand rule. Point your index finger at $\hat{K}$ and your middle finger at $\vec{h}$. Your thumb will point in the direction of $\vec{n}$.

Because Ω is the angle between $\hat{I}$ and $\vec{n}$, we can use the dot product relationship again to find the longitude of the ascending node

$$\boxed{\Omega = \cos^{-1}\frac{\hat{I} \cdot \vec{n}}{In}}$$
(5-9)

where

Ω = longitude of ascending node (deg or rad)

$\hat{I}$ = unit vector in the principle direction

$\hat{n}$ = ascending node vector (unitless, points at ascending node)

I = magnitude of $\hat{I}$ = 1

n = magnitude of $\hat{n}$ (unitless)

The longitude of ascending node can range between 0° and 360°, so a quadrant check is necessary. How do we decide which quadrant Ω belongs in? Looking at Figure 5-20, we see the equatorial plane and the location of the ascending node vector, $\hat{n}$. Notice the $\hat{J}$ component of $\hat{n}$, n_J, tells us which side of the $\hat{I}$ axis $\hat{n}$ is on. This, in turn, can tell us how big an angle Ω is. Thus, we can write a logic statement for our quadrant check

If $n_J \geq 0$ then $0 \leq \Omega \leq 180°$

If $n_J < 0$ then $180° < \Omega < 360°$

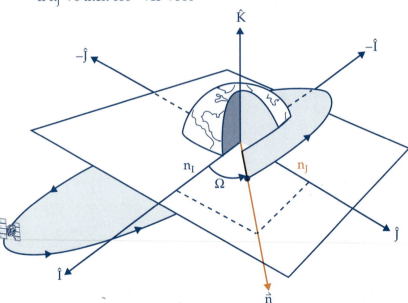

Figure 5-20. Quadrant Check for Ω. We can find the quadrant for the longitude of ascending node, Ω, by looking at the sign of the $\hat{J}$ component of $\hat{n}$, n_J. If n_J is greater than zero, Ω is between 0 and 180°. If n_J is less than zero, Ω is between 180° and 360°.

Finding Argument of Perigee, ω

Now let's look at argument of perigee, ω. It locates perigee in the orbit plane. Remember, it was defined as the angle between the ascending node and perigee. We already know where the ascending node is from the ascending node vector, $\hat{n}$. We also know the eccentricity vector, $\hat{e}$, points at perigee. Using our dot product relationship once again, we can solve for argument of perigee, ω.

157

$$\omega = \cos^{-1}\frac{\vec{n} \cdot \vec{e}}{ne} \qquad\qquad (5\text{-}10)$$

where

ω = argument of perigee (deg or rad)
$\vec{n}$ = ascending node vector (unitless, points at ascending node)
$\vec{e}$ = eccentricity vector (unitless, points at perigee)
n = magnitude of $\vec{n}$ (unitless)
e = eccentricity (unitless)

Once more, we have two possible answers that satisfy the equation, so we have another quadrant-identity crisis. How do we know which quadrant ω belongs in? In Figure 5-21, we can see that if ω is between $0°$ and $180°$, perigee will be above the equator; if ω is between $180°$ and $360°$, perigee will be below the equator. Luckily, we have the trusty $\vec{e}$ vector to tell us exactly where perigee is. If we look at just the $\hat{K}$ component of $\vec{e}$ (e_K), we can tell if perigee is in the Northern or Southern Hemisphere (positive $\hat{K}$ for Northern, negative $\hat{K}$ for Southern). We can write this as a logic statement

If $e_K \geq 0$ then $0° \leq \omega \leq 180°$
If $e_K < 0$ then $180° < \omega < 360°$

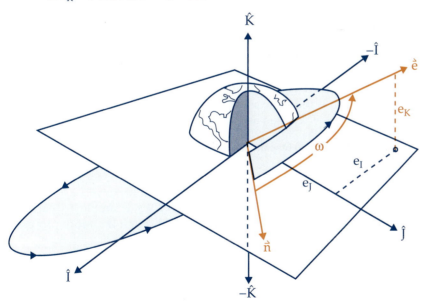

Figure 5-21. Finding ω. We can find the quadrant for argument of perigee, ω, by looking at the $\hat{K}$ component of the eccentricity vector, $\vec{e}$. If e_K is greater than zero, perigee lies above the equator; thus, ω is between $0°$ and $180°$. If e_K is less than zero, perigee lies in the Southern Hemisphere; thus, ω is between $180°$ and $360°$.

Finding True Anomaly, ν

Finally, it's time to find out what's nu. That is, we must find true anomaly, ν, to locate the satellite's position in its orbit. Because ν is defined to be the angle between perigee and the position vector, $\vec{R}$, we can start from our last point of reference, the perigee direction (using the eccentricity vector again), and measure to the position vector, $\vec{R}$. Applying our dot product relationship one last time, we arrive at

$$\nu = \cos^{-1} \frac{\vec{e} \cdot \vec{R}}{eR} \tag{5-11}$$

where
ν = true anomaly (deg or rad)
$\vec{e}$ = eccentricity vector (unitless, points at perigee)
$\vec{R}$ = position vector (km)
e = eccentricity (unitless)
R = magnitude of $\vec{R}$ (km)

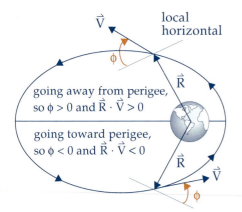

going away from perigee, so $\phi > 0$ and $\vec{R} \cdot \vec{V} > 0$

going toward perigee, so $\phi < 0$ and $\vec{R} \cdot \vec{V} < 0$

Figure 5-22. Finding ν. To resolve the quadrant for true anomaly, ν, check the sign on the flight-path angle, φ. If φ is positive, the satellite is moving away from perigee, so true anomaly is between 0 and 180°. If φ is negative, the satellite is moving toward perigee, so true anomaly is between 180° and 360°.

To sort out the quadrant for this angle, we want to tell whether we are heading away from or toward perigee. Recall from Chapter 4 our discussion of the flight-path angle, φ. If φ is positive, we're gaining altitude and heading away from perigee ("the houses are getting smaller"). If φ is negative, we're losing altitude and heading toward perigee ("the houses are getting bigger") as seen in Figure 5-22.

So all we have to do is find the sign on φ. No problem. Remember that φ is the angle between the local horizontal and the velocity vector. By applying a little bit of trigonometry, we can show that the sign of the quantity $(\vec{R} \cdot \vec{V})$ is the same as the sign of φ! Thus, if we know $(\vec{R} \cdot \vec{V})$, we know what's nu, ν. Written as a logic statement, this boils down to

If $(\vec{R} \cdot \vec{V}) \geq 0$ ($\phi \geq 0$) then $0° \leq \nu \leq 180°$

If $(\vec{R} \cdot \vec{V}) < 0$ ($\phi < 0$) then $180° < \nu < 360°$

Now we've been through all the steps needed to convert a lowly set of $\vec{R}$ and $\vec{V}$ vectors into the extremely useful COEs. Example 5-1 parts 1, 2, and 3 take some real-life numbers and goes through the entire process so you can see how it really works.

Section Review

Key Terms

ascending node vector, $\vec{n}$
eccentricity vector, $\vec{e}$

Key Equations

$$\varepsilon = \frac{V^2}{2} - \frac{\mu}{R}$$

$$\vec{e} = \frac{1}{\mu}\left[\left(V^2 - \frac{\mu}{R}\right)\vec{R} - (\vec{R}\cdot\vec{V})\,\vec{V}\right]$$

$$i = \cos^{-1}\frac{\hat{K}\cdot\vec{h}}{Kh}$$

$$\vec{n} = \hat{K}\times\vec{h}$$

$$\Omega = \cos^{-1}\frac{\hat{I}\cdot\vec{n}}{n}$$

$$\omega = \cos^{-1}\frac{\vec{n}\cdot\vec{e}}{ne}$$

$$\nu = \cos^{-1}\frac{\vec{e}\cdot\vec{R}}{eR}$$

Key Concepts

➤ You can compute all six classical orbital elements for an orbit using just one position vector, $\vec{R}$, and velocity vector, $\vec{V}$—a "snap shot" of the satellite at any point in time.

➤ You can find semi-major axis, a, using the magnitudes of $\vec{R}$ and $\vec{V}$ to determine the orbit's specific mechanical energy, which is a function of semi-major axis

➤ You can find eccentricity using the eccentricity vector, $\vec{e}$, which points at perigee

➤ Because all the remaining COEs are angles, you can find them using simple vector dot products. In general, for any two vectors $\vec{A}$ and $\vec{B}$, the angle between them, θ, is

$$\theta = \cos^{-1}\frac{\vec{A}\cdot\vec{B}}{AB}$$

➤ Because inclination, i, is the angle between the unit vector, $\hat{K}$, and the specific angular momentum vector, $\vec{h}$, you can find it using the dot product relationship.
  • Remember: $0 \leq i \leq 180°$

➤ Because longitude of ascending node, Ω, is the angle between the principle direction, $\hat{I}$, and the ascending node (represented by $\vec{n}$), you can find Ω using the dot product relationship.
  • Remember: If $n_J \geq 0$, then $0° \leq \Omega \leq 180°$
         If $n_J < 0$, then $180° < \Omega < 360°$

➤ Because argument of perigee, ω, is the angle between the ascending node, $\vec{n}$, and perigee (represented by the eccentricity vector, $\vec{e}$), you can find it using the dot product relationship.
  • Remember: If $e_K \geq 0$, then $0° \leq \omega \leq 180°$
         If $e_K < 0$, then $180° < \omega < 360°$

➤ Because true anomaly, ν, is the angle between perigee (represented by the eccentricity vector $\vec{e}$) and the satellite's position vector, $\vec{R}$, you can find it using the dot product relationship.
  • Remember: If $(\vec{R}\cdot\vec{V}) \geq 0\,(\phi \geq 0)$, then $0° \leq \nu \leq 180°$
         If $(\vec{R}\cdot\vec{V}) < 0\,(\phi < 0)$, then $180° < \nu < 360°$

Example 5-1 (Part 1)

Problem Statement

Space Operations Officers at Air Force Space Command have given you this set of position ($\vec{R}$) and velocity ($\vec{V}$) vectors for a new European Space Agency (ESA) satellite.

$$\vec{R} = 8228\ \hat{I} + 389.0\ \hat{J} + 6888\ \hat{K}\ \text{km}$$

$$\vec{V} = -0.7000\ \hat{I} + 6.600\ \hat{J} - 0.6000\ \hat{K}\ \frac{\text{km}}{\text{s}}$$

Determine the size (semi-major axis) and shape (eccentricity) for this satellite's orbit.

Problem Summary

Given: $\vec{R} = 8228\ \hat{I} + 389\ \hat{J} + 6888\ \hat{K}\ \text{km}$

$\vec{V} = -0.7000\ \hat{I} + 6.600\ \hat{J} - 0.6000\ \hat{K}\ \dfrac{\text{km}}{\text{s}}$

Find: a and e

Conceptual Solution

1) Determine magnitudes of the vectors, R and V

2) Solve for the semi-major axis

Determine the orbit's size using the relationships shown in Equation (5-3) and Equation (5-1).

$$\varepsilon = \frac{V^2}{2} - \frac{\mu}{R}$$

$$\varepsilon = \frac{-\mu}{2a}$$

3) Solve for the eccentricity vector, $\hat{e}$, and its magnitude, e.

$$\hat{e} = \frac{1}{\mu}\left[\left(V^2 - \frac{\mu}{R}\right)\vec{R} - (\vec{R} \cdot \vec{V})\vec{V}\right]$$

Analytical Solution

1) Determine magnitudes of the vectors, $\vec{R}$ and $\vec{V}$

$$R = \sqrt{(8228)^2 + (389)^2 + (6888)^2} = 10738\ \text{km}$$

$$V = \sqrt{(-0.7000)^2 + (6.600)^2 + (-0.6000)^2}$$

$$= 6.664\ \frac{\text{km}}{\text{s}}$$

2) Solve for the semi-major axis

$$\varepsilon = \frac{V^2}{2} - \frac{\mu}{R}$$

$$\varepsilon = \frac{(6.664\ \text{km/s})^2}{2} - \frac{3.986005 \times 10^5\ \dfrac{\text{km}^3}{\text{s}^2}}{10{,}738\ \text{km}}$$

$$= -14.916\ \frac{\text{km}^2}{\text{s}^2}$$

$$\varepsilon = \frac{-\mu}{2a}$$

$$a = \frac{-\mu}{2\varepsilon} = \frac{-3.986005 \times 10^5\ \dfrac{\text{km}^3}{\text{s}^2}}{2\left(-14.916\ \dfrac{\text{km}^2}{\text{s}^2}\right)} = 1.336 \times 10^4\ \text{km}$$

3) Solve for the eccentricity vector, $\hat{e}$, and its magnitude, e.

$$\hat{e} = \frac{1}{\mu}\left[\left(V^2 - \frac{\mu}{R}\right)\vec{R} - (\vec{R} \cdot \vec{V})\vec{V}\right]$$

We can start by finding the dot product between $\vec{R}$ and $\vec{V}$.

$$\vec{R} \cdot \vec{V} = (8228)(-0.7000) + (389.0)(6.600) +$$

$$(6888)(-0.6000) = -7325\ \frac{\text{km}^2}{\text{s}}$$

$$\hat{e} = \frac{1}{3.986005 \times 10^5} \times$$

$$\left[\left((6.66)^2 - \frac{3.986005 \times 10^5\ \dfrac{\text{km}^3}{\text{s}^2}}{10{,}738\ \text{km}}\right)\vec{R} - (-7325)\vec{V}\right]$$

Example 5-1 (Part 1) Continued

$$\vec{e} = (2.508778 \times 10^{-6}) \, [\, (7.288) \, \vec{R} - (-7325) \, \vec{V} \,]$$

$$\vec{e} = (1.8284 \times 10^{-5}) \, [8228 \, \hat{I} + 389.0 \, \hat{J} + 6888 \, \hat{K}] -$$
$$(-0.018377) \, [-0.7000 \, \hat{I} + 6.600 \, \hat{J} - 0.6000 \, \hat{K}]$$

$$\vec{e} = 0.15044 \, \hat{I} + 0.0071125 \, \hat{J} + 0.12594 \, \hat{K}$$
$$- [0.012864 \, \hat{I} - 0.12129 \, \hat{J} + 0.011026 \, \hat{K}]$$

$$\vec{e} = 0.1376 \, \hat{I} + 0.1284 \, \hat{J} + 0.1149 \, \hat{K}$$

Now that we have the eccentricity vector, we can solve for the magnitude, which tells us the shape of the orbit.

$$e = \sqrt{(0.1376)^2 + (0.1284)^2 + (0.1149)^2} = 0.2205$$

Interpreting the Results

We can see that the semi-major axis of this orbit is 13,360 km (8296 miles). Because e = 0.2205 (0 < e < 1), we also know the orbit is an ellipse (not a circle, parabola, or hyperbola).

Example 5-1 (Part 2)

Problem Statement

Using the same position and velocity vector as in Example 5-1 (Part 1), determine the inclination of the orbit.

Problem Summary

Given: $\vec{R} = 8228\ \hat{I} + 389.0\ \hat{J} + 6888\ \hat{K}$ km

$\vec{V} = -0.7000\ \hat{I} + 6.600\ \hat{J} - 0.6000\ \hat{K}\ \dfrac{km}{s}$

Find: i

Conceptual Solution

1) Solve for the specific angular momentum vector, $\vec{h}$, and its magnitude, h.

$$\vec{h} = \vec{R} \times \vec{V}$$

2) Solve for the inclination angle, i.

$$i = \cos^{-1}\dfrac{\hat{K} \cdot \vec{h}}{Kh} = \cos^{-1}\dfrac{\hat{K} \cdot \vec{h}}{h}$$

Analytical Solution

1) Solve for the specific angular momentum vector, $\vec{h}$, and its magnitude, h.

$$\vec{h} = \vec{R} \times \vec{V}$$

$$\vec{R} \times \vec{V} = \begin{vmatrix} \hat{I} & \hat{J} & \hat{K} \\ 8228 & 389.0 & 6888 \\ -0.7000 & 6.600 & -0.6000 \end{vmatrix} \dfrac{km^2}{s}$$

$\vec{h} = [\,(389.0)\,(-0.6000) - (6.600)\,(6888)\,]\,\hat{I} -$

$\qquad [\,(8228)\,(-0.6000) - (-0.7000)\,(6888)\,]\,\hat{J} +$

$\qquad [\,(8228)\,(6.600) - (-0.7000)\,(389.0)\,]\,\hat{K}\ \dfrac{km^2}{s}$

$\vec{h} = [\,(-233.4) - (45,460.8)\,]\,\hat{I} -$

$\qquad [\,(-4936.8) + (4821.6)\,]\,\hat{J} +$

$\qquad [\,(54,304.8) + (272.3)\,]\,\hat{K}\ \dfrac{km^2}{s}$

$\vec{h} = -45,694.2\ \hat{I} + 115.2\ \hat{J} + 54,577.1\ \hat{K}\ \dfrac{km^2}{s}$

$h = \sqrt{(45,694.2)^2 + (115.2)^2 + (54,577.1)^2}$

$\qquad = 71,180.3\ \dfrac{km^2}{s}$

2) Solve for the inclination angle, i.

$$i = \cos^{-1}\dfrac{\hat{K} \cdot \vec{h}}{Kh} = \cos^{-1}\dfrac{\hat{K} \cdot \vec{h}}{h}$$

$$\hat{K} \cdot \vec{h} = h_k = 54,577.1$$

$$i = \cos^{-1}\dfrac{h_K}{h} = \cos^{-1}\dfrac{54,577.1}{71,180.3} = \cos^{-1}(0.76674)$$

At this point we need to pull out our calculator and take the inverse cosine of 0.76674 (unless you can figure things like that out in your head!). **But be careful!** When you take inverse trigonometric functions your calculator will give you only *one* of the possible correct angles. For an inverse cosine we must subtract this result from 360° to get the second possible answer. For our result from above we get two possible answers for inclination:

$$i = 39.94° \text{ or } (360°-39.94°) = 320.1°$$

To resolve this ambiguity, we must return to the definition of inclination. Because i must be between 0° and 180°, our answer must be i = 39.94°.

Interpreting the Results

The inclination of this orbit is 39.94°.

Example 5-1 (Part 3)

Problem Statement

Using the same position and velocity information from Example 5-1 (Part 1), determine the longitude of ascending node, argument of perigee, and true anomaly.

Problem Summary

Given: $\vec{R} = 8228\ \hat{I} + 389.0\ \hat{J} + 6888\ \hat{K}$ km

$\vec{V} = -0.7000\ \hat{I} + 6.600\ \hat{J} - 0.6000\ \hat{K}\ \dfrac{km}{s}$

Find: Ω, ω, and v

Conceptual Solution

1) Solve for the ascending node vector, $\vec{n}$, and its magnitude, n.

$$\vec{n} = \hat{K} \times \vec{h}$$

2) Solve for the longitude of the ascending node, Ω—**Do quadrant check**.

$$\Omega = \cos^{-1}\frac{\hat{I} \cdot \vec{n}}{In} = \cos^{-1}\frac{\hat{I} \cdot \vec{n}}{n}$$

3) Solve for the argument of perigee, ω—**Do quadrant check**.

$$\omega = \cos^{-1}\frac{\vec{n} \cdot \vec{e}}{ne}$$

4) Solve for the true anomaly angle, v—**Do quadrant check**.

$$v = \cos^{-1}\frac{\vec{e} \cdot \vec{R}}{eR}$$

Analytical Solution

1) Solve for the ascending node vector, $\vec{n}$, and its magnitude, n.

$$\vec{n} = \hat{K} \times \vec{h} = \begin{vmatrix} \hat{I} & \hat{J} & \hat{K} \\ 0.0 & 0.0 & 1.0 \\ -45{,}694.2 & 115.2 & 54{,}577.1 \end{vmatrix}$$

$$= -115.2\ \hat{I} - 45{,}694.2\ \hat{J} + 0\ \hat{K}$$

Solving for the magnitude of $\vec{n}$ we get n = 45,694.3

2) Solve for the longitude of ascending node angle, Ω —**Do quadrant check**.

$$\Omega = \cos^{-1}\frac{\hat{I} \cdot \vec{n}}{In} = \cos^{-1}\frac{\hat{I} \cdot \vec{n}}{n}$$

$$\hat{I} \cdot \vec{n} = n_I = -115.2$$

$$\Omega = \cos^{-1}\frac{-115.2}{45{,}694.2} = \cos^{-1}(-0.0025211)$$

$$\Omega = 90.14° \text{ or } (360° - 90.14°) = 269.9°$$

Once again we must figure out which of the two possible answers is the correct one. Using the logic we developed earlier, we must check n_J.

If $n_J \geq 0$ then $0° \leq \Omega \leq 180°$

If $n_J < 0$ then $180° < \Omega < 360°$

Because $n_J = -45{,}694.2$, $n_J < 0$, which means $180° < \Omega < 360°$. Thus, we can conclude $\Omega = 269.9°$.

3) Solve for the argument of perigee angle, ω—**Do quadrant check**.

$$\omega = \cos^{-1}\frac{\vec{n} \cdot \vec{e}}{ne}$$

$$\vec{n} \cdot \vec{e} = (-115.2)(0.1376) + (-45{,}694.2)(0.1284)$$
$$+ (0.0)(0.1149) = -5882.99$$

$$\omega = \cos^{-1}\left[\frac{-5882.9}{(45{,}694.2)(0.2205)}\right] = \cos^{-1}(-0.58389)$$

$$125.7° \text{ or } (360° - 125.7) = 234.3$$

To resolve which ω is right we use the logic developed earlier and check the value of e_K.

If $e_K \geq 0$ then $0° \leq \omega \leq 180°$

If $e_K < 0$ then $180° < \omega < 360°$

Because $e_K = +0.1140$, $e_K > 0$, which means $0° \leq \omega \leq 180°$. Thus, we can conclude

$$\omega = 125.7°$$

Example 5-1 (Part 3) Continued

4) Solve for the true anomaly angle, v—**Do quadrant check**.

$$v = \cos^{-1}\frac{\hat{e} \cdot \vec{R}}{eR}$$

$$\hat{e} \cdot \vec{R} = (0.1376)\,(8228) + (0.1284)\,(389.0) +$$

$$(0.1149)\,(6888) = 1974.05$$

$$v = \cos^{-1}\frac{1974.05}{(0.2205)\,(10{,}738)} = \cos^{-1} 0.83373$$

$$v = 33.52° \text{ or } (360° - 33.78°) = 326.52°$$

Finally, we resolve one last ambiguity between possible answers. Using our logic developed earlier, we check the sign on $(\vec{R} \cdot \vec{V})$.

If $(\vec{R} \cdot \vec{V}) \geq 0$ $(\phi > 0)$ then $0° \leq v \leq 180°$

If $(\vec{R} \cdot \vec{V}) < 0$ $(\phi < 0)$ then $180° < v < 360°$

The value for $(\vec{R} \cdot \vec{V})$ we found earlier was -7325. Because $(\vec{R} \cdot \vec{V}) < 0$, $180° < v < 360°$. Therefore, we can conclude

$$v = 326.5°$$

Interpreting the Results

We started with:

$$\vec{R} = 8228\ \hat{I} + 389.0\ \hat{J} + 6888\ \hat{K}\ \text{km}$$

$$\vec{V} = -0.7000\ \hat{I} + 6.600\ \hat{J} - 0.6000\ \hat{K}\frac{\text{km}}{\text{s}}$$

And found the following COEs:

$a = 13{,}360$ km	$\Omega = 269.9°$
$e = 0.2205$	$\omega = 125.7°$
$i = 39.94°$	$v = 326.5°$

5.3 Satellite Ground Tracks

In This Section You'll Learn to...

- ☞ Explain why satellite ground tracks look the way they do
- ☞ Use ground tracks to determine the inclination and period for direct orbits

The six classical orbital elements allowed us to visualize how an orbit looks from way out in space. Now let's beam back to Earth to see how an orbit looks from our perspective here on the ground.

A satellite user needs to know what part of the Earth the satellite is passing over at any given time. For instance, remote-sensing satellites must be over the location they need to observe. As we'll see, we can learn a lot about a satellite's orbit and mission by examining the track it makes over the Earth.

To understand ground tracks, imagine you're going to take a trip in your car from San Francisco to Omaha. You'll drive east out of San Francisco on Interstate 80 for a couple thousand miles. If you get out a road map of the western United States, you can trace your route on the map by drawing a line along I-80 as shown in Figure 5-23. This is your ground track from San Francisco to Omaha.

Now imagine you're going to take the same trip, but this time in an airplane. Once again, you could trace your air route on a map of the U.S. from San Francisco east to Omaha (only this time you wouldn't need roads on the map to go by). This, too, would be your ground track.

A satellite ground track is really no different from either of these examples. It's simply the trace of the path the satellite takes over the Earth's surface. The main complications arise when we have to consider that the satellite covers the entire circumference of the Earth (more than 40,000 km or 25,000 miles) each orbit and that the Earth is spinning on its axis at a speed of more than 1600 km/hr (1000 m.p.h.) at the equator while this happens, as we can see in Figure 5-24.

So what does a ground track look like? To make things easy, we'll start out by pretending the Earth doesn't rotate. (Try not to get dizzy—we'll throw the rotation back in soon.) Picture an orbit above this non-rotating Earth. The ground track will follow what is called a great circle route around the Earth. A *great circle* is any circle which "slices through" the center of a sphere. For example, lines of longitude, as shown in Figure 5-25, are great circles because they slice through the center of the Earth, but lines of latitude are not great circles (except for 0° latitude at the equator) because they don't. An orbit trace must be a great circle because the satellite is in orbit around the center of the Earth; thus, the orbit plane also passes through the Earth's center.

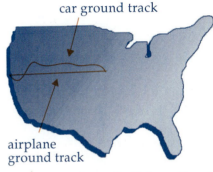

Figure 5-23. Car and Airplane Ground Tracks. Ground tracks for a trip by car and air from San Francisco to Omaha.

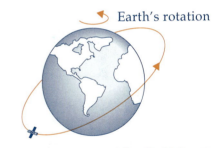

Figure 5-24. Earth and Satellite Motion. The Earth spins on its axis at nearly 1000 m.p.h. at the equator while a satellite goes around it.

Figure 5-25. Great Circles. A great circle is any circle around a sphere which bisects it (cuts it exactly in half). Lines of longitude are great circles whereas lines of latitude (except for the equator) are not.

When the Earth is stretched out to a flat-map projection (called a Mercator projection), things start to look a little different. To get a feel for how this affects the shape we see for the ground track, imagine the Earth is shaped like a soda can. A trace of the orbit on the soda can, as in Figure 5-26, would look like a circle slicing through the center of the can. But what if we were to flatten the can out and look at the orbit trace? It would look like a sine wave!

Figure 5-26. Orbiting Around a Soda Can. Imagine an orbit around a soda can. It draws a circle around the can which, when flattened out, looks like a sine wave.

Now imagine yourself on the ground watching the orbit pass overhead. Because we've stopped the Earth from rotating, the ground track will always stay the same, and the satellite will continue to pass overhead orbit after orbit, as shown in Figure 5-27. Even if we change the size and shape of the orbit, the ground track would look the same.

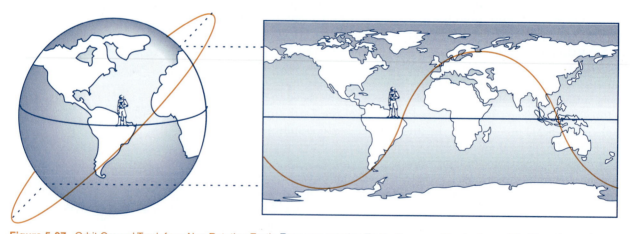

Figure 5-27. Orbit Ground Track for a Non-Rotating Earth. For a non-rotating Earth, the ground track of an orbit will continuously repeat.

But suppose we start the Earth rotating again. What happens? If you're standing on the ground watching the satellite pass directly overhead on one orbit, something different happens on the next orbit. The satellite will appear to pass to the *west* of you! How can this be? Because the orbit plane is fixed in inertial space, the satellite will always return to the same point in inertial space. But you're fixed to the Earth as it rotates to the east. So as the satellite moves around in its orbit, you move away from the orbit, making it look as if the satellite moved, as seen in Figure 5-28.

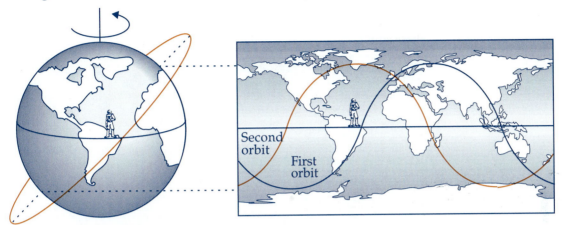

Figure 5-28. A Normal Satellite Ground Track. As the Earth rotates, successive ground tracks appear to shift to the west from an Earth-based observer's viewpoint.

Can we learn something about the orbit from all of this? Sure! Because the Earth rotates at a fixed rate of about 15° per hr (360° in 24 hrs = 15°/hr) or 0.25°/min., we can use this rotation as a "clock" to tell us the orbit's period. All we have to do is measure how much the orbit's ground track moves to the west from one orbit to the next. We can quantify this movement using a new parameter, *node displacement, ΔN*, which is measured along the equator from one ascending node to the next. ΔN is defined to be positive in the direction of satellite motion. Thus, the shift over one orbit is the difference between 360° and ΔN.

We can put this node displacement to work in finding the period because the shift is simply the Earth's rotation rate times the period of the orbit. For example, if the period of the orbit were two hours, the Earth would rotate 30° (2 hr × 15°/hr) during one orbit revolution, producing a node displacement of 330° (360° – 30°). In terms of ΔN, the period can be found from

$$\text{Period (hours)} \; = \; \frac{360° - \Delta N}{15°/\text{hr}} \; \text{(for direct orbits)} \qquad (5\text{-}12)$$

[Note: As is, this equation applies only to direct orbits with period less than 24 hours. For retrograde and super-synchronous orbits, the same concept applies but the equation would change.] If we can determine the period, we can also determine the semi-major axis of the orbit using the equation for orbit period from Chapter 4.

$$P \; = \; 2\pi \sqrt{\frac{a^3}{\mu}} \qquad (5\text{-}13)$$

where
P = period (s)
a = semi-major axis (km)
μ = gravitational parameter (km^3/s^2)

So, by finding ΔN from the ground track, we can find the semi-major axis. For example, in the ground track in Figure 5-29, ΔN is 315°. We can find the orbit period using Equation (5-12) and the semi-major axis using Equation (5-13). But **be careful!** You must watch your units when using these equations.

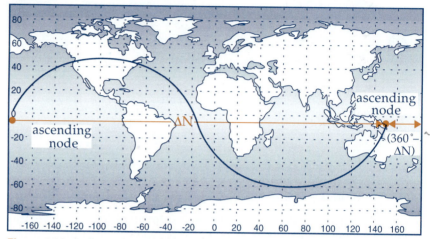

Figure 5-29. Ascending Node Shift Due to Rotating Earth. ΔN is measured along the equator from one ascending node to the next and is positive in the direction of satellite motion. Thus, 360° − ΔN represents the amount the Earth rotates during one orbit.

As the orbit size increases, the semi-major axis gets bigger, so ΔN gets smaller. This happens because the satellite takes longer to make one revolution as the Earth rotates beneath it (the bigger the semi-major axis, a, the longer the period). As the orbit size gets bigger and bigger, the ΔN gets smaller and smaller, so the ground tracks appear to compress or "scrunch" together. Recall, we define a geosynchronous orbit as one with a period of exactly 24 hours. For such an orbit, the ΔN is 0°. This means the satellite period exactly matches the rotational period of the Earth. Thus, the orbit appears to come back on itself and form a figure 8 as shown in Figure 5-30,

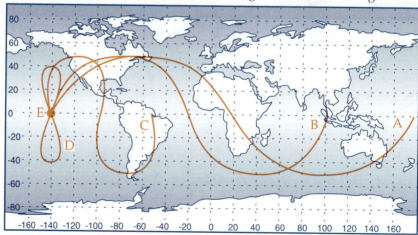

Figure 5-30. Orbit Ground Tracks. Orbit A: Period of 2.67 hours. Orbit B: Period of 8 hours. Orbit C: Period of 18 hours. Orbit D: Period of 24 hours. Orbit E: Period of 24 hours.

orbit D. If the orbit lies in the plane of the equator (that is, it has an inclination angle of 0°), the ground track would be just a dot on the equator, orbit E, in Figure 5-30. A satellite with a period of 24 hours and an inclination of 0° is said to be in a geostationary orbit. This means the satellite appears to be stationary to Earth-based observers which makes these orbits very useful for communication satellites. Once you point the ground receiving antenna, you don't have to move it as the Earth rotates.

Besides determining the semi-major axis, we can also find the orbit's inclination by looking at the ground track. Imagine a satellite in a 50° inclined orbit. From our definition of inclination, we know in this case the angle between the equatorial plane and the orbit plane would be 50°. What's the highest latitude the spacecraft will pass directly over? 50°! The highest latitude any satellite passes over is equal to its inclination. Let's see why. Remember that latitude is the Earth-centered angle measured from the equator north or south to the point in question. But the orbit plane also passes through the center of the Earth, and the angle it forms with the equatorial plane is the inclination, as you can see in Figure 5-31. Thus, for direct or prograde orbits, if you're on the satellite and you look straight down when it reaches the northernmost point in the orbit, you'll see the latitude line equal to your inclination.

In this way, we can use the ground track to tell us the orbit's inclination.

- For a direct orbit (0 < i < 90°), all we have to do is look at the "crest of the wave" of the orbit ground track and read off the latitude. This latitude will be equal to the orbit inclination.

- For a retrograde orbit (90 < i < 180°), we subtract the maximum latitude from 180°.

The Earth coverage a satellite requires will affect how we select the orbit's inclination. For example, if a remote-sensing satellite needs to be able to view the entire surface of the Earth at some time during the mission, it needs an inclination of near 90°. In Figure 5-32 we see several satellites with the same period but with varying inclinations.

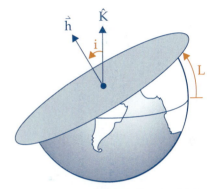

Figure 5-31. Inclination Equals Highest Latitude, L. Because inclination relates the angle between the orbit plane and the equator, the highest latitude reached by a satellite orbit is equal to its inclination (for direct orbits).

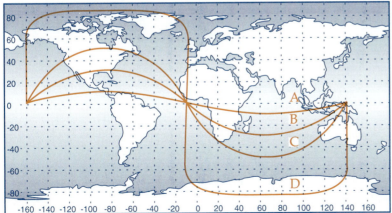

Figure 5-32. Changing Inclination. All four ground tracks represent orbits with a period of 4 hours. We can find the inclination of these orbits by looking at the highest latitude reached. Orbit A has an inclination of 10°. Orbit B has an inclination of 30°. Orbit C has an inclination of 50°. Orbit D has an inclination of 85°.

So far we've looked only at circular orbits. Now let's look at how eccentricity and the location of perigee affect the shape of the ground track. If an orbit is circular, its ground track will be symmetrical. If an orbit is not circular, its ground track will be lopsided. That is, it will not look the same above and below the equator. Remember, a satellite moves fastest at perigee, so it will cover the most real estate near perigee and make the ground track look spread out. But, near apogee it's going slower, so the ground track will be more scrunched up. You can see this in the two ground tracks in Figure 5-33. Orbit A has perigee in the Northern Hemisphere; orbit B has perigee in the Southern Hemisphere.

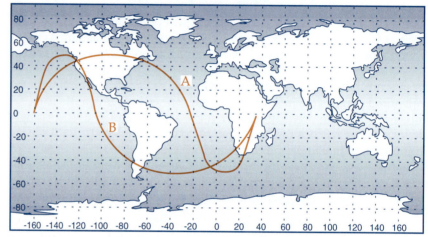

Figure 5-33. Changing Perigee Location. Both ground tracks represent orbits with periods of 11.3 hours and inclinations of 50°. Both orbits are highly eccentric. Orbit A has perigee over the Northern Hemisphere. Orbit B has perigee over the Southern Hemisphere.

▰ Section Review

Key Terms

great circle
node displacement, ΔN

Key Concepts

➤ A ground track is the path a satellite traces on the Earth's surface as it orbits. Because a satellite orbits around the center of the Earth, the orbit plane slices through the center, so the ground track is a great circle.

➤ When the spherically shaped Earth is spread out on a two-dimensional, Mercator-projection map, the orbit ground track resembles a sine wave for orbits with periods less than 24 hours

➤ Because orbit planes are fixed in inertial space and the Earth rotates beneath them, ground tracks appear to shift westward during successive orbits

➤ From the ground track, you can find many interesting things:
  • The orbit period—by measuring the westward shift of the ground track
  • The inclination of a satellite's orbit—by looking at the highest latitude reached on the ground track (for direct orbits)
  • Eccentricity of the orbit—by noting that nearly circular orbits appear to be symmetric whereas eccentric orbits appear lopsided
  • The location of perigee—by looking at the point where the ground track is spread out the most

References

Bate, Roger R., Donald D. Mueller, and Jerry E. White. *Fundamentals of Astrodynamics*. New York, NY: Dover Publications, 1971.

Wertz, James R. and Wiley J. Larson. *Space Mission Analysis and Design*. Netherlands: Kluwer Academic Publishers, 1991.

Mission Problems

5.1   Orbital Elements

1   Why do we prefer classical orbital elements over a set of $\vec{R}$ and $\vec{V}$ vectors for describing an orbit?

2   How many initial conditions (ICs) are needed for solving the two-body equation of motion? Give an example of one set of ICs.

3   If a satellite has a high specific mechanical energy, what does this tell us about the size of the orbit? Why?

4   What is the specific mechanical energy, ε, of an orbit with a semi-major axis of 42,160 km?

5   What four things do classical orbital elements (COEs) tell us about a satellite orbit?

6   What are the six COEs?

7   What are the two major categories of inclination relative to the Earth's motion?

8   How can we look only at the $\vec{R}$ and $\vec{V}$ vectors and tell if i = 0. What about i = 90° or 180°?

9   An Atlas IV launches a satellite from Vandenburg AFB (34.6° N latitude, 120.6° W longitude). What's the most northerly point (latitude) the satellite can view directly below it?

5.2   Computing Orbital Elements

10   After maneuvering, the satellite from the previous question has these orbital elements

semi-major axis = 8930 km
eccentricity = 0.3
inclination = 53°
longitude of the ascending node = 165°
argument of perigee = 90°
true anomaly = 90°

a)   Sketch a picture of the satellite's orbit

b)   Is this a direct or retrograde orbit?

c)   Where is the satellite located in its orbit?

d)   Is perigee in the Northern or Southern Hemisphere?

e)   Is this a circular, elliptical, parabolic, or hyperbolic orbit?

11   Discuss the use of inclination for different satellite missions.

12   Why don't we use vectors in the orbit and equatorial planes to measure inclination?

13 Complete the following table:

	a	e	ε
Circle			
Ellipse			
Parabola			
Hyperbola			

14 In general, how do we measure the additional orbital elements when some of the reference vectors are undefined?

15 As a program manager for a major corporation, determine which classical orbital elements (COEs) describe these various orbits

a) Circular, $i > 0°$

b) Equatorial, $0 < e < 1$

c) Circular-equatorial

16 A satellite has these orbital elements

semi-major axis = 5740 km
eccentricity = 0.1
inclination = 53°
longitude of the ascending node = 345°
argument of perigee = 270°
true anomaly = 183°

What is peculiar about this orbit?

17 A ground-based tracking station observes that a new Russian satellite has the following $\vec{R}$ and $\vec{V}$ vectors:

$$\vec{R} = 7016\ \hat{I} + 5740\ \hat{J} + 638\ \hat{K}\ km$$

$$\vec{V} = 0.24\ \hat{I} - 0.79\ \hat{J} - 7.11\ \hat{K}\ km/s$$

a) What is the specific angular momentum of the satellite?

b) What is the satellite's inclination?

c) Calculate the ascending node vector.

d) What is the satellite's longitude of the ascending node?

18 The above satellite was supposedly put into a Molniya type orbit.

a) Compute the eccentricity vector.

b) What is the satellite's argument of perigee? Is this the value needed for a Molniya orbit? See Table 5-4.

c) Where is the satellite? In other words, what is its true anomaly?

5.3 Satellite Ground Tracks

19 Given a non-rotating Earth, if the inclination stays the same but the orbit size increases or decreases, does the ground track change? Why or why not? Describe what, if anything, happens to the ground track.

20 Given a rotating Earth, if the inclination stays the same but the orbit size increases or decreases, does the ground track change? Why or why not? Describe what, if anything, happens to the ground track.

21 What type of ground track does a geostationary satellite have? How about a geosynchronous satellite?

22 Can we "hang" a reconnaissance satellite over Moscow? Why or why not?

23 Sketch the ground track of a satellite with the following elements:

period = 480 min.
eccentricity = 0.0
inclination = 25°

a) What is the longitude shift of the ground track?

b) What is the highest and lowest latitude that the satellite's ground track will reach?

c) What shape will the ground track have? Will it be symmetrical?

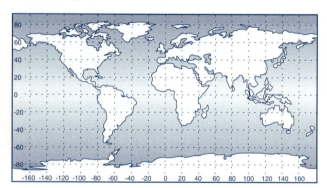

24 Given the ground track below,

a) Identify the inclination

b) Determine the longitude shift and then compute the period of the orbit

c) Suggest a possible mission(s) for the satellite

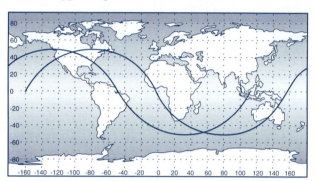

Mission Profile—Skylab

Soon after the first successful lunar landing, NASA realized that without clear goals for its manned space program, Congress and the public would quickly lose interest in funding future spaceflights. NASA proposed the Apollo Applications Program (AAP), which would use surplus Apollo hardware to bridge the gap between Apollo, and the proposed shuttle program, and a permanent space station. After cuts to Apollo (from ten lunar landings to six), the AAP became Skylab.

Mission Overview

Skylab was a prefabricated space station launched on a Saturn V. Three crews conducted long-duration manned missions to the station that included solar and Earth science experiments. Future shuttle missions were to use it until a permanent station was built. Unfortunately, Skylab reentered the atmosphere on July 11, 1979—two years before the Space Shuttle's first flight.

Mission Data

✓ Skylab 1, May 14, 1973: Launch of Skylab station atop a Saturn V. Damaged during launch.

✓ Skylab 2, May 25, 1973 (Crew: Conrad, Kerwin, Weitz): Mission lasted for 28 days and included extensive extra-vehicular activity to repair damage to the station, most notably to install a new heat shield.

✓ Skylab 3, July 29, 1973, (Crew: Bean, Garriott, Lousma): Mission lasted for 59 days. Emphasis was on solar, Earth, and life sciences.

✓ Skylab 4, November 16, 1973 (Crew: Carr, Gibson, Pogue): Mission lasted for 84 days (U.S. record). Observations of comet Kahoutek were a notable addition to the mission.

Mission Impact

Skylab was valuable because it was the United States' first space station. It discovered the effects of long spaceflights on astronauts, as well as valuable scientific data about the Earth and Sun. (Ironically, sunspot activity contributed to Skylab's orbital decay before it could be rescued).

Skylab orbiting the Earth. *(Courtesy of NASA)*

For Discussion

- What did we learn from Skylab?
- How will current proposals for a space station differ from Skylab?

Contributor

Todd Lovell, United States Air Force Academy

References

Baker, David. *The History of Manned Spaceflight.* New York: Crown, 1981.

Yenne, Bill. *The Encyclopedia of U.S. Spacecraft.* New York: Exeter, 1985.

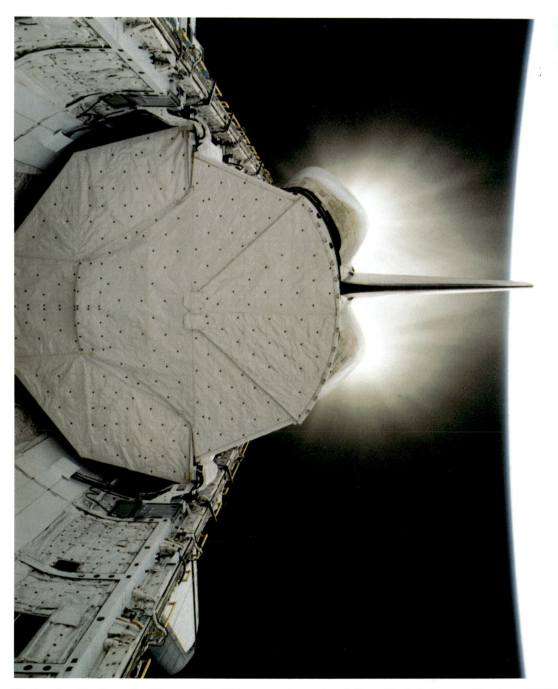

The flare from the Space Shuttle's powerful orbital maneuvering engines lights up the dark of space. *(Courtesy of NASA)*

Maneuvering In Space

6

In This Chapter You'll Learn to...

☛ Explain the most energy-efficient means of transferring between two orbits—the Hohmann Transfer

☛ Determine the velocity change (ΔV) needed to perform a Hohmann Transfer between two orbits

☛ Explain plane changes and how to determine the required ΔV

☛ Explain orbital rendezvous and how to determine the required ΔV and wait time needed to start one

You Should Already Know...

❑ Basic orbit concepts (Chapter 4)

❑ Classical orbital elements (Chapter 5)

Space...is big. Really big. You just won't believe how vastly, hugely, mind-boggling big it is. I mean, you may think it's a long way down the road to the chemist's [druggist's], but that's just peanuts to space.

Douglas Adams
The Hitch-hiker's Guide to the Galaxy
1979

A satellite is seldom happy with the orbit it's in. On nearly every space mission, we must change one or more of the classical orbital elements. Communication satellites, for instance, never directly assume their geostationary positions. They first go into a low-altitude (300 km or so) "parking orbit" before transferring out to geosynchronous altitude (about 35,000 km). At the same time this large change in semi-major axis occurs, another maneuver takes the satellite's inclination from that of the parking orbit to 0°. On other missions additional maneuvers are required to bring two satellites together in orbit as was done when the Apollo lunar module docked with the command module.

Unfortunately, as we'll see in this chapter, these orbital maneuvers aren't as simple as "motor boating" from one point to another. Because we're usually in the gravitational field of some central body (like the Earth or the Sun), we have to be concerned with orbital motion in getting from one place to another. In this chapter we'll use our understanding of the two-body problem to see what it takes to get around in space. We'll discover the most economical way to move from one orbit to another, find out how and when to change our orbit plane, and finally, describe the intricate ballet needed to bring spacecraft together in orbit.

6.1 Hohmann Transfers

In This Section You'll Learn to...

- ☞ Describe the steps in the Hohmann Transfer, the most efficient way to get from one orbit to another orbit in the same plane

- ☞ Determine the velocity change (ΔV) needed to complete a Hohmann Transfer

One of the very first problems faced by space-mission designers was figuring out how to go from one orbit to another. Refining this process for eventual missions to the Moon was one of the objectives of the Gemini program in the 1960s, shown in Figure 6-1. Let's say we're in one orbit and we want to go to another orbit. We'll assume for the moment that the initial and final orbits are in the same plane to keep things simple. Such coplanar maneuvers are often used to move satellites from their initial parking orbits to their final mission orbits. Because fuel is critical for all orbital maneuvers, let's look at the most fuel-efficient way to do this, known as the Hohmann Transfer.

Figure 6-1. The Gemini Program. During the Gemini program in the 1960s, NASA developed the procedures for all of the orbital maneuvers we'll discuss in this chapter, including orbit transfers, plane changes, and rendezvous. *(Courtesy of NASA)*

In 1925 a German engineer, Walter Hohmann, theorized a fuel-efficient way to transfer between orbits. (It's amazing someone was thinking about this considering satellites didn't exist at the time.) This method, called the *Hohmann Transfer*, uses an elliptical transfer orbit tangent to both the initial and final orbits.

To better understand this idea, let's pretend you're driving a fast car around a racetrack, as shown in Figure 6-2. The effort needed to exit the track depends on the off-ramp's location and orientation. For instance, if the off-ramp is on the same level as the racetrack (in the same plane) and tangent to it, your exit will be quite easy—all you have to do is straighten out the wheel. But if the off-ramp is perpendicular to the track (even though it's in the same plane), you'll have to slow down a lot, and maybe even stop, to negotiate the turn. Why the difference? With the tangential exit you have to change only the *magnitude* of your velocity, so you just hit the brakes. With the perpendicular exit, you have to change both the *magnitude* and *direction* of your velocity. This is hard to do at high speed without rolling your car!

Figure 6-2. Maneuvering. One way to think about maneuvering in space is to imagine driving around a racetrack. It takes more effort to exit at a sharp turn than to exit tangentially.

The Hohmann Transfer applies this simple racetrack example to a cosmic scale. By sticking to orbital "on/off-ramps" tangential to our initial and final orbits, we get from here to there as cheaply as possible. For rocket scientists, "cheap" means saving weight, which in this case is fuel.

By definition, Hohmann Transfers are limited to orbits in the same plane, called *co-planer orbits*, and those with their major axes (line of apsides) aligned, called *co-apsidal orbits*. Technically, then, we could do a Hohmann Transfer between two elliptical orbits. But, to keep things simple, we'll always assume we start and finish in circular orbits. In addition, all velocity changes, called *delta-Vs* or ΔVs, during the transfer are tangent to the initial and final orbits. Thus, the velocity vector changes magnitude but not direction. If you're flying the spaceship, this means you want to fire your thrusters or "burn" in a direction parallel to your velocity vector. As we've seen, these tangential ΔVs are the real secret to the economy of the Hohmann Transfer.

Now let's look at what these velocity changes are doing to the orbit. By assuming all ΔVs occur nearly instantaneously (sometimes called an "impulsive burn"), we can continue to use the results from the two-body problem we developed in Chapter 4 to help us here. (Otherwise, we'd have to integrate the thrust over time and that would be too messy.) Whenever we add or subtract velocity, we change the orbit's specific mechanical energy, ε, and hence its size, or semi-major axis, a. Remember these quantities are related by

$$\varepsilon = -\frac{\mu}{2a}$$
(6-1)

where

ε = specific mechanical energy (km^2/s^2)
μ = gravitational parameter = 3.986005×10^5 (km^3/s^2) for Earth
a = semi-major axis (km)

If we want to move a satellite to a higher orbit, we have to increase the semi-major axis. To do so, we need to add energy to the orbit by increasing the satellite's speed. On the other hand, if we want to move the satellite to a lower orbit, we decrease the semi-major axis by decreasing the velocity and thus the energy.

So now we know that to go from one orbit to another we must change our velocity by some specific amount. But by how much? Before we answer this question, let's step back and look at the big picture.

We want to go from one orbit (orbit 1) to another (orbit 2). We must follow some path connecting the two orbits called a *transfer orbit*, shown in Figure 6-3. To get from orbit 1 to the transfer orbit, we need to change the energy. Then, when the transfer orbit gets us to orbit 2, we must change our energy again. As a result, the complete maneuver taking us from orbit 1 to orbit 2 requires two separate velocity changes (ΔV_1 and ΔV_2). If you don't do ΔV_2, you'll be stuck on the transfer orbit and return right back to where you started from at orbit 1.

Any ΔV represents a change from the velocity you have to the velocity you want. For a tangential burn, we can write this

$$\Delta V = \left| V_{want} - V_{have} \right|$$

Notice we normally take the absolute value of this difference because we want to know the amount of energy or fuel consumed rather than the direction of the ΔV. We could be going from orbit 1 to orbit 2 or from orbit 2 to orbit 1 and still use the same amount of fuel. In this case we'll assume ΔV_1 is the change in velocity which will take us from orbit 1 onto the transfer orbit.

$$\Delta V_1 = \left| V_{transfer\ at\ orbit\ 1} - V_{orbit\ 1} \right|$$

where
ΔV_1 = ΔV to move from orbit 1 onto transfer orbit (km/s)
$V_{transfer\ at\ orbit\ 1}$ = velocity on transfer orbit at orbit 1 radius (km/s)
$V_{orbit\ 1}$ = velocity in orbit 1 (km/s)

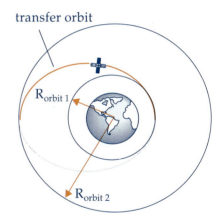

Figure 6-3. Getting From One Orbit to Another. The problem in orbital maneuvering is getting from orbit 1 to orbit 2. Here we see a satellite moving from a lower orbit to a higher orbit on a transfer orbit. If a second ΔV is not done when the satellite reaches orbit 2, it will remain in the transfer orbit.

ΔV_2 is the change to get us from the transfer orbit onto orbit 2. Both of these ΔVs are shown in Figure 6-4.

$$\Delta V_2 = \left| V_{orbit\ 2} - V_{transfer\ at\ orbit\ 2} \right|$$

where

$\Delta V_2 = \Delta V$ to move from transfer orbit to orbit 2 (km/s)

As a mission planner budgeting all the fuel needed for the trip from orbit 1 to orbit 2, you add the ΔV from each burn to get a total ΔV.

$$\Delta V_{total} = \Delta V_1 + \Delta V_2 \tag{6-2}$$

where

ΔV_{total} = total ΔV needed for entire transfer (km/s)

When we cover the rocket equation in Chapter 14, we'll see how to convert this number into the amount of fuel required.

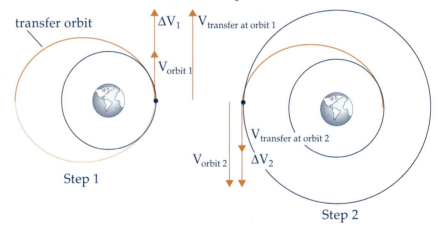

Figure 6-4. Hohmann Transfer. Step 1: The first burn or ΔV of a Hohmann Transfer takes you out of your first circular orbit and puts you on an elliptical transfer orbit. Step 2: The second burn of a Hohmann Transfer takes you from the transfer orbit and places you in the desired final, circular orbit.

Now that we've seen the big picture, we can determine how big ΔV_{total} is. Again, we return to our earlier discussion of orbital mechanics. Because we're trying to find orbital velocities, we start by digging out our trusty energy equations. As it turns out, everything we need to know to solve our orbital-maneuvering problem comes from these two humble relationships, as we'll see in Example 6-1.

$$\boxed{\varepsilon = \frac{V^2}{2} - \frac{\mu}{R}} \tag{6-3}$$

where

ε = specific mechanical energy (km²/s²)
V = velocity (km/s)
μ = gravitational parameter (km³/s²)
R = position (km)

$$\boxed{\varepsilon = \frac{-\mu}{2a}} \qquad (6\text{-}4)$$

where

ε = specific mechanical energy (km^2/s^2)

μ = gravitational parameter (km^3/s^2)

a = semi-major axis (km)

Let's review the steps in the transfer process to see how all this fits together. Referring to Figure 6-4,

- Step 1: ΔV_1 takes you from orbit 1 and puts you on the transfer orbit

- Step 2: ΔV_2 occurs when the transfer orbit is tangent to orbit 2, and puts you onto orbit 2

To solve for these ΔVs, you'll need to find the energy in each orbit. The semi-major axis, a, of both orbit 1 and orbit 2 are known and, for circular orbits, are equal to the orbit radius. The major axis of the transfer orbit is equal to the sum of the two orbit radii, as shown in Figure 6-5.

The Hohmann Transfer is energy efficient, but it can take a long time. To find the time of flight, look at the diagram of the maneuver. The transfer covers exactly one half of an ellipse. Recall the total period for any closed orbit is found by:

$$P = 2\pi\sqrt{\frac{a^3}{\mu}} \qquad (6\text{-}5)$$

So, our transfer orbit time of flight (TOF) will be half of the period

$$\boxed{TOF = \frac{P}{2} = \pi\sqrt{\frac{a_{transfer}^3}{\mu}}} \qquad (6\text{-}6)$$

where

TOF = time of flight (s)

P   = period (s)

a   = semi-major axis of the transfer orbit (km)

μ = gravitational parameter (km^3/s^2)

Example 6-1 shows how to find time of flight for a Hohmann Transfer.

Now that we've gone through the Hohmann Transfer, let's step back to see what went on here. In the example, we've gone from a low orbit to a higher orbit. To do this, we had to speed up twice: ΔV_1 and ΔV_2. But notice the velocity in the higher circular orbit is lower than the velocity in the lower circular orbit. Thus, we sped up twice and ended up slowing down! Does this make sense?

ΔV_1 increases our velocity, taking us out of orbit 1 and putting us on the transfer orbit. On the transfer orbit, our velocity gradually decreases as we gain altitude and trade kinetic energy for potential, just as a baseball thrown into the air slows down as it gets higher. When the spacecraft reaches the altitude of orbit 2, the velocity increases again, with ΔV_2

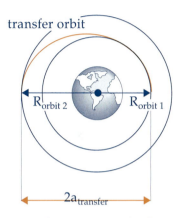

Figure 6-5. Size of the Transfer Orbit. The major axis of the transfer orbit is equal to the sum of the radii of the initial and final orbits.

allowing us to enter the final orbit. Even though the velocity in orbit 2 is lower than in orbit 1, the *total energy* is higher because it's at a higher altitude. Remember, energy is the sum of kinetic and potential energy. Thus, we use our engines to add kinetic energy, which allows us to gain potential energy. In the end, we have higher total energy.

▬ Section Review

Key Terms

co-apsidal orbits
co-planer orbits
delta-Vs, ΔV
Hohmann Transfer
magnitude
transfer orbit
total energy

Key Equations

$$\varepsilon = -\frac{\mu}{2a}$$

$$\varepsilon = \frac{V^2}{2} - \frac{\mu}{R}$$

$$TOF = \frac{P}{2} = \pi\sqrt{\frac{a^3}{\mu}}$$

Key Concepts

➤ The Hohmann Transfer moves a spacecraft from one orbit to another in the same plane. It's the simplest kind of orbital maneuver because it focuses only on changing the satellite's specific mechanical energy.

➤ The Hohmann Transfer is the cheapest way to get from one orbit to another. It's based on these assumptions
  • Initial and final orbits are in the same plane (coplanar)
  • Velocity changes (ΔVs) are aimed tangential to the initial and final orbits. Thus, the orbit velocity changes size but not direction
  • ΔVs occur instantaneously—impulsive burns

➤ The Hohmann Transfer consists of two separate ΔVs
  • The first, ΔV_1, moves the spacecraft from its initial orbit into an elliptical transfer orbit
  • The second, ΔV_2, moves the spacecraft from the elliptical transfer orbit into the final orbit

Example 6-1

Problem Statement

NASA wants to place a communications satellite into a geosynchronous orbit from a low-Earth parking orbit.

$R_{orbit\ 1}$ = 6570 km

$R_{orbit\ 2}$ = 42,160 km

What is the ΔV for this transfer and how long will it take?

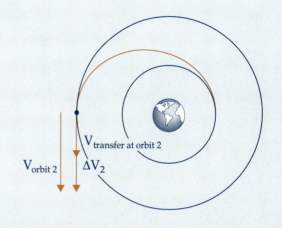

Problem Summary

Given: $R_{orbit\ 1}$ = 6570 km

$R_{orbit\ 2}$ = 42,160 km

Find: ΔV_{total} and TOF

Problem Diagram

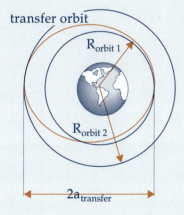

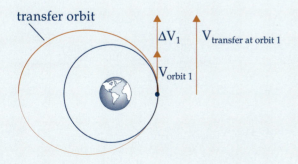

1) Compute the semi-major axis of the transfer orbit

$$a_{transfer} = \frac{R_{orbit\ 1} + R_{orbit\ 2}}{2}$$

2) Solve for the specific mechanical energy of the transfer orbit

$$\varepsilon_{transfer} = \frac{-\mu}{2a_{transfer}}$$

3) Solve for the energy and velocity in orbit 1

$$\varepsilon_{orbit\ 1} = \frac{-\mu}{2a_{orbit\ 1}}$$

$$a_{orbit\ 1} = R_{orbit\ 1} \text{ (circular orbit)}$$

$$\varepsilon = \frac{V^2}{2} - \frac{\mu}{R}$$

$$\therefore V_{orbit\ 1} = \sqrt{2\left(\frac{\mu}{R_{orbit\ 1}} + \varepsilon_{orbit\ 1}\right)}$$

4) Solve for $V_{transfer\ at\ orbit\ 1}$

$$V_{transfer\ at\ orbit\ 1} = \sqrt{2\left(\frac{\mu}{R_{orbit\ 1}} + \varepsilon_{transfer}\right)}$$

5) Find ΔV_1

$$\Delta V_1 = \left|V_{transfer\ at\ orbit\ 1} - V_{orbit\ 1}\right|$$

Example 6-1 (Continued)

6) Solve for $V_{\text{transfer at orbit 2}}$

$$V_{\text{transfer at orbit 2}} = \sqrt{2\left(\frac{\mu}{R_{\text{orbit 2}}} + \varepsilon_{\text{transfer}}\right)}$$

7) Solve for the energy and velocity in orbit 2

$$\varepsilon_{\text{orbit 2}} = \frac{-\mu}{2a_{\text{orbit 2}}}$$

$$a_{\text{orbit 2}} = R_{\text{orbit 2}} \text{ (circular orbit)}$$

$$\varepsilon = \frac{V^2}{2} - \frac{\mu}{R}$$

$$\therefore V_{\text{orbit 2}} = \sqrt{2\left(\frac{\mu}{R_{\text{orbit 2}}} + \varepsilon_{\text{orbit 2}}\right)}$$

8) Find ΔV_2

$$\Delta V_2 = \left|V_{\text{orbit 2}} - V_{\text{transfer at orbit 2}}\right|$$

9) Solve for ΔV_{total}

$$\Delta V_{\text{total}} = \Delta V_1 + \Delta V_2$$

10) Compute TOF

$$\text{TOF} = \pi\sqrt{\frac{a_{\text{transfer}}^3}{\mu}}$$

Analytical Solution

1) Compute the semi-major axis of the transfer orbit

$$a_{\text{transfer}} = \frac{R_{\text{orbit 1}} + R_{\text{orbit 2}}}{2} = \frac{6570\,\text{km} + 42{,}160\,\text{km}}{2}$$

$$a_{\text{transfer}} = 24{,}365\,\text{km}$$

2) Solve for the specific mechanical energy of the transfer orbit

$$\varepsilon_{\text{transfer}} = \frac{-\mu}{2a_{\text{transfer}}} = \frac{-3.986005 \times 10^5 \dfrac{\text{km}^3}{\text{s}^2}}{2\,(24{,}365\,\text{km})}$$

$$\varepsilon_{\text{transfer}} = -8.1798\frac{\text{km}^2}{\text{s}^2}$$

(Note the energy is negative, which implies the transfer orbit is an ellipse; that's what we'd expect.)

3) Solve for energy and velocity of orbit 1.

$$\varepsilon_{\text{orbit 1}} = \frac{-\mu}{2a_{\text{orbit 1}}}$$

$$= \frac{-3.986005 \times 10^5 \dfrac{\text{km}^3}{\text{s}^2}}{2\,(6570\,\text{km})} = -30.33\frac{\text{km}^2}{\text{s}^2}$$

$$V_{\text{orbit 1}} = \sqrt{2\left(\frac{\mu}{R_{\text{orbit 1}}} + \varepsilon_{\text{orbit 1}}\right)}$$

$$\sqrt{2\left(\frac{3.986005 \times 10^5 \dfrac{\text{km}^3}{\text{s}^2}}{6570\,\text{km}} - 30.33\frac{\text{km}^2}{\text{s}^2}\right)} = 7.789\frac{\text{km}}{\text{s}}$$

4) Solve for $V_{\text{transfer at orbit 1}}$

$$V_{\text{transfer at orbit 1}} = \sqrt{2\left(\frac{\mu}{R_{\text{orbit 1}}} + \varepsilon_{\text{transfer}}\right)}$$

$$\sqrt{2\left(\frac{3.986005 \times 10^5 \dfrac{\text{km}^3}{\text{s}^2}}{6570\,\text{km}} - 8.1798\frac{\text{km}^2}{\text{s}^2}\right)}$$

$$V_{\text{transfer at orbit 1}} = 10.246\frac{\text{km}}{\text{s}}$$

5) Find ΔV_1

$$\Delta V_1 = \left|V_{\text{transfer at orbit 1}} - V_{\text{orbit 1}}\right|$$

$$\left|10.246\frac{\text{km}}{\text{s}} - 7.789\frac{\text{km}}{\text{s}}\right|$$

$$\Delta V_1 = 2.457\frac{\text{km}}{\text{s}}$$

6) Solve for $V_{\text{transfer at orbit 2}}$

$$V_{\text{transfer at orbit 2}} = \sqrt{2\left(\frac{\mu}{R_{\text{orbit 2}}} + \varepsilon_{\text{transfer}}\right)}$$

Example 6-1 (Continued)

$$\sqrt{2\left(\cfrac{3.986005 \times 10^5 \cfrac{km^3}{s^2}}{42,160\ km} - 8.1798 \cfrac{km^2}{s^2}\right)}$$

$$V_{\text{transfer at orbit 2}} = 1.597 \frac{km}{s}$$

7) Solve for energy and velocity in orbit 2

$$\varepsilon_{\text{orbit 2}} = \frac{-\mu}{2a_{\text{orbit 2}}}$$

$$= \frac{-3.986005 \times 10^5 \cfrac{km^3}{s^2}}{2\,(42,160\ km)} = -4.727 \frac{km^2}{s^2}$$

$$V_{\text{orbit 2}} = \sqrt{2\left(\frac{\mu}{R_{\text{orbit 2}}} + \varepsilon_{\text{orbit 2}}\right)}$$

$$\sqrt{2\left(\cfrac{3.986005 \times 10^5 \cfrac{km^3}{s^2}}{42,160\ km} - 4.727 \cfrac{km^2}{s^2}\right)} = 3.075 \frac{km}{s}$$

8) Find ΔV_2

$$\Delta V_2 = \left| V_{\text{orbit 2}} - V_{\text{transfer at orbit 2}} \right|$$

$$\left| 3.075 \frac{km}{s} - 1.597 \frac{km}{s} \right|$$

$$\Delta V_2 = 1.478 \frac{km}{s}$$

9) Solve for ΔV_{total}

$$\Delta V_{\text{total}} = \Delta V_1 + \Delta V_2 = 2.457 \frac{km}{s} + 1.478 \frac{km}{s}$$

$$= 3.935 \frac{km}{s}$$

10) Compute TOF

$$TOF = \pi \sqrt{\frac{a_{\text{transfer}}^3}{\mu}}$$

$$= \pi \sqrt{\cfrac{(24,365\ km)^3}{3.986005 \times 10^5 \cfrac{km^3}{s^2}}}$$

$$TOF = 18,925\ s \cong 315\ \text{min} = 5\ \text{hrs}\ 15\ \text{min}$$

Interpreting the Results

To move the communication satellite from its low-altitude (192 km) parking orbit out to geosynchronous altitude, the engines must provide a total velocity change of about 4 km/s (about 8800 m.p.h.). The transfer will take more than five hours to complete.

6.2 Plane Changes

▰ In This Section You'll Learn to...

- ☞ Explain when to use a simple plane change and how a simple plane change can modify an orbit plane

- ☞ Explain how to use a plane change combined with a Hohmann Transfer to efficiently change an orbit's size and orientation

- ☞ Determine the ΔV needed for simple and combined plane changes

So far we've seen how to change the size of an orbit using a Hohmann Transfer. However, this transfer is confined to coplanar orbits. To change the orbit plane, we must point our velocity changes (ΔVs) out of plane. By changing the orbit plane, we also alter the orbit's tilt (inclination, i) or its swivel (longitude of ascending node, Ω), depending on where in the orbit the ΔV is done. For plane changes, we must consider the direction and magnitude of our initial and final velocities.

To understand plane changes let's return to our racetrack example from the last section. What if the off-ramp were like those on a freeway—causing you not only to change your velocity *within* the plane of the racetrack but also to go *above* or *below* the level of the track? This "out of plane" maneuver causes you to use even more energy because you now have to speed up to make it up the ramp or brake as you go down. As a result, out-of-plane maneuvers require considerably more energy than in-plane maneuvers.

We'll look at two types of plane changes—simple and combined. The difference between the two depends on what is changing. With a *simple plane change* only the velocity vector's direction changes. The *combined plane change* alters the vector's direction and magnitude. We'll take on the simple plane change first.

Simple Plane Change

Let's imagine we have a spacecraft in an orbit with an inclination of 28.5°. (As we'll see in Chapter 9, this is the inclination you'd get if you launched due East out of the Kennedy Space Center, as the Shuttle often does.) Assume you want to change into an equatorial orbit (i = 0°). Once again we have a situation where we're in one orbit and we want to change to another. As before, we must change our velocity to do this, but we want to change the orbit's orientation and not its size. This means the velocity vector's magnitude stays the same, that is $\left|\vec{V}_{initial}\right| = \left|\vec{V}_{final}\right|$, but its direction must change.

How do we change just the direction of our velocity vector? Look at the situation in Figure 6-6. You can see we have an initial inclined orbit with a velocity $\vec{V}_{initial}$ and we want to rotate the orbit by an angle θ to reach a final

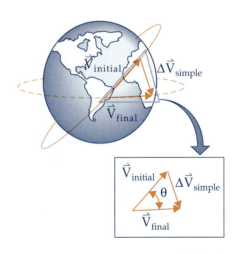

Figure 6-6. Simple Plane Change. A simple plane change affects only the direction and not the magnitude of the original velocity vector.

velocity $\vec{V}_{final}$. The vector triangle shown in Figure 6-6 summarizes this problem. It's an isosceles triangle (meaning two sides of equal length). Using a little geometry we can get a relationship for ΔV_{simple}—the change in velocity needed to rotate our plane.

$$\Delta V_{simple} = 2\, V_{initial} \sin\left(\frac{\theta}{2}\right) \qquad (6\text{-}7)$$

where
ΔV_{simple} = ΔV for simple plane change (km/s)
$V_{initial} = V_{final}$ = velocities in initial and final orbits (km/s)
θ = plane change angle (deg or rad)

If we're interested in changing only the orbit's inclination, we must change the velocity at either the ascending node or the descending node. When the ΔV is done at one of these points, the orbit will pivot about a line connecting the two nodes, thus changing only the inclination.

We can also use a plane change to change the longitude of the ascending node, Ω. This might be useful if we want a remote-sensing satellite to pass over a certain point on the Earth at a certain time of day. When we consider a polar orbit ($i = 90°$), we can see that a ΔV_{simple} at the north or south pole will change just the longitude of the ascending node, as illustrated in Figure 6-7. It is also possible to change Ω alone for inclinations other than 90°. The trick is to perform the ΔV_{simple} where the initial and final orbits intersect. (Think of this maneuver as pivoting around a line connecting the burn point to the center of the Earth.) We won't go into the details of these cases because the spherical trigonometry gets out of hand.

The amount of velocity change we need to re-orient the orbit plane depends on two things—the angle we're turning through and the initial velocity. As the angle we're turning through increases, so does ΔV_{simple}. For example, when this angle is 60°, our vector triangle becomes equilateral (all sides equal). Thus, ΔV_{simple} is equal to the initial velocity, which is the amount of velocity you needed to get into the orbit in the first place! That's why we want our initial parking orbit to have an inclination as close as possible to the final orbit.

Also notice that ΔV_{simple} will increase as the initial velocity increases. Therefore, we can lower ΔV_{simple} by reducing the initial velocity. For a circular orbit the velocity is constant throughout the orbit, but we know a satellite in an elliptical orbit slows down as it approaches apogee. Thus, if we can choose where to do a simple plane change in an elliptical orbit, we should do it at apogee, where the satellite's velocity is slowest. Remember our earlier analogy about changing speeds and directions on a racetrack. It's easier to change direction when you're going slower (even if you're a stunt driver). Example 6-2 demonstrates simple plane change.

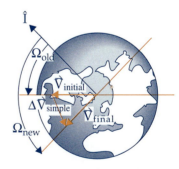

Figure 6-7. Changing Ω. A simple plane change as you cross the pole in a polar orbit ($i = 90°$) will change only the longitude of the ascending node, Ω. Imagine the orbit plane pivoting about the Earth's poles.

Combined Plane Change

Suppose our satellite is in a low-altitude parking orbit with i = 28.5° and it needs to transfer to a geostationary orbit (R = 42,160 km, i = 0°). This presents us with two separate problems: changing the size of the orbit and changing the orientation of the orbit plane. You might be tempted to tackle this problem in two parts—a Hohmann Transfer followed by a simple plane change. Based on what we've discussed so far, this gets the job done in three separate burns. But we can do the job in two burns rather than three and save fuel. How? By combining the plane-change burn with one of the Hohmann Transfer burns to get a maneuver we call a combined plane change.

If we draw a diagram of this problem, as in Figure 6-8, we can see that $\Delta\vec{V}_{combined}$ is really the vector sum of doing a simple plane change ($\Delta\vec{V}_{simple}$) and changing the orbit size where $\Delta\vec{V}_{increase}$ is one of the two Hohmann Transfer burns. These three ΔVs form a triangle with $\Delta\vec{V}_{combined}$ as the third side. You may recall from geometry that the sum of any two sides of a triangle is greater than the third side. That is:

$$\left|\Delta\vec{V}_{simple}\right| + \left|\Delta\vec{V}_{increase}\right| > \left|\Delta\vec{V}_{combined}\right|$$

This means it's always cheaper (in terms of ΔV) to do a combined plane change than to do a simple plane change followed by one of the Hohmann Transfer burns.

To solve for the needed velocity change let's apply the ever-popular law of cosines to get

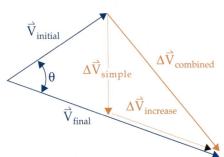

Figure 6-8. Vector Diagram of the Combined Plane Change. For a combined plane change, $\Delta\vec{V}_{combined}$ is always less than a simple plane change, $\Delta\vec{V}_{simple}$, followed by a tangential velocity increase, $\Delta\vec{V}_{increase}$.

$$\Delta V_{combined} = \sqrt{\left(\left|\vec{V}_{initial}\right|\right)^2 + \left(\left|\vec{V}_{final}\right|\right)^2 - 2\left|\vec{V}_{initial}\right|\left|\vec{V}_{final}\right|\cos\theta} \qquad (6\text{-}8)$$

where

$\Delta V_{combined}$ = ΔV for combined plane change (km/s)

$\left|\vec{V}_{initial}\right|$ = magnitude of the velocity in initial orbit (km/s)

$\left|\vec{V}_{final}\right|$ = magnitude of the velocity in final orbit (km/s)

θ = plane change angle (deg or rad)

Working through this equation, we find that it's usually cheaper to do a combined plane change at slower velocities (when farther from the Earth) just as we found for the simple plane change.

So what's the cheapest way to do a Hohmann Transfer with a plane change? For the case of going from a smaller to a larger orbit, we should begin the Hohmann Transfer (ΔV_1) and then do the combined plane change at apogee of the transfer orbit.

Table 6-1 summarizes the four options for a transfer from a low-Earth orbit with inclination of 28° to a geostationary orbit. It shows that starting the Hohmann Transfer closest to the Earth and finishing with the combined plane change at apogee (Case 4) is the most economical in terms of ΔV. The results in Table 6-1 are based on the following example

Given: $R_{orbit\ 1} = 6570$ km
$R_{orbit\ 2} = 42,160$ km
$i_{orbit\ 1} = 28°$
$i_{orbit\ 2} = 0°$
Find: ΔV_{total}

Table 6-1.   Plane Change and Hohmann Transfer Options. Case 4 requires the least amount of ΔV.

Case 1	Case 2	Case 3	Case 4
Do 28° inclination change using a simple plane change. Then do the Hohmann Transfer, ΔV_1 and ΔV_2.	Do the Hohmann Transfer, ΔV_1 and ΔV_2. Then do the 28° inclination change using a simple plane change.	Do combined plane change at perigee of transfer orbit. Do ΔV_2 of Hohmann.	Do ΔV_1 of Hohmann Transfer. Do combined plane change at apogee of transfer orbit.
$\Delta V_{simple} = 3.77$ km/s (in orbit 1)	$\Delta V_{Hohmann} = 3.94$ km/s	$\Delta V_{combined} = 4.98$ km/s (at perigee)	$\Delta V_1 = 2.46$ km/s
$\Delta V_{Hohmann} = 3.94$ km/s	$\Delta V_{simple} = 1.49$ km/s (in orbit 2)	$\Delta V_2 = 1.47$ km/s	$\Delta V_{combined} = 1.82$ km/s (at apogee)
$\Delta V_{total} = 7.70$ km/s	$\Delta V_{total} = 5.43$ km/s	$\Delta V_{total} = 6.46$ km/s	$\Delta V_{total} = 4.29$ km/s

Section Review

Key Terms

combined plane change
simple plane change

Key Equations

$$\Delta V_{simple} = 2\ V_{initial} \sin\left(\frac{\theta}{2}\right)$$

$$\Delta V_{combined}$$
$$= \sqrt{(|\vec{V}_{initial}|)^2 + ((|\vec{V}_{final}|)^2 - 2|\vec{V}_{initial}||\vec{V}_{final}|\cos\theta)}$$

Key Concepts

➤ We need maneuvers to change from one orbit plane to another:

• Simple plane changes alter only the direction, not the magnitude, of the velocity vector for the original orbit

$$|\vec{V}_{initial}| = |\vec{V}_{final}|$$

- A simple plane change at either the ascending or descending node will change only the orbit inclination. On a polar orbit a simple plane change made over the north or south pole will change only the longitude of the ascending node. A simple plane change made anywhere else will change both inclination and longitude of ascending node.

• A combined plane change alters the magnitude and direction of the original velocity vector

- It's always cheaper (in terms of ΔV) to do a combined plane change than to do a simple plane change followed by a Hohmann Transfer burn

➤ It's always cheaper (in terms of ΔV) to change planes when the orbit velocity is *slowest*, which is at apogee for elliptical transfer orbits

Example 6-2

Problem Statement

A GPS satellite is in a circular orbit at an altitude of 250 km. It needs to move from its current inclination of 28° to an inclination of 57°. What ΔV is required?

Problem Summary

Given: Altitude = 250 km

$i_{initial} = 28.0°$

$i_{final} = 57.0°$

Find: ΔV_{simple}

Conceptual Solution

1) Solve for the orbit's energy and velocity

$$\varepsilon = \frac{-\mu}{2a}$$

$$= \frac{-\mu}{2R} \text{ (circular orbit)}$$

$$\varepsilon = \frac{V^2}{2} - \frac{\mu}{R}$$

$$V = \sqrt{2\left(\frac{\mu}{R} + \varepsilon\right)}$$

2) Solve for the inclination change

$$\theta = \left|i_{final} - i_{initial}\right|$$

3) Find the change in velocity for a simple plane change

$$\Delta V_{simple} = 2\ V_{initial} \sin\frac{\theta}{2}$$

Analytical Solution

1) Solve for the energy and velocity of the orbit

$$\varepsilon = \frac{-\mu}{2R} = \frac{-3.986005 \times 10^5 \frac{km^3}{s^2}}{2\ (6378 + 250\ km)}$$

$$= -30.069\ \frac{km^2}{s^2}$$

$$V_{initial} = \sqrt{2\left(\frac{\mu}{R} + \varepsilon\right)}$$

$$V_{initial} = \sqrt{2\left(\frac{3.986005 \times 10^5 \frac{km^3}{s^2}}{6628\ km} - 30.069\ \frac{km^2}{s^2}\right)}$$

$$= 7.755\ \frac{km}{s}$$

2) Solve for the inclination change

$$\theta = \left|i_{final} - i_{initial}\right| = \left|57° - 28°\right|$$

$$\theta = 29°$$

Find ΔV for the simple plane change

$$\Delta V_{simple} = 2\ V_{initial} \sin\frac{\theta}{2} = 2\left(7.755\frac{km}{s}\right)\sin\frac{29°}{2}$$

$$\Delta V_{simple} = 3.88\ km/s$$

Interpreting the Results

To change the inclination of the GPS satellite by 29°, we must apply a ΔV of 3.88 km/s. This is 50% of the velocity we needed to get the GPS satellite into space in the first place. Plane changes are *very expensive* (in terms of ΔV.)

6.3 Rendezvous

▬ In This Section You'll Learn to...

- ☞ Describe orbit rendezvous
- ☞ Determine the ΔV and wait time to execute a rendezvous

For the previous maneuvers, we've discussed how to move a satellite around without considering where it is in relation to other spacecraft. However, several types of space missions require a satellite to connect or *rendezvous* with another satellite. This places an additional requirement on the maneuver: when going from one orbit to another, the first satellite must arrive at the same place at the same time as the second satellite. The Gemini program in the 1960s perfected this maneuver.

Rendezvous was also a critical requirement for the Apollo mission to the Moon. Two astronauts returning from the lunar surface had to rendezvous with their companion in the Command/Service Module in lunar orbit for the trip back to Earth. Future Space Shuttle missions will need to rendezvous with the Space Station routinely to transfer people and equipment.

Astro Fun Fact

Docking Drama on Gemini 8

The first man to walk on the Moon almost didn't survive his first flight into space! On March 16, 1966, Gemini 8 was launched from Cape Canaveral, Florida. The mission was to rendezvous and perform the first-ever docking with another spacecraft. A modified Agena upperstage (shown in the photo) had been launched 100 minutes before.

Following a smooth rendezvous, astronauts Neil Armstrong and Dave Scott started the docking sequence and, for the first time, two spacecraft were joined together in orbit. Thirty minutes later, the two spacecraft began to spin uncontrollably. Thinking the problem was with the Agena, Armstrong undocked. Unfortunately, this made the spinning worse as the Gemini reached rates of 360° per second (60 R.P.M.). Nearing blackout, the two astronauts finally regained control by shutting down power to the primary thrusters and firing the re-entry thrusters. The mission was aborted early to a safe splashdown, but Neil Armstrong would fly again—next time to the Moon.

David Baker, PhD, <u>The History of Manned Space Flight</u>. New York, NY: Crown Publishers Inc., 1981.

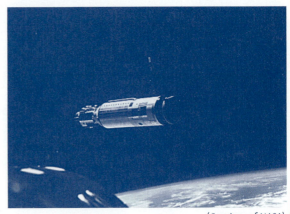

(Courtesy of NASA)

Coplanar Rendezvous

The simplest type of rendezvous uses a Hohmann Transfer between coplanar orbits. The key to this maneuver is timing. Deciding when to fire your engines involves leading the target satellite just as a quarterback leads a receiver. After the ball is snapped, the receiver starts running straight down the field toward the goal line, as Figure 6-9 shows. The quarterback must mentally calculate how fast the receiver is running and how long it will take the ball to get to a certain spot on the field. When the ball is released, it will take a specific amount of time to reach its destination. Over this same period, the receiver will go from where he was when the ball was released to the "rendezvous" point with the ball.

rendezvous point

Figure 6-9. Orbit Rendezvous and Football. The spacecraft rendezvous problem is similar to the problem a quarterback faces when trying to pass to a receiver. The quarterback must time the pass just right so the ball and the receiver are in the same place at the same time.

Let's take a closer look at this football analogy to see how the quarterback decides when to throw the ball so it will "rendezvous" with the receiver. We have a quarterback who will throw a 20-yard pass travelling at 10 yd/s and a wide receiver who will run at 4 yd/s. How long must the quarterback wait from the snap (assuming the receiver starts running immediately) before throwing the ball? To analyze the problem, let's define the following symbols

$V_{receiver}$ = velocity of receiver
    = 4 yd/s
V_{ball} = velocity of the ball
    = 10 yd/s

We know we need to "lead" the receiver; that is, we know the receiver will travel some distance while the ball is in the air. But how long will the ball take to travel 20 yards from the quarterback to the receiver? Let's define:

TOF_{ball} = time of flight of the ball
    = distance ball travels/V_{ball}
    = 20 yd/(10 yd/s)
    = 2 s

The lead distance is then the receiver's velocity times the ball's time of flight.

$$\alpha = \text{lead distance}$$
$$= V_{receiver} \times TOF_{ball}$$
$$= (4 \text{ yd/s}) \times 2 \text{ s}$$
$$= 8 \text{ yd}$$

This means the receiver will run 8 yards while the ball is in the air. From this we can easily figure out how much of a head start the receiver needs before the quarterback throws the ball. If the receiver runs 8 yards while the ball is in the air, and the ball is being thrown 20 yards, the receiver then needs a head start of:

$$\phi_{head\ start} = \text{head start distance needed by receiver}$$
$$= 20 \text{ yd} - \alpha$$
$$= 20 \text{ yd} - 8 \text{ yd}$$
$$= 12 \text{ yd}$$

So before the quarterback throws the ball, the receiver should be 12 yards down the field. We can now determine how long it will take the receiver to go 12 yards down field. This is the time the quarterback must wait before throwing the ball to ensure the receiver will be at the rendezvous point when the ball arrives.

$$\text{W.T.} = \text{wait time}$$
$$= \phi_{head\ start} / V_{receiver}$$
$$= 12 \text{ yd}/(4 \text{ yd/s})$$
$$= 3 \text{ s}$$

That's all well and good for footballs, but what about satellites trying to rendezvous in space? It turns out the approach is exactly the same as in the football problem. Let's look at the geometry of the problem shown in Figure 6-10. We have a target spacecraft (say a disabled communication satellite the crew of the shuttle plans to fix) and an interceptor (the Space Shuttle). In this case, the target satellite is in a higher orbit than the shuttle, but we'd do the same thing if it were in a lower orbit. To rendezvous, the shuttle must initiate a ΔV to transfer to the rendezvous point. But we must do this ΔV at just the right time to make sure the target arrives at the same time as the shuttle.

To see how to solve this problem, remember that the quarterback first had to know the velocities of the interceptor (the ball) and the target (the receiver). Because footballs move in straight lines, their velocities are easy to see. However, when things move around in orbits, velocities aren't so straightforward. Instead of using straight-line velocities (in meters per second or miles per hour), we'll use rotational velocities measured in radians per second or degrees per hour. We normally call this rotational velocity "angular velocity" and use the Greek letter omega, ω, to represent it. How do we find this angular velocity for spacecraft? Because any orbiting spacecraft moves through 360° (or 2π radians) in one orbital period, we can find its angular velocity from

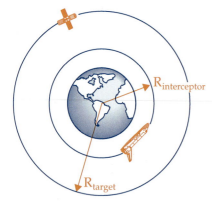

Figure 6-10. The Rendezvous Problem. The Space Shuttle commander must do a Hohmann Transfer at precisely the right moment to link up with another satellite.

$$\omega = \frac{2\pi \, (\text{radians})}{2\pi \sqrt{\frac{a^3}{\mu}}}$$

$$\boxed{\omega = \sqrt{\frac{\mu}{a^3}}} \tag{6-9}$$

where
ω = angular velocity (rad/s)
μ = gravitational parameter (km^3/s^2)
a = semi-major axis (km)

For circular orbits, $a = R$, so this angular velocity is constant.

To solve the football problem, we next had to find the ball's time of flight. For the spacecraft rendezvous problem, the time of flight is the same as the Hohmann Transfer's time of flight which we found earlier

$$TOF = \pi \sqrt{\frac{a_{transfer}^3}{\mu}} \tag{6-10}$$

where
TOF = time of flight (s)
$a_{transfer}$ = semi-major axis of transfer orbit (km)
μ = gravitational parameter (km^3/s^2)

Finally, we need to get our timing right. We begin by finding the amount by which the interceptor should lead the target or its *lead angle*, α_{lead}, when the shuttle starts the Hohmann Transfer. This lead angle, shown in Figure 6-11, represents the angular distance covered by the target during the shuttle's time of flight. We find it by multiplying the target's angular velocity by the shuttle's time of flight.

$$\boxed{\alpha_{lead} = \omega_{target} TOF} \tag{6-11}$$

where
α_{lead} = amount by which interceptor should lead the target (rad)
ω_{target} = angular target velocity of target (rad/s)
TOF = time of flight (s)

We can now determine how big of a head start we need to give the target. For satellites, we call this a *phase angle*, ϕ, measured from the interceptor's radius vector to the target's radius vector in the direction of motion. Because we're using a Hohmann Transfer, the shuttle will travel 180° (π radians) during the transfer. This means we can easily compute the phase angle we need, ϕ_{final}, if we know the lead angle.

$$\boxed{\phi_{final} = \pi - \alpha_{lead}} \tag{6-12}$$

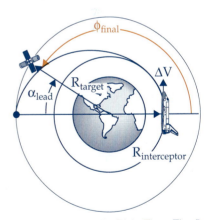

Figure 6-11. ΔV at the Right Time. The first ΔV of the rendezvous Hohmann Transfer is started when the interceptor is at an angle ϕ_{final} from the target.

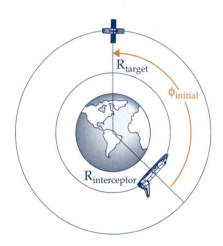

Figure 6-12. Rendezvous Initial Condition. At the start of the rendezvous problem, the target is some angle, $\phi_{initial}$, away from the interceptor.

where

ϕ_{final} = phase angle between interceptor and target as transfer begins (rad)

α_{lead} = amount by which interceptor should lead the target (rad)

Chances are, when we're first ready to start the rendezvous, the target won't be in the right position, as seen in Figure 6-12. So what do we do? We wait until the relative position between the target and interceptor are just right as in Figure 6-11. But how long do we wait? To answer this we have to relate where the target is now (relative to the interceptor), $\phi_{initial}$, to where we want it to be, ϕ_{final}. Because the shuttle and target are moving in circular orbits at constant velocities, $\phi_{initial}$ and ϕ_{final} are related by

$$\phi_{final} = \phi_{initial} + (\omega_{target} - \omega_{interceptor}) \times \text{wait time} \quad (6\text{-}13)$$

Solving for wait time gives

$$\boxed{\text{wait time} = \frac{\phi_{final} - \phi_{initial}}{\omega_{target} - \omega_{interceptor}}} \quad (6\text{-}14)$$

where

wait time	= time until rendezvous is initiated (s)
$\phi_{final}, \phi_{initial}$	= initial and final phase angles (rad)
$\omega_{target}, \omega_{interceptor}$	= target and interceptor angular velocities (rad/s)

So far, so good. But if you look at the wait time equation, you can see that sometimes wait time can be less than zero. Does this mean we have to go back in time? Luckily, no. Because both the shuttle and the target are going around in circles, the correct angular relationship will repeat itself periodically. When the difference between ϕ_{final} and $\phi_{initial}$ changes by 2π radians (360°), you're right back where you started. To calculate the next available opportunity, we either add or subtract 2π from the numerator in Equation (6-14)—whichever it takes—to make the resulting wait time positive. In fact, we can continue to determine future rendezvous opportunities by adding or subtracting multiples of 2π.

Co-orbital Rendezvous

Another twist to the rendezvous problem occurs when the satellites are co-orbital. This means both the target and interceptor are in exactly the same orbit, with one ahead or behind the other.

Whenever the target is ahead, as shown in Figure 6-13, the interceptor must somehow catch up to the target. For example, suppose you live five miles away from school, and your friend lives three miles from school, and you both leave at the same time. If so, you'll have to go faster to get to school at the same time as your friend. To catch up, the interceptor needs to move into a waiting or *phasing orbit*, which will return it to the same spot one orbit later, in the time it takes the target to move around to that spot. Notice the angular distance the target has to travel is less than 360° while

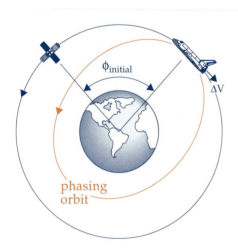

Figure 6-13. Slow Down to Speed Up. To catch up with another spacecraft ahead of you in the same orbit, you *slow down*, entering a smaller phasing orbit with a shorter period. This allows the interceptor to catch up to the target.

the travel distance for the interceptor is exactly 360°. How can we do this in an orbit? By *slowing down*! What?! Does this make sense? Go back to orbital mechanics. When you slow down (decrease energy) you enter a smaller orbit. Smaller orbits have a shorter period, so you cover 360° faster. If you slow down the right amount, you'll get back to where you started just as the target gets there.

To determine the right amount to slow down, first we find out how far the target must travel to get to our current position. If the target is ahead of us by an amount $\phi_{initial}$, it must travel through an angle, ϕ_{travel}, to reach the rendezvous spot found from

$$\phi_{travel} = 2\pi - \phi_{initial} \qquad (6\text{-}15)$$

where

ϕ_{travel} = angle through which target travels to reach the rendezvous location (rad)

$\phi_{initial}$ = initial angle between interceptor and target (rad)

Now, if we know the angular velocity of the target, we can find the time it will take to cover this angle ϕ_{travel}

$$TOF = \frac{\phi_{travel}}{\omega_{target}} \qquad (6\text{-}16)$$

Remember we found the angular velocity of the target from Equation (6-9)

$$\omega_{target} = \sqrt{\frac{\mu}{a_{target}^3}}$$

Because the time of flight is equal to the period of the phasing orbit, we equate this to our trusty equation for the period of an orbit, producing

$$TOF = \frac{\phi_{travel}}{\omega_{target}} = 2\pi \sqrt{\frac{a_{phasing}^3}{\mu}}$$

We can now solve for the required size of the phasing orbit

$$\boxed{a_{phasing} = \sqrt[3]{\mu \left(\frac{\phi_{travel}}{\omega_{target} (2\pi)} \right)^2}}$$

where

$a_{phasing}$ = semi-major axis of phasing orbit (km)

μ = gravitational parameter (km^3/s^2)

ϕ_{travel} = angular distance target must travel to get to rendezvous location (rad)

ω_{target} = angular velocity of target (rad/s)

Knowing the size of the phasing orbit, we can compute the necessary ΔVs for the rendezvous. The first ΔV slows us down and puts us in the phasing orbit. The second ΔV returns us to the original orbit right next to

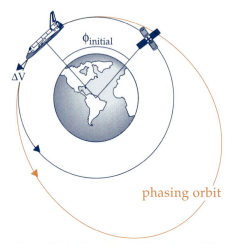

$\phi_{initial}$

ΔV

phasing orbit

Figure 6-14. Speed Up to Slow Down. If the target is behind the interceptor in the same orbit, the interceptor can speed up and enter a higher, slower orbit, thereby allowing the target to catch up.

the target. A nice feature of this maneuver is that the ΔV to go into the phasing orbit is the same as the ΔV to return from the phasing orbit into the original circular orbit.

We must also know how to rendezvous whenever the target is behind the interceptor. In this case, the angular distance the target must cover to get to the rendezvous spot is greater than 360°. Thus, the phasing orbit for the interceptor will have a period greater than the current circular orbit period. To get into the phasing orbit, the interceptor *speeds up*. It then enters a higher, slower orbit, allowing the target to catch up, as Figure 6-14 shows.

Section Review

Key Terms

lead angle, α_{lead}
phase angle, ϕ
phasing orbit
rendezvous

Key Equations

$$\omega = \sqrt{\frac{\mu}{a^3}}$$

$$\alpha_{lead} = \omega_{target} TOF$$

$$\phi_{final} = \pi - \alpha_{lead}$$

$$\text{wait time} = \frac{\phi_{final} - \phi_{initial}}{\omega_{target} - \omega_{interceptor}}$$

$$a_{phasing} = \sqrt[3]{\mu \left(\frac{\phi_{travel}}{\omega_{target}(2\pi)} \right)^2}$$

Key Concepts

➤ Rendezvous is getting two or more spacecraft to arrive at the same point in an orbit at the same time

➤ The rendezvous problem is very similar to the problem quarterbacks face when they must "lead" a receiver before throwing a pass. But because the interceptor and target spacecraft are going around in circular orbits, the proper relative positions for rendezvous will repeat periodically.

➤ We'll assume spacecraft rendezvous uses a Hohmann Transfer

➤ The lead angle, α_{lead}, is the angular distance the target spacecraft will travel during the interceptor time of flight, TOF

➤ The final phase angle, ϕ_{final}, is the "headstart" the target spacecraft needs

➤ The wait time is the time between some initial starting time and the time when the geometry is right to begin the Hohmann Transfer for a rendezvous

   • Remember, for negative wait times, we must modify the numerator in the wait time equation by adding multiples of 2π radians

Example 6-3

Problem Statement

A next-generation Space Shuttle in low-Earth orbit needs to rendezvous with a target satellite in a geosynchronous orbit. If the initial angle between the two satellites is 180°, how long must the interceptor wait before starting the rendezvous?

$R_{interceptor}$ = 6570 km

R_{target} = 42,160 km

Problem Summary

Given: $R_{interceptor}$ = 6570 km

R_{target} = 42,160 km

$\phi_{initial}$ = 180° = π radians

Find: wait time

Problem Diagram

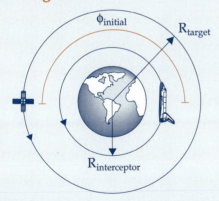

Conceptual Solution

1) Compute the semi-major axis of the transfer orbit

$$a_{transfer} = \frac{R_{interceptor} + R_{target}}{2}$$

2) Find the time of flight (TOF) of the transfer orbit

$$TOF = \pi \sqrt{\frac{a_{transfer}^3}{\mu}}$$

3) Find the angular velocities of the interceptor and target

$$\omega_{interceptor} = \sqrt{\frac{\mu}{R_{interceptor}^3}}$$

$$\omega_{target} = \sqrt{\frac{\mu}{R_{target}^3}}$$

4) Compute the lead angle

$$\alpha_{lead} = (\omega_{target})(TOF)$$

5) Solve for the final phase angle

$$\phi_{final} = \pi - \alpha_{lead}$$

6) Find the wait time

$$Wait\ Time = \frac{\phi_{final} - \phi_{initial}}{\omega_{target} - \omega_{interceptor}}$$

Analytical Solution

1) Compute the semi-major axis of the transfer orbit

$$a_{transfer} = \frac{R_{interceptor} + R_{target}}{2}$$

$$= \frac{6570\ km + 42,160\ km}{2}$$

$a_{transfer}$ = 24,365 km

2) Find the TOF of the transfer orbit

$$TOF = \pi \sqrt{\frac{a_{transfer}^3}{\mu}} = \pi \sqrt{\frac{(24,365\ km)^3}{3.986005 \times 10^5 \frac{km^3}{s^2}}}$$

TOF = 18,925 s = 315 min

3) Find the angular velocities of the interceptor and target

$$\omega_{interceptor} = \sqrt{\frac{\mu}{R_{interceptor}^3}} = \sqrt{\frac{3.986005 \times 10^5 \frac{km^3}{s^2}}{(6570\ km)^3}}$$

Example 6-3 Continued

$\omega_{interceptor} = 0.0012 \text{ rad/s}$

$$\omega_{target} = \sqrt{\frac{\mu}{R_{target}^3}} = \sqrt{\frac{3.986005 \times 10^5 \frac{km^3}{s^2}}{(42,160 \text{ km})^3}}$$

$\omega_{target} = 0.000073 \text{ rad/s}$

4) Compute the lead angle

$\alpha_{lead} = (\omega_{target})(TOF)$

$$\left(0.000073 \frac{rad}{s}\right)(18,925 \text{ s})$$

$\alpha_{lead} = 1.38 \text{ rad}$

5) Solve for the final phase angle

$\phi_{final} = \pi - \alpha_{lead} = \pi - 1.38 \text{ rad}$

$\phi_{final} = 1.76 \text{ rad}$

6) Find the wait time

$$\text{wait time} = \frac{\phi_{final} - \phi_{initial}}{\omega_{target} - \omega_{interceptor}}$$

$$\text{wait time} = \frac{1.76 \text{ rad} - \pi}{0.000073 \frac{rad}{s} - 0.0012 \frac{rad}{s}}$$

$\text{wait time} = 1226 \text{ s} = 20.4 \text{ min}$

Interpreting the Results

From the initial separation of 180°, the interceptor must wait 20.4 minutes before starting the Hohmann Transfer to rendezvous with the target.

References

Bate, Robert R., Donald D. Mueller, Jerry E. White. *Fundamentals of Astrodynamics*. New York, NY: Dover Publications, Inc., 1971.

Escobal, Pedro R. *Methods of Orbit Determination*. Malabar, FL: Krieger Publishing Company, Inc., 1976.

Kaplan, Marshall H. *Modern Spacecraft Dynamics and Control*. New York, NY: Wiley & Sons. 1976.

Vallado, Capt. David. *Orbit Maneuvering and Orbital Phasing*. USAFA Astro 321 Course Handout. 1991.

Mission Problems

6.1   Hohmann Transfers

1   What assumptions allow us to use a Hohmann Transfer?

2   What makes a Hohmann Transfer the most energy-efficient maneuver between coplanar orbits?

3   When going from a smaller circular orbit to a larger one, why do we speed up twice but end up with a slower velocity in the final orbit?

4   Why do we take the absolute value of the difference between the two orbit velocities when we compute total ΔV?

5   NASA wants to move a malfunctioning satellite from a circular orbit at 500 km altitude to one at 150 km altitude, so a shuttle crew can repair it.

a) What is the energy of the transfer orbit?

b) What is the velocity change (ΔV_1) needed to go from the initial circular orbit into the transfer orbit?

c) What is the velocity change (ΔV_2) needed to go from the transfer orbit to the final circular orbit?

d) What is the time (TOF) required for the transfer?

6   What is the orbital velocity of a circular orbit at 220 km altitude?

6.2   Plane Changes

7   What orbital elements can a simple plane change alter?

8   For changing inclination only, where do we do the ΔV? Why?

9   Why do we prefer to use a combined plane change when going from a low-Earth parking orbit to a geostationary orbit rather than a Hohmann Transfer and a simple plane change?

10 Why does Case 3 in Table 6-1 (doing a combined plane change at perigee followed by ΔV_2 of Hohmann Transfer) have a higher total ΔV than Case 2 (doing a Hohmann Transfer and then a simple plane change)?

11 A satellite deployed into an orbit inclined 57° at 130 km altitude needs to change to a polar orbit at the same altitude. What ΔV is required?

12 Now that the satellite from Problem 11 is in a polar orbit, what ΔV is required to change the longitude of the ascending node by 35°?

13 NASA wants to send a newly repaired satellite from its circular orbit at 150 km altitude (28° inclination) to a circular orbit at 20,000 km altitude (inclination of 45°).

a) What is the energy of the transfer orbit?

b) What is the velocity change (ΔV_1) needed to go from the initial circular orbit to the transfer orbit?

c) What is the combined plane change ΔV to go from the transfer orbit to the final circular orbit and change the inclination?

6.3   Rendezvous

14 Describe a rendezvous for an interceptor in a high orbit to a target satellite in a lower orbit.

15 You are in charge of a rescue mission. The satellite in distress is in a circular orbit at 240 km altitude.

The shuttle (rescue vehicle) is in a coplanar circular orbit at 120 km altitude. The shuttle is 135° behind the target satellite.

a) What is the TOF of the transfer orbit to rendezvous with the target satellite?

b) What is the shuttle's angular velocity? The target satellite's?

c) What is the lead angle?

d) What is the final phase angle?

e) How long must the shuttle wait before starting the rendezvous maneuver?

16 In the above rescue mission, the shuttle engines misfired, placing the shuttle in the same 240 km circular orbit as the target satellite, but 35° ahead of the target.

a) What is the TOF (and, therefore, the period) of the rendezvous phasing orbit?

b) What is the semi-major axis of the phasing orbit?

c) Compute the ΔV necessary for the shuttle to move into the phasing orbit.

For Discussion

17 What extra steps must you add for a rendezvous between non-coplanar satellites?

18 What types of space missions use rendezvous?

Mission Profile—Gemini

In December 1961, NASA let a contract to the McDonnell Corporation to build the "two-man spacecraft." This was the beginning of Project Gemini, the second U.S. manned space program. It was conceived as an extension of the Mercury program. But President Kennedy's goal of putting a man on the Moon by the end of the decade required solutions for many of the technical problems in a lunar landing.

Ed White performs the first U.S. spacewalk. *(Courtesy of NASA)*

Mission Overview

The Gemini spacecraft carried 2 astronauts and was launched by a Titan 2 missile. Between April 1964 and November 1966, the program completed 10 manned launches and 2 unmanned launches. Major goals for the program included proving rendezvous and docking capabilities that would be required for Apollo, extending the endurance of U.S. astronauts in space, and proving the ability to do extravehicular activity (EVA) or "spacewalking."

Mission Data

✓ Gemini 1 and 2 were unmanned tests to prove the performance of the launch vehicle and spacecraft.

✓ Gemini 3 (Crew: Grissom, Young): First manual control of space maneuver and first manual reentry.

✓ Gemini 4 (Crew: McDivitt, White): First U.S. citizen (White) to spacewalk. 11 scientific experiments completed.

✓ Gemini 5 (Crew: Conrad, Cooper): 17 scientific experiments completed.

✓ Gemini 6 (Unmanned) Failed at launch

✓ Gemini 6A (Crew: Schirra, Stafford): Performed the first successful orbital rendezvous with Gemini 7.

✓ Gemini 7 (Crew: Borman, Lovell): Established an endurance record of 206 orbits in 330 hrs 36 mins, which was longer than any of the Apollo missions.

✓ Gemini 8 (Crew: Armstrong, Scott): Completed the first successful docking in space with an Atlas Agena upperstage. Failure of the spacecraft's attitude maneuvering system caused wild gyrations of the spacecraft and was one of the worst emergencies of the program.

✓ Gemini 9/9A (Crew: Cernan, Stafford): Failure of the target vehicle resulted in delay and redesignation as Gemini 9A two weeks later. Completed rendezvous with new target but aborted docking because the docking apparatus had mechanical problems.

✓ Gemini 10 (Crew: Collins, Young): Rendezvoused and docked but used up twice as much fuel as planned.

✓ Gemini 11 (Crew: Conrad, Gordon): Docking achieved on first orbit

✓ Gemini 12 (Crew: Aldrin, Lovell): Conducted the first visual docking (due to a radar failure). More than five hours of EVA by Aldrin.

Mission Impact

Gemini accomplished many "firsts" and showed manned space-flight missions could overcome major problems. Despite some set-backs, the program succeeded beyond anyone's expectations and moved NASA toward more flexible operations.

For Discussion

- Do the lessons learned from Project Gemini affect how we work in space today?
- How might our current operations in space be different if we had not learned to walk in space or rendezvous and dock?

Contributor

Todd Lovell, United States Air Force Academy

References

Baker, David. *The History of Manned Spaceflight.* New York: Crown, 1981.

Yenne, Bill. *The Encyclopedia of US Spacecraft.* New York, NY: Exeter, 1985.

Earth rise over the lunar horizon. *(Courtesy of NASA)*

Interplanetary Travel

7

▬ In This Chapter You'll Learn to...

☛ Describe the basic steps involved in getting from one planet in the solar system to another

☛ Determine the required velocity change, ΔV, needed for interplanetary transfer

☛ Explain how we can use the gravitational pull of planets to get "free" velocity changes, making interplanetary transfer faster and cheaper

▬ You Should Already Know...

❑ Definition and use of coordinate systems (Chapter 4)

❑ The limitations on a restricted two-body problem and its solution (Chapter 4)

❑ Definitions of specific mechanical energy for various conic sections (Chapter 4)

❑ How to use the Hohmann Transfer to get from one orbit to another (Chapter 6)

❑ The rendezvous problem (Chapter 6)

▬ Outline

Greetings from the children of the planet Earth.

> *Anonymous greeting*
> *placed on the Voyager spacecraft*
> *in case it encounters aliens*

The wealth of information from interplanetary missions such as the Pioneer, Voyager, and Magellan has given us insight into the history of the solar system and a better understanding of the basic mechanisms at work in Earth's atmosphere and geology. Our quest for knowledge of the solar system continues. Perhaps in the not-too-distant future, we'll undertake manned missions back to the Moon, to Mars, and beyond.

How do we get from Earth to these exciting new worlds? That's the problem of interplanetary transfer. In Chapter 4 we laid the foundation for understanding orbits. In Chapter 6 we developed the Hohmann Transfer. Using this as a tool, we saw how to transfer between two orbits around the same body, such as the Earth. The problem of interplanetary transfer is really just an extension of the Hohmann Transfer. Only now, the central body is the Sun. In addition, as we'll see, we must be concerned with orbits around our departure and destination planets.

We'll begin by looking at the basic equation of motion for interplanetary transfer and then learn how we can greatly simplify the problem using a technique called the "patched-conic approximation." We'll see an example of how to use this simple method to plot a path from Earth to Mars. Finally, we'll look at gravity assist or "slingshot" trajectories to see how to use them for "free" ΔV, making interplanetary missions faster and cheaper.

7.1 Equation of Motion

≡ In This Section You'll Learn to...

☛ Describe the coordinate systems and equation of motion used for interplanetary transfer

☛ Describe the basic concept of the patched-conic approximation and why it's needed

To develop an understanding of interplanetary transfer, we'll start by dusting off our trusty Motion Analysis Process Checklist shown in Figure 7-1. For our analysis, we'll deal with only the first three items on the checklist because looking at initial conditions, error analysis, and model testing would get far too involved for our simplified approach.

Coordinate Systems

When we developed the two-body equation of motion to analyze satellite motion around the Earth in Chapter 4, we assumed

- There are only two bodies—the spacecraft and the Earth
- Earth's gravitational pull is the only force acting on the spacecraft

For Earth-based problems, the Geocentric-Equatorial frame is suitable. Once we cross the boundary into interplanetary space, however, the Earth's gravitational pull becomes less significant and the Sun's pull becomes the dominant force. Therefore, because the Sun is central to interplanetary transfer, we must develop a sun-centered or heliocentric coordinate system. By definition, *heliocentric* means the origin is the center of the Sun. In choosing a fundamental plane, we'll use the plane of the Earth's orbit about the Sun, which is called the *ecliptic plane*. Next, because we still need a principal direction fixed with respect to the universe, we bring the vernal equinox direction (♈) back for an encore performance. With the fundamental plane fixed and the principle direction chosen, the $\hat{J}$ and $\hat{K}$ directions complete our right-handed system. Now we can relate our trajectory from the Earth to another planet, or even to the edge of the solar system, to this *heliocentric-ecliptic coordinate system* defined in Figure 7-2.

Equation of Motion

Now that we have a coordinate frame, we can get an equation to describe the motion of spacecraft around the Sun by returning to Newton's Second Law. First we must identify the forces a spacecraft can expect to encounter in a flight from the Earth to another planet. As always, we have the gravity of the Earth at the start of the mission. Quite early on,

MOTION ANALYSIS PROCESS (MAP) CHECKLIST

☐ Coordinate System

☐ Equation of Motion

☐ Simplifying Assumptions

☐ Initial Conditions

☐ Error Analysis

☐ Testing the Model

Figure 7-1. Motion Analysis Process Checklist.

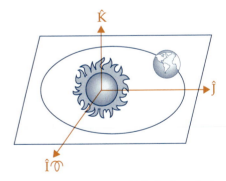

Figure 7-2. Heliocentric-Ecliptic Coordinate System for Interplanetary Transfer. Origin—center of the Sun; fundamental plane—ecliptic plane (Earth's orbit plane around the Sun); principle direction—vernal equinox direction.

however, when we get far enough away from the Earth, the Sun's gravitational pull begins to dominate. Finally, at journey's end, we must also consider the gravitational attraction of the target planet. This could range from Mercury's slight tug to Jupiter's immense pull. As before, we can throw in "other" forces to cover anything we might have forgotten such as solar pressure or pull from asteroids. When we consider all these forces, our equation of motion gets cumbersome

$$\Sigma \vec{F}_{external} = m\ddot{\vec{R}} = \vec{F}_{gravity\ Sun} + \vec{F}_{gravity\ Earth} + \vec{F}_{gravity\ target} + \vec{F}_{other} \quad (7\text{-}1)$$

Simplifying Assumptions

Luckily, we can assume that the forces of gravity are much greater than all "other" forces acting on the spacecraft. This leaves us with only the force of gravity, but gravity from three different sources!

$$\Sigma \vec{F}_{external} = m\ddot{\vec{R}} = \vec{F}_{gravity\ Sun} + \vec{F}_{gravity\ Earth} + \vec{F}_{gravity\ target} \quad (7\text{-}2)$$

Thus, as you can see in Figure 7-3, we have a four-body problem—spacecraft, Earth, Sun, and target planet. Trying to solve for the spacecraft's motion under the influence of all these bodies could give you nightmares! Remember, gravity depends on the distance from the central body to the spacecraft. To calculate all of these gravitational forces at once, you'd have to know the spacecraft's position and the positions of the planets as they orbit around the Sun. This may not sound too tough, but the

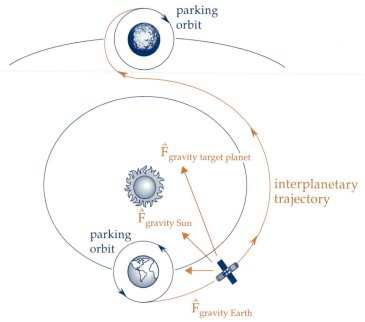

Figure 7-3. Gravitational Forces on an Interplanetary Spacecraft. Consider the forces on an interplanetary spacecraft as it makes its way from Earth to the target planet. We have the gravitational forces due to the Earth, the Sun, and the target planet making it a four-body problem—Earth, Sun, target planet, and spacecraft.

equation of motion becomes a highly non-linear, vector, differential equation which is very hard to solve. In fact, there's no closed-form solution to even a three-body problem, let alone four bodies. So how do we solve it? We use the old "divide and conquer" approach, taking one big problem and splitting it into three little ones. Which little problems can we solve? Two-body problems. For interplanetary transfers this approach is called the patched-conic approximation. The *patched-conic approximation* breaks the interplanetary trajectory into three separate regions. In each region, only the gravitational attraction on the spacecraft from one body is significant.

By looking at the problem with respect to one attracting body at a time, we're back to our good ol' two-body problem. Its equation of motion is

$$\ddot{\vec{R}} + \frac{\mu}{R^2}\hat{R} = 0 \tag{7-3}$$

where
$\ddot{\vec{R}}$ = acceleration vector (km/s^2)
μ = gravitational parameter of central body (km^3/s^2)
R = position (km)
$\hat{R}$ = unit vector in $\vec{R}$ direction

As you may remember, the solution to this equation describes a conic section (circle, ellipse, parabola, or hyperbola). Thus, the individual pieces of our trajectory will all be conic sections. By solving one two-body problem at a time, we splice one conic trajectory onto another, arriving at the patched-conic approximation. In the next section we'll see how all these pieces fit together.

▬ Section Review

Key Terms	Key Concepts
ecliptic plane heliocentric heliocentric-ecliptic coordinate system patched-conic approximation	➤ The coordinate system for Sun-centered or interplanetary transfers is the heliocentric-ecliptic system: • The origin is the center of the Sun • The fundamental plane is the ecliptic plane (the Earth's orbit plane) • The principle direction ($\hat{I}$) is the vernal equinox direction ➤ Taken together, the interplanetary transfer problem involves four separate bodies: • The spacecraft • The Earth (or departure planet) • The Sun • The target or destination planet ➤ Because the four-body problem is difficult to solve, we split it into three, two-body problems using a method called patched-conic approximation

7.2 The Patched-Conic Approximation

▬ In This Section You'll Learn to...

- ☛ Describe how to solve interplanetary transfers with the patched-conic approximation
- ☛ Determine the velocity change (ΔV) needed to go from one planet to another
- ☛ Determine the time of flight for interplanetary transfer and discuss the problem of planetary alignment

The patched-conic approximation is a way of breaking the interplanetary trajectory into pieces we can handle, using methods we already know. We divide the trajectory into pieces so we have to deal with the gravity from only one body at a time. In Figure 7-4 you can see the three separate regions of the trajectory for interplanetary transfer:

- Region 1—*Sun-centered transfer from Earth to target planet.* In this region the Sun's gravitational pull dominates
- Region 2—*Earth departure.* In this region the Earth's gravitational pull dominates
- Region 3—*Arrival at the target planet.* In this region the target planet's gravitational pull dominates

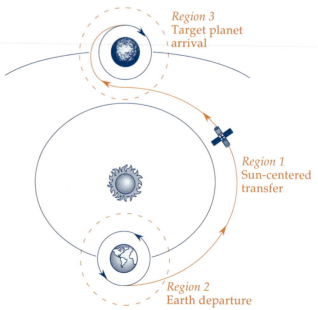

Region 3
Target planet arrival

Region 1
Sun-centered transfer

Region 2
Earth departure

Figure 7-4. Three Regions of the Patched-Conic Approximation. We can break the trajectory for interplanetary transfer into three distinct regions in which the gravitational pull of only one body dominates the spacecraft.

By considering each of these three regions separately, we get three distinct two-body problems that we can solve individually and then "patch" together to get a final solution. Our ultimate goal is to determine the total ΔV we need to leave Earth orbit and get into orbit around another planet. (In Chapter 14, we'll learn how to use this total ΔV requirement to determine the amount of rocket propellant needed for the trip.) Our three distinct problems are

❏ **Problem 1**: *Sun-centered transfer from Earth to the target planet.* This is a Hohmann Transfer around the Sun (heliocentric), shown in Figure 7-5. Thus, the orbit for interplanetary transfer is an ellipse. Because the Sun's gravitational pull is the only force on the spacecraft, we use the heliocentric-ecliptic coordinate system to describe this motion. In this problem, we must identify several different velocities, *all referenced to the Sun*.

- V_{Earth} is the Earth's velocity around the Sun
- $V_{transfer\ at\ Earth}$ is the velocity the spacecraft needs near Earth to transfer to the target planet
- $V_{\infty\ Earth}$ is the velocity "at infinity" with respect to Earth. This is the "extra" velocity the spacecraft needs to enter the transfer orbit.
- V_{target} is the target planet's velocity around the Sun
- $V_{transfer\ at\ target}$ is the velocity the spacecraft has on the transfer orbit near the target planet
- $V_{\infty\ target}$ is the velocity "at infinity" with respect to target planet. This is the "extra" velocity the spacecraft has with respect to the target planet.

Note: The nomenclature is important here! Remember when we introduced orbits we used R_{Earth} as the radius of the planet Earth. In this chapter we'll use $R_{to\ Earth}$ as the distance from the center of the Sun to the center of the Earth. V_{Earth} is the velocity of the Earth about the Sun. $R_{to\ target}$ is the distance from the center of the Sun to the center of the target planet. V_{target} is the velocity of the target planet around the Sun.

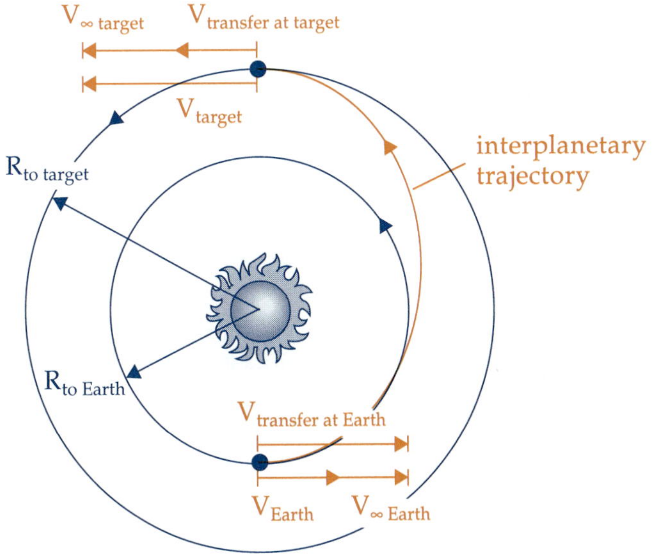

Figure 7-5. Problem 1. The first problem in the interplanetary patched-conic approximation is the heliocentric Hohmann Transfer from Earth to the target planet.

❑ **Problem 2**: *Earth departure*. We assume Earth's gravitational pull is the only force on the spacecraft and use the geocentric-equatorial coordinate system to describe motion. Recall from Chapter 4, to escape a planet entirely, we must be on either a parabolic or hyperbolic trajectory. As we'll see, the parabolic trajectory doesn't really take us anywhere, so we depart Earth on a hyperbolic trajectory. Here we must also understand several different velocities, *all referenced to the Earth*, as shown in Figure 7-6.

- $V_{park\ at\ Earth}$ is the spacecraft's original velocity in a parking orbit around the Earth

- $V_{hyperbolic\ at\ Earth}$ is the velocity the spacecraft must have at the parking orbit altitude to enter a hyperbolic departure trajectory

- ΔV_{boost} is the change in velocity needed to propel the spacecraft from its parking orbit onto the hyperbolic departure trajectory

- $V_{\infty\ Earth}$ is velocity "at infinity" with respect to Earth. This is also known as the "hyperbolic excess velocity" because it represents the "extra" velocity the spacecraft has as it leaves the Earth.

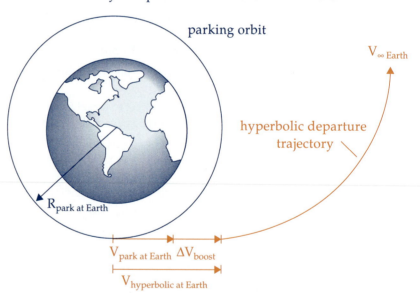

Figure 7-6. Problem 2. The second problem in the patched-conic approximation is the hyperbolic escape from Earth. This problem uses an Earth-centered perspective and requires the spacecraft to increase its velocity by an amount ΔV_{boost}.

❑ **Problem 3**: *Arrival at the target planet*. We assume the target planet's gravitational pull is the only force on the spacecraft and use a coordinate frame centered at the target planet (similar to the geocentric-equatorial system) to describe motion. Because the spacecraft's velocity with respect to the target planet is so high, it's once again on a hyperbolic trajectory with respect to the planet. From the planet's perspective, it looks like the spacecraft arrives ahead of it and the planet runs into it, like when you're going 65 m.p.h. on the highway and someone pulls in front of you going 50 m.p.h. To slow down into orbit around the planet, the spacecraft needs to do another ΔV maneuver. To compute the required change in velocity, we must understand several different velocities, *all referenced to the target planet*, as shown in Figure 7-7.

- $V_{park\ at\ target}$ is the velocity the spacecraft needs to enter into a parking orbit around the planet

- $V_{hyperbolic\ at\ target}$ is the velocity the spacecraft has on the hyperbolic trajectory when it reaches parking-orbit altitude

- ΔV_{retro} is the change in velocity needed to slow the spacecraft from its hyperbolic trajectory into a circular parking orbit around the target planet

- $V_{\infty\ target}$ is velocity "at infinity" with respect to the target planet. This is also known as the "hyperbolic excess velocity" because it represents the "extra" velocity the spacecraft has with respect to the target planet at arrival.

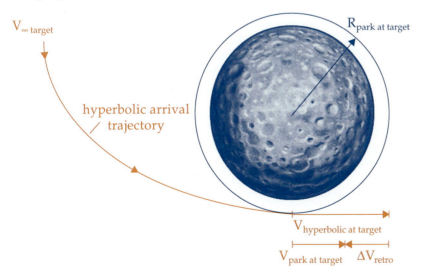

Figure 7-7. Problem 3. The third problem in the patched-conic approximation focuses on the target planet. In this case, the spacecraft must slow down by an amount ΔV_{retro} to drop into orbit.

As we've seen, these three separate problems require us to keep track of many velocities. These velocities relate to one particular reference frame, but the reference frame keeps changing! That is, the spacecraft's velocity with respect to the Earth is not the same as its velocity with respect to the Sun. This is a very important distinction to understand. As we'll see, the only velocity common to both Problem 1 and Problem 2 is $V_{\infty\text{ Earth}}$; $V_{\infty\text{ target}}$ is common to both Problem 1 and Problem 3. This commonality that allows us to patch the trajectories together. To see this more clearly, let's use the patched-conic approach to analyze a problem more down to Earth. Imagine you're driving down the interstate at 85 m.p.h. Your friend is chasing you in another car going 95 m.p.h., as shown in Figure 7-8. With respect to a stationary observer on the side of the road, the two cars are moving at 85 m.p.h. and 95 m.p.h., respectively. But your friend's velocity, with respect to you, is only 10 m.p.h. (she's gaining on you at 10 m.p.h.)

$V_{\text{relative to you}} = 10$ m.p.h.

$V_{\text{friend}} = 95$ m.p.h.

$V_{\text{you}} = 85$ m.p.h.

Figure 7-8. Relative Velocity. From your perspective at 85 m.p.h., you see your friend at a speed of 95 m.p.h. gaining on you at a relative speed of 10 m.p.h.

Now suppose your friend throws a water balloon at your car as shown in Figure 7-9. How fast is the balloon going? Well, that depends on your perspective. From your friend's perspective, if the balloon is thrown hard enough, it will appear to her to move ahead of her car at 20 m.p.h. From the viewpoint of our observer standing on the side of the highway, your friend's car is going 95 m.p.h., and the balloon leaves the car going 115 m.p.h. What do you see? Looking behind, you see the balloon rushing toward you at a closing speed of 30 m.p.h. (10 m.p.h. closing speed for your friend's car plus 20 m.p.h. closing speed for the balloon.)

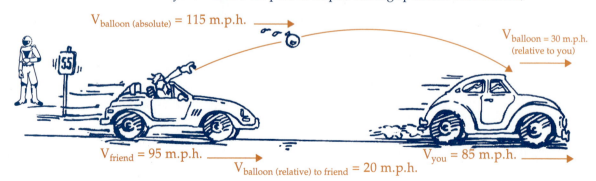

$V_{\text{balloon (absolute)}} = 115$ m.p.h.

$V_{\text{balloon}} = 30$ m.p.h. (relative to you)

$V_{\text{friend}} = 95$ m.p.h.

$V_{\text{you}} = 85$ m.p.h.

$V_{\text{balloon (relative) to friend}} = 20$ m.p.h.

Figure 7-9. Transfer from Car to Car. If your friend throws a water balloon at you at the speed of 20 m.p.h. relative to your friend, it will be going 115 m.p.h. relative to a fixed observer and will appear to you to be gaining on you at 30 m.p.h.

Analyzing the motion of the balloon, you can see the three problems used in the patched-conic approximation.

❏ **Problem 1**: *Stationary observer watching your friend throw a water balloon.* The observer sees your friend's car going 95 m.p.h., your car going 85 m.p.h., and a balloon travelling from one car to the other at 115 m.p.h. The reference frame is a stationary frame at the side of the road.

❏ **Problem 2**: *Water balloon depart in your friend's car.* The balloon leaves your friend's car with a relative speed of 20 m.p.h. as shown in Figure 7-9. The reference frame is your friend's car.

❏ **Problem 3**: *Water balloon landing in your car!* The balloon catches up to your car at a relative speed of 30 m.p.h. The reference frame is your car.

As we apply this approach to an interplanetary mission, keep in mind these three problems.

Elliptical Hohmann Transfer Between Planets—Problem 1

We'll begin by looking at the "big picture" because information from this problem allows us to solve the other two. Starting with Problem 1, we remove the planets and examine our interplanetary trajectory as going from one orbit around the Sun to another as in Figure 7-10. This is just a Hohmann Transfer on a larger scale. As before, we'll assume we start and end in circular orbits. For most planets, this assumption is fine, because their orbital eccentricities are small.

For the interplanetary transfer we follow nearly the same steps as we did for the Earth-centered Hohmann Transfer in Chapter 6. We first find our spacecraft's initial velocity around the Sun. We'll assume the spacecraft starts out in a parking orbit around the Earth. Because this orbit moves along with the Earth as the Earth orbits the Sun, the spacecraft's initial velocity with respect to the Sun is essentially the same as Earth's. (In reality, this isn't strictly true but it's close enough for a good approximation.)

Recall that we can find the specific mechanical energy for any orbit from

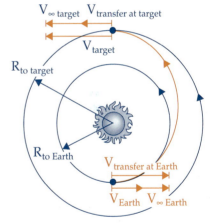

Figure 7-10. Problem 1. To enter the heliocentric-elliptical transfer orbit, the spacecraft must have a velocity of $V_{transfer}$ at Earth relative to the Sun. To achieve this, it must change its current heliocentric velocity, V_{Earth}, by an amount $V_{\infty\ Earth}$.

$$\boxed{\varepsilon = \frac{V^2}{2} - \frac{\mu}{R}} \qquad (7\text{-}4)$$

where
ε = specific mechanical energy (km^2/s^2)
V = velocity (km/s)
μ = gravitational parameter (km^3/s^2)
R = position (km)

and

$$\varepsilon = \frac{-\mu}{2a}$$ (7-5)

where
ε = specific mechanical energy (km^2/s^2)
μ = gravitational parameter (km^3/s^2)
a = semi-major axis (km)

We can rearrange Equation (7-4) to get a useful relationship for velocity

$$V = \sqrt{2\left(\frac{\mu}{R} + \varepsilon\right)}$$ (7-6)

The velocity of Earth around the Sun can then be found using Equation (7-6)

$$V_{Earth} = \sqrt{2\left(\frac{\mu_{Sun}}{R_{to\ Earth}} + \varepsilon_{Earth}\right)}$$ (7-7)

where
μ_{Sun} = gravitational parameter of the Sun
          = 1.327×10^{11} km^3/s^2
$R_{to\ Earth}$ = distance from Sun to Earth
          = 1 astronomical unit (AU), see Appendix B
          = 1.459×10^8 km (about 91 million miles)
ε_{Earth} = specific mechanical energy of Earth's orbit (km^2/s^2)

Notice we use μ of the Sun because we're referencing our motion to the Sun. This gives us Earth's velocity around the Sun, as well as the spacecraft's velocity with respect to the Sun while it's in orbit around Earth.

Next we find the velocity we need to enter the transfer ellipse. As before, we start with specific mechanical energy

$$\varepsilon_{transfer} = -\frac{\mu_{Sun}}{2a_{transfer}}$$ (7-8)

where
$\varepsilon_{transfer}$ = specific mechanical energy of heliocentric transfer orbit (km^2/s^2)
μ_{Sun} = gravitational parameter of the Sun (km^3/s^2)
$a_{transfer}$ = major axis of transfer orbit (km)

The major axis of the transfer orbit is found from

$$a_{transfer} = \frac{R_{to\ Earth} + R_{to\ target}}{2}$$ (7-9)

where
$a_{transfer}$ = semi-major axis of the transfer orbit (km)
$R_{to\ Earth}$ = radius from the Sun to the Earth (km)
$R_{to\ target}$ = radius from the Sun to the target planet (km)

The velocity for a transfer orbit at Earth can then be found by using

$$V_{\text{transfer at Earth}} = \sqrt{2\left(\frac{\mu_{\text{Sun}}}{R_{\text{to Earth}}} + \varepsilon_{\text{transfer}}\right)} \qquad (7\text{-}10)$$

where

$V_{\text{transfer at Earth}}$	= velocity the spacecraft needs near Earth to transfer to the target planet (km/s)
μ_{Sun}	= gravitational parameter of the Sun (km^3/s^2)
$R_{\text{to Earth}}$	= radius from the Sun to the Earth (km)
$\varepsilon_{\text{transfer}}$	= specific mechanical energy of heliocentric transfer orbit (km^2/s^2)

The difference between these two velocities, V_{Earth} and $V_{\text{transfer at Earth}}$, is the velocity relative to the Earth which we must have as we leave the Earth.   For the patched-conic approximation, this velocity difference is called the Earth-departure velocity, $V_{\infty\,\text{Earth}}$ or "V infinity at Earth." (Why "V infinity"? As we'll see in a bit, this is the spacecraft's velocity at an "infinite" distance from Earth.)

$$V_{\infty\,\text{Earth}} = \left|V_{\text{transfer at Earth}} - V_{\text{Earth}}\right| \qquad (7\text{-}11)$$

where

$V_{\infty\,\text{Earth}}$	= velocity "at infinity" with respect to Earth (km/s)
$V_{\text{transfer at Earth}}$	= velocity the spacecraft needs near Earth to transfer to the target planet (km/s)
V_{Earth}	= Earth's velocity around the Sun

We can relate this to our previous experience with the Hohmann Transfer by thinking of $V_{\infty\,\text{Earth}}$ as the ΔV_1 for the heliocentric transfer even though no actual burn takes place here.

Let's review what this all means. The Earth (and any spacecraft in orbit around the Earth) is going around the Sun at V_{Earth}. To enter a heliocentric transfer orbit to the target planet, the spacecraft needs to increase its velocity with respect to the Sun by an amount $V_{\infty\,\text{Earth}}$, as shown in Figure 7-11.

Continuing with our heliocentric transfer, let's see what happens at the other end when we get near the target planet. Remember from the "big picture" of the Sun-centered Hohmann Transfer, the spacecraft coasts 180° around the Sun from Earth to the target planet. We can compute the spacecraft's velocity when it arrives in the region of the target planet as we did earlier.

$$V_{\text{transfer at target}} = \sqrt{2\left(\frac{\mu_{\text{Sun}}}{R_{\text{to target}}} + \varepsilon_{\text{transfer}}\right)} \qquad (7\text{-}12)$$

where

$V_{\text{transfer at target}}$	= velocity the spacecraft has on transfer orbit near target planet (km/s)

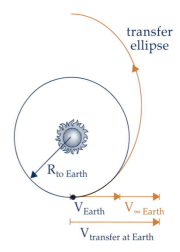

Figure 7-11. Starting the Transfer. To enter the heliocentric transfer orbit, the spacecraft must change its velocity by an amount $V_{\infty\,\text{Earth}}$.

μ_{Sun} = gravitational parameter of Sun (km^3/s^2)
$R_{to\ target}$ = distance from the Sun to the target planet (km)
$\varepsilon_{transfer}$ = specific mechanical energy of the transfer orbit (km^2/s^2)

As with any Hohmann Transfer, the spacecraft must change its velocity on the transfer ellipse to match the transfer planet's circular velocity around the Sun. (If it doesn't, the probe will fly by the planet and head back toward Earth's orbit.)

$$V_{target} = \sqrt{2\left(\frac{\mu_{Sun}}{R_{to\ target}} + \varepsilon_{target}\right)} \qquad (7\text{-}13)$$

where
V_{target} = velocity of target planet around the Sun (km/s)
μ_{Sun} = gravitational parameter of Sun (km^3/s^2)
$R_{to\ target}$ = distance from the Sun to the target planet (km)
ε_{target} = specific mechanical energy of target planet's orbit (km^2/s^2)

We now know the heliocentric velocity we have and the velocity we want. All that remains is to determine the difference between the two, which we'll call "V infinity target, $V_{\infty\ target}$."

$$V_{\infty\ target} = \left|V_{transfer\ at\ target} - V_{target}\right| \qquad (7\text{-}14)$$

where
$V_{\infty\ target}$ = velocity "at infinity" with respect to target planet (km/s)
$V_{transfer\ at\ target}$ = velocity the spacecraft has on transfer orbit near target planet (km/s)
V_{target} = velocity of target planet around the Sun (km/s)

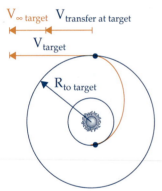

Figure 7-12. Arriving at the Target Planet. From the Sun-centered perspective, the spacecraft must change its velocity by an amount $V_{\infty\ target}$ to match orbits with the target planet.

Here again, using our Hohmann Transfer experience, you can think of $V_{\infty\ target}$ as ΔV_2. Keep in mind, however, no actual burn takes place at this point. From the perspective of an observer standing on the Sun, we see the spacecraft arriving at $V_{transfer\ at\ target}$ and the target planet moving at V_{target} with respect to the Sun. The spacecraft must change its velocity by an amount $V_{\infty\ target}$, as shown in Figure 7-12. The method for determining both $V_{\infty\ Earth}$ and $V_{\infty\ target}$ is in Example 7-1 (Part 1).

How and when do we actually fire our rockets to achieve $V_{\infty\ Earth}$ and $V_{\infty\ target}$? To find out, we need to examine the other two problems in the patched-conic approximation.

Example 7-1 (Part 1)

Problem Statement

The Jet Propulsion Lab (JPL) wants to send a probe from Earth ($R_{\text{to Earth}} = 1.496 \times 10^8$ km) to Mars ($R_{\text{to Mars}} = 2.278 \times 10^8$ km) to map landing sites for future manned missions. The probe will leave Earth from a parking orbit of 6697 km and arrive at Mars in another parking orbit of 3580 km.

- Part 1: What is the "extra" velocity the spacecraft needs to leave Earth ($V_{\infty\ \text{Earth}}$) and that it has at Mars ($V_{\infty\ \text{Mars}}$)?

- Part 2: What ΔV is needed in a parking orbit around Earth to begin the transfer?

- Part 3: What ΔV is needed to inject into Mars orbit and what is the total mission ΔV?

Problem Summary—Part 1

Given: $R_{\text{to Earth}} = 1.496 \times 10^8$ km
$R_{\text{to Mars}} = 2.278 \times 10^8$ km
$R_{\text{park at Earth}} = 6697$ km
$R_{\text{park at Mars}} = 3580$ km

Find: $V_{\infty\ \text{Earth}}$, $V_{\infty\ \text{Mars}}$ (Part 1)

Problem Diagram

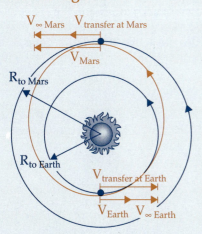

Conceptual Solution

Elliptical Hohmann Transfer—Problem 1

1) Find semi-major axis of transfer orbit, a_{transfer}

2) Find energy of transfer orbit, $\varepsilon_{\text{transfer}}$

3) Find velocity of Earth around the Sun, V_{Earth}

4) Find velocity in transfer orbit at Earth, $V_{\text{transfer at Earth}}$

5) Find velocity at infinity near Earth, $V_{\infty\ \text{Earth}}$

6) Find velocity of Mars around the Sun, V_{Mars}

7) Find velocity in transfer orbit at Mars, $V_{\text{transfer at Mars}}$

8) Find velocity at infinity near Mars, $V_{\infty\ \text{Mars}}$

Analytical Solution

1) Find a_{transfer}

$$a_{\text{transfer}} = \frac{R_{\text{to Earth}} + R_{\text{to Mars}}}{2}$$

$$= \frac{1.496 \times 10^8 \text{ km} + 2.278 \times 10^8 \text{ km}}{2}$$

$$a_{\text{transfer}} = 1.887 \times 10^8 \text{ km}$$

2) Find $\varepsilon_{\text{transfer}}$

$$\varepsilon_{\text{transfer}} = \frac{-\mu_{\text{Sun}}}{2a_{\text{transfer}}} = \frac{-1.327 \times 10^{11} \frac{\text{km}^3}{\text{s}^2}}{2\,(1.887 \times 10^8 \text{ km})}$$

$$\varepsilon_{\text{transfer}} = -351.6 \frac{\text{km}^2}{\text{s}^2}$$

(Note: negative energy because the transfer orbit is elliptical)

3) Find velocity of Earth around the Sun, V_{Earth}

$$\varepsilon_{\text{Earth}} = \frac{-\mu_{\text{Sun}}}{2a_{\text{Earth}}} = \frac{-\mu_{\text{Sun}}}{2R_{\text{to Earth}}}$$

Example 7-1 (Part 1) Continued

$$= \frac{-1.327 \times 10^{11} \dfrac{km^3}{s^2}}{2\,(1.496 \times 10^8\ km)}$$

$= -443.5\ km^2/s^2$ (Note: negative energy because the Earth is in a circular orbit around the Sun)

$$V_{Earth} = \sqrt{2\left(\frac{\mu_{Sun}}{R_{to\ Earth}} + \varepsilon_{Earth}\right)}$$

$$= \sqrt{2\left(\frac{1.327 \times 10^{11} \dfrac{km^3}{s^2}}{1.496 \times 10^8\ km} - 443.5 \dfrac{km^2}{s^2}\right)}$$

$$= 29.78\ km/s$$

4) Find $V_{transfer\ at\ Earth}$

$$V_{transfer\ at\ Earth} = \sqrt{2\left(\frac{\mu_{Sun}}{R_{to\ Earth}} + \varepsilon_{transfer}\right)}$$

$$= \sqrt{2\left(\frac{1.327 \times 10^{11} \dfrac{km^3}{s^2}}{1.496 \times 10^8\ km} - 351.6 \dfrac{km^3}{s^2}\right)}$$

$$\doteq 32.72\ km/s$$

5) Find $V_{\infty\ Earth}$

$$V_{\infty\ Earth} = \left|V_{transfer\ at\ Earth} - V_{Earth}\right|$$
$$= \left|32.72\ km/s - 29.78\ km/s\right|$$
$$= 2.94\ km/s$$

6) Find velocity of Mars around the Sun, V_{Mars}

$$\varepsilon_{Mars} = \frac{-\mu_{Sun}}{2a_{Mars}} = \frac{-\mu_{Sun}}{2R_{to\ Mars}}$$

$$= \frac{-1.327 \times 10^{11} \dfrac{km^3}{s^2}}{2\,(2.278 \times 10^8\ km)}$$

$= -291.3\ km^2/s^2$ (Note: negative energy)

$$V_{Mars} = \sqrt{2\left(\frac{\mu_{Sun}}{R_{to\ Mars}} + \varepsilon_{Mars}\right)}$$

$$= \sqrt{2\left(\frac{1.327 \times 10^{11} \dfrac{km^3}{s^2}}{2.278 \times 10^8\ km} - 291.3 \dfrac{km^2}{s^2}\right)}$$

$$= 24.14\ km/s$$

7) Find $V_{transfer\ at\ Mars}$

$$V_{transfer\ at\ Mars} = \sqrt{2\left(\frac{\mu_{Sun}}{R_{to\ Mars}} + \varepsilon_{transfer}\right)}$$

$$= \sqrt{2\left(\frac{1.327 \times 10^{11} \dfrac{km^3}{s^2}}{2.278 \times 10^8\ km} - 351.6 \dfrac{km^2}{s^2}\right)}$$

$$= 21.49\ km/s$$

8) Find $V_{\infty\ Mars}$

$$V_{\infty\ Mars} = \left|V_{Mars} - V_{transfer\ at\ Mars}\right|$$
$$= \left|24.14 \frac{km}{s} - 21.49 \frac{km}{s}\right|$$
$$= 2.65\ km/s$$

Interpreting the Results

To leave Earth and enter an interplanetary trajectory to Mars, our probe needs to gain 2.94 km/s with respect to the Sun. To match orbits with Mars, the probe must gain another 2.65 km/s with respect to Mars. We'll see how these velocities are achieved in Parts 2 and 3.

Hyperbolic Earth Departure—Problem 2

Remember, we broke the interplanetary problem into three problems based on which attracting body (Sun, Earth, or target planet) was the major player in the satellite's trajectory. We just described how to get onto the Sun-centered transfer ellipse. In Problem 2 we now back up to see how we get from an Earth-centered trajectory to the Sun-centered one.

As we know, a mass in space exerts a certain gravitational pull on other bodies. Newton's law of gravitation describes this force as decreasing inversely with the square of the distance from the body. Theoretically, the gravitational attraction from any body reaches out to infinity. From a practical standpoint, however, the gravitational pull of a given body becomes insignificant when you get far enough away. As you leave the Earth, for instance, at some point its gravitational pull becomes insignificant and the pull of other bodies, such as the Moon and Sun, begins to dominate. Thus, a planet's gravitational pull is effective only within a specific volume of space. We will define the *sphere of influence (SOI)* to be the volume of space around a planet or attracting body where that body causes the dominant gravitational force, as shown in Figure 7-13.

This means the Earth's gravity dominates a satellite travelling close to the Earth. The size of the sphere of influence depends on the planet's mass (the more massive the planet the farther it can extend its gravitational reach) and how close it is to the Sun (the closer the Sun the harder it will pull). The SOI's size is approximated by

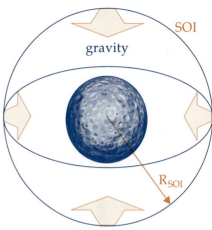

Figure 7-13. Sphere of Influence. A planet's Sphere of Influence (SOI) is the volume of space within which the planet's gravitational force is dominant.

$$\text{radius of SOI} = a_{planet}\left(\frac{m_{planet}}{m_{Sun}}\right)^{\frac{2}{5}} \qquad (7\text{-}15)$$

where

a_{planet} = semi-major axis of planet's orbit around the Sun (km)
m_{planet} = mass of planet (kg)
m_{Sun} = mass of Sun = 1.989×10^{30} kg

Earth's sphere of influence is approximately 1,000,000 km in radius. To put this in perspective, imagine the Earth being the size of a baseball, as in Figure 7-14. Its SOI would extend out 78 times its radius or 7.9 m (26 ft.) (well beyond the orbit of the Moon). Appendix D lists the sizes of the spheres of influence for other planets in the solar system.

Now let's see how we relate our motion within this SOI to motion outside. From Problem 1, we know we want to leave the SOI with some velocity we called $V_{\infty \, Earth}$. Let's think about what kind of trajectory we must be on within the SOI to do this. Recall from our discussion of orbital trajectories that parabolic and hyperbolic trajectories allow a satellite to escape from an attracting body. Will the parabolic trajectory work? Remember, a parabola has a specific mechanical energy equal to zero.

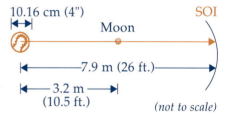

Figure 7-14. The Earth's Sphere of Influence Extends Well Beyond the Orbit of the Moon. To put this in perspective, imagine the Earth were the size of a baseball; then the Moon would be 3.2 m (10.5 ft.) away and the SOI 7.9 m (26 ft.).

$$\varepsilon = 0 \text{ (for parabola)}$$

This means if we depart the Earth on a parabolic trajectory we'll go some large distance from the Earth (out to the boundary of the sphere of influence) and arrive with zero velocity relative to the Earth.

$$V_{\text{relative to Earth}} = \sqrt{2\left(\frac{\mu}{R_{\text{"at infinity"}}} + \varepsilon\right)} = 0 \text{ (for a parabola)}$$

This would place our satellite in an orbit around the Sun exactly like the Earth's. Relative to the Earth, we'd be stationary, so we wouldn't go anywhere. To illustrate this, imagine you're driving down the interstate at 55 m.p.h. If your friend pulls up "at infinity" next to you also going 55 m.p.h., the relative velocity between the two cars is zero. This would be the case of the velocity at the sphere of influence on a parabolic trajectory. A spacecraft arriving at the SOI on a parabolic trajectory would have the same velocity as the Earth going around the Sun or 29.78 km/s. But relative to the Earth its velocity would be zero, so it wouldn't really go anywhere! Realize, of course, that a parabolic trajectory is a "special" case that can't be achieved in real-life. However, by understanding this special case, we can better understand what trajectories we really need.

So, what trajectory do we use to go somewhere? A hyperbolic trajectory! As we increase our satellite's velocity to the value for a hyperbola, our specific mechanical energy goes from negative (for an ellipse) to zero (for a parabola) to positive for the hyperbola. Now when we arrive at the boundary of the sphere of influence we have some *excess* velocity. How much excess velocity do we need to get to our destination? Our solution of the heliocentric transfer gave us the answer: $V_{\infty \text{ Earth}}$. This excess put us onto the elliptical transfer orbit relative to the Sun which we need to get to our destination planet.

Now we can start at the launch pad and work our way out to the sphere of influence to see how we reach the SOI with this required excess velocity. For technical as well as operational reasons, an interplanetary probe is never launched directly into the transfer orbit. Instead, the launch vehicle will first put our spacecraft into a circular parking orbit close to the Earth. This allows ground controllers time to check out all systems and wait for the right moment to start the upperstage. In a circular parking orbit, we can find the probe's velocity with respect to Earth as before

$$V_{\text{park at Earth}} = \sqrt{2\left(\frac{\mu_{\text{Earth}}}{R_{\text{park at Earth}}} + \varepsilon_{\text{park at Earth}}\right)} \tag{7-16}$$

where

$V_{\text{park at Earth}}$ = velocity in Earth parking orbit (km/s)
μ_{Earth} = gravitational parameter of Earth
= 3.986×10^5 km^3/s^2
$R_{\text{park at Earth}}$ = radius of parking orbit (km)
$\varepsilon_{\text{park at Earth}}$ = energy in parking orbit around Earth (km^2/s^2)

$$= \frac{-\mu_{\text{Earth}}}{2a_{\text{park at Earth}}} = \frac{-\mu_{\text{Earth}}}{2R_{\text{park at Earth}}}$$

225

Next, the spacecraft has to be boosted from the circular orbit onto the hyperbolic trajectory. It will then coast on this hyperbolic escape trajectory from the parking-orbit altitude to the sphere of influence, arriving with the needed excess velocity, as shown in Figure 7-15.

To see how much we must increase the spacecraft's velocity in the parking orbit to change to the hyperbolic orbit, we consider specific mechanical energy, ε. Using our energy relationship for this hyperbolic trajectory with respect to Earth, we get

$$\varepsilon_{\infty \text{ Earth}} = \frac{V^2_{\infty \text{ Earth}}}{2} - \frac{\mu_{\text{Earth}}}{R_{\infty \text{ Earth}}}$$

where

$\varepsilon_{\infty \text{ Earth}}$ = specific mechanical energy on the hyperbolic departure trajectory (km^2/s^2)

$V_{\infty \text{ Earth}}$ = velocity of the satellite at the SOI relative to the Earth (km/s)

$R_{\infty \text{ Earth}}$ = "infinite" distance to the SOI from the center of the Earth (km)

Because the distance to the sphere of influence is so big ($R_{\infty \text{Earth}} \approx \infty$), the term for potential energy effectively goes to zero. Thus, the energy at the SOI becomes

$$\boxed{\varepsilon_{\infty \text{ Earth}} = \frac{V^2_{\infty \text{ Earth}}}{2}} \tag{7-17}$$

where

$\varepsilon_{\infty \text{ Earth}}$ = specific mechanical energy on the hyperbolic departure trajectory (km^2/s^2)

$V_{\infty \text{ Earth}}$ = velocity of the satellite at the SOI relative to the Earth (km/s)

(Note: This energy is positive, which, as we learned in Chapter 4, means the trajectory is hyperbolic). We can now use this relationship to find the velocity the spacecraft must achieve at R_{park} to enter the hyperbolic escape trajectory. Solving the energy equation for velocity we get

$$\boxed{V_{\text{hyperbolic at Earth}} = \sqrt{2\left(\frac{\mu_{\text{Earth}}}{R_{\text{park at Earth}}} + \varepsilon_{\infty \text{ Earth}}\right)}} \tag{7-18}$$

where

$V_{\text{hyperbolic at Earth}}$ = velocity on hyperbolic departure trajectory at parking-orbit altitude (km/s)

μ_{Earth} = gravitational parameter of Earth (km^3/s^2)

$R_{\text{park at Earth}}$ = radius of parking orbit (km)

$\varepsilon_{\infty \text{ Earth}}$ = specific mechanical energy on the hyperbolic departure trajectory (km^2/s^2)

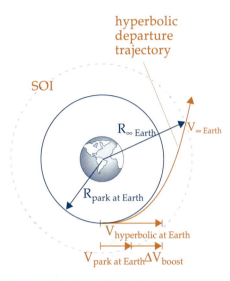

Figure 7-15. Escaping Earth. To escape the Earth on a hyperbolic trajectory and arrive at the SOI with the required velocity to enter into the heliocentric transfer orbit, $V_{\infty \text{ Earth}}$, a satellite needs to increase its velocity in the parking orbit, $V_{\text{park at Earth}}$, by an amount ΔV_{boost}.

Finally, we can determine the change in velocity, ΔV, we need to go from our circular parking orbit onto the hyperbolic departure trajectory away from Earth.

$$\Delta V_{boost} = \left| V_{hyperbolic\ at\ Earth} - V_{park\ at\ Earth} \right| \qquad (7\text{-}19)$$

where

ΔV_{boost} = change in velocity required to go from parking orbit around Earth onto hyperbolic departure trajectory (km/s)

$V_{hyperbolic\ at\ Earth}$ = velocity on hyperbolic departure trajectory at parking-orbit altitude (km/s)

$V_{park\ at\ Earth}$ = velocity in parking orbit around Earth (km/s)

ΔV_{boost} represents the change in velocity the spacecraft must generate to start its interplanetary journey. An upperstage—with a rocket engine, fuel tanks, and guidance system—normally provides this velocity. Once ΔV_{boost} is applied, we're on our way to another planet!

To recap, let's see how we've "patched" Problems 1 and 2 together. We started in a circular parking orbit around the Earth at a radius $R_{park\ at\ Earth}$ and a velocity $V_{park\ at\ Earth}$. We then fired our rocket engines to increase our velocity by an amount ΔV_{boost} and to give us a velocity of $V_{hyperbolic\ at\ Earth}$. This velocity puts us on a hyperbolic trajectory away from Earth. Upon arrival at the SOI, the spacecraft has the necessary velocity, $V_{\infty\ Earth}$, to escape the Earth and enter a heliocentric elliptical transfer orbit. By design, our geocentric hyperbolic trajectory blends smoothly onto the heliocentric elliptical orbit, so the two are now "patched" together. Example 7-1 (Part 2) shows how to determine ΔV_{boost}.

Example 7-1 (Part 2)

Problem Summary—Part 2

Given: $R_{to\ Earth} = 1.496 \times 10^8$ km
$R_{to\ Mars} = 2.278 \times 10^8$ km
$R_{park\ at\ Earth} = 6697$ km
$R_{park\ at\ Mars} = 3580$ km
$V_{\infty\ Earth} = 2.94$ km/s (from Part 1)

Find: ΔV_{boost} (Part 2)

Problem Diagram

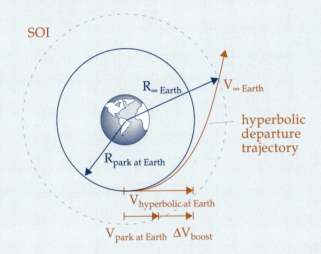

Conceptual Solution

Hyperbolic Earth Departure—Problem 2

1) Find the energy on the hyperbolic escape trajectory, $\varepsilon_{\infty\ Earth}$

2) Find the velocity in the circular parking orbit around Earth, $V_{park\ at\ Earth}$

3) Find the velocity to enter the hyperbolic escape trajectory from the parking orbit, $V_{hyperbolic\ at\ Earth}$

4) Find the velocity change needed to enter the hyperbolic escape trajectory, ΔV_{boost}

Analytical Solution

1) Find energy on the hyperbolic escape trajectory. Energy is the same everywhere on the trajectory, so we'll find it at the SOI.

$$\varepsilon_{\infty\ Earth} = \frac{V_{\infty Earth}^2}{2} = \frac{\left(2.94\frac{km}{s}\right)^2}{2} = 4.323\frac{km^2}{s^2}$$

(Note: positive energy on hyperbolic trajectory)

2) Find $V_{park\ at\ Earth}$

$$\varepsilon_{park\ at\ Earth} = \frac{-\mu_{Earth}}{2a_{park\ at\ Earth}} = \frac{-\mu_{Earth}}{2R_{park\ at\ Earth}}$$

$$= \frac{-3.986005 \times 10^5 \frac{km^3}{s^2}}{2\,(6697\ km)}$$

$$= -29.76\frac{km^2}{s^2}$$

(negative energy in circular orbit around Earth)

$$V_{park\ at\ Earth} = \sqrt{2\left(\frac{\mu_{Earth}}{R_{park\ at\ Earth}} + \varepsilon_{park\ at\ Earth}\right)}$$

$$= \sqrt{2\left(\frac{3.986005 \times 10^5 \frac{km^3}{s^2}}{6697\ km} - 29.76\frac{km^2}{s^2}\right)}$$

$$= 7.71\ km/s$$

3) Find $V_{hyperbolic\ at\ Earth}$

$$V_{hyperbolic\ at\ Earth} = \sqrt{2\left(\frac{\mu_{Earth}}{R_{park\ at\ Earth}} + \varepsilon_{\infty\ Earth}\right)}$$

$$= \sqrt{2\left(\frac{3.986005 \times 10^5 \frac{km^3}{s^2}}{6697 km} + 4.323\frac{km^2}{s^2}\right)}$$

$$= 11.30\ km/s$$

4) Find ΔV_{boost}

$$\Delta V_{boost} = \left|V_{hyperbolic\ at\ Earth} - V_{park\ at\ Earth}\right|$$

Example 7-1 (Part 2) Continued

$$= \left| 11.30 \frac{km}{s} - 7.71 \frac{km}{s} \right|$$

$$= 3.59 \text{ km/s}$$

Interpreting the Results

From our circular parking orbit around Earth, we must fire our upperstage engines to increase our velocity by 3.59 km/s, to can achieve a hyperbolic departure trajectory which will start us on our way to Mars.

Hyperbolic Planetary Arrival—Problem 3

At the other end of our transfer ellipse, we arrive at the target planet's sphere of influence. The spacecraft's velocity with respect to the planet will be

$$V_{\infty target} = \left| V_{target} - V_{transfer\ at\ target} \right|$$ (7-20)

where

$V_{\infty\ target}$ = velocity of the satellite at the SOI with respect to target planet (km/s)

V_{target} = velocity of target planet with respect to the Sun (km/s)

$V_{transfer\ at\ target}$ = velocity in heliocentric transfer orbit with respect to the Sun at the target planet (km/s)

The velocity, $V_{\infty\ target}$, marks the crossing of another boundary in the patched conic. It represents the excess hyperbolic velocity at the sphere of influence. Just as we left the Earth on a hyperbolic trajectory, we arrive at the target on a hyperbolic trajectory, as shown in Figure 7-16. Applying the same energy technique, we can find the probe's energy at an "infinite" radius from the target planet:

$$\varepsilon_{\infty\ target} = \frac{V_{\infty\ target}^2}{2}$$ (7-21)

where

$\varepsilon_{\infty\ target}$ = specific mechanical energy on hyperbolic arrival trajectory (km^2/s^2)

$V_{\infty\ target}$ = velocity of the satellite at the SOI with respect to target planet (km/s)

In a reflection of the way we left the Earth, we want to coast on the hyperbolic trajectory and then perform the ΔV or "burn" at a certain distance away from the target planet ($R_{park\ at\ target}$). This maneuver will move our satellite into a circular parking orbit around the target planet. Solving for this velocity

$$V_{hyperbolic\ at\ target} = \sqrt{2\left(\frac{\mu_{target}}{R_{park\ at\ target}} + \varepsilon_{\infty\ target}\right)}$$ (7-22)

where

$V_{hyperbolic\ at\ target}$ = velocity of spacecraft when it reaches parking orbit altitude (km/s)

μ_{target} = gravitational parameter of target planet (km^3/s^2)

$R_{park\ at\ target}$ = radius of parking orbit around target planet (km)

$\varepsilon_{\infty\ target}$ = specific mechanical energy on hyperbolic arrival trajectory (km^2/s^2)

If we didn't change our velocity we'd speed around the planet and back out into space on the other leg of the hyperbolic trajectory. To avoid this, we do a ΔV to put ourselves into a circular orbit at the parking orbit's radius.

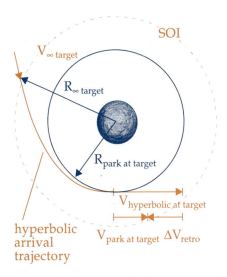

Figure 7-16. Arriving at the Target Planet. If the spacecraft does nothing as it approaches the target planet, it will swing by on a hyperbolic trajectory and depart the SOI on the other side. To slow down enough to be captured into orbit at a radius R_{park}, it must change its velocity by an amount ΔV_{retro}.

$$V_{\text{park at target}} = \sqrt{2\left(\frac{\mu_{\text{target}}}{R_{\text{park at target}}} + \varepsilon_{\text{park at target}}\right)} \qquad (7\text{-}23)$$

$V_{\text{park at target}}$ = velocity in parking orbit around target planet (km/s)
μ_{target} = gravitational parameter of target planet (km³/s²)
$R_{\text{park at target}}$ = radius of parking orbit around target planet (km)
$\varepsilon_{\text{park at target}}$ = energy in parking orbit around target (km²/s²)

This velocity change to enter the parking orbit is

$$\Delta V_{\text{retro}} = \left| V_{\text{park at target}} - V_{\text{hyperbolic at target}} \right| \qquad (7\text{-}24)$$

where
ΔV_{retro} = change in velocity required to go from a hyperbolic arrival trajectory or to the parking orbit around the target planet (km/s)

$V_{\text{park at target}}$ = velocity in parking orbit around target planet (km/s)

$V_{\text{hyperbolic at target}}$ = velocity on a hyperbolic arrival trajectory at parking-orbit altitude (km/s)

ΔV_{retro} represents how much we must slow down to be captured into orbit around the target planet. The total velocity change our rockets must provide for the mission is then

$$\Delta V_{\text{mission}} = \Delta V_{\text{boost}} + \Delta V_{\text{retro}} \qquad (7\text{-}25)$$

where
$\Delta V_{\text{mission}}$ = total velocity change required for the mission (km/s)
ΔV_{boost} = change in velocity required to go from parking orbit around Earth onto a hyperbolic departure trajectory (km/s)

ΔV_{retro} = change in velocity required to go from a hyperbolic arrival trajectory to the parking orbit around the target planet (km/s)

The propulsion system on the satellite must provide $\Delta V_{\text{mission}}$ to leave the Earth and arrive into orbit around the target planet. Example 7-1 (Part 3) shows how to calculate ΔV_{retro}.

To review, interplanetary flight involves connecting or "patching" three conic sections to approximate our path. We fire our engines once to produce ΔV_{boost}, which allows us to depart from a circular parking orbit around the Earth on a hyperbolic trajectory, arriving at the Earth's SOI with some excess velocity, $V_{\infty \text{ Earth}}$. This excess velocity is enough to put us on a heliocentric, elliptical transfer orbit from Earth to the target planet. After travelling half of the ellipse, we cross the target planet's SOI, arriving at the target planet on a hyperbolic trajectory. We then fire our engines a second time to produce ΔV_{retro}, which captures us into orbit. Table 7-1 summarizes the three regions of the problem and the necessary equations.

Example 7-1 (Part 3)

Problem Summary—Part 3

Given: $R_{to\ Earth} = 1.496 \times 10^8$ km
$R_{to\ Mars} = 2.278 \times 10^8$ km
$R_{park\ at\ Earth} = 6697$ km
$R_{park\ at\ Mars} = 3580$ km
$\mu_{Mars} = 43{,}050$ km^3/s^2
$V_{\infty\ Mars} = 2.65$ km/s (from Part 1)
$\Delta V_{boost} = 3.59$ km/s (from Part 2)

Find: $\Delta V_{mission}$ (Part 3)

Problem Diagram

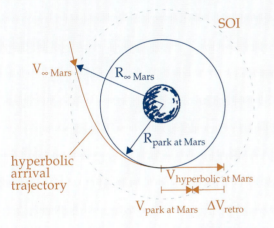

hyperbolic
arrival
trajectory

$V_{\infty\ Mars}$ $R_{\infty\ Mars}$ SOI
$R_{park\ at\ Mars}$
$V_{hyperbolic\ at\ Mars}$
$V_{park\ at\ Mars}$ ΔV_{retro}

Conceptual Solution

Hyperbolic Planetary Arrival—Problem 3

1) Find the energy on the hyperbolic arrival trajectory, $\varepsilon_{\infty\ Mars}$

2) Find the velocity on the hyperbolic arrival trajectory at parking-orbit altitude, $V_{hyperbolic\ at}$
   $_{Mars}$

3) Find the velocity in the circular parking orbit, $V_{park\ at\ Mars}$

4) Find the velocity change needed to capture into the parking orbit, ΔV_{retro}

5) Find the total velocity change needed for the

Analytical Solution

1) Find energy at Mar's SOI

$$\varepsilon_{\infty\ Mars} = \frac{V^2_{\infty\ Mars}}{2} = \frac{\left(2.65\frac{km}{s}\right)^2}{2} = 3.51\frac{km^2}{s^2}$$

2) Find $V_{hyperbolic\ at\ Mars}$

$$V_{hyperbolic\ at\ Mars} = \sqrt{2\left(\frac{\mu_{Mars}}{R_{park\ at\ Mars}} + \varepsilon_{\infty\ Mars}\right)}$$

$$= \sqrt{2\left(\frac{43{,}050\frac{km^3}{s^2}}{3580\ km} + 3.51\frac{km^2}{s^2}\right)}$$

$$= 5.57\ km/s$$

3) Find $V_{park\ at\ Mars}$

$$\varepsilon_{park\ at\ Mars} = \frac{-\mu_{Mars}}{2a_{park\ at\ Mars}} = \frac{-\mu_{Mars}}{2R_{park\ at\ Mars}}$$

$$= \frac{-43{,}050\frac{km^3}{s^2}}{2\,(3580\ km)}$$

$$= -6.0123\ km^2/s^2$$

$$V_{park\ at\ Mars} = \sqrt{2\left(\frac{\mu_{Mars}}{R_{park\ at\ Mars}} + \varepsilon_{park\ at\ Mars}\right)}$$

$$= \sqrt{2\left(\frac{43{,}050\ km^3/s^2}{3580\ km} - 6.0123\,km^2/s^2\right)}$$

$$= 3.47\ km/s$$

4) Find ΔV_{retro}

$$\Delta V_{retro} = \left|V_{park\ at\ Mars} - V_{hyperbolic\ at\ Mars}\right|$$

$$= \left|3.47\frac{km}{s} - 5.57\frac{km}{s}\right|$$

$$= 2.10\ km/s$$

Example 7-1 (Part 3) Continued

5) Find $\Delta V_{mission}$

$$\Delta V_{mission} = \Delta V_{boost} + \Delta V_{retro}$$
$$= 3.59 \text{ km/s} + 2.10 \text{ km/s}$$
$$= 5.69 \text{ km/s}$$

Interpreting the Results

To be captured into an orbit around Mars, we need to slow down by 2.10 km/s. Thus, to go from a parking orbit of 6697 km radius at Earth to a parking orbit of 3580 km radius at Mars requires a total velocity change from our rockets of 5.69 km/s.

Table 7-1. Summary of Interplanetary Transfer Problem.

Region	Reference Frame	Energy	Velocities
1: From Earth to target planet (elliptic trajectory)	Heliocentric-ecliptic	$\varepsilon_{transfer} = -\dfrac{\mu_{Sun}}{2a_{transfer}}$ $a_{transfer} = \dfrac{R_{to\ Earth} + R_{to\ target}}{2}$	$V_{Earth} = \sqrt{2\left(\dfrac{\mu_{Sun}}{R_{to\ Earth}} + \varepsilon_{Earth}\right)}$ $V_{target} = \sqrt{2\left(\dfrac{\mu_{Sun}}{R_{to\ target}} + \varepsilon_{target}\right)}$ $V_{transfer\ at\ Earth} = \sqrt{2\left(\dfrac{\mu_{Sun}}{R_{to\ Earth}} + \varepsilon_{transfer}\right)}$ $V_{transfer\ at\ target} = \sqrt{2\left(\dfrac{\mu_{Sun}}{R_{to\ target}} + \varepsilon_{transfer}\right)}$ $V_{\infty Earth} = \left|V_{transfer\ at\ Earth} - V_{Earth}\right|$ $V_{\infty target} = \left|V_{target} - V_{transfer\ at\ target}\right|$
2: Departure from Earth (hyperbolic trajectory)	Geocentric-equatorial	$\varepsilon_{\infty Earth} = \dfrac{V_{\infty Earth}^2}{2}$	$V_{\infty\ Earth} = \text{from above}$ $V_{hyperbolic\ at\ Earth} = \sqrt{2\left(\dfrac{\mu_{Earth}}{R_{park\ at\ Earth}} + \varepsilon_{\infty\ Earth}\right)}$ $V_{park\ at\ Earth} = \sqrt{2\left(\dfrac{\mu_{Earth}}{R_{park\ at\ Earth}} + \varepsilon_{park\ at\ Earth}\right)}$ $\Delta V_{boost} = \left|V_{hyperbolic\ at\ Earth} - V_{park\ at\ Earth}\right|$
3: Arrival at target planet (hyperbolic trajectory)	Planet-centered equatorial	$\varepsilon_{\infty target} = \dfrac{V_{\infty target}^2}{2}$	$V_{\infty\ target} = \text{from above}$ $V_{hyperbolic\ at\ target} = \sqrt{2\left(\dfrac{\mu_{target}}{R_{park\ at\ target}} + \varepsilon_{\infty target}\right)}$ $V_{park\ at\ target} = \sqrt{2\left(\dfrac{\mu_{target}}{R_{park\ at\ target}} + \varepsilon_{park\ at\ target}\right)}$ $\Delta V_{retro} = \left|V_{park\ at\ target} - V_{hyperbolic\ at\ target}\right|$

Transfer Time of Flight

So far we've spent all our time trying to figure how much ΔV we need to get between planets. But before we pack our bags, we'd like to know how long we'll be gone. Let's see how long an interplanetary transfer might take. The Hohmann Transfer ellipse approximates the time for our interplanetary journey. Because the Hohmann Transfer is one-half of an ellipse, we use one-half the period of the transfer orbit to determine the time of flight

$$TOF = \pi \sqrt{\frac{a_{transfer}^3}{\mu_{Sun}}}$$

(7-26)

where
TOF = time of flight (s)
$a_{transfer}$ = semi-major axis of transfer ellipse (km)
μ_{Sun} = gravitational parameter of the Sun = 1.327×10^{11} km^3/s^2

(This does neglect the hyperbolic departure and arrival trajectories, but those times are insignificant compared to the long journey around the Sun.) Using information on a trip to Mars presented in Example 7-1 (Part 1, 2, and 3), we can determine the time of flight.

$$a_{transfer} = 1.887 \times 10^8 \text{ km}$$

$$TOF = \pi \sqrt{\frac{a_{transfer}^3}{\mu_{Sun}}} = \pi \sqrt{\frac{(1.887 \times 10^8 \text{ km})^3}{1.327 \times 10^{11} \dfrac{\text{km}^3}{\text{s}^2}}}$$

$$= 2.235 \times 10^7 \text{ s}$$
$$= 6208 \text{ hours}$$

$$TOF = 258.7 \text{ days or about } 8.5 \text{ months}$$

That's a long time to be stuck in a tiny spaceship. Very long manned missions, such as the trip to Mars, put significant demands on mission planners to sustain the crew by protecting them from the space environment and providing them life support. Consequently, all interplanetary missions have been unmanned so far.

Phasing of Planets for Rendezvous

Another item we've omitted from our discussion is the proper phasing of the planets for the transfer. Recall from Chapter 6 how we related the rendezvous problem to the way a quarterback synchronizes the flight of a football to a receiver down field. Similarly, a crew of astronauts going to Mars would like to find the planet there when they arrive! The way we solve this problem is the same as the rendezvous of two spacecraft in Earth orbit. We need to find the lead angle, α_{lead}, and then the final phase angle, ϕ_{final}. For our Mars example, the angular velocity, ω, is

$$\omega = \sqrt{\frac{\mu}{R^3}} \tag{7-27}$$

$$\omega_{Mars} = \sqrt{\frac{\mu_{Sun}}{R^3_{to\ Mars}}} = \sqrt{\frac{1.327 \times 10^{11} \dfrac{km^3}{s^2}}{(2.278 \times 10^8\ km)^3}} = 1.06 \times 10^{-7} \frac{rad}{s}$$

We would then use the TOF from Equation (7-26) to calculate the lead angle:

$$\alpha_{lead} = \omega\ TOF \tag{7-28}$$

$$\alpha_{lead} = (\omega_{Mars})\ (TOF) = \left(1.06 \times 10^{-7} \frac{rad}{s}\right)(2.235 \times 10^7\ s)$$

$$\alpha_{lead} = 2.37\ rad = 135.8°$$

So the final phase angle is

$$\phi_{final} = 180° - \alpha_{lead} \tag{7-29}$$

$$\phi_{final} = 180° - 135.8° = 44.2°$$

A final angle of 44.2° means that, when we start on the interplanetary Hohmann Transfer, Mars needs to be 44.2° ahead of the Earth, as shown in Figure 7-17. If the Earth is 50° behind Mars, the phase angle, $\phi_{initial}$, is 50°. Using the equation we developed in Chapter 6, we can solve for the amount of time to wait before we need to have our spacecraft ready to go.

But if this were a manned mission, we'd have to worry not only about getting out to Mars but also getting back. We know from Chapter 6 that the proper configuration for rendezvous recurs periodically, but we'd like to know how long we need to wait between opportunities. This wait time between successive opportunities is called the *synodic period*. Using the 2π relationship we discussed in Chapter 6, we can develop a relationship for synodic period as

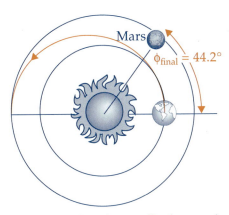

Figure 7-17. Interplanetary Rendezvous. A spacecraft launched from Earth to rendezvous with Mars should be 44.2° behind Mars at launch.

$$Synodic\ Period = \frac{2\pi}{\left|\omega_{Earth} - \omega_{target\ planet}\right|} \tag{7-30}$$

For a trip to Mars using a Hohmann Transfer, the proper alignment between the two planets repeats itself about every two years. This means if you have a spacecraft sitting on the launch pad ready to go and you somehow miss your chance to launch, you must wait two more years before you get another chance! Appendix D gives the synodic periods between Earth and the other planets.

▆ Section Review

Key Terms

sphere of influence (SOI)
synodic period

Key Equations

$$\varepsilon = \frac{V^2}{2} - \frac{\mu}{R}$$

$$\varepsilon = \frac{-\mu}{2a}$$

$$\Delta V_{mission} = \Delta V_{boost} + \Delta V_{retro}$$

$$TOF = \pi \sqrt{\frac{a_{transfer}^3}{\mu_{Sun}}}$$

See Table 7-1 for other Key
Equations

Key Concepts

➤ The patched-conic approximation breaks interplanetary transfer into three regions and their associated problems:

- **Problem 1:** From Earth to target planet. This is a heliocentric transfer on an elliptical trajectory from the Earth to the target planet:
  - The velocity needed to change from the Earth's orbit around the Sun to the elliptical transfer orbit is $V_{\infty \, Earth}$
  - The velocity needed to change from the elliptical transfer orbit to the target planet's orbit around the Sun is $V_{\infty \, target}$

- **Problem 2:** Earth departure. The spacecraft leaves Earth's vicinity on a hyperbolic trajectory:
  - The Earth's Sphere of Influence (SOI) defines an imaginary boundary in space within which the Earth's gravitational pull dominates. When a spacecraft goes beyond the SOI, it has effectively left the Earth. Earth's SOI extends out to about 1,000,000 km.
  - To begin interplanetary transfer, the spacecraft needs a velocity relative to the Earth of $V_{\infty \, target}$ at the SOI. This is achieved with ΔV_{boost}.

- **Problem 3:** Arrival at the target planet. The spacecraft arrives at the target planet on a hyperbolic trajectory:
  - The planet's Sphere of Influence (SOI) defines an imaginary boundary in space within which the planet's gravitational pull dominates
  - The velocity the spacecraft has at the SOI, relative to the planet is $V_{\infty \, target}$
  - To capture into orbit around the target planet, the spacecraft performs a ΔV_{retro} burn.

➤ Table 7-1 summarizes all equations for the interplanetary transfer

➤ Practically speaking, a spacecraft will begin the interplanetary transfer in a parking orbit around the Earth and end in a final mission orbit around the target planet. To transfer between these two orbits, we must fire our engines twice to get two separate velocity changes:

- **First burn:** ΔV_{boost} transfers the spacecraft from a circular parking orbit around Earth to a hyperbolic trajectory with respect to Earth. This trajectory "patches" to an elliptical orbit around the Sun, taking the spacecraft to the target planet.

- **Second burn:** ΔV_{retro} slows the spacecraft from a hyperbolic trajectory with respect to the target planet to a final mission orbit around the target planet

Continued on next page

Section Review (Continued)

Key Concepts (Continued)

➤ The total change in velocity the rocket must provide for the mission is the sum of ΔV_{boost} and ΔV_{retro}

➤ The time of flight (TOF) for an interplanetary transfer is approximately one-half the period of the transfer ellipse

➤ To ensure the target planet is there when you arrive, you must consider phasing of the planets:
  • The phasing problem for interplanetary transfer is identical to the rendezvous problem from Chapter 6

➤ The synodic period of two planets is the time between successive chances to launch

7.3 Gravity-Assist Trajectories

▬ In This Section You'll Learn to...

☞ Explain the concept of gravity-assist trajectories and how they can help spacecraft travel between the planets

In the previous sections we saw how to get to other planets using an interplanetary Hohmann Transfer. Even this efficient maneuver requires a tremendous amount of rocket propellant, significantly driving up the mission's cost. Often, a mission relying solely on rockets to get the required ΔV simply can't be justified. For example, if the Voyager probes sent on a "grand tour" of the solar system had to rely totally on rockets to steer between the planets, they would never have gotten off the ground.

Fortunately, spacecraft can sometimes get "free" velocity changes with gravity assist as they travel through the solar system. This *gravity assist* technique uses a planet's gravitational field and orbital velocity to "sling shot" the spacecraft, changing its velocity (in magnitude and direction) with respect to the Sun.

Gravity-assist trajectories often make the difference between possible and impossible missions. After the Challenger accident, the Galileo spacecraft's missions were in trouble. Liquid-fueled upperstages were banned from the payload bay of the shuttle, and solid fuel rockets simply weren't powerful enough to send it directly to Jupiter. Fortunately, mission designers hit on the idea of going to Jupiter by way of Venus and Earth. They used one gravitational assist from Venus and two from Earth to speed the spacecraft on its way; hence, the name VEEGA (Venus, Earth, Earth Gravity Assist).

Of course, gravity assisted velocity changes aren't totally free. Actually, the spacecraft "steals" velocity from the planet, causing the planet to speed up or slow down ever so slightly in its orbit around the Sun. Gravity assist can also bend the satellite's trajectory to allow it to travel closer to some other point of interest. The Ulysses spacecraft got a gravity assist from Jupiter to change planes, sending it out of the ecliptic into a polar orbit around the Sun.

How does gravity assist work? As the spacecraft enters a planet's sphere of influence (SOI), the planet pulls it in the direction of the planet's motion. Of course, as Isaac Newton said, for every action there is an equal but opposite reaction. Because the spacecraft is insignificantly small compared to the size of the planet, the same force which can radically change the spacecraft's trajectory has no significant affect on the planet. (Imagine a mosquito landing on a dinosaur; the dinosaur would never notice!)

As we saw in the previous sections, a spacecraft's velocity depends on the eye of the beholder. During a gravity assist, we want to change the spacecraft's velocity with respect to the Sun; this puts it on a different heliocentric orbit so it can go somewhere else.

Let's start by thinking about what's going on from the planet's perspective. As a spacecraft flies by on a hyperbolic trajectory, the planet pulls it. If the spacecraft passes behind the planet, as shown in Figure 7-18, it's pulled in the direction of the planet's motion and thus gains velocity (and hence energy) with respect to the Sun. This alters the spacecraft's original orbit around the Sun, as shown in Figure 7-19, sending it off to a different part of the solar system to rendezvous with another planet.

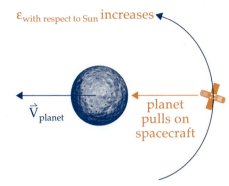

Figure 7-18. Spacecraft Passing Behind Planet. During a gravity-assist maneuver, a spacecraft's energy will increase with respect to the Sun if it passes behind the planet.

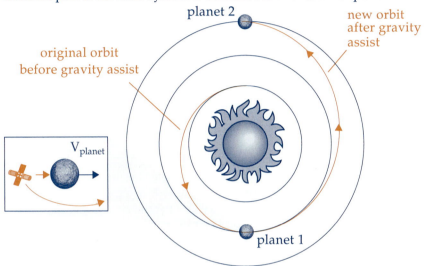

Figure 7-19. Gravity Assist. During a gravity assist, a spacecraft is pulled by a planet, changing its velocity with respect to the Sun and thus altering its orbit around the Sun. The planet's orbit also changes, but very little.

If a spacecraft passes in front of a planet as in Figure 7-20, it's pulled in the opposite direction, slowing the spacecraft down and lowering its orbit with respect to the Sun.

A gravity assist which changes the magnitude of the spacecraft's velocity is called *orbit pumping*. Using a planet's gravity to change the direction of travel is called *orbit cranking*. The gravity of Jupiter "cranked" the Ulysses solar-polar satellite out of the ecliptic plane into an orbit around the Sun's poles.

Realistically, mission requirements will constrain which planets we can use for gravity assist or whether it's even possible. Voyager's flyby past Neptune did not allow it to change its orbit to go to Pluto, for example. Doing so would have required a trip beneath the surface of Neptune (which would have been a bit hard on the spacecraft!).

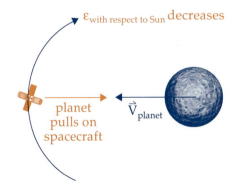

Figure 7-20. Spacecraft Passing in Front of Planet. During a gravity-assist maneuver, a spacecraft's energy will decrease with respect to the Sun if it passes in front of the planet.

Astro Fun Fact
Slowing Down the Earth

Feeling a little slow today? Maybe you should. Following the Challenger accident, mission designers for the Galileo spacecraft had a problem. The high-energy Centaur upperstage they'd planned to use to boost the spacecraft to Jupiter wasn't available due to safety concerns. Instead, they had to use the safer, but less powerful, Inertial Upper Stage (IUS). Unfortunately, the IUS couldn't provide the necessary ΔV_{boost} to begin the transfer to Jupiter. Faced with this dilemma, they came up with a unique solution—not one, not two, but three gravity assists! To achieve the necessary ΔV, they planned to "steal" energy from Venus once and Earth twice, hence the name VEEGA—Venus, Earth, Earth Gravity Assist. The entire maneuver is in the figure on the right. Launched on October 18, 1989, Galileo began its journey to Jupiter by going to Venus,

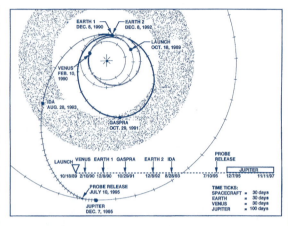

flying by on February 10, 1990, and gaining 2.2 km/s.   On December 8, 1990, Galileo returned to Earth on a hyperbolic trajectory that increased its velocity by another 5.2 km/s. Exactly two years later, it made a second pass by Earth, gaining the additional 3.7 km/s needed to take it out to Jupiter, where it will arrive on 7 December 1995. But this extra ΔV from the gravity assists was not totally "free." Energy must be conserved, so for Galileo to speed up, the Earth had to slow down. But don't worry, the result of both assists slowed Earth down by a grand total of 4.3×10^{-21} km/s. That's about five inches in one billion years!

Information and diagram courtesy of NASA/JPL

▬ Section Review

Key Terms

gravity assist
orbit cranking
orbit pumping

Key Concepts

➤ Gravity-assist trajectories allow a spacecraft to get "free" velocity changes by using a planet's gravity to change a spacecraft's trajectory. This changes the spacecraft's velocity with respect to the Sun and slows the planet (but by a very small amount).

References

Bate, Roger R., Donald D. Miller, and Jerry E. White. *Fundamentals of Astrodynamics.* New York, NY: Dover Publications, Inc., 1971.

G.A. Flandro. *Fast Reconnaissance Missions to the Outer Solar System Utilizing Energy Derived from the Gravitational Field of Jupiter.* Astronautica Acta, Vol. 12 No. 4, 1966.

Wilson, Andrew. *Space Directory.* Alexandria, VA: Jane's Information Group Inc., 1990.

Mission Problems

7.1 Equation of Motion

1   How do we define the heliocentric-ecliptic coordinate frame?

2   Why can't we include all the appropriate gravitational forces for an interplanetary trajectory in a single equation of motion and solve it directly?

7.2 The Patched-Conic Approximation

3   What are the three regions used in the patched-conic approximation of an interplanetary trajectory and what coordinate frame does each use?

4   What "head start" does a spacecraft have in achieving the large velocities needed to travel around the Sun?

5   What does a planet's sphere of influence (SOI) represent?

6   What does the size of the SOI depend on?

7   Why do we escape from the Earth (or any planet) on a hyperbolic trajectory versus a parabolic trajectory?

8   A new research spacecraft is designed to measure the environment in the tail of Earth's magnetic field. It's in a circular parking orbit with a radius of 12,756 km. What ΔV will take the spacecraft to the Earth's SOI such that it arrives there with zero velocity with respect to Earth?

9   To continue the studies of Venus begun by the Magellan probe, NASA is sending a remote-sensing satellite to that planet.

Given:

$R_{\text{to Earth}} = 1.496 \times 10^8$ km

$R_{\text{to Venus}} = 1.081 \times 10^8$ km

$R_{\text{park at Earth}} = 6600$ km

$R_{\text{park at Venus}} = 6400$ km

$\mu_{\text{Earth}} = 3.986005 \times 10^5$ km^3/s^2

$\mu_{\text{Venus}} = 3.257 \times 10^5$ km^3/s^2

$\mu_{\text{Sun}} = 1.327 \times 10^{11}$ km^3/s^2

a) Find the semi-major axis and specific mechanical energy of the transfer orbit.

b) Find $V_{\infty \text{ Earth}}$

c) Find $V_{\infty \text{ Venus}}$

d) Find ΔV_{boost}

e) Find ΔV_{retro}

f) Find $\Delta V_{\text{mission}}$

10 Find the time of flight (TOF) for the above mission to Venus.

11 Adjust Equation (7-15) and compute the radius of the moon's sphere of influence relative to the Earth. Assume:

$$a_{Moon} = 3.844 \times 10^5 \text{ km and } \left(\frac{mass_{Moon}}{mass_{Earth}}\right) = \frac{1}{81.3}$$

12 What is the needed phase angle for the rendezvous of a flight from Earth to Saturn?

Given:

$$R_{to\ Earth} = 1.497 \times 10^8 \text{ km}$$

$$R_{to\ Saturn} = 1.426 \times 10^9 \text{ km}$$

7.3   Gravity-Assist Trajectories

13 Explain how gravity assist can be used to get "free ΔV" for interplanetary transfers.

14 Is ΔV from a gravity assist really "free?"

15 Can a flyby of the Sun help us change a satellite's interplanetary trajectory?

Projects

16 Write a computer program or spreadsheet to compute $\Delta V_{mission}$ for trips to all of the other eight planets in the solar system.

Mission Profile—Magellan

On May 4, 1989, the Magellan Space Probe launched from the Space Shuttle Atlantis—the first interplanetary spacecraft launched from a space shuttle and the first U.S. interplanetary mission since 1978. Magellan was designed to produce the first high-resolution images of the surface of Venus using a synthetic aperture radar (SAR). The project's manager was NASA, with coordinators at the Jet Propulsion Laboratory and the primary builders at Martin Marietta and Hughes Aircraft Company. By studying our neighboring planet with a spacecraft equipped with radar, researchers hope to learn more about how the solar system and the Earth were formed.

Mission Overview

The primary objectives for Magellan were

✓ Map at least 70% of the surface of Venus through the SAR

✓ Take altitude readings of its surface, take its "temperature" (radiometry), and chart changes in its gravity field.

Mission Data

✓ Launched by the Space Shuttle and placed into an interplanetary transfer orbit by the inertial upperstage

✓ The interplanetary transfer orbit took Magellan around the Sun and placed it into a highly elliptical mapping orbit around Venus. The mapping orbit was elliptical to allow Magellan to map close to perigee and give it enough time to transmit data back to Earth near apogee.

✓ Magellan was built from left-over and back-up parts of previous satellites such as Voyager, Galileo, and Ulysses. Its configuration included

  • Antennas with high, medium, and low gain plus an altimeter

  • A forward equipment module, housing the communications and radar electronics and reaction wheels for attitude control

  • A ten-sided spacecraft bus

  • Two solar panels

  • A propulsion module

✓ The main payload included the synthetic aperture radar and the altimeter

✓ To withstand harsh conditions close to the Sun, Magellan was covered in thermal blankets and heat-reflecting inorganic paint. Louvers were also used to dissipate heat created by the SAR and other electronic components.

✓ The high-gain antenna provides communications and data gathering. Deep Space Network antennas on Earth receive scientific data from the payload. This data then goes to the Jet Propulsion Laboratory for interpretation. The deep space network is also used to send transmissions to Magellan from the Earth.

✓ Because Magellan used the high-gain antenna for both radar mapping and data transmission to Earth, the spacecraft had to rotate four times in each orbit. A combination of reaction wheels, sun-sensors, and thrusters provided enough pointing accuracy for these maneuvers.

Mission Impact

Magellan performed better than mission planners expected, mapping more than 95% of Venus's surface with images better than any of the Earth. This single mission gathered more data than in all other NASA exploratory missions combined. Magellan's funding was cut for fiscal year 1993, the mission was to end on May 14, 1993.

For Discussion

• What information can we learn from studying Venus?

• How might the lessons learned from Magellan affect future unmanned space probes?

Contributors

Luciano Amutan and Scott Bell, United States Air Force Academy

References

Wilson, Andrew. *Interavia Space Directory*. Alexandria, VA: Jane's Information Group, 1990.

Young, Carolynn (ed.). *The Magellan Venus Explorer's Guide*. Pasadena, CA: Jet Propulsion Laboratory, 1990.

What a dish! Antennas scan the heavens to monitor the paths of satellites through space. *(Courtesy of NASA)*

Predicting Orbits

8

Robert B. Giffen
United States Air Force Academy

In This Chapter You'll Learn to...

☛ Determine the time of flight between two satellite positions in a given orbit

☛ Determine the satellite's future position using Kepler's Equation

☛ Describe the effects of perturbations on orbits and explain their practical applications

☛ Describe the overall problem of tracking satellites and predicting orbits

You Should Already Know...

❑ The assumptions of the restricted two-body problem (Chapter 4)

❑ The two-body equation of motion and its solution (Chapter 4)

❑ The definition and use of the classical orbital elements, including argument of latitude, u (Chapter 5)

What goes around comes around.

Anonymous

By now you should have a pretty good feel for orbits: how they look, how they're defined, and how they're used. So far, everything we've done with orbits has been the result of a few basic equations developed back in Chapter 4. Now it's time to take the next step. In this chapter we'll turn our attention to predicting orbits. To track satellites through space, we need to know where they are now and where they'll be later, so we can point our antennas at them to perform key functions. Although we can easily predict this motion when the orbit is a circle, the problem becomes more complicated when the orbit is an ellipse, and most orbits are at least slightly elliptical.

In this chapter we'll begin by looking at the orbit-prediction problem solved by Johannes Kepler over 300 years ago. We'll see how Kepler developed a method to analyze the motion of a satellite on an elliptical orbit. Next, we'll reexamine what we assumed about orbit motion back in Chapter 4. We'll see that, in the "real world," our assumptions break down for low-Earth-orbiting satellites. We'll find both the Earth's atmosphere and its non-spherical shape disturb or "perturb" orbits from the path we explained in Chapter 4. Finally, we'll combine Kepler's method with an understanding of how orbits are perturbed to predict orbits in the real world.

8.1 Predicting an Orbit (Kepler's Problem)

▰ In This Section You'll Learn to...

☛ Use Kepler's Equation to calculate a satellite's time of flight

☛ Use Kepler's Equation to predict a satellite's position at some future time

Let's take a look at the "big picture" of tracking and predicting shown in Figure 8-1. Pretend the Space Shuttle is in orbit and we've just received a position update on it from our tracking site located on an island in the Indian Ocean. The site provides us with the Shuttle's *range* (the distance from the tracking site), *azimuth* (the angle from true north), and *elevation* (the angle between the local horizon and the Shuttle). We can then convert these observations into a position vector, $\vec{R}_{initial}$, and, with at least one more set of observations, we can also find a velocity vector, $\vec{V}_{initial}$. Using the techniques developed in Chapter 5, we can then convert $\vec{R}_{initial}$ and $\vec{V}_{initial}$ into a set of classical orbital elements (COEs).

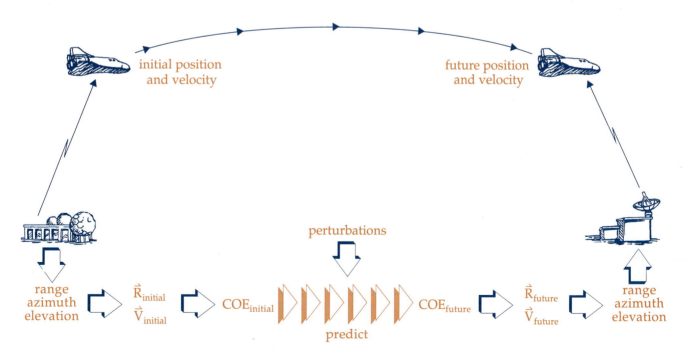

Figure 8-1. The Tracking Problem. To track and predict a satellite's orbit, we take tracking data, convert it first to $\vec{R}_{initial}$ and $\vec{V}_{initial}$ and then to COEs, and move these COEs to a future time (including perturbations). We then reconvert the COEs back to $\vec{R}_{future}$ and $\vec{V}_{future}$ vectors, and, finally, back to range, azimuth, and elevation angles.

For this experiment, we can pretend the astronauts aboard the Shuttle are going to deploy a large mirror and we're going to bounce a low-power laser off the mirror's surface and back to the ground. To aim our laser, we have to know *precisely* when the Shuttle will pass over our school and where it will be at that time. If all we know is the Shuttle's position and velocity some time in the past, we'll have to *predict* when it will be overhead. To predict or *propagate* any orbit into the future (COE_{future}), we have to develop a prediction method and understand how environmental factors such as drag and the Earth's oblate shape affect this prediction. Once we know COE_{future}, we can then re-convert these updated orbital elements back into $\vec{R}_{future}$ and $\vec{V}_{future}$, and then back into range, elevation, and azimuth. This will tell us how to point our laser.

In this section we'll determine the time of flight between two orbit positions. Then we'll predict a satellite's position within its orbit at some future time. We'll tackle the effects of drag and the oblate Earth in the next section.

Kepler's Equation and Time of Flight

In nice circular orbits, determining how long a satellite takes to get from an initial position to a future position is simple, because the satellite is moving at a constant speed. Using Figure 8-2, let's define this angular speed as the *mean motion, n*, which tells us the mean, or average, speed on the orbit. The mean motion can be found in terms of the period, P, and the radius, which in the case of circular orbits is the same as the semi-major axis, a. The mean motion is

$$n = \frac{angle}{time} = \frac{2\pi}{P} = \sqrt{\frac{\mu}{a^3}} \qquad (8\text{-}1)$$

where
n = mean motion (rad/s)
P = period (s)
μ = gravitational parameter (km^3/s^2)
a = semi-major axis (km)

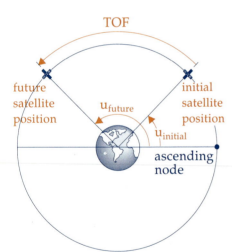

Figure 8-2. Time of Flight on a Circular Orbit. To find the time of flight between two satellite positions in a circular orbit, divide the angle between these two positions by the orbital mean motion, n. Conversely, to find the satellite's location some time in the future, just multiply the time of flight by the mean motion and add the result to the initial position.

Remember from Chapter 5 for a circular orbit, position is defined by the *argument of latitude, u*, which is the angular distance from the ascending node to the satellite. So, as Figure 8-2 shows, the time of flight in a circular orbit is

$$\text{Time of flight} = TOF = \frac{u_{future} - u_{initial}}{n} \qquad (8\text{-}2)$$

where
$u_{initial}$ = initial argument of latitude (rad)
u_{final} = final argument of latitude (rad)
n = mean motion (rad/s)

And the satellite's position at some future time is equal to its initial position plus the angle it travels through during the time of flight

$$u_{future} = u_{initial} + n\,(TOF) \qquad (8\text{-}3)$$

where

$$n\,(TOF) = \frac{angle}{time} \times time = angle$$

In an elliptical orbit, however, the satellite motion is not uniform. We don't know how the true anomaly, v, changes with time because it doesn't change uniformly. Here's where Johannes Kepler came to the rescue. Remember from Chapter 2, he was trying to make his theory match Tycho Brahe's meticulous observations of the planet Mars. To solve this problem, he figured out how to move v to a time in the future and, conversely, given a future v, how to find out how long Mars would take to get there.

Kepler's approach was purely geometrical—he related motion on a circle to motion on an ellipse. He also defined a satellite's mean motion, n, to be the average angular rate a satellite travels in one orbital period

$$n = \sqrt{\frac{\mu}{a^3}} \qquad (8\text{-}4)$$

He then defined a new angle called the *mean anomaly, M*

$$M = nT \qquad (8\text{-}5)$$

where
M = mean anomaly (rad)
n = mean motion (rad/s)
T = the time since last perigee passage (s)

Mean anomaly is an angle that has no physical meaning and can't be drawn in a picture. We'll have to describe it mathematically. Expressing this equation in terms of two points in the same orbit

$$M_{future} - M_{initial} = n\,(t_{future} - t_{initial}) - 2k\pi \qquad (8\text{-}6)$$

where

M_{future}	= mean anomaly when the satellite is in the future position (rad)
$M_{initial}$	= mean anomaly when the satellite is in the initial position (rad)
$t_{future} - t_{initial}$	= time of flight (TOF)
t_{future}	= time when the satellite is in the final position (e.g., 3:47 A.M.)
$t_{initial}$	= time when the satellite is in the initial position (e.g., 3:30 A.M.)
k	= the number of times the satellite passes perigee

It's important to keep track of time when working these problems. Figure 8-3 shows the relationship between the time since last perigee passage, T, and the time of the clock, t.

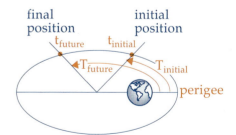

Figure 8-3. Keeping Track of Time in Kepler's Equation. We use T to represent the amount of time since the satellite passed perigee. t represents the actual time (shown on your watch) that the satellite is of a particular location. $T_{initial}$ is the amount at time elapsed since the satellite located in the initial position was at perigee (e.g., 30 min). $t_{initial}$ is the actual time when the satellite was at the initial position (e.g., 3:30 A.M.).

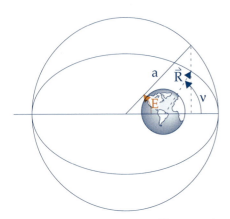

Figure 8-4. Eccentric Anomaly. The eccentric anomaly, E, is defined geometrically by circumscribing an elliptical orbit with a circle and relating E to M.

To relate elliptical motion to circular motion, he defined another new angle called the *eccentric anomaly, E*, so he could relate M to E and then E to ν, as seen in Figure 8-4. With all these things defined, Kepler was able to develop his now-famous equation, commonly called Kepler's Equation. (For this equation to work, all angles must be in radians.)

$$M = E - e \, \sin E \qquad (8\text{-}7)$$

where
E = eccentric anomaly (rad)
e = eccentricity

He then related E to ν using

$$\cos E = \frac{e + \cos \nu}{1 + e \, \cos \nu} \qquad (8\text{-}8)$$

where
ν = true anomaly (rad)

And ν to E through

$$\cos \nu = \frac{\cos E - e}{1 - e \, \cos E} \qquad (8\text{-}9)$$

Now we have the equations needed to solve two problems. The first problem, and the easiest, is finding the time of flight between two points in an orbit. Given $\nu_{initial}$ and ν_{future}, we simply go through the following steps:

- Use Equation (8-8) to solve for $E_{initial}$ and E_{future}
- Use Equation (8-7) to solve for $M_{initial}$ and M_{future}
- Use Equation (8-6) to solve for the time of flight ($t_{future} - t_{initial}$)

Remember in Chapter 5, when we solved for the orbital elements, we had to check the quadrant when taking the inverse cosine? We have to do that here, too, but it's easier. Look at Figure 8-4 and you can see ν and E are always in the same half plane. It turns out mean anomaly follows the same rule. That means if ν is between 0° and 180°, so are E and M.

Kepler's Equation and Future Position

The second problem we can solve using Kepler's method is far more practical. This involves determining a satellite's position at some future time, t_{future}, as shown in Figure 8-5. This second problem is trickier. We assume we know where the satellite is at time $t_{initial}$, so we know $\nu_{initial}$. We start by finding $E_{initial}$, using Equation (8-8). Then we find $M_{initial}$ using Kepler's Equation (8-7).

$$M_{initial} = E_{initial} - e \, \sin E_{initial}$$

Now, because we know t_{future}, using Equation (8-6), we can find M_{future}.

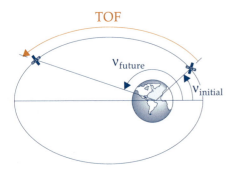

Figure 8-5. Time of Flight on an Elliptical Orbit. Predicting a satellite's future position at some time, t_{future}, is the second application of Kepler's Equation.

$$M_{future} - M_{initial} = n(t_{future} - t_{initial}) - 2k\pi$$

Great. We're now on our way to finding v_{future}, which tells us where the satellite will be. So we go to Kepler's Equation again to find E_{future}. Let's rearrange this equation and put E on the left side

$$E_{future} = M_{future} + e \sin E_{future} \qquad (8\text{-}10)$$

OOPS! E_{future} is on both sides of the equation. This is called a *transcendental equation* and can't be solved for E_{future} directly. In fact, almost every notable mathematician over the past 300 years has tried to find a direct solution to this form of Kepler's Equation without success. So we must resort to "math tricks" to solve for E_{future}. The "math trick" we'll use is called *iteration*. To see how iteration works think about the kids' game Twenty Questions. In this game, your partner thinks of a person, place, or thing and you must guess what he's thinking of. You get 20 questions (guesses) to which your partner can answer only "yes" or "no." In seeking the right answer, a good player will systematically eliminate all other possibilities until only the correct answer remains.

A mathematical application of this can be seen using another transcendental equation

$$y = \cos y$$

Because we can't solve for y using algebra (we can't get the y out of the cosine function to put all the y's on the left side), we must iterate. Begin by taking a guess at the value for y, and take the cosine to see how close you were. Then take this as the new value of y and use it for the next guess, and keep doing this iteration until the new y equals the old y (or is, say, within 0.000001 radians of the old value).

Let's try it to see what the answer for y really is. Get out your calculator and use $\pi/4$ radians as your first guess for y. (Remember to set your calculator to use radians, not degrees.) Keep punching the cosine function button and you'll see the value slowly converges to 0.739085 radians (about 43°). Presto—you've now solved the transcendental equation y = cos y using iteration!

We can use this same iterative technique to solve Equation (8-10) for E_{future}. It turns out the values for M and E are always pretty close together, even for the most eccentric orbits, so let's use M_{future} for our first guess at E_{future}. Here's the algorithm

- Use M_{future} for the first E_{future}

- Solve Equation (8-10) for a new E_{future}

- Use this new E_{future} for the next guess for Equation (8-10)

- Keep doing the previous step until E_{future} doesn't change by much (less than about 0.0001 rad). At this point the solution is said to have *converged*.

This brute force iteration method will solve Equation (8-10), but there are better methods. The most notable is Newton's Iteration Method (see Bate, et al, pages 220–221).

Let's quickly summarize what we've learned. If we know where we are in an orbit and where we want to be, we can use Kepler's Equation to solve for the time of flight it takes to get there. The solution is very straightforward. If, however, we know where we are and want to know where we'll be at some future time, we can use Kepler's Equation to find that location only by iterating a transcendental equation for eccentric anomaly.

Astro Fun Fact
Astronomy vs. Astrology

Johannes Kepler was also a dabbler in astrology. From an early age, Kepler was interested in the study of how the movements of the heavenly bodies affected peoples' lives. The beginnings of astrology go back to around 2000 B.C. in Babylonia. It purports to tell how people are affected by the positions of the Earth, the planets (including the Sun and Moon), the zodiac, and the "houses" (similar to the zodiac except located on the Earth). At first, Kepler told fortunes for family members, but later he made calendars of predictions to make more money. Into his third and final year of seminary, Kepler was appointed to become a math instructor at a Lutheran school in Graz, Austria, after the death of the professor. After arriving in Graz, his successful prediction of cold weather, peasant uprisings, and invasion by the Turks made Kepler's calendars a hot item. Kepler was split on his feelings about astrology. He referred to it as the "foolish little daughter of respectable astronomy" and stated "if astrologers sometimes do tell the truth, it ought to be attributed to luck." However, Kepler's firm belief in the harmony and sense of order in the universe kept him involved in astrology. He was also able to provide food for his family and pay the bills doing prediction calendars. Kepler was a bridge between the mysticism of astrology and the realism of astronomy.

Dictionary of Scientific Biography. New York, NY: Scribner's Sons, 1992.

Contributed by Steve Crumpton, United States Air Force Academy

▬ Section Review

Key Terms

argument of latitude, u
azimuth
eccentric anomaly, E
elevation
iteration
mean motion, n
mean anomaly, M
range
transcendental equation

Key Equations

$$M = E - e \sin E$$

$$\cos E = \frac{e + \cos v}{1 + e \cos v}$$

$$\cos v = \frac{\cos E - e}{1 - e \cos E}$$

Key Concepts

➤ Kepler's Equation gives us the solution to two problems:
  • Finding the time of flight between two known orbital positions
  • Finding a future orbital position, given the time of flight

➤ Mean motion is the average speed of a satellite in orbit

➤ Mean anomaly is related to mean motion through the time, T, since passing by perigee

➤ Eccentric anomaly relates motion on an ellipse to motion on a circumscribed circle

➤ Satellite position defined by true anomaly, v, can be related to eccentric anomaly

➤ Given a new satellite position, v, we can find E, M, and finally T. Or given T (or some future time, t_{future}), we can find M, solve for E using substitution and iteration, and then find v.

Example 8-1

Problem Statement

The Space Shuttle is conducting Spacelab experiments in a slightly elliptical orbit (e = 0.1) with a semi-major axis of 7000 km. Data from the Tracking and Data Relay Satellite tell us the Shuttle's current true anomaly is 270°. How long until it reaches a true anomaly of 50°?

Problem Summary

Given: a = 7000 km
        e = 0.10
        $v_{initial}$ = 270°

Find: Time of flight from $v_{initial}$ = 270° to v_{future} = 50°

Problem Diagram

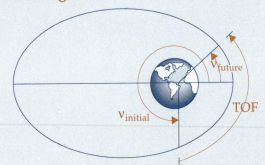

Conceptual Solution

1) Find mean motion, n

$$n = \frac{2\pi}{P} = \sqrt{\frac{\mu}{a^3}}$$

2) Find $E_{initial}$ and E_{future}

$$\cos E = \frac{e + \cos v}{1 + e \cos v}$$

3) Find $M_{initial}$ and M_{future}

$$M = E - e \sin E$$

4) Find time of flight

$$M_{future} - M_{initial} = n\,(t_{future} - t_{initial}) - 2k\pi$$

$$t_{future} - t_{initial} = \frac{M_{future} - M_{initial} + 2k\pi}{n}$$

Note from the diagram we must pass perigee once to get from $v_{initial}$ to v_{final}

Analytical Solution

1) Find mean motion, n

$$n = \sqrt{\frac{\mu}{a^3}} = \sqrt{\frac{3.986005 \times 10^5 \dfrac{km^3}{s^2}}{(7000 \text{ km})^3}}$$

$$= 0.001078 \text{ rad/s} = 14.82 \text{ rev/day}$$

2) Find $E_{initial}$, E_{future}

$$\cos E_{initial} = \frac{e + \cos v_{initial}}{1 + e \cos v_{initial}}$$

$$= \frac{0.1 + \cos 270°}{1.0 + 0.1 \cos 270°} = 0.1$$

$E_{initial}$ = 84.26° or 275.74°

Remember, v, M, and E must all lie in the same half plane. Therefore, because $v_{initial}$ = 270°, $E_{initial}$ = 275.74° = 4.81 rad

$$\cos E_{future} = \frac{e + \cos v_{future}}{1 + e \cos v_{future}}$$

$$= \frac{0.1 + \cos 50°}{1.0 + 0.1 \cos 50°} = 0.6979$$

E_{future} = 45.74° or 314.26°

E_{future} = 45.74° (same half-plane as v_{future}) = 0.798 rad

3) Find $M_{initial}$, M_{future}

$$M_{initial} = E_{initial} - e \sin E_{initial}$$

Note: Here $E_{initial}$ must be in **radians**

$$M_{initial} = 4.813 - 0.1 \sin 4.813 = 281.437° = 4.912 \text{ rad}$$

Example 8-1 (Continued)

$M_{future} = E_{future} - e \sin E_{future}$

$M_{future} = 0.798 - 0.1 \sin 0.798 = 41.597° = 0.726 \, rad$

4) Find Time of Flight

$$t_{future} - t_{initial} = \frac{M_{future} - M_{initial} + 2k\pi}{n}$$

Because we must pass perigee once, $k = 1$

$$= \frac{(0.726 - 4.912 + 2\pi) \, rad}{0.001078 \dfrac{rad}{s}}$$

$$= 1945.44 \, s = 32.42 \, min$$

Interpreting the Results

Using Kepler's equation, we found it takes about 32 minutes for the Space Shuttle to travel from $v_{initial} = 270°$ to $v_{future} = 50°$. We pass perigee one time so we must add in the factor 2π to the equation for time of flight. This answer makes sense because the period of a low-Earth orbit is about 90 minutes, and we're travelling 140°, or a little more than a third of the way around the orbit.

Example 8-2

Problem Statement

An Earth-observation satellite is in a slightly eccentric orbit (e = 0.1) with a semi-major axis of 7000 km and an inclination of 50°. If its current true anomaly is 270°, what will be the true anomaly six hours from now?

Problem Summary

Given: a = 7000 km
        e = 0.1
        i = 50°
        $v_{initial}$ = 270°

Find: The satellite's true anomaly six hours from now

Problem Diagram

$v_{initial} = 270°$

Conceptual Solution

1)   Find mean motion, n

$$n = \frac{2\pi}{P} = \sqrt{\frac{\mu}{a^3}}$$

2)   Find $E_{initial}$

$$\cos E = \frac{e + \cos v}{1 + e \ \cos v}$$

3)   Find $M_{initial}$

$$M = E - e \sin E$$

4)   Move mean anomaly to the desired time

$$M_{future} = M_{initial} + n \left(t_{future} - t_{initial} \right) - 2k\pi$$

5)   Solve for E_{future} using Kepler's equation (iterative solution required)

$$E = M + e \sin E$$

6)   Find v_{future}

$$\cos v = \frac{\cos E - e}{1 - e \ \cos E}$$

Analytical Solution

1)   Find mean motion, n

$$n = \sqrt{\frac{\mu}{a^3}} = \sqrt{\frac{398600.5 \ \dfrac{km^3}{s^2}}{(7000 \ km)^3}}$$

$$= 0.001078 \ rad/s = 14.82 \ rev/day$$

2)   Find $E_{initial}$

$$\cos E_{initial} = \frac{e + \cos v_{initial}}{1 + e \ \cos v_{initial}}$$

$$= \frac{0.1 + \cos 270°}{1.0 + 0.1 \ \cos 270°} = 0.1$$

$E_{initial} = 84.26°, 275.74° = 275.74° = 4.813 \ rad$

Remember, v, M, and E must all lie in the same half plane.

3)   Find $M_{initial}$

$M_{initial} = E_{initial} - e \sin E_{initial}$

Note: Here $E_{initial}$ must be in **Radians**

$M_{initial} = 4.813 - 0.1 \sin 4.813 = 4.912 \ rad = 281.437°$

4)   Move mean anomaly to the desired time

$M_{future} = M_{initial} + n(t_{future} - t_{initial}) - 2k\pi$

$M_{future} = 4.912 + (0.001078 \ rad/s)(6 \ hr \cdot 3600 \ s/hr) - 0$

$$= 28.197 \ rad$$

Because this value is greater than 2π, we need to keep subtracting 2π until M_{future} is less than 2π.

Example 8-2 (Continued)

$M_{future} = 28.197 - 4(2\pi) = 3.064$ rad

Physically, this means our satellite passes perigee four times in the next six hours

5) Solve for E_{future} using Kepler's equation (iterative solution required)

First, guess $E_{future} = M_{future} = 3.064$ rad

$E_{future} = M_{future} + e \sin E_{future}$

$\qquad = 3.064 + 0.1 \sin 3.064 = 3.072$ rad

Now, use $E_{future} = 3.072$ rad for your next guess

$E_{future} = M_{future} + e \sin E_{future}$

$\qquad = 3.064 + 0.1 \sin 3.072 = 3.071$ rad

Again, use $E_{future} = 3.071$ rad next

$E_{future} = M_{future} + e \sin E_{future}$

$\qquad = 3.064 + 0.1 \sin 3.071 = 3.071$ rad $= 176.0°$

Okay, within the accuracy we've chosen here, our solution has converged.

6) Find v_{future}

$$\cos v_{future} = \frac{\cos E_{future} - e}{1 - e \ \cos E_{future}} = \frac{\cos 3.071 - 0.1}{1 - 0.1 \ \cos 3.071}$$

$v_{future} = 176.34°$ or $185.66° = 176.34°$ (since $E_{future} = 176.0°$)

Interpreting the Results

In six hours our satellite will pass perigee four times and end up at $v = 176.34°$, which is about 94° behind where it is now. The period of our orbit is about 97 minutes, so in six hours we should go around the orbit about four times.

8.2 Orbit Perturbations

▬ In This Section You'll Learn to...

☛ Explain and determine how Earth's atmosphere changes a spacecraft's orbit

☛ Explain and determine how Earth's non-spherical shape changes a spacecraft's orbit

☛ Explain how sun-synchronous and Molniya orbits take advantage of the Earth's non-spherical shape

☛ Describe other sources of orbit perturbations

Back in Chapter 4 we developed an equation of motion for our satellite that looked like this

$$\sum \vec{F}_{external} = \vec{F}_{gravity} + \vec{F}_{drag} + \vec{F}_{thrust} + \vec{F}_{3rd\ body} + \vec{F}_{other} \qquad (8\text{-}11)$$

We then assumed

- Gravity was the only force
- The Earth's mass was much greater than the satellite's mass
- The Earth was spherically symmetric with uniform density, so it could be treated as a point mass

These assumptions led us to the restricted two-body equation of motion

$$\ddot{\vec{R}} + \frac{\mu}{R^2}\hat{R} = 0 \qquad (8\text{-}12)$$

The solution to this equation gave us the six classical orbital elements (COEs)

> a—semi-major axis
> e—eccentricity
> i—inclination
> Ω—longitude of the ascending node
> ω—argument of perigee
> ν—true anomaly

Under our assumptions, the first five of these elements remain constant for a given orbit. Only the true anomaly, ν, varies with time as the satellite travels around its fixed orbit. What happens if we now change some of our original assumptions? Other COEs besides ν will begin to change as well. Any changes to these COEs due to other forces are called *perturbations*. To see which COEs will change and by how much, let's look at our first assumption—gravity is the only force.

Atmospheric Drag

Gravity really isn't the only force acting on a satellite. In Chapter 3 we talked about the space environment and the effect the Earth's atmosphere has on a satellite's lifetime. Recall the Earth's atmosphere gets thinner with altitude but still has some effect up to altitudes as high as 1000 km (620 miles). Because most of our Earth-sensing satellites orbit at altitudes lower than 1000 km, this very thin air causes drag on these satellites. Let's look at how drag affects our orbital elements.

Drag is a non-conservative force—it takes energy away from the orbit in the form of friction on the satellite. Thus, we expect the semi-major axis, a, to get smaller. The eccentricity also decreases, since the orbit becomes more circular. Let's see why this is so. When a satellite in an elliptical orbit is at perigee, it has a greater speed then it would if the orbit were circular at that same altitude. The drag decreases the speed, making it closer to the circular orbit speed. That's exactly what we see in Figure 8-6. It's like drag is giving the satellite a small negative ΔV (slowing it down) each time it passes perigee.

Drag is very difficult to model because of the many factors affecting the Earth's upper atmosphere and the satellite's attitude. The Earth's day-night cycle, seasonal tilt, variable solar distance, the fluctuation of Earth's magnetic field, the Sun's 27-day rotation, and the 11-year cycle for Sun spots make the modeling task nearly impossible. The force of drag, which we'll see in greater detail in Chapter 10, also depends on the satellite's coefficient of drag and frontal area, which can also vary widely, further complicating the modeling problem.

The uncertainty in these variables is the main reason Skylab decayed and burned up in the atmosphere several years earlier than first predicted. For a given orbit, however, we can approximate how the semi-major axis and the eccentricity change with time, at least in the short term. Satellite-tracking organizations use complex techniques to determine these values and make them available on request for each satellite.

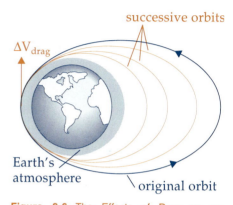

Figure 8-6. The Effects of Drag on an Eccentric Low-Earth Orbit. As the satellite passes through the atmosphere at perigee, the effect is like a series of small ΔVs, which lower apogee altitude, circularizing the orbit, until it decays and the satellite reenters.

Astro Fun Fact
Lagrange Points

Five points (L1–L5) near the Earth and Moon are within both bodies' influence and rotate at the same rate as the Moon. French scientist Joseph Lagrange discovered this curious fact in 1764 while attempting to solve the complex three-body problem. In his honor these points of equilibrium are now known as Lagrange Libration Points (LLP). LLPs exist for any multi-body system. The LLPs for the Earth-Moon system are shown at the right. These points would be an ideal location for future space stations or Lunar docking platforms, but no one has yet used them.

Cousins, Frank W. _The Solar System_. New York, NY: Pica Press, 1972.

Szebehely, Victor. _Theory of Orbits: The Restricted Problem of Three Bodies_. Yale University, New Haven, CT: Academic Press, Inc., 1967.

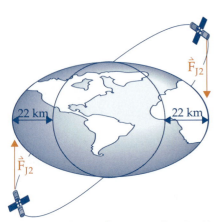

Figure 8-7. Earth's Oblateness. The Earth's equatorial bulge, shown here greatly exaggerated, causes a twisting force on a satellite's orbit.

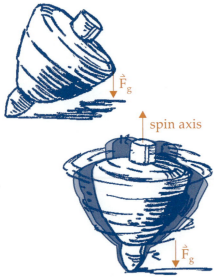

Figure 8-8. Precession of a Top. A non-spinning top will fall over if placed upright. But a spinning top wobbles about its spin axis due to the pull of gravity, $\vec{F}_g$. This motion is called precession.

Earth's Oblateness

Our second assumption about a satellite's mass being much less than Earth's mass is still true, but what about the third assumption? Columbus was wrong! The Earth isn't really round. From space, it looks like a big, blue spherical marble, but if you take a closer look, it's really kind of squashed. Thus, it can't really be treated as a point mass. We call this squashed shape *oblateness*. What exactly does an oblate Earth look like? Imagine spinning a ball of jello around its axis and you can visualize how the middle (or equator) of the spinning jello would bulge out—the Earth is fatter at the equator than at the poles. This bulge is often modeled by complex mathematics (which we won't do here) and is frequently referred to as the *J2 effect*. J2 is a constant describing the size of the bulge in the mathematical formulas used to model the oblate Earth.

What effect does this have on the orbit? Let's look at Figure 8-7. Here it's shown exaggerated; actually the bulge is only about 22 km thick. That is, the Earth's radius is about 22 km longer along the equator than through the poles.

Let's see if we can reason out what this bulge will do to the orbital elements. The force caused by the equatorial bulge is still gravity. Recall from Chapter 4 that gravity is a conservative force—the total mechanical energy must be conserved. So, one of the constants of orbital motion we defined—the specific mechanical energy, ε—will not change. That means the semi-major axis, a, will remain constant over the long term. It turns out the eccentricity, e, also doesn't change, although the explanation for this is beyond the scope of our discussion here. The bulge does pull on our orbit, so you would expect the inclination to change, but it doesn't! Because the satellite is going around in the orbit, the effect is to change the longitude of ascending node, Ω, and to move the argument of perigee, ω, within that plane. That's not very intuitive, but it's like a force acting on a spinning gyroscope, which we'll discuss in detail in Chapter 12. A similar analogy is a toy top. If you stand a non-spinning top up on its point, gravity will cause it to fall over. But if you spin the top first, gravity still tries to make it fall over, but because of its angular momentum, it begins to swivel—this motion is called *precession*, as shown in Figure 8-8. Let's examine this effect on Ω and ω more closely.

How J2 Affects the Longitude of the Ascending Node, Ω

The gravitational effect of this equatorial bulge slightly perturbs the satellite because the force is no longer coming from the center of the Earth. This causes the plane of the orbit to precess (like the spinning top), resulting in a movement of the ascending node, $\Delta\Omega$. This motion is westward for direct orbits (inclination < 90°), and eastward for retrograde orbits (inclination > 90°).

Figure 8-9 shows this *nodal regression rate*, $\dot{\Omega}$, as a function of inclination and orbital altitude. Let's look more closely at this figure. What it says is that the higher the satellite is, the less effect the bulge has on the orbit. This makes sense because the gravitational pull of the bulge decreases according to the inverse square law (μ/R^2). It also says that if the satellite is in a polar orbit (center of the graph), the bulge will have no effect. The greatest effect occurs at low altitudes with low inclinations. This makes sense, too, because the satellite will be traveling much closer to the bulge during its orbit, and thus be pulled more by the bulge. For low-altitude and low-inclination orbits, the ascending node can move as much as 9° per day (lower left corner and upper right corner of the graph).

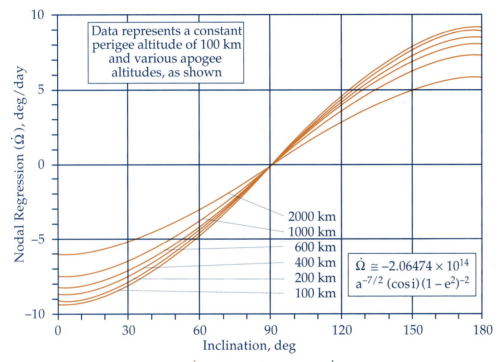

Figure 8-9. Nodal Regression Rate, $\dot{\Omega}$. The nodal regression rate, $\dot{\Omega}$, caused by the Earth's equatorial bulge. Positive numbers represent eastward movement; negative numbers represent westward movement. The less inclined an orbit is to the equator, the greater the effect of the bulge. The higher the orbit, the smaller the effect. An appropriate equation is given for finding $\dot{\Omega}$ in Earth orbit where a = semi-major axis, i = inclination, and e = eccentricity.

How J2 Affects the Argument of Perigee, ω

Figure 8-10 shows how perigee location rotates for an orbit with a perigee altitude of 100 km depending on the inclination for various apogee altitudes. This *perigee rotation rate, $\dot{\omega}$,* is difficult to explain physically, but it could be derived mathematically from the equation for J2 effects on perigee location. With this perturbation, the major axis, or line of nodes, will rotate in the direction of satellite motion if the inclination is less than 63.4° or greater than 116.6°. It will rotate opposite to satellite motion for inclinations between 63.4° and 116.6°.

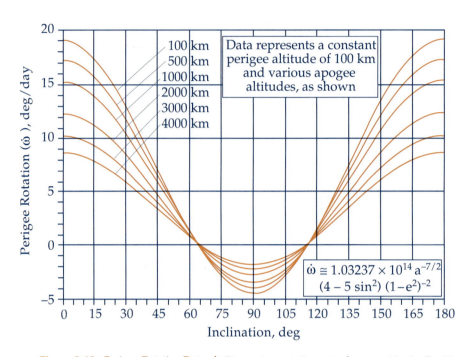

Figure 8-10. Perigee Rotation Rate, $\dot{\omega}$. The perigee rotation rate, $\dot{\omega}$, caused by the Earth's equatorial bulge depends on inclination and altitude at apogee. An approximate equation for $\dot{\omega}$ finding in Earth orbit is given where a = semi-major axis, i = inclination, and e = eccentricity.

In the figure:

Data represents a constant perigee altitude of 100 km and various apogee altitudes, as shown

100 km
500 km
1000 km
2000 km
3000 km
4000 km

$$\dot{\omega} \cong 1.03237 \times 10^{14}\, a^{-7/2} (4 - 5 \sin^2)(1 - e^2)^{-2}$$

Perigee Rotation ($\dot{\omega}$), deg/day — Inclination, deg

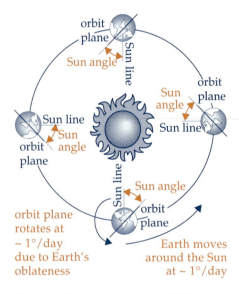

Figure 8-11. Sun-synchronous Orbits. Sun-synchronous orbits take advantage of the rate of change in longitude of ascending node, $\dot{\Omega}$, caused by the Earth's oblateness. By carefully selecting the proper inclination and altitude, we can match the rotation of Ω with the movement of the Earth around the Sun. In this case, the same angle between the orbit plane and the Sun will be maintained. Such orbits are useful for remote-sensing applications because shadows cast by targets on Earth then stay the same.

Sun-Synchronous and Molniya Orbits

The effects of the Earth's oblateness on the node and perigee positions give rise to two unique orbits that have very practical applications. The first of these, the *sun-synchronous orbit*, takes advantage of eastward nodal regression at inclinations greater than 90°. Looking at Figure 8-9, we see at an inclination of about 98° (depending on the satellite's altitude), the ascending node moves eastward about 1° per day. Coincidentally, the Earth also moves around the Sun about 1° per day (360° in 365 days), so at this sun-synchronous inclination, the satellite's *orbital plane* will always maintain the same orientation to the Sun, as shown in Figure 8-11. This means the satellite will see the same Sun angle when it passes over a particular point on the Earth's surface. As a result, the Sun shadows cast by features on the Earth's surface will not change when pictures are taken days or even weeks apart. This is important for remote-sensing missions such as reconnaissance, weather, and monitoring of the Earth's resources, because they use shadows to measure an object's height. By maintaining the same Sun angle day after day, observers can better track changes in weather, terrain, or man-made features.

The second unique orbit is the *Molniya orbit*, named after the Russian word for lightning (as in "quick-as-lightning"). This is a 12-hour orbit with high eccentricity (about e = 0.7) and a perigee location in the Southern Hemisphere. The inclination is 63.4°—why? Because at this inclination the perigee doesn't rotate, as Figure 8-10 shows, so the satellite will "hang"

263

over the Northern Hemisphere for nearly 11 hours of its 12-hour period before it whips "quick as lightning" through perigee in the Southern Hemisphere. Figure 8-12 shows the ground track for a Molniya orbit. The Russians used this orbit for their communication satellites because they didn't have launch vehicles large enough to put them into geosynchronous orbits from their far northern launch sites. Remember from Chapter 6, plane changes require large ΔVs and, as we'll see in Chapter 9, far northern launch sites don't get much extra kick from the Earth's rotation. These Molniya orbits also better covered the polar and high latitudes above 80° North.

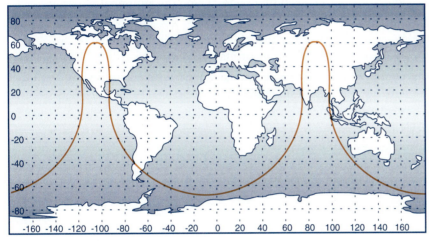

Figure 8-12. Molniya Orbit. Molniya orbits take advantage of the fact that $\dot{\omega}$ due to Earth's oblateness is zero at an inclination of 63.4°. Thus, apogee stays over the Northern Hemisphere, covering high latitudes 11 hours out of the orbit's 12-hour period.

Other Perturbations

Other perturbative forces can affect a satellite's orbit and its orientation within that orbit. These forces usually are much smaller than the J2 (oblate Earth) and drag forces, but depending on the required accuracy, satellite planners may have to anticipate their effects. These forces include

- Solar radiation pressure—can cause long-term orbit perturbations and unwanted satellite rotation
- Third-body gravitational effects (Moon, Sun, planets, etc.)—can perturb orbits at high altitudes and on interplanetary trajectories
- Unexpected thrusting—caused by either outgassing or malfunctioning thrusters; can perturb orbits or cause satellite rotation

▌ Section Review

Key Terms

J2 effect
Molniya orbit
nodal regression rate, $\dot{\Omega}$
oblateness
perigee rotation rate, $\dot{\omega}$
perturbations
precession
sun-synchronous orbit

Key Concepts

➤ Perturbations resulting from small disturbing forces cause our two-body orbit to vary

➤ Atmospheric drag causes orbital decay by decreasing both the semi-major axis, a, and the eccentricity, e.

➤ Equatorial bulge of the oblate Earth (J2) causes both the longitude of the ascending node, Ω, and the argument of perigee, ω, to change in a predictable way

➤ We use oblateness perturbations to practical advantage in sun synchronous and Molniya orbits

➤ Other perturbations may also have long-term effects on a satellite's orbit and orientation
  • Solar wind
  • Third body
  • Unexpected thrust

Example 8-3

Problem Statement

A remote-sensing satellite has the following orbital elements

$a_{initial} = 7303$ km

$e_{initial} = 0.001$

$i_{initial} = 50°$

$\Omega_{initial} = 90°$

$\omega_{initial} = 45°$

$v_{initial} = 0°$

U. S. Space Command has told you the semi-major axis is decreasing by 2 km/day. You would like nature to move the ascending node to a $\Omega = 30°$ before you command a satellite maneuver. You would also like to estimate your satellite's lifetime in case your orbit-correction thrusters stop operating. Assume your satellite will reenter almost immediately if your semi-major axis drops below 6500 km. How long will it take for the ascending node to move to a position of 30°? How long can you expect your satellite to remain in orbit without further thrusting?

Problem Summary

Given: $a_{initial} = 7303$ km $\dot{a} = 2$ km/day

$e_{initial} = 0.001$ Re-enter when

$i_{initial} = 50°$ $a = 6500$ km

$\Omega_{initial} = 90°$

$\omega_{initial} = 45°$

$v_{initial} = 0°$

Find: Time until $\Omega = 30°$

Time until $a = 6500$ km

Conceptual Solution

1) The orbit is essentially circular so we can use Figure 8-9 to find $\dot{\Omega}$, then

$$\text{wait time} = \frac{\Delta\Omega}{\dot{\Omega}}$$

2) Find decay time

$$\text{decay time} = \frac{\text{current } a - \text{minimum } a \text{ for decay}}{\text{decay rate}}$$

Analytical Solution

1) Using Figure 8-9, find $\dot{\Omega}$, then

$$\text{wait time} = \frac{\Delta\Omega}{\dot{\Omega}} = \frac{30° - 90°}{-5°/\text{day}} = 12 \text{ days}$$

2) Find decay time

$$\text{decay time} = \frac{\text{current } a - \text{minimum } a \text{ for decay}}{\text{decay rate}}$$

$$= \frac{7303 \text{ km} - 6500 \text{ km}}{2 \text{ km/day}} = 401.5 \text{ days}$$

Interpreting the Results

The Earth's equatorial bulge will move your ascending node naturally to 30° in 12 days, thus saving precious fuel. Your satellite will reenter the Earth's atmosphere in about 400 days if you can't thrust it back into a higher orbit.

8.3 Predicting Orbits in the Real World

▰ In This Section You'll Learn to...

☞ Combine what you've learned about Kepler's Problem and orbital perturbations to predict a satellite's future position

Let's put what we've learned about orbit perturbations together with the solution to Kepler's Problem and discuss in more detail how to predict a satellite's position in the real world. Assume we're tracking a satellite and have determined the orbital elements, $COE_{initial}$, at time $t_{initial}$.

Now let's step through the process of predicting the orbital elements, COE_{future}, at some time in the future, t_{future}. First, we need to find how these elements change with time due to the perturbations caused by atmospheric drag and the oblate Earth. We learned earlier that the oblate Earth (J2) affects both Ω and ω, so we can use Figures 8-9 and 8-10 back in Section 8.2 to find $\dot{\Omega}$ and $\dot{\omega}$. Inclination, i, isn't affected by either the oblate Earth or drag, so $i_{initial} = i_{future}$. We'll need to find out from our tracking organization how drag affects our orbit's semi-major axis and eccentricity. They'll give us the time rate of change of semi-major axis, $\dot{a}$, and the time rate of change in eccentricity, $\dot{e}$.

Now we know how the first five elements change with time, so let's update them by multiplying the rate of change by the time interval and adding this to the initial value of the orbital element.

$$a_{future} = a_{initial} + \dot{a}(t_{future} - t_{initial})$$

$$e_{future} = e_{initial} + \dot{e}(t_{future} - t_{initial})$$

$$i_{future} = i_{initial}$$

$$\Omega_{future} = \Omega_{initial} + \dot{\Omega}(t_{future} - t_{initial})$$

$$\omega_{future} = \omega_{initial} + \dot{\omega}(t_{future} - t_{initial})$$

(8-13)

where

$a_{initial}, a_{future}$	= initial and future values of semi-major axis (km)
$\dot{a}$	= time rate of change of semi-major axis (km/day)
$t_{initial}, t_{future}$	= initial and future time (days)
$e_{initial}, e_{future}$	= initial and future values of eccentricity
$\dot{e}$	= time rate of change of eccentricity (1/day)
$i_{initial}, i_{future}$	= initial and future values of inclination (deg)
$\Omega_{initial}, \Omega_{future}$	= initial and future values of longitude of the ascending node (deg)

$\dot{\Omega}$ = time rate of change of longitude of the ascending node (deg/day)

$\omega_{initial}$, ω_{future} = initial and future values of argument of perigee (deg)

$\dot{\omega}$ = time rate of change of argument of perigee (deg/day)

We have only the true anomaly, ν, left to update. We know true anomaly, ν, changes with time, even without perturbations, but atmospheric drag also affects it, because drag decreases the semi-major axis, a, and therefore shortens the period. This means the satellite speeds up and so does the rate of change of ν. As we learned in the last section, we need to use Kepler's equation to update ν by using the mean anomaly, M, and the eccentric anomaly, E. Let's update the mean anomaly first, (using Equation (8-6)).

$$M_{future} - M_{initial} = n\,(t_{future} - t_{initial}) - 2k\pi$$

or

$$M_{future} = M_{initial} + n\,(t_{future} - t_{initial}) - 2k\pi \qquad (8\text{-}14)$$

where

M_{future}, $M_{initial}$ = future and initial values of mean anomaly (rad)

n = mean motion (rad/s)

k = number of complete orbits from initial position

We know this equation works when there are no perturbations, but what happens when we add drag? What changes? Remember from Equation (8-1), we can find the mean motion

$$n = \frac{2\pi}{P} = \sqrt{\frac{\mu}{a^3}}$$

where

P = period (s)

μ = gravitational parameter (km^3/s^2)

a = semi-major axis (km)

Recall the semi-major axis, a, changes due to drag ($\dot{a}$). This means the mean motion, n, also changes, so we can find $\dot{n}$. What value for n should we use in Equation (8-14) to solve for M_{future}? Let's look at how the mean motion changes. At $t_{initial}$, the mean motion is $n_{initial}$. At t_{future}, the mean motion is $n_{initial} + \dot{n}\,(t_{future} - t_{initial})$, so the *average mean motion, $\bar{n}$*, is

$$\bar{n} = \frac{n_{initial} + [\,n_{initial} + \dot{n}\,(t_{future} - t_{initial})\,]}{2}$$

$$= n_{initial} + \frac{\dot{n}}{2}\,(t_{future} - t_{initial}) \qquad (8\text{-}15)$$

where

$\bar{n}$ = average mean motion (rad/s)

$\dot{n}$ = time rate of change of mean motion (rad/s^2)

We just added the initial and future values for mean motion and divided by 2. If we now substitute this value of $\bar{n}$ for n back into Equation (8-14), we get

$$M_{future} = M_{initial} + n_{initial}\,(t_{future} - t_{initial})$$

$$+ \frac{\dot{n}}{2}\,(t_{future} - t_{initial})^2 - 2k\pi \qquad (8\text{-}16)$$

You math majors may notice this equation just represents the first three terms of a Taylor Series approximation. We use the $2k\pi$ to keep subtracting 2π from M_{future} until M_{future} is less than 2π. Now we use Kepler's Equation (8-10) to find E_{future}

$$E_{future} = M_{future} + e_{future}\sin E_{future}$$

Remember, this is a transcendental equation, so we'll have to iterate to solve for E_{future}. Finally, we can use Equation (8-9) to solve for the true anomaly at t_{future}

$$\cos v_{future} = \frac{\cos E_{future} - e_{future}}{1 - e_{future}\cos E_{future}} \qquad (8\text{-}17)$$

We now have all the orbital elements (a_{future}, e_{future}, i_{future}, Ω_{future}, ω_{future}, and v_{future}) for the future time, t_{future}. In real life we would then convert these elements back to position and velocity vectors, $\vec{R}$ and $\vec{V}$, and then back to range, azimuth, and elevation for our tracking site to know where to point. Look back at Figure 8-1 and you'll see we've just completed the entire tracking and prediction process.

In summary, we've looked at how our original restricted two-body problem changes by adding in the real-world effects of a non-spherical Earth and atmospheric drag. Then, using Kepler's Equations, we've learned to update our orbit to some future time, so we can use its position for our mission planning.

▬ Section Review

Key Terms

average mean motion, $\bar{n}$

Key Equations

$a_{future} = a_{initial} + \dot{a}\,(t_{future} - t_{initial})$

$e_{future} = e_{initial} + \dot{e}\,(t_{future} - t_{initial})$

$i_{future} = i_{initial}$

$\Omega_{future} = \Omega_{initial} + \dot{\Omega}(t_{future} - t_{initial})$

$\omega_{future} = \omega_{initial} + \dot{\omega}\,(t_{future} - t_{initial})$

$\bar{n} = \dfrac{n_{initial} + [\,n_{initial} + \dot{n}\,(t_{future} - t_{initial})\,]}{2}$

$= n_{initial} + \dfrac{\dot{n}}{2}\,(t_{future} - t_{initial})$

$M_{future} = M_{initial} + n_{initial}\,(t_{future} - t_{initial})$
$+ \dfrac{\dot{n}}{2}\,(t_{future} - t_{initial})^2 - 2k\pi$

Key Concepts

➤ Using our knowledge of perturbations, we can update the orbital elements from time $t_{initial}$, to time t_{future}

➤ Drag will cause the semi-major axis, and hence mean motion, to change with time

➤ To find v_{future}, we
  • Determine average mean motion, $\bar{n}$
  • Determine M_{future}

$M_{future} = M_{initial} + n_{initial}\,(t_{future} - t_{initial}) + \dfrac{\dot{n}}{2}\,(t_{future} - t_{initial})^2 - 2k\pi$

  • Compute E_{future} using iteration

$$E_{future} = M_{future} + e_{future}\,\sin E_{future}$$

  • Solve for v_{future}

$$\cos v_{future} = \frac{\cos E_{future} - e_{future}}{1 - e_{future}\cos E_{future}}$$

Example 8-4

Problem Statement

A remote-sensing satellite has the following COEs and perturbations. Determine the COEs ten days from now.

$a_{initial} = 7000$ km $\dot{a} = -0.7$ km/day

$\hspace{3.8cm} \dot{n} = 0.00003$ rad/day^2

$e_{initial} = 0.1$ $\dot{e} = -0.00003$/day

$i_{initial} = 50°$

$\Omega_{initial} = 90°$ $\dot{\Omega} = -5.0°$/day

$\omega_{initial} = 45°$ $\dot{\omega} = 4.0°$/day

$\nu_{initial} = 270°$

Problem Summary

Given: $a_{initial} = 7000$ km $\dot{a} = 0.7$ km/day

$\hspace{1.5cm} e_{initial} = 0.1$ $\dot{n} = 0.00003$ rad/day^2

$\hspace{1.5cm} i_{initial} = 50°$ $\dot{e} = -0.00003$/day

$\hspace{1.5cm} \Omega_{initial} = 90°$ $\dot{\Omega} = -5.0°$/day

$\hspace{1.5cm} \omega_{initial} = 45°$ $\dot{\omega} = 4.0°$/day

$\hspace{1.5cm} \nu_{initial} = 270°$

Find: The satellite's COEs and position ten days from now

Problem Diagram

$\nu_{initial}$

Conceptual Solution

1) Update the orbital elements to the new time, t_{future}

$$a_{future} = a_{initial} + \dot{a}\,(t_{future} - t_{initial})$$

$$e_{future} = e_{initial} + \dot{e}\,(t_{future} - t_{initial})$$

$$i_{future} = i_{initial}$$

$$\Omega_{future} = \Omega_{initial} + \dot{\Omega}\,(t_{future} - t_{initial})$$

$$\omega_{future} = \omega_{initial} + \dot{\omega}\,(t_{future} - t_{initial})$$

2) Find $n_{initial}$

$$n_{initial} = \frac{2\pi}{p} = \sqrt{\frac{\mu}{a_{initial}^3}}$$

3) Find $E_{initial}$

$$\cos E = \frac{e + \cos\nu}{1 + e\,\cos\nu}$$

4) Find $M_{initial}$

$$M = E - e\sin E$$

5) Move mean anomaly to the desired time

$$M_{future} = M_{initial} + n_{initial}\,(t_{future} - t_{initial}) +$$

$$\frac{\dot{n}}{2}\,(t_{future} - t_{initial})^2 - 2k\pi$$

6) Solve for E_{future} using Kepler's Equation and iteration

$$E = M + e\sin E$$

7) Find ν_{future}

$$\cos\nu = \frac{\cos E - e}{1 - e\,\cos E}$$

Analytical Solution

1) Update the orbital elements

$$a_{future} = a_{initial} + \dot{a}\,(t_{future} - t_{initial}) = 7000\text{ km} -$$

$$(0.7\text{ km/day})\,(10\text{ days}) = 6993\text{ km}$$

271

Example 8-4 (Continued)

$e_{future} = e_{initial} + \dot{e}\,(t_{future} - t_{initial}) = 0.1 -$

$\qquad (0.00003/day)\,(10\ days) = 0.0997$

$i_{future} = i_{initial} = 50°$

$\Omega_{future} = \Omega_{initial} + \dot{\Omega}\,(t_{future} - t_{initial})$

$\qquad = 90° - (5°/day)\,(10\ days) = 40°$

$\omega_{future} = \omega_{initial} + \dot{\omega}\,(t_{future} - t_{initial})$

$\qquad = 45° + (4°/day)\,(10\ days) = 85°$

2) Find $n_{initial}$

$$n_{initial} = \sqrt{\frac{\mu}{a_{initial}^3}} = \sqrt{\frac{398600.5\ \dfrac{km^3}{s^2}}{(7000\ km)^3}}$$

$$= 0.001078\ rad/s = 14.82\ rev/day$$

3) Find $E_{initial}$

$$\cos E_{initial} = \frac{e_{initial} + \cos v_{initial}}{1 + e_{initial}\cos v_{initial}}$$

$$= \frac{0.1 + \cos 270°}{1.0 + 0.1\cos 270°} = 0.1$$

$E_{initial} = 84.261°,\ 275.739° = 275.739° = 4.813\ rad$

Remember, v, M, and E must all lie in the same half plane

4) Find $M_{initial}$

$M_{initial} = E_{initial} - e_{initial}\sin E_{initial}$

  NOTE: Here $E_{initial}$ must be in **radians**

$M_{initial} = 4.813 - 0.1\sin 4.813 = 4.912\ rad = 281.440°$

5) Move mean anomaly to the desired time

$M_{future} = M_{initial} + n_{initial}\,(t_{future} - t_{initial}) + \dfrac{\dot{n}}{2}\,(t_{future}$

$\qquad - t_{initial})^2 - 2k\pi$

$M_{future} = 4.912 + 0.001078\ rad/s$

$(10\ day \times 86{,}400\ s/day) + \dfrac{0.00003\ rad/day^2}{2}$

$(10\ day)^2 = 936.312\ rad$

Because this value is greater than 2π, we need to keep subtracting 2π until M_{future} is less than 2π

$M_{future} = 0.11759\ rad = 6.74°$

Physically, this means our satellite passes perigee 149 times in the next ten days

6) Solve for E_{future} using Kepler's Equation (iterative solution required)

First, guess $E_{future} = M_{future}$

$E_{future} = M_{future} + e_{future}\sin E_{future} = 0.11759 +$

$\qquad 0.0997\sin 0.11759 = 0.1293\ rad$

Now, use this value of E_{future} for your next guess

$E_{future} = M_{future} + e_{future}\sin E_{future} = 0.11759 +$

$\qquad 0.0997\sin 0.1293 = 0.1304\ rad$

Again, use this new value of E_{future} next

$E_{future} = M_{future} + e_{future}\sin E_{future} = 0.11759 +$

$\qquad 0.0997\sin 0.1304 = 0.1306\ rad$

Again,

$E_{future} = M_{future} + e_{future}\sin E_{future} = 0.11759 +$

$\qquad 0.0997\sin 0.1306 = 0.1306\ rad$

Okay, within the accuracy we've chosen here, our solution has converged.

7) Find v_{future}

$$\cos v_{future} = \frac{\cos E_{future} - e_{future}}{1 - e_{future}\cos E_{future}}$$

$$= \frac{\cos 0.1306 - 0.0997}{1 - 0.0997\cos 0.1306}$$

$\cos v_{future} = 0.9896\quad v_{future} = 0.1443\ rad = 8.27°$

Interpreting the Results

In ten days, our satellite will make 149 trips around its orbit, which itself has been perturbed by drag and the Earth's oblateness. Both the eccentricity and the semi-major axis will decrease during this period. Inclination will remain unchanged while longitude of ascending node will regress 50° and argument of perigee will advance to 85°. True anomaly will be 8.27°.

References

Bate, Roger R., Mueller, Donald D., and White, Jerry E., *Fundamentals of Astrodynamics*. New York: Dover Publications, Inc., 1971.

Mission Problems

8.1   Predicting an Orbit (Kepler's Problem)

1   Describe how to track and predict an orbit.

2   For a satellite in a circular polar orbit with an altitude of 400 km, find its time of flight between the equator and the North Pole.

3   A satellite in an elliptical polar orbit, with a semi-major axis of 6778.135 km and an eccentricity of 0.1, is at perigee as it passes the ascending node. How long will the satellite take to reach the North Pole?

4   For a satellite in a circular polar orbit with an altitude of 400 km, find the satellite's location, v, in 40 minutes if the satellite is at the ascending node.

5   A satellite in an elliptical polar orbit, with a semi-major axis of 6778.135 km and an eccentricity of 0.1, is at perigee as it passes the equator. Find the true anomaly 40 minutes later.

8.2   Orbit Perturbations

6   Describe the effects of atmospheric drag on satellites in low-Earth orbit.

7   Describe the effects of an oblate Earth on satellites in low-Earth orbit.

8   List three other sources of orbital perturbations.

9   How long will a satellite in elliptical orbit with a perigee altitude of 650 km take to decay to an altitude of 160 km if the average decay rate is 0.6 km/day?

10   A satellite is in a circular orbit at an altitude of 200 km and an inclination of 28°. How far will the ascending node move in one day? In which direction will it move?

8.3   Predicting Orbits in the Real World

11   An Earth-resource satellite has the following orbital parameters

$a_{initial} = 6900$ km	$\dot{a} = -0.8$ km/day
$e_{initial} = 0.12$	$\dot{e} = -0.00004$/day
$i_{initial} = 75°$	$\dot{i} = 0.0$
$\Omega_{initial} = 90°$	$\dot{\Omega} = -3.0°$/day
$\omega_{initial} = 45°$	$\dot{\omega} = -1.5°$/day
$v_{initial} = 90°$	$\dot{n} = 0.00003$ rad/day^2

Find the future position of the satellite in 30 days.

Projects

12   Pick a clear night and try to observe satellites in low-Earth orbit right after sunset or right before sunrise (this is the only time they are illuminated by the Sun and not yet in the Earth's shadow). Time their passage and approximate position in the sky and see if you can estimate their inclinations and longitudes of ascending node.

Mission Profile—Viking

In 1968, with the Apollo Moon program in full swing, NASA set its sights on another target—Mars. Dubbed Project Viking, it would continue the Mars exploration begun by the Soviet Mars 1-7 spacecraft and by the U.S. Mariner spacecraft. The ultimate objective of Viking was to answer an age-old question—Is there life on Mars?

Mission Overview

The Viking program consisted of two spacecraft containing an orbiter and a lander. The spacecraft were launched on Titan III boosters using Centaur upperstages. The orbiters were first used in mapping the Martian surface to select a final landing site and then in cataloging much of the planet's surface and observing the two Martian moons, Phobos and Demos. The landers collected data on the Martian atmosphere and searched for signs of life.

Mission Data

✓ Viking 1: Launched on August 20, 1975. Arrived in Martian orbit on June 19, 1976. Viking 2 followed close behind using the same launch opportunity on September 9, 1975, arriving on August 7, 1976.

✓ The payload on the orbiters consisted of visual and infrared imaging systems as well as atmospheric water detectors.

✓ Viking 1 landed on the Chryse Panitia (22.4° N, 48° W) seven years to the day after the Apollo 11 Moon landing, July 20, 1976. Viking 2 landed more than one month later on September 3, in the Utopia Planitia area (48° N, 45.7° W).

✓ The landers conducted three primary experiments on Martian soil to search for life. The Pyrolytic Release Experiment heated a soil sample to 650°C to detect the release of CO_2 by organic material. The Labeled Release Experiment incubated a soil sample along with several nutrients and looked for reactions which would indicate the presence of organic life. The third experiment looked for signs of life by analyzing the gas exchange between a soil sample and the gas contained in the experiment module.

Mission Impact

The results of the lander's search for life were ambiguous and so, for now, the question remains unanswered. However, the extensive photographic and other data collected by the Viking orbiters have revolutionized our understanding of Mars' evolution. Viking data indicates that at one time in the distant past, Mars had water flowing on its surface. In fact, early conditions on Mars may have been quite similar to Earth four billion years ago or so. Some scientists therefore speculate that life may no longer exist on Mars but may have flourished and then died out as Mars' water and atmosphere evaporated.

Viking orbiter together with the lander on the surface studied the Martian environment and searched for signs of life. *(Courtesy of NASA)*

For Discussion

• What would have been the impact on human society if Viking had discovered evidence of life on Mars?

• Viking show that robots could carry out significant science on other worlds. Does this mean humans have no place in interplanetary exploration?

• What should the next step be in our search for life on Mars?

References

Rycroft, Michael (ed.), *The Cambridge Encyclopedia of Space*. New York, N.Y.: Cambridge University Press, 1990.

It's launch time. *(Courtesy of NASA)*

Getting To Orbit

9

In This Chapter You'll Learn to...

☛ Describe launch windows and how they constrain when we launch a booster into orbit

☛ Determine when and where you must launch, as well as the required velocity, to reach a given orbit

☛ Demonstrate how mission planners determine when, where, and with what velocity to launch spacecraft into their desired orbits

You Should Already Know...

❑ Constants of orbit motion (Chapter 4)

❑ Definition of the Geocentric-Equatorial Coordinate System (Chapter 4)

❑ Definitions of the Classical Orbital Elements (Chapter 5)

❑ The difference between direct and retrograde orbits (Chapter 5)

❑ How to determine orbit inclination from the orbit ground track (Chapter 5)

If you don't know where you're going, you'll probably end up somewhere else.

Yogi Berra
former New York Yankees catcher

Few scenes are as spectacular as the launch of a Space Shuttle. The three main engines ignite, followed a few seconds later by the powerful solid-rocket boosters. Then, the 20-story vehicle arcs into the clear Florida skies on a plume of white exhaust. But how do we know when to light the engines, when to shut them off, and perhaps most important (at least for the people who live down range), where to point the thing? In this chapter we'll answer these questions.   We'll first examine when the time is right to start the rocket motors on our journey into space. Then we'll dig into what velocity we need and in what direction we must launch to get a payload into a desired orbit.

9.1 Launch Windows and Time

▅ In This Section You'll Learn to...

☛ Define a launch window

☛ Calculate time using the Earth's rotation

For most space missions, the spacecraft must get into a specific orbit. For example, remote-sensing satellites normally use polar orbits, and communication satellites typically go into geostationary orbits. To meet these requirements, we need to launch the satellite from a specific launch site at a particular time and in a particular direction. Let's see how to meet these requirements.

Launch Windows

As part of mission planning we must select the orbit we want. The most common way to specify that orbit is to define a set of classical orbital elements that satisfy mission objectives. This may sound like no big deal, but trying to achieve a specific set of orbital elements along with other mission constraints can severely limit when and in what direction to launch. In some cases, certain orbit conditions may be impossible to achieve because of the launch-site location or booster size. The *launch window* is the period when you can launch a satellite directly into a specified orbit from a given launch site. Notice we said "directly" into the desired orbit. It's always possible to launch into some parking orbit and then go through Hohmann Transfers and plane changes to end up in the desired orbit, but this is much more complicated and expensive.

One way to understand launch windows is to think about bus schedules. Suppose you've made a date to meet a friend in a particular time and place and you need to catch a bus to get there. The time and place specified for your meeting are like the desired orbital elements. Only certain buses will get you to your meeting on time. If the bus you need is scheduled to leave at 11:13, you'd better be at the bus stop at 11:13. If you miss this bus you may have to wait quite a while until the next bus going your way comes along.

We'll define a launch window at an exact time—e.g., 11:13. In practice however, a launch window normally covers a period of time during which you can launch—usually several minutes or even hours around this exact time—just as a bus scheduled to leave at 11:13 could leave anytime between 11:10 and 11:15. Actual mission planners have some flexibility in the orbital elements they'll accept, and booster rockets usually can steer enough to expand the length of the window.

In addition, we'll deal with only one major constraint on launch time—the physical alignment of the launch site and desired orbit. In practice, constraints on launch time include availability of launch-site sensors, weather, lighting for aborts (for manned launches), and political considerations.

Because a rocket must follow trajectories governed by Newton's laws of motion, a launch window restricts us much more than a bus schedule. Let's begin our investigation of launch windows by looking at the problem we face in relating what goes on in an orbit to what goes on at a launch site. As we've learned so far, orbit planes are fixed in inertial space while the Earth (and the launch site along with it) rotate *underneath* this orbit plane. As a result, a launch site at a particular point on the Earth will intersect the orbit plane only periodically as the Earth rotates. (In some cases, it may not intersect it at all.) When the launch site and the orbit plane intersect, we have a launch window and can launch directly into the orbit.

Launch Time

How do we know what time we can launch? Because we're dealing with two periodic events (the rotating Earth and the orbiting spacecraft), it's helpful to think about a car driving around a one-mile, circular racetrack. Assume the car is 3/4 of a mile past the starting line, and the turnoff for the pit is 1/8 of a mile in front of the starting line, as seen in Figure 9-1. Using simple math, you can see the car must be 1/8 of a mile from the pit turnoff. If we also know the car's speed, we can easily determine how long before the car reaches the pit. If the car stays at a constant speed, we can also predict several laps in advance when it will reach the pit turn-off.

What does all this have to do with launch windows? Well, launch windows also repeat periodically, so we can use almost the same approach to find out when launch windows occur. Remember, we defined the geometry for Earth-orbiting spacecraft in terms of the Geocentric-Equatorial Coordinate System. In this system the principle direction, $\hat{I}$, is the vernal equinox direction. We can use this same point fixed in inertial

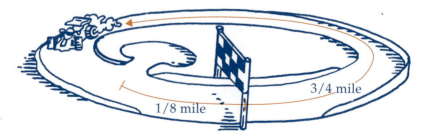

Figure 9-1. Orbital Racetrack. To visualize how we relate periodic events for launch windows, imagine a car going around a one-mile, circular racetrack. If the car is 3/4 mile past the starting line, and the pit turnoff is 1/8 mile before the starting line, then the car must be 1/8 mile away from the pit. Given the car's speed we could then determine how long before the car reaches the pit.

space as a convenient reference from which to measure the angular distance between the orbit plane and our launch site as the Earth (and the launch site along with it) rotates. Thus, it's like using the starting line on the racetrack to reference the location of the car and the pit. In effect, we're going to use the Earth itself as a big clock. To visualize how, imagine a polar orbit around the Earth, as shown in Figure 9-2. We know from Chapter 5 that the angle between the $\hat{I}$ axis and the ascending node of the orbit is the longitude of ascending node, Ω. We also know that in the ideal case an orbit is fixed in inertial space while the Earth rotates beneath it. To use the Earth as a clock, we must use a new angle to relate the distance between a launch site on the Earth's surface to the location of the orbit plane defined by Ω. But unlike Ω, this new angle will be changing as the Earth rotates. We call this angle the *local sidereal time (LST)*, the angle from the vernal equinox direction, $\hat{I}$, to the launch site. Sidereal means "related to the stars" and, because we're referencing the first point in the constellation Aries, this is a good description.

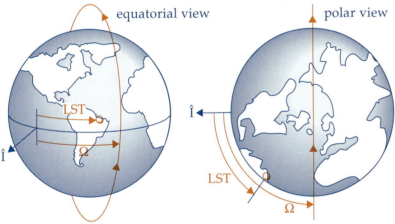

Figure 9-2. Defining Local Sidereal Time. (LST). We used the longitude of ascending node, Ω, to represent the angle from the vernal equinox direction, $\hat{I}$, to the ascending node. Similarly, we can use an angle called the Local Sidereal Time (LST) to reference some point on the Earth's surface, such as a launch site, to the vernal equinox direction. Unlike Ω, LST will change as the Earth rotates.

Now, we're normally used to telling time in hours, minutes, and seconds. But here we've defined an angle, LST, as a "time." Does this make sense? If you think about it, we can just as easily tell time in degrees as in hours. Because the Earth rotates 360° in 24 hours, we can say the Earth rotates 15°/hr (360°/24 hr). (Later we'll see it's actually a little more than 15°.) This explains why time zones around the world span about 15° of longitude. For example, we can say 1:00 A.M. is the same thing as 15°. (The Earth has rotated 1 hour or 15° since midnight, which we take to be 0 o'clock or 0°.) 1:00 P.M., which is 1300 using a 24-hour clock (as the military and most of Europe does), is 195° (15°/hr × 13 hr). A standard 12-hour clock face and the corresponding angles are in Figure 9-3. The relationship between time measured in hours and time measured in degrees is

Figure 9-3. Telling Time. We can tell time in degrees as easily as in hours. Because we use a 12-hour clock instead of a 24-hour clock, 3:00 A.M. or 3 hours past midnight is 45° of Earth rotation. Noon is 180°, etc.

$$\text{Time (degrees)} = \text{Time (hours)} \times \omega_{Earth}$$

where

$$\omega_{Earth} = \text{rotation rate of Earth} = 15°/hr$$

Time in degrees may seem kind of strange at first, but you'll see later that it allows us to relate launch time to launch constraints, especially longitude of ascending node, Ω, which is always expressed in degrees.

We'll define successive passages of the vernal equinox over a certain longitude as a *sidereal day*. The time since the vernal equinox last passed over a certain longitude is *sidereal time*. If the longitude you're using as a reference is the local longitude (say of your home town or a launch site), the time since the vernal equinox last passed over the local longitude is the LST.

▅ Section Review

Key Terms	Key Concepts
launch window local sidereal time (LST) sidereal day sidereal time	➤ A *launch window* is the period during which we can launch directly into a desired orbit from a particular launch site ➤ Local sidereal time is the angle measured from the vernal equinox to the local longitude ➤ We can measure time in degrees as easily as in hours

9.2   When and Where to Launch

In This Section You'll Learn to...

☛ Explain how many chances exist to launch from a given launch site into a particular orbit

☛ Draw a diagram representing launch-window geometry and use it to determine launch-window parameters

Let's take a closer look at how we determine launch-window times and directions. But before we begin, you must remember one very important thing—*launch-window calculations always depend on geometry*! This means you *must* draw a picture to fully understand what is going on! In Chapter 5 we saw that, as an orbit plane slices through the center of the Earth, it extends North and South to latitude lines equal to its inclination as shown in Figure 9-4. Thus, to launch directly from a given launch site into a given orbit plane, the launch site must pass through the orbit plane.

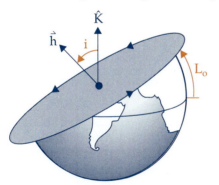

Figure 9-4. Inclination Versus Latitude. An orbit plane slices through the center of the Earth and extends North and South to latitude lines equal to the orbit inclination. To launch directly from a given launch site on the Earth, the launch site must pass through the orbit plane.

This means the orbit's inclination, i, must be equal to or greater than the launch site's latitude, L_o. In other words, a launch window can exist only for the following conditions

$$L_o \leq i \text{ (direct orbits)}$$

$$\text{or } L_o \leq 180° - i \text{ (retrograde orbits)}$$

For example, we can't launch a satellite from the Kennedy Space Center (KSC) ($L_o = 28.5°$) *directly* into an equatorial orbit ($i = 0°$). Instead, we must launch *indirectly* into this orbit by first launching into an orbit with $i = 28.5°$ and then doing a plane change to $i = 0°$. But as we saw in Chapter 6, plane changes are very expensive in terms of ΔV.

Unless indicated, all of our discussion of launch-window geometry will focus on direct orbits ($i < 90°$). The geometry for retrograde orbits is the same, but the relationships are slightly different. Assuming a launch window does exist ($L_o \leq i$ for a direct orbit), let's look at the number of launch windows we can have. Consider the two cases shown in Figure 9-5. In Case 1, the launch-site latitude is equal to the orbit inclination, that is $L_o = i$. In this case, the orbit plane is tangent to the launch site's latitude only once per orbit. Thus, we have only one launch opportunity per day. For example, if we want to launch into orbit with $i = 28.5°$ from KSC ($L_o = 28.5°$), we have exactly one launch window per day. A polar view of this case is in Figure 9-6.

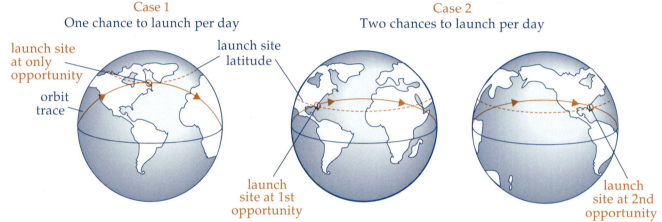

Case 1
One chance to launch per day

launch site at only opportunity

launch site latitude

orbit trace

Case 2
Two chances to launch per day

launch site at 1st opportunity

launch site at 2nd opportunity

Figure 9-5. Launch Opportunity. Chances to launch fall into two possible cases. First, if the launch-site latitude is equal to the orbit inclination, we have one chance per day. Second, if the launch-site latitude is less than the orbit inclination, we have two chances per day: once near the ascending node and once near the descending node. If the launch-site latitude is greater than the inclination, we have no chance to launch.

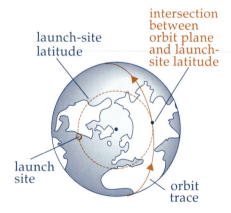

launch-site latitude

intersection between orbit plane and launch-site latitude

launch site

orbit trace

Figure 9-6. Case 1. If we look down on the Earth from the North Pole, we can sketch the latitude line of the launch site and the arc the orbit plane makes. If the orbit inclination is equal to the launch-site latitude, the two intersect each other at only one point. (Keep in mind the orbit plane is fixed, and the Earth [and the launch site along with it] rotate underneath.) As the Earth rotates, the launch site moves closer to the point where it will intersect the orbit plane.

Now consider Case 2, where $L_o < i$. To understand this case, you must remember the orbit plane is fixed in inertial space, and the launch site is fixed to the rotating Earth. As the Earth rotates, it will carry the launch site under the orbit to intersect it twice each day. This gives us one launch opportunity near the ascending node and another near the descending node.

Now, let's determine when to launch. As we saw earlier, the local sidereal time (LST) for the launch site is measured from the vernal equinox direction ($\hat{I}$ unit vector) to the longitude of the site. We will define the *launch window sidereal time (LWST)* as the angle measured from the vernal equinox direction ($\hat{I}$ unit vector) to the point where the launch site passes through the orbit plane. When these two angles are equal (LST = LWST), we can launch into our desired orbit. In other words, if you're waiting for the 11:13 bus, and your watch says 11:13, it's time to go!

Figure 9-7 illustrates the relationship between LWST and LST, which we can use to find the LWST for a particular opportunity. We start by drawing the orbit ground track on the sphere of the Earth. We then sketch in a dotted line showing the latitude for the launch site (L_o) and a figure representing the launch site. We'll use Case 1 for the example. Notice the launch site latitude intersects the orbit trace at only one point. As the Earth rotates, the launch site moves closer to this point.

Next, we determine the current LST at the site, the LWST, and the longitude of the ascending node of the orbit, again referring to Figure 9-7. The launch site will rotate along with the Earth until it intersects the orbit plane. When this happens, LST = LWST, it's time to launch. For Case 1, the point of intersection between the orbit plane and the launch-site latitude occurs 90° from the ascending node. Thus, LWST = Ω + 90°.

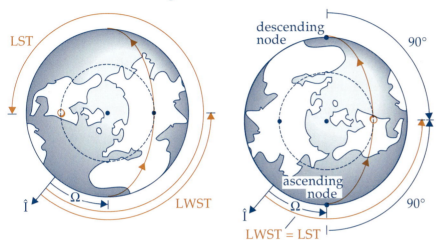

Figure 9-7. *Relating LST to LWST.* The Local Sidereal Time (LST) at the launch site is defined as the angle from the vernal equinox direction, $\hat{I}$, to the launch site as shown in the figure on the left. (Keep in mind the orbit plane is fixed and the Earth [along with launch site] rotates *underneath*.) The launch window opens when the Earth has rotated enough to cause the launch site to intersect the orbit plane (LST = LWST), as shown in the figure on the right.

For Case 1, this is pretty straightforward, but for Case 2, things get a little more complicated. Because in this case the angle from the ascending node to either of the two points of intersection is *not* 90°, we must take a closer, 3-dimensional look at the geometry of the problem to find exactly what these angles are. We know that in Case 2 the orbit plane intersects the launch-site latitude at two points. One of these points is closer to the ascending node and is called the *ascending-node opportunity*, the other is closer to the descending node and is called the *descending-node opportunity*.

Let's draw the local longitude line on the Earth's surface from the North Pole, through the launch site at the ascending-node opportunity, to the equator, as seen in Figure 9-8. This line works with the orbit trace and the equator to form a triangle on the Earth's surface. Because this triangle will be so important in finding launch time and direction, let's take a closer look at it. We'll define two auxiliary angles, α and γ, in this triangle. We call the first angle the *inclination auxiliary angle, α,* and define it at the node between the equator and the ground trace of the orbit. Notice α is equal to inclination, i, (for direct orbits). The second angle, called the *launch direction auxiliary angle, γ,* is measured at the intersection of the orbit ground trace and the longitude line. The side opposite γ we'll call the *launch window location angle, δ,* and measure it along the equator, between the node closest to the launch opportunity being considered and the longitude where the orbit crosses the launch-site latitude. The side

Figure 9-8. Case 2. Taking a three-dimensional view of the problem, we can draw launch-site latitude and longitude lines on the Earth to show where they intersect the orbit plane for the ascending-node opportunity. Notice this method forms a triangle on the Earth's surface. Because the Earth is a sphere, this is a spherical triangle, giving it properties different from the plane triangles you may be used to.

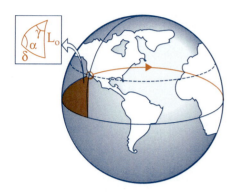

Figure 9-9. Auxiliary Triangle for a Launch Window. Look at the spherical triangle formed on the Earth's surface for the ascending-node opportunity. You can see the right side represents the launch-site latitude, L_O. The angle between the orbit trace and the equator is the inclination auxiliary angle, α. The angle between the orbit trace and the longitude line is the auxiliary angle, γ. The side opposite γ is the angle δ. [Note that in spherical triangles, *sides* are actually *angles* measured from the center of the Earth].

opposite α is the *launch-site latitude, L_o*. Figure 9-9 shows another view of this auxiliary triangle.

Because this triangle is drawn on the surface of a sphere, it is naturally called a *spherical triangle*. Spherical triangles are quite different from the good old plane triangles you're used to. For instance, the sum of the angles in a spherical triangle can be more than 180°. Furthermore, the sides are measured as angles. By using the law of cosines for spherical trigonometry, we can get a relationship between the two sides and two angles in the triangle (See Appendix B for an explanation of spherical trigonometry.)

$$\cos\alpha = -\cos 90° \cos\gamma + \sin 90° \sin\gamma \cos L_o$$

$$\cos\alpha = \sin\gamma \cos L_o \qquad (9\text{-}1)$$

To find γ we rearrange Equation (9-1) to get

$$\boxed{\sin\gamma = \frac{\cos\alpha}{\cos L_o}} \qquad (9\text{-}2)$$

where
γ = launch direction auxiliary angle (deg or rad)
α = inclination auxiliary angle (deg or rad)
L_o = launch-site latitude (deg or rad)

We can find δ by using spherical trigonometry again to get

$$\sin\alpha \cos\delta = \cos\gamma \sin 90° + \sin\gamma \cos 90° \cos L_o \qquad (9\text{-}3)$$

Rearranging Equation (9-3), we get

$$\boxed{\cos\delta = \frac{\cos\gamma}{\sin\alpha}} \qquad (9\text{-}4)$$

where
δ = launch window location angle (deg or rad)

Finally, we'll define one more important angle—the *launch azimuth, β* which tells us what direction to launch. It is measured from true North at the launch site clockwise to the launch direction, as shown in Figure 9-10. If you've used a compass, note that β is measured the same way as magnetic heading, with East 90°, South 180°, etc. Notice for a launch at the ascending-node opportunity from a site in the Northern Hemisphere, $\beta = \gamma$. Table 9-1 summarizes all these angles.

Figure 9-10. Launch Azimuth, β. The launch azimuth, β, is the angle measured from true North (the launch-site longitude line) clockwise to the launch direction. When only one chance to launch exists, the launch azimuth is exactly 90°.

With all these definitions out of the way, we can return to finding the launch-window sidereal time (LWST) for Case 2 in Figure 9-5. LWST will always be a function of the desired orbit's longitude of ascending node, Ω, and the launch-window location angle, δ. Unfortunately, LWST depends totally on the launch situation, so it's not convenient to write one equation to cover all cases. Therefore, to determine LWST, you must draw a diagram of the launch site!

By now you may be confused about the difference between α and LWST and all these other angles. To clear things up, let's look at an example to see how it all works. Computing LWST and β depends on three choices:

- Direct or retrograde orbit
- Opportunity near the ascending-node or descending-node
- Launch-site in Northern or Southern Hemisphere

Table 9-1. Key Angles for Launch Geometry.

Angle	Name	Definition
α	inclination auxiliary angle	measured from the equator to the ground track at the node—same as inclination for direct orbits
γ	launch-direction auxiliary angle	measured from ground track to the launch site local longitude line
δ	launch-window location angle	measured from the node closest to the launch opportunity to the launch site local longitude line
β	launch azimuth	measured from due North clockwise to the launch direction

As an example, let's pick a direct orbit near the descending node in the Northern Hemisphere with $i > L_o$. To see how to compute LWST for this situation, we begin by drawing two diagrams, as shown in Figure 9-11. One is in three dimensions showing the spherical launch geometry and another is a two-dimensional polar view showing the relationship between LWST, Ω, and δ.

By inspection, you can see the inclination auxiliary angle, α, is equal to the inclination, i. The launch azimuth, β, is equal to $180°$ minus the auxiliary angle, γ ($\beta = 180° - \gamma$). We can find the launch-window location angle, δ, knowing α and γ and using Equation (9-4). To find LWST for this situation, we look again at our diagram to see

$$LWST_{DN} = \Omega + (180° - \delta) \text{ for direct orbit, Northern Hemisphere site}$$

Let's review where this came from. Ω takes us from $\hat{I}$ to $\hat{n}$ (the ascending node vector); $180°$ moves us to the descending node; then, we subtract δ to come back to the launch-site longitude.

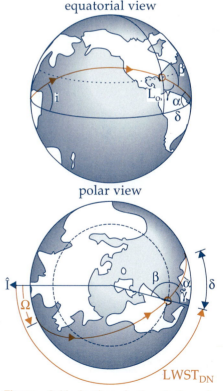

Figure 9-11. Another Look at Launch Geometry. We can analyze the launch problem for a launch at the descending node opportunity from a launch site in the Northern Hemisphere. We begin by drawing the 3-D view showing the spherical Earth and the auxiliary triangle. We then sketch a 2-D polar view of the problem showing Ω and LWST.

If we know the current local sidereal time (LST) for the launch site, we can find out how long we must wait until it's time to launch by subtracting from the LWST. For example, if the LWST is 1330 and LST is 1200, we must wait 1 1/2 hours before the launch window opens. (As we'll see later, these are sidereal hours, which are slightly shorter than the solar hours we keep on our watches.)

In all, there are eight launch-window situations. Table 9-2 summarizes the four direct-orbit cases, but the four retrograde-orbit cases aren't shown because we seldom launch into retrograde orbits.

Table 9-2. Summary of Direct-Orbit Cases for Launch Opportunities.

Near Ascending Node	Direct Orbits	Near Descending Node
LWST = $\Omega + \delta$ i = α $\beta = \gamma$	Northern Hemisphere	LWST = $\Omega + 180° - \delta$ i = α $\beta = 180° - \gamma$
Near Descending Node	**Direct Orbits**	**Near Ascending Node**
LWST = $\Omega + 180° + \delta$ i = α $\beta = 180° - \gamma$	Southern Hemisphere	LWST = $\Omega - \delta$ i = α $\beta = \gamma$

▦ Section Review

Key Terms

inclination auxiliary angle, α
launch azimuth, β
launch direction auxiliary angle, γ
launch-site latitude, L_o
launch window sidereal time (LWST)
launch window location angle, δ
spherical triangle

Key Equations

$$\sin\gamma = \frac{\cos\alpha}{\cos L_o}$$

$$\cos\delta = \frac{\cos\gamma}{\sin\alpha}$$

3-D view at launch time

Key Concepts

➤ For a launch window to exist at a given launch site, the latitude of the launch site, L_o, must be less than or equal to the inclination of the desired orbit ($L_o \le i$).

➤ Computing launch-window sidereal time (LWST) and launch azimuth, β, depends on geometry. You must draw a diagram to clearly visualize all angles.

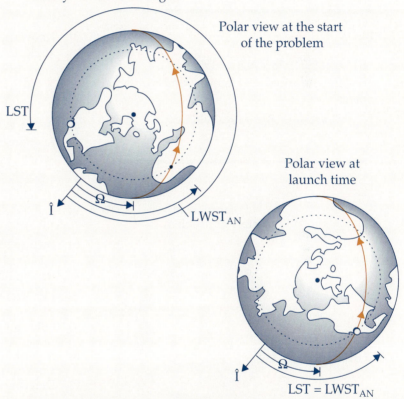

Polar view at the start of the problem

Polar view at launch time

- Launch-window geometry depends on spherical trigonometry
- After sketching launch-window geometry, we can see an auxiliary triangle. Table 9-1 defines auxiliary angles.
- Launch-window sidereal time (LWST) is a function of the desired longitude of ascending node, Ω, and the launch-window location angle, δ
- Launch azimuth, β, is defined as the direction to launch from a given site to achieve a desired orbit. β is measured clockwise from due North at the launch site.

➤ Table 9-2 summarizes launch-window geometry for the four most common of the eight possible cases where $L_o < i$

Example 9-1

Problem Statement

An interplanetary probe bound for Saturn will be deployed from the Space Shuttle. It requires a parking orbit with a longitude of the ascending node, Ω, of 195° and an inclination of 41°. If the current LST at the launch site is 0100, how long before the next launch window opens for a launch from Kennedy Space Center (28.5°N, 80°W)? What is the launch azimuth for this opportunity?

Problem Summary

Given: LST = 0100, L_o=28.5°N, Ω = 195°, i= 41°

Find: Time in hours until next launch window opens

Launch azimuth, β, for next window

Problem Diagram

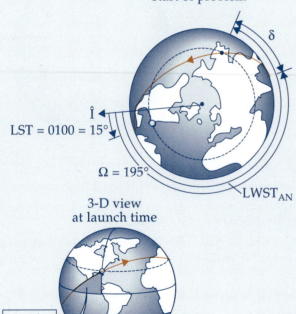

Polar view at start of problem

δ

$\hat{I}$

LST = 0100 = 15°

Ω = 195°

$LWST_{AN}$

3-D view at launch time

α γ L_o

δ

Conceptual Solution

Time until next launch window opens

1) By inspection, find the inclination auxiliary angle, α

$$\alpha = i$$

2) Knowing α and L_o, find the launch-direction auxiliary angle, γ

$$\sin\gamma = \frac{\cos\alpha}{\cos L_o}$$

3) Knowing γ, find the launch-window location angle, δ

$$\cos\delta = \frac{\cos\gamma}{\sin\alpha}$$

4) By inspecting launch-site geometry, determine the relationship between LWST, Ω, and δ for the ascending-node opportunity

$$LWST_{AN} = \Omega + \delta$$

5) Solve for LWST for ascending-node opportunity

6) By inspecting launch-site geometry, determine the relationship between LWST, Ω, and δ for the descending-node opportunity

$$LWST_{DN} = \Omega + 180° - \delta$$

7) Solve for LWST for descending-node opportunity

8) Determine which LWST is next

9) Determine time until next opportunity

10) Knowing γ and inspecting launch-site geometry for the next opportunity, find β

Analytical Solution

1) By inspection, find the inclination auxiliary angle, α

$$\alpha = i = 41°$$

2) Knowing α and L_o, find the launch-direction auxiliary angle, γ

289

Example 9-1 (Continued)

$$\sin\gamma = \frac{\cos\alpha}{\cos L_o} = \frac{\cos 41°}{\cos 28.5°} = 0.8588$$

$$\gamma = 59.18°$$

3) Knowing γ, find the launch-window location angle, δ

$$\cos\delta = \frac{\cos\gamma}{\sin\alpha}$$

$$\cos\delta = \frac{\cos 59.18°}{\sin 41°} = 0.7809$$

$$\delta = 38.65°$$

4) By inspecting launch-site geometry, determine the relationship between LWST, Ω, and δ for the ascending-node opportunity

$$LWST_{AN} = \Omega + \delta$$

5) Solve for LWST for ascending-node opportunity

$$LWST_{AN} = 195° + 38.65° = 233.65°$$

$$LWST_{AN}\,(hrs) = \frac{LWST_{AN}\,(°)}{\omega_{Earth}} = \frac{233.65°}{15°/hr}$$

$$= 15.58\ hr$$

$0.58\ hr = 34.8\ min$

so the time is 1534.8

6) By inspecting launch-site geometry, determine the relationship between LWST, Ω, and δ for the descending-node opportunity

$$LWST_{DN} = \Omega + (180° - \delta)$$

7) Solve for LWST for descending-node opportunity

$$LWST_{DN} = 195° + (180° - 38.65°) = 336.35°$$

$$LWST_{DN}\,(hrs) = \frac{LWST_{DN}\,(°)}{\omega_{Earth}} = \frac{336.35°}{15°/hr}$$

$$= 22.42\ hr$$

so the time is 2225.4

8) Determine which LWST is next in hours

Because LST = 0100 and $LWST_{DN}$ = 2225.4, we have missed the descending-node opportunity (0100 in the morning is after 2225.4 at night). Thus, we must wait for the ascending-node opportunity, and the next LWST is at 1534.8 in the afternoon.

9) Determine time until next opportunity

Time until next opportunity is time from current LST (0100 = 1.00) until $LWST_{AN}$ (1534.8 = 15.58) or 15.58 - 1.0 = 14.58 hours = 14 hours and 34.8 minutes.

10) Knowing γ and inspecting launch-site geometry for the next opportunity, find β.

For the ascending-node opportunity, $\beta = \gamma$, thus $\beta_{AN} = 59.18°$

Interpreting the Results

We must wait 14 hours and 34.8 minutes until the ascending-node opportunity to launch the Space Shuttle into a parking orbit with $\Omega = 195°$ and $i = 41°$. We will launch when LST is 1534.8 hours, at an azimuth of 59.18° (northeast).

9.3 What Time Is It?

▬ In This Section You'll Learn to...

☛ Explain the difference between the sidereal time we use to compute launch windows and the solar time we keep on our watches

So far, our discussion about when to launch has depended on knowing LST, which is referenced to the vernal equinox direction. However, since the beginning of time (literally), people have used a more obvious reference—the Sun. By placing a stick in the ground, people found they could measure the passage of time based on the shadow cast by the Sun. Thus, the time between the Sun's successive passages above a certain point became known as an *apparent solar day*.

Because the Earth's orbit around the Sun is slightly elliptical (e = 0.017), the apparent solar day varies slightly throughout the year. To compensate for this, we take the annual average of apparent solar days to get a *mean solar day*. This is the time we keep on our watch. There are 24 mean solar hours in a mean solar day. Because people all around the world like to keep their time in synch with the Sun, we have 24 time zones. To avoid confusion between people trying to coordinate time across several time zones, we usually choose the Greenwich or 0° longitude line (Prime Meridian) as an international reference point. The local mean solar time at the Prime Meridian is called *Greenwich Mean Time (GMT)*. This is also known as "Zulu time" and is used by the military and other international organizations.

Astro Fun Fact
Prime Meridian

(Courtesy of British Royal Observatory)

The Royal Observatory in Greenwich, England was constructed in the late seventeenth century to track the movement of the Moon and location of the stars. This information (published in the Nautical Almanac of 1767) was essential for mariners and seamen who had no other way to locate their east-west position (longitude). They discovered their latitude by using this data coupled with an onboard device called a sextant. With the introduction of the railroads in the nineteenth century, a universal time standard became necessary. In England, local time had to be adjusted almost every time the train would enter another town! The Greenwich site, with it's accurate time measurement and famous observatory, became the natural choice for a universal standard time. In time, the world accepted the Greenwich locale as 0° longitude, employing this Meridian as a timekeeping and navigation standard for land, air, space, and sea.

The Old Royal Observatory: The Story of Astronomy & Time. Courtesy British Royal Observatory, Greenwich, England.

Contributed by Troy Kitch, United States Air Force Academy

So if our launch window is based on sidereal time, and we have only solar watches, how do we know when to punch the button to launch the booster? To understand the difference between solar time and sidereal time, imagine a line extending above your head out to infinity. As this line sweeps by the Sun (as the Earth rotates on its axis), it counts off solar days. This is fine for most events we keep track of on Earth.

Now suppose, as an amateur astronomer, you want to keep track of some sidereal event such as the position of the constellation Aquarius (it will be the "age of Aquarius" in only about 400 years). If you keep track of the time Aquarius rises in the night sky, you'll notice something strange. Each night it appears about 4 minutes earlier than the previous night, and, eventually, it no longer appears in the night sky at all (until next year).

What's going on? Is your watch broken? No! The difference between solar and sidereal time causes this time shift. As the world *turns* it also *moves* around the Sun in its orbit, causing a shift in the apparent position of the Sun due to the Earth's movement along its orbit. Any shift in an object's apparent position due to the motion of the observer is called *parallax*. Because of the Earth's motion around the Sun, it must actually rotate slightly more than 360° to bring the Sun directly overhead again. On the other hand, a sidereal day uses the vernal equinox direction, ♈, as a reference, which eliminates parallax. This difference between a solar and a sidereal day is shown in Figure 9-12.

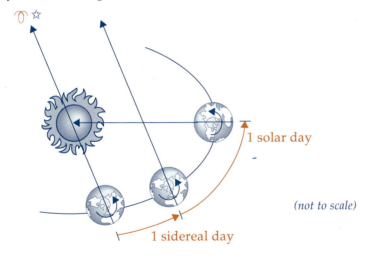

1 solar day

(not to scale)

1 sidereal day

Figure 9-12. Solar Versus Sidereal Day. A solar day is longer than a sidereal day because the Earth must rotate slightly more than 360° to bring the Sun back over a certain point on the Earth to compensate for Earth's movement around the Sun. The vernal equinox direction used to define a sidereal day is so far away that the movement of the Earth around the Sun is insignificant.

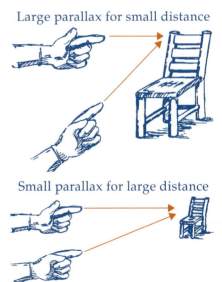

Large parallax for small distance

Small parallax for large distance

Figure 9-13. Parallax. You can see parallax by looking at how your angle of reference changes as you move relative to a nearby object versus how it changes for a distant object.

This parallax is easy to visualize if you do a little experiment. Pick out a chair about 10 feet away and point at it. Now start walking in a straight line at right angles to the line between you and the chair, as seen in Figure 9-13. Notice you must move your arm to continue to point directly at the chair. This is because of parallax caused by the change in apparent position

---

Astro Fun Fact
Mean Solar Time

Mean Solar Time is calculated by taking the annual average of the Sun's successive passes over a point on the Earth. This period of time must be averaged because the Earth's orbit is slightly elliptical (which makes the Earth's speed vary with its distance from the Sun.) As this method correlates with the Earth's rotation it has some inherent inaccuracy. As you can imagine, many scientists and engineers want extreme accuracy regardless of the movement of the Earth and Sun. In response to this need, they developed a nearly perfect time-keeping device. This clock uses hydrogen atoms to produce a frequency which vibrates with amazing consistency. If this clock operated for a human's average life span, its error would be less than one microsecond. If it began operating when the solar system began, it would now be less than two minutes off the actual time!

Goudsmit, Samuel A. and Robert Claiborne, ed. Time. New York: Time Incorporated, 1966.

Cowan, Harrison J. Time and It's Measurement. Cleveland: The World Publishing Company, 1958.

Contributed by Troy Kitch, United States Air Force Academy

---

of the chair with respect to you as you move. Now imagine the same chair but this time several blocks away, again as shown in Figure 9-13. If you point at the chair now and then start to move, you'll notice your arm moves very little to keep pointing at it. Your movement is small relative to the much greater distance to the more distant chair.

Thus, by using a distant star as a reference, we can avoid the parallax problem we have with the Sun. This is why we chose the vernal equinox direction as our reference for LST. Using this as a reference, you can picture the rotating Earth as a big clock. Imagine sitting on the second hand of a watch. You'll move 360° from the 12 o'clock position back to the 12 o'clock position every minute. Similarly, put yourself on the surface of the Earth. As the Earth rotates through 360°, you can watch the vernal equinox sweep by overhead, ticking off sidereal days.

Now let's compare the difference between a solar day and a sidereal day. Remember a sidereal day is exactly one rotation of the Earth but a solar day is made up of the Earth's rotation on its axis and its revolution about the Sun. Because the Earth revolves 360° around the Sun in about 365 days this means we move about 1° per day along our orbit. Thus, the Earth must rotate slightly *more* than 360° to bring the Sun back overhead. This means a solar day is longer than a sidereal day.

1 mean solar day = 1.0027 sidereal days

or

Note: the rotation rate of the Earth on its axis, ω_{Earth}, is 360° in 24 sidereal hours or 15° per sidereal hour. ω_{Earth} is also 15.04107° per mean solar hour.

$23^{hr} 56^{min} 04^{sec}$ mean solar time = 24 sidereal hours = 1 sidereal day

But converting from LST to local solar time is not simply a matter of subtracting four minutes. Actually, the conversion is quite involved and it requires you to know the position of the vernal equinox at some specific solar time in the past. You then have to "propagate," or account for the Earth's rotation and revolution about the Sun, from that time up to the LST you're interested in. This is what mission planners do. They work in LST, then convert LST to solar time or, most often, GMT for launch controllers. Fortunately, we won't worry about doing this conversion here.

▰ Section Review

Key Terms

apparent solar day
Greenwich Mean Time (GMT)
mean solar day
parallax

Key Concepts

➤ A mean solar day is the average time between the Sun's successive passages over a given point on the Earth:

- Mean solar time is the time we keep on our clocks and watches
- Greenwich Mean Time (GMT) is the mean solar time at Greenwich, England, which is on the Prime Meridian or 0° longitude

➤ Solar time is measured with respect to the Sun. Because the Earth is in motion about the Sun, solar time isn't a good inertial time reference for launching satellites. Instead, sidereal time (having to do with the stars) is used, with the vernal equinox direction as a reference:

- A sidereal day is defined as the time between successive passages of the vernal equinox over a given point on the Earth or vice versa
- Local Sidereal Time (LST) is the time since the vernal equinox direction was last over a given local longitude

➤ The Earth must rotate slightly more than 360° to bring a given point back directly under the Sun because of Earth's revolution about the Sun. Thus, to bring the Sun back over a given longitude, a solar day is slightly longer than a sidereal day.

9.4 Launch Velocity

▬ In This Section You'll Learn to...

☞ Determine the change in velocity a booster must deliver to put a
spacecraft into a given orbit

The launch window tells us when and in what direction to launch our
booster in order to achieve some desired orbit. Now let's examine how
much velocity the booster needs to place a payload into this orbit.

During liftoff, a booster goes through several distinct phases on its way
into orbit, as shown in Figure 9-14.

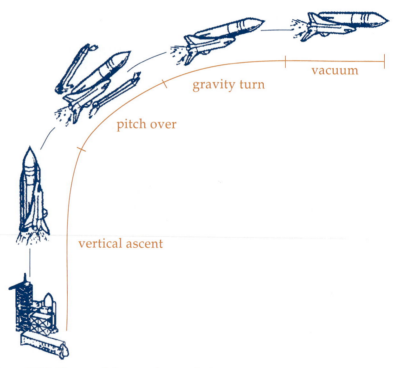

Figure 9-14. Phases of Booster Ascent. During ascent a booster goes through four
phases—vertical ascent, pitch over, gravity turn, and vacuum.

❏ Phase one—vertical ascent (pronounced áss-sent, not ass-śent, by
real rocket scientists)

• During this phase, the booster needs to gain altitude quickly to get up
out of the dense atmosphere which slows the vehicle down because
of drag. Also during this phase, you see the distinctive roll maneuver
as the Space Shuttle and other boosters leave the pad. This roll
properly aligns the launch azimuth so the booster gets into the correct
orbit.

295

❏ Phase two—pitch over

- Once the booster has gained enough altitude, the vehicle must pitch over slightly so it can begin to gain velocity downrange (horizontally). As we saw in Chapter 4, horizontal velocity keeps a vehicle in orbit.

❏ Phase three—gravity turn

- During this phase, gravity bends the velocity vector more and more toward horizontal.

❏ Phase four—vacuum phase

- During this phase, the booster is effectively out of the Earth's atmosphere and accelerates to gain the necessary velocity to achieve orbit. In this final phase of powered flight, the control system concentrates on delivering the vehicle to the desired burnout conditions: velocity ($\vec{V}_{burnout}$), altitude ($Alt_{burnout}$), flight-path angle ($\phi_{burnout}$), and downrange angle ($\theta_{burnout}$) as shown in Figure 9-15.

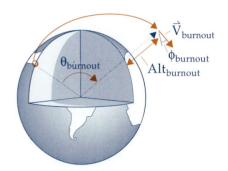

Figure 9-15. Booster Burnout Conditions. During the vacuum phase the control system is trying to deliver the booster to a specified set of burnout conditions including velocity, flight-path angle, altitude, and downrange angle.

In this section we'll concern ourselves with the burnout conditions a booster needs to achieve to attain orbit. Specifically, we'll focus on burnout velocity ($\vec{V}_{burnout}$) and see how it can be used to determine how much velocity a booster must be designed to deliver.

As you sit there in your chair, you're moving. "Doesn't feel like it," you say? Well, remember all the trouble we went through in defining an inertial coordinate frame? The Earth is spinning on its axis (our giant clock), so any point on it (except on the spin axis) has some velocity due to this rotation. You don't feel it, but the Earth's surface (and you along with it) is moving at nearly 1000 miles per hour! Sitting on the launch pad, a booster is moving eastward just like us. Instantaneous tangential velocity for a spinning object is equal to its angular velocity, ω, times the distance from the spin axis, R. If you look at a record spinning on an old-fashioned record player, as shown in Figure 9-16, you can see that the farther out from the center you look, the faster the tangential velocity. The entire record is spinning at the same angular velocity (revolutions per minute), but the tangential velocity increases as you move out to a larger radius. This can be expressed as

$$V = R\, \omega_{Earth} \qquad (9\text{-}5)$$

Figure 9-16. Tangential Velocity Increases with Radius. If you look at a spinning record you can see that the tangential velocity increases as you look farther out from the center.

The same thing happens on the Earth. ω_{Earth} is the Earth's rotation rate (15.04107°/mean solar hr) converted to rad/s, and R is the radius from the spin axis out to the launch site. This radius would be the Earth's radius (6378 km) if our launch site were on the equator. In this case, the velocity would be:

$$V_{at\ equator} = (15.04107°/\text{hr})\ (6378.137\ \text{km})$$

$$V_{at\ equator} = (0.000072921165\ \text{rad/s})\ (6{,}378{,}137\ \text{m})$$

$$V_{at\ equator} = 465.1\ \text{m/s}$$

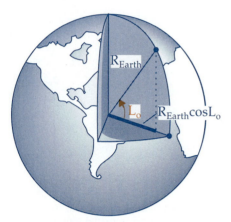

Figure 9-17. Higher Latitude Means Lower Velocity. The tangential velocity of a point on the surface of the Earth is a function of latitude. As you increase in latitude (either North or South) your perpendicular distance to the spin axis decreases, so the tangential velocity decreases.

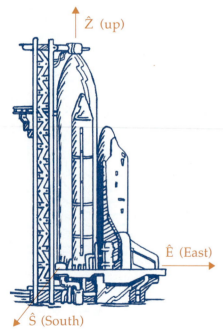

Figure 9-18. The SEZ Frame. We use the South, East, Up (zenith) or SEZ coordinate system as our launch-site coordinate frame.

Unfortunately, we don't have any launch sites on the equator (although the European Space Agency (ESA) site at Kourou French Guyana is close, with a latitude of 4° N). To find the radius from the spin axis to a site at some other location, we must multiply by the cosine of the latitude. Thus, as you increase latitude, R (distance from the spin axis) decreases and so the tangential velocity decreases, as shown in Figure 9-17.

$$V_{\text{launch site}} = (R_{\text{Earth}} \cos L_o) \omega_{\text{Earth}}$$

and because

$$V_{\text{at equator}} = R_{\text{Earth}} \, \omega_{\text{Earth}} = 465.1 \text{ m/s}$$

we can say

$$V_{\text{launch site}} = (465.1 \text{ m/s}) \cos L_o$$

Let's consider the direction of this motion so we can put it into vector form. To express the velocity as a vector, we need to choose a new coordinate frame. Previously, we picked inertial frames because we were writing equations of motion and needed to apply Newton's laws. Now, because we want to know the velocity a rocket must deliver from a given launch site, we pick an Earth-fixed reference such as the *topocentric-horizon frame*. As the name implies, the origin for this frame is at the site on the Earth, with the horizontal (a plane tangent to the Earth at the launch site) as the fundamental plane. If we choose the vector pointing due South from the site as the principle direction ($\hat{S}$) and the straight-up or zenith direction ($\hat{Z}$) as the out-of-plane vector, the East direction ($\hat{E}$) completes the right-hand rule. This is sometimes called the *South-East-Zenith (SEZ)* system, as shown in Figure 9-18. The velocity due to the Earth's rotation can now be expressed as a vector in the East direction.

$$\vec{V}_{\text{launch site}} = (465.1 \, \text{m/s}) \cos L_o \, \hat{E} \qquad (9\text{-}6)$$

So what does this mean in terms of putting a payload into orbit? The closer our launch site is to the equator the greater assist we get when launching into a direct orbit. In other words, because the launch site is already moving eastward at some velocity, it has a "head start" for launches in the easterly direction. For example, the European Space Agency's launch site at Kourou (4° latitude) gets an assist of 464 m/s versus 409 m/s for the Kennedy Space Center at 28.5° latitude. This means that for a given booster you can launch a larger payload from the launch site at a lower latitude. This is one of the reasons we picked Florida instead of, say, New York for the U.S. launch site on the East coast.

What about launching into a retrograde orbit? You won't get any help because you'd be launching in a westerly direction while the launch site is already moving eastward. In fact, you're coming from behind because you have to make up this difference to get into orbit. Thus, it's more costly (in terms of velocity) to launch into a retrograde orbit. Both eastward and westward launches are shown in Figure 9-19.

To achieve some specified orbital elements the satellite must reach a certain inertial velocity at some required altitude. It must reach this velocity when the booster shuts down, so it's called the *velocity at burnout,* $\vec{V}_{burnout}$. The satellite starts off at the launch site with the velocity due to the Earth's rotation and must use the $\Delta\vec{V}$ supplied by the booster to achieve the required velocity at burnout. The *needed velocity,* $\Delta\vec{V}_{needed}$, is the difference between the *velocity of the launch site,* $\vec{V}_{launch\ site}$, and the velocity at burnout, $\vec{V}_{burnout}$.

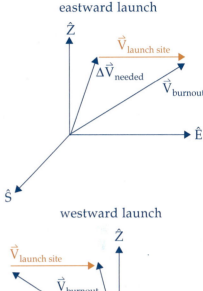

eastward launch

$$\Delta\vec{V}_{needed} = \vec{V}_{burnout} - \vec{V}_{launch\ site} \qquad (9\text{-}7)$$

where
$\Delta\vec{V}_{needed}$ = velocity change needed by the satellite to get it from the launch site to orbit (m/s)

$\vec{V}_{burnout}$ = velocity vector of the booster at burnout which will put the satellite in the required orbit (m/s)

$\vec{V}_{launch\ site}$ = velocity vector of the launch site due to Earth's rotation (m/s)

Because this is a vector equation, we must be concerned with the individual components of each vector. To keep things consistent, we'll define these vectors with respect to the topocentric-horizon frame, just as we did for the launch-site velocity. Using the geometry of the SEZ coordinate system, we can develop relationships for the three components of $\Delta\vec{V}_{needed}$. We must also consider the flight-path angle, ϕ, (measured from the horizon to the velocity vector) and the launch azimuth angle, β, (measured clockwise from due North to the projection of the velocity on the horizon plane). With these values, we can derive the components for the burnout velocity (see Appendix C).

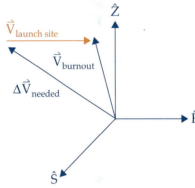

Figure 9-19. Eastward Versus Westward Launches. For eastward launches the Earth's rotation, $\vec{V}_{launch\ site}$, gives the booster a "head start," so ΔV_{needed} is less than $V_{burnout}$. For westward launches the opposite is true.

$$\Delta V_{needed_{South}} = V_{burnout_{South}} = -V_{burnout}\cos\phi\cos\beta \qquad (9\text{-}8)$$

$$\Delta V_{needed_{East}} = V_{burnout}\cos\phi\sin\beta - \left(465.1\frac{m}{s}\right)\cos L_o \qquad (9\text{-}9)$$

$$\Delta V_{needed_{Zenith}} = V_{burnout_{Zenith}} = V_{burnout}\sin\phi \qquad (9\text{-}10)$$

where
$\Delta V_{needed_{South,\ East,\ Zenith}}$ = components of the needed velocity and the South, East, and Zenith directions (m/s)

$V_{burnout_{South,\ Zenith}}$ = components of the booster velocity vector at burnout in the South and Zenith directions (m/s)

$V_{burnout}$ = magnitude of booster velocity vector at burnout (m/s)

ϕ = flight-path angle of satellite orbit at burnout (deg or rad)
= 0° for circular orbits

β = launch azimuth (deg or rad)
L_o = launch-site latitude (deg or rad)

The magnitude of the needed velocity is

$$\Delta V_{needed} = \left| \Delta \vec{V}_{needed} \right| = \sqrt{\Delta V_{needed_{South}}^2 + \Delta V_{needed_{East}}^2 + \Delta V_{needed_{Zenith}}^2}$$

This is the velocity needed to take the satellite from the launch-site velocity to the required velocity at burnout. Later, when we examine booster control systems, we'll see how the booster must continually measure its current velocity, compare it to the desired velocity, and then adjust accordingly, until $\vec{V}_{burnout}$ is achieved.

So far we've considered only the effects of the Earth's rotation in getting our satellite off the ground and into orbit. But, the booster is also under the influence of gravity and drag as it moves off the launch pad and into space. To determine the actual amount of velocity the booster requires, we have to add the needed velocity to the estimated velocity losses from gravity and drag. Modeling or historical data give us estimates of these losses, usually about 1000 m/s or so. Summing these values gives us the minimum total velocity our booster must be designed to deliver to achieve orbit, which we'll call ΔV_{design}

$$\Delta V_{design} = \left| \Delta \vec{V}_{needed} \right| + \Delta V_{losses} \qquad (9\text{-}11)$$

where

ΔV_{design} = velocity the booster must be designed to deliver to achieve the desired orbit (m/s)

$\Delta \vec{V}_{needed}$ = velocity change needed by the satellite to get it from the launch site to orbit (m/s)

ΔV_{losses} = velocity losses during booster ascent due to gravity and drag (m/s)
$\cong 1000$ m/s

▬ Section Review

Key Terms

ΔV_{design}
needed velocity, $\Delta \vec{V}_{needed}$
South-East-Zenith (SEZ)
topocentric-horizon frame
velocity at burnout, $\vec{V}_{burnout}$
velocity of the launch site, $\vec{V}_{launch\ site}$

Key Equations

$$\Delta \vec{V}_{needed} = \vec{V}_{burnout} - \vec{V}_{launch\ site}$$

$$\Delta V_{needed_{South}} = V_{burnout_{South}} = -V_{burnout}\cos\phi\ \cos\beta$$

$$\Delta V_{needed_{East}} = V_{burnout}\cos\phi\sin\beta - \left(465.1\frac{m}{s}\right)\cos L_o$$

$$\Delta V_{needed_{Zenith}} = V_{burnout_{Zenith}} = V_{burnout}\sin\phi$$

$$\Delta V_{design} = \left|\Delta \vec{V}_{needed}\right| + \Delta V_{losses}$$

Key Concepts

➤ A booster system is designed to launch from a given launch site and deliver a spacecraft of a certain size into space. It does this through four phases:
   • vertical ascent
   • pitch over
   • gravity turn
   • vacuum

➤ Because the Earth is rotating eastward, a booster sitting on the launch pad already has some velocity in the eastward direction. Thus,
   • A booster has a "head start" for launching into direct orbits
   • A booster must overcome the Earth's rotation to get into a retrograde orbit
   • The velocity of a launch site depends on the launch-site latitude and is in the East direction

➤ The velocity the booster needs to provide to get a satellite into orbit is the difference between the desired burnout velocity and the velocity of the launch site

➤ To determine the velocities needed to get into orbit, we define the topocentric-horizon coordinate system (or SEZ)

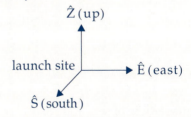

➤ When actually designing the size of a booster, we must also account for losses due to gravity and drag. These losses are typically about 1000 m/s. ΔV_{design} is the velocity the booster must actually deliver to put the payload into orbit.

Example 9-2

Problem Statement

A reconnaissance satellite needs to be launched with a velocity of 7900 m/s, a flight-path angle of 40°, and an azimuth angle of 280°. If the estimated velocity losses for our booster are 900 m/s, how much overall velocity do we design for the booster? (We are launching at a latitude of 28.5°.) Note: The launch azimuth is 280°, so we are launching West into a retrograde orbit.

Problem Summary

Given: $V_{burnout} = 7900$ m/s
$\phi_{burnout} = 40°$
$\beta_{burnout} = 280°$
$L_o = 28.5°$
$\Delta V_{losses} = 900$ m/s

Find: ΔV_{design}

Conceptual Solution

1) Determine the components of

$\Delta \vec{V}_{needed}$

$\Delta V_{needed_{South}} = -V_{burnout} \cos\phi_{burnout} \cos\beta_{burnout}$

$\Delta V_{needed_{East}} = V_{burnout} \cos\phi_{burnout} \sin\beta_{burnout}$
$- (465.1 \text{ m/sec}) \cos L_o$

$\Delta V_{needed_{Zenith}} = V_{burnout} \sin\phi_{burnout}$

2) Find the magnitude of $\vec{V}_{needed}$

$\left| \Delta \vec{V}_{needed} \right|$

$= \sqrt{\Delta V_{needed_{South}}^2 + \Delta V_{needed_{East}}^2 + \Delta V_{needed_{Zenith}}^2}$

3) Solve for ΔV_{design}

$\Delta V_{design} = \left| \Delta \vec{V}_{needed} \right| + \Delta V_{losses}$

Analytical Solution

1) Determine $\Delta \vec{V}_{needed}$ components:

$\Delta V_{needed_{South}} = -V_{burnout} \cos\phi_{burnout} \cos\beta_{burnout}$

$\Delta V_{needed_{South}} = -(7900 \text{ m/s}) \cos 40° \cos 280°$

$\Delta V_{needed_{South}} = -1051 \text{ m/s}$

$\Delta V_{needed_{East}} = V_{burnout} \cos\phi_{burnout} \sin\beta_{burnout}$
$- (465.1 \text{ m/s}) \cos L_o$

$\Delta V_{needed_{East}} = (7900 \text{ m/s}) \cos 40° \sin 280°$
$- (465.1 \text{ m/s}) \cos 28.5°$

$\Delta V_{needed_{East}} = -5960 \text{ m/s} - 409 \text{ m/s}$

$\Delta V_{needed_{East}} = -6369 \text{ m/s}$ (Note: This value is negative because we are launching to the West.)

$\Delta V_{needed_{Zenith}} = V_{burnout} \sin\phi_{burnout}$
$= (7900 \text{ m/s}) \sin 40°$

$\Delta V_{needed_{Zenith}} = 5078 \text{ m/s}$

2) Find $\left| \Delta \vec{V}_{needed} \right|$

$\left| \Delta \vec{V}_{needed} \right| = \sqrt{\Delta V_{needed_{South}}^2 + \Delta V_{needed_{East}}^2 + \Delta V_{needed_{Zenith}}^2}$

$\left| \Delta \vec{V}_{needed} \right| = \sqrt{\left(-1051 \frac{m}{s}\right)^2 + \left(-6369 \frac{m}{s}\right)^2 + \left(5078 \frac{m}{s}\right)^2}$

$\left| \Delta \vec{V}_{needed} \right| = \sqrt{67,449,067 \frac{m^2}{s^2}} = 8213 \frac{m}{s}$

3) Solve for ΔV_{design}

$\Delta V_{design} = \left| \Delta \vec{V}_{needed} \right| + \Delta V_{losses}$

$\Delta V_{design} = 8213 \text{ m/s} + 900 \text{ m/s} = 9113 \text{ m/s}$

Interpreting the Results

Our booster needs a ΔV_{design} of 9113 m/s to achieve the required burnout conditions and overcome the losses of gravity and drag.

References

Bate, Roger R., Donald D. Mueller, and Jerry E. White. *Fundamentals of Astrodynamics*. New York: Dover Publications, Inc., 1971.

Mission Problems

9.1   Launch Windows and Time

1   What is a launch window?

2   How do mission planners specify desired orbits?

3   What do we mean when we say orbit planes are fixed in inertial space?

4   If an orbit has an inclination of 45°, what is the highest northern latitude it will pass over? The highest southern latitude?

5   Define longitude of ascending node, Ω.

6   What is Local Sidereal Time (LST)?   Draw a diagram to illustrate your answer. What is meant by "sidereal"?

7   How does LST change as the Earth rotates? How does Ω change as the Earth rotates?

8   If LST is 45°, what is it in hours, minutes, and seconds? Draw a diagram to illustrate this time.

9   If your current location has rotated 50° past the Vernal Equinox direction, what is LST in hours, minutes, and seconds?

9.2   When and Where to Launch

10   Mission planners want to launch the Space Shuttle from Kennedy Space Center ($L_o = 28.5°$) into an orbit with an inclination of 28.5°. How many launch windows will there be each day? Draw a diagram to illustrate this case. How would this change if the desired inclination were 57°? Draw a diagram to illustrate this case.

11   Define launch window sidereal time (LWST)? What is the difference between $LWST_{AN}$ and $LWST_{DN}$? Draw a diagram to illustrate your answers. How does LWST change as the Earth rotates?

12   Mission planners want to launch the Space Shuttle from Kennedy Space Center ($L_o = 28.5°$) into an orbit with an inclination of 28.5° and a longitude of ascending node of 45°.

a)   What is LWST for this launch in degrees?

b)   What is LWST for this launch in hours, minutes, seconds?

c)   If current LST at Kennedy Space Center is 1200 hrs, how long until the launch window opens?

13 Why do we need to determine angles on a spherical auxiliary triangle to determine LWST for cases with two opportunities per day?

14 Sketch the spherical auxiliary triangle we use to compute launch windows and define all angles.

15 Mission planners at the European Space Agency want to launch their Ariane IV booster from French Guyana ($L_o = 4°$) into a low-Earth orbit with an inclination of $30°$ and a longitude of ascending node of $135°$. LST at the launch site is 1430.

a) How many launch opportunities will there be per day?

b) Draw a 2-D polar view of this launch geometry.

c) Draw a 3-D side view of this launch geometry.

d) Draw the auxiliary triangle for the ascending-node and descending-node opportunities.

e) What is the inclination auxiliary angle, α, for the ascending- and descending-node opportunities?

f) What is the launch-direction auxiliary angle, γ, for the ascending- and descending-node opportunities?

g) What is the launch-window location angle, δ, for both opportunities?

h) What is $LWST_{AN}$?

i) How long until the $LWST_{AN}$?

j) What is the launch azimuth, β, for the ascending node?

k) What is $LWST_{DN}$?

l) How long until $LWST_{DN}$?

m) What is the launch azimuth, β, for the descending node?

16 Mission planners want to launch the Energia booster from Baikanur cosmodrome ($L_o = 51°$) into an orbit with an inclination of $63°$ and a longitude of ascending node of $270°$. If the LST is 0945, when will the next launch window open? What direction will they launch?

17 A new launch site is being planned for the northern coast of Australia ($L_o = 13°S$).

a) What is the lowest inclination for an orbit from this site?

b) If planners want to use this site to resupply the Space Station ($i = 28.5°$ and $\Omega = 35°$), how much time elapses between launch opportunities each day?

9.3   What Time is It?

18 Why do we use sidereal rather than solar time for computing launch windows?

19 What is the difference between solar and sidereal time? Draw a diagram to illustrate which is longer and why.

9.4   Launch Velocity

20   What are the four phases of booster ascent?

21   How fast is Kennedy Space Center (L_o = 28.5°) moving?

22   Explain why eastward launches get a "head start" but westward launches don't.

23   Define the following:

a) $\vec{V}_{burnout}$

b) $\vec{V}_{launch\ site}$

c) $\vec{V}_{needed}$

d) ΔV_{losses}

e) ΔV_{design}

f)   The SEZ coordinate system

24   Mission planners want to launch a new satellite to monitor hurricanes in the Pacific ocean. The booster will launch from Kennedy Space Center into a circular orbit at an altitude of 400 km, with an inclination of 28.5° and longitude of ascending node of 25°.

a)   What is the magnitude of $\vec{V}_{burnout}$?

b)   What is $\vec{V}_{launch\ site}$?

c)   What is $\vec{V}_{needed}$ in SEZ coordinates?

d)   What is the magnitude of $\vec{V}_{needed}$?

e)   If ΔV_{losses} are assumed to be 1000 m/s for this launch, what is ΔV_{design}?

25   Launch-process teams are preparing the new Ariane V booster for launch from French Guyana (L_o = 4°). For propellant loading, what ΔV_{design} is needed to achieve a circular orbit at 500 km, with an inclination of 4°, if the total losses from drag and gravity are 800 m/s?

Mission Profile—Salyut

One of the most ambitious goals of space exploration is establishing a permanent human presence beyond the cradle of Earth. From 1971–1986, the Soviet Salyut program logged a staggering number of human hours in space, established procedures for resupplying cosmonauts, and laid the foundation of technology and experience for eventually colonizing of space.

Mission Overview

From 1971 to 1986, the Soviet Union designed, built, launched, and operated seven Salyut space stations of increasing capability. Cosmonauts conducted experiments in life sciences, astronomy, Earth observation, and materials processing. They developed procedures for automated resupply, extravehicular activities (EVA or space walks), and in-flight maintenance and repair.

Mission Data

- ✓ Salyut 1: Launched April 19, 1971. Manned for more than 23 days beginning June 16, 1971. Cosmonauts Dobrovolsky, Volkov, and Patsaiev died returning to Earth due to a sudden loss of cabin pressure.

- ✓ Salyut 2: Launched April 3, 1973. Station broke up and decayed shortly after launch. No cosmonauts ever visited the station.

- ✓ Salyut 3: Launched June 25, 1974. Cosmonauts Popvich and Artyukhin spent more than 15 days onboard. Station re-entered in January 1975 after seven months of occupied and unoccupied operation.

- ✓ Salyut 4: Launched December 26, 1974. Cosmonauts Goubarev and Grechko logged more than 29 days onboard followed by Klimauk and Sevastianov with a record-setting 63-day stay. Unmanned Soyuz 20 mission demonstrated automated rendezvous and docking. Station deorbited February 3, 1977.

- ✓ Salyut 5: Launched June 22, 1976. Cosmonauts Volynov and Jolobov spent more than 49 days in the station. Soyuz 23 failed to dock. Soyuz 24 with Gorbatko and Glazkov, successfully docked in February 1977, spending 17 days onboard. Station deorbited August 8, 1977.

- ✓ Salyut 6: Launched September 29, 1977. Significantly advanced over its predecessors, it consisted of five modules and two docking ports. The crews used a shower and self-contained space suits. Salyut 6 was home to 33 cosmonauts who kept it occupied for 676 days—far longer than Skylab's 171 days. Deorbited July 1982.

- ✓ Salyut 7: Launched April 19, 1982. Ten crews spent a total of 812 days onboard. It received more than 37,000 kg (81,400 lb) of cargo from twelve Progress and three Cosmos vehicles. Cosmonauts Dzhanibekov and Savinykh rescued the station after a nearly catastrophic power failure. Salyut 7 was visited one last time by Kizim and Soloviev to salvage equipment for use onboard the new Mir space station in 1986. Salyut 7 finally succumbed to atmospheric drag, reentering in 1992.

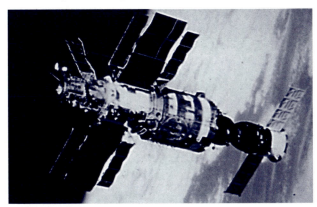

Salyut 7 was the capstone of the Soviet Salyut program. *(Courtesy of Energiya NPO)*

Mission Impact

The Salyut program showed the importance of incremental improvement in a total program, each success setting the stage for the next. Building on their Salyut experience, in 1986 the Soviets launched the Mir station, which can expand by adding on modules.

For Discussion

- The U.S. built the Space Shuttle while the Soviets built Salyut. Which was the better choice?

- How could the experience gained during Salyut apply toward planning a manned Mars mission?

References

Wilson, Andrew (ed.) *Space Directory 1990-91*. Coulsdon, U.K.: Jane's Information Group, 1990.

Rycroft, Michael (ed.) *The Cambridge Encyclopedia of Space*. New York, N.Y.: Cambridge University Press, 1990.

The Space Shuttle orbiter streaks into the atmosphere blazing a trail through the sky. *(Courtesy of NASA)*

Returning From Space: Entry

10

▰ In This Chapter You'll Learn to...

- ☛ Describe the competing design requirements of an entry vehicle
- ☛ Describe the process for analyzing entry motion
- ☛ Describe the basic trajectory options and trade-offs in entry design
- ☛ Describe the basic vehicle options and trade-offs in entry design
- ☛ Describe how a lifting vehicle changes the entry problem

▰ You Should Already Know...

- ❏ The Motion Analysis Process checklist (Chapter 4)
- ❏ Conservation of energy (Chapter 4)
- ❏ Newton's Second Law of Motion (Chapter 4)
- ❏ Basic concepts of calculus (Appendix A)
- ❏ Basic approach to interplanetary transfer (Chapter 7)

▰ Outline

All around him glows the brilliant orange color. Behind, visible through the center of the window is a bright yellow circle. He sees that it is the long trail of glowing ablation material from the heat shield, stretching out behind him and flowing together. "This is Friendship 7. A real fireball outside!"

Astronaut, John Glenn
during entry of Mercury-Atlas 6
February 20, 1962
[Voas, 1962]

Walking along the shore of a tranquil lake on a sunny spring day, most of us have indulged in one of life's simplest pleasures: skipping stones. When the wind is calm, the mirror-like surface of the water practically begs you to try your skill. Searching through pebbles on the sandy bank, you find the perfect skipping rock: round and flat and just big enough for you to get a good grip. You take careful aim because you want the stone to strike the water's surface at the precise angle and speed that will allow its wide, flat bottom to take the full force of impact, causing it to skip. If you have great skill (and a good bit of luck), it may skip three or four times before finally losing its momentum and plunging beneath the water. You know from experience that, if the rock is not flat enough or the angle of impact is too steep, you'll make only a noisy splash rather than a quiet and graceful skip.

Returning from space, astronauts face a similar challenge. The Earth's atmosphere presents to them a dense, fluid medium which, at orbital velocities, is not all that different from a lake's surface. They must plan to hit the atmosphere at the precise angle and speed for a safe landing. If they hit too steeply or too fast, they risk making a big "splash," which would mean a fiery end. If their impact is too shallow, they may literally skip off the atmosphere and back into the cold of space. This subtle dance between fire and ice is the science of atmospheric entry.

In this chapter we'll explore the mission requirements of vehicles entering an atmosphere—whether returning to Earth or trying to land on another planet. We'll discover what engineers must trade off in designing missions which must plunge into dense atmospheres. When we're through, you may never skip rocks the same way again!

10.1 Entry Design

▬ In This Section You'll Learn to...

- ☞ List and discuss the competing requirements of entry design
- ☞ Define an entry corridor and discuss its importance
- ☞ Apply the Motion Analysis Process (MAP) checklist to entry motion and discuss the results
- ☞ Describe the process for entry design

Trade-Offs for Entry Design

All space-mission planning begins with a set of requirements we must meet to achieve mission objectives. The entry phase of a mission is no different. We must delicately balance four, often competing, requirements:

- Deceleration
- Heating
- Accuracy of landing or impact
- Size of the entry corridor

The structure and payload of a vehicle limits the maximum deceleration or "g's" it can pull. (1 g is the gravitational acceleration at the surface of the Earth, or 9.798 m/s^2.) When subjected to enough g's, even steel and aluminum can crush like paper. Fortunately, the structural g limit for a vehicle can be quite high, perhaps hundreds of g's. But a fragile human payload would be crushed to death long before that. Humans can withstand a maximum deceleration of about 12 g's for only a few minutes at a time. Stop for a moment to think about what this would feel like. Imagine eleven other people with your same weight all stacked on top of you. You'd be lucky to breathe! Just as a chain is only as strong as its weakest link, the maximum deceleration a vehicle experiences during entry must be low enough to prevent damage or injury to the weakest part of the spacecraft.

But maximum g's aren't the only deceleration concern of entry designers. Too little deceleration can also cause serious problems. Like a rock skipping off a pond, a vehicle that doesn't slow down enough may literally bounce off the atmosphere and back into the cold reaches of space.

Another limitation during entry is heating. If you've ever seen the fiery trail of a meteor streaking across the night sky, you know that entry can get hot! This intense heat is a result of friction between the speeding meteor and the air. When you rub your hands together vigorously, you produce heat in much the same way. How hot can something get during entry? Think about the energies involved. The Space Shuttle in orbit has a mass of 100,000 kg (220,000 lb.), an orbital velocity of 7700 m/s (17,300 m.p.h.), and an altitude of 300 km (186 miles). We can find its total mechanical energy using the relationship we developed in Chapter 4.

$$E = \frac{1}{2}mV^2 + mgh \tag{10-1}$$

where
E = total mechanical energy ($kg \cdot m^2/s^2 = J$)
m = mass (kg)
V = velocity (m/s)
g = acceleration due to gravity (m/s^2)
   = 8.94 m/s^2
h = altitude (m)

Substituting the above values and converting to standard units of energy, we get

$$E = 3.23 \times 10^{12} \text{ Joules} = 3.06 \times 10^9 \text{ BTU}$$

Let's put this number in perspective. The average house in Colorado needs about 73.4×10^7 BTU/year to keep it heated. This means the Shuttle dissipates enough heat during entry to heat the average home in Colorado for 41 years! All of this energy must be lost in only about one-half hour to bring the Shuttle to a full stop on the runway. But, remember, energy is conserved! Where does all the energy go? Heat! Parts of the Shuttle can reach temperatures of up to 1477° C (2690° F). We must design the entry trajectory, and the vehicle itself, to withstand these tremendous heat loads. As we'll see, we must contend with not only the total heating over the entry but also the peak heating rate.

A third mission requirement is accuracy. Beginning its descent from over 4000 miles away, the Space Shuttle must land on a runway only 100 m (300 ft.) wide. The reentry vehicle (RV) of an Intercontinental Ballistic Missile (ICBM) may have even tighter accuracy requirements. To meet these constraints, we must again adjust the trajectory and vehicle design.

As you can see from all these constraints, an entry vehicle must walk a tightrope between being squashed and skipping out, between fire and ice, and between hitting and missing the target. This tightrope is actually a three-dimensional *entry corridor*, shown in Figure 10-1, through which an entering vehicle must pass to avoid skipping out or burning up.

The size of the corridor depends on three competing constraints—deceleration, heating, and accuracy. For example, if the vehicle strays below the lower, or undershoot, boundary of the corridor, it'll experience too much drag, slowing it down rapidly and heating it up too quickly. On the other hand, if the vehicle enters above the upper, or overshoot, boundary of the corridor, it won't experience enough drag and may literally skip off the atmosphere back into space. If we're not careful, all these competing requirements may literally constrain us to an entry corridor too narrow for the vehicle to drive through!

Whereas the above three constraints fix the entry corridor's size, the vehicle's control system determines its ability to steer through the entry corridor. In this chapter we'll concentrate on understanding what affects the corridor's size. We'll discuss limits on the control system in Chapter 12.

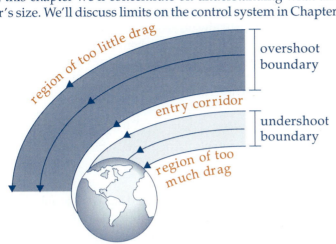

Figure 10-1. Entry Corridor. The entry corridor is a narrow region in space an entering vehicle must drive through. If the vehicle strays above the corridor, it may skip out. If it strays below the corridor, it may burn up.

The Motion Analysis Process

Before we can understand how to juggle all these entry trade-offs, we need to develop a way of analyzing entry motion to see how various trajectories and vehicle shapes affect these entry constraints. Whether it's a rock hitting the water or a spacecraft hitting the atmosphere, we still have a problem in dynamics which we can solve by dusting off our tried-and-true Motion Analysis Process (MAP) checklist, as shown in Figure 10-2.

First on the list is a coordinate system. We still need an inertial reference frame (so Newton's laws will be valid), which we'll call the entry coordinate system. To make things easier, we'll place the origin of the *entry coordinate system* at the vehicle's center of mass *at the start of entry*. We'll then analyze the motion with respect to this fixed center.

Our fundamental plane is the orbit plane of the entering vehicle. Within this plane, we can pick a convenient principle direction which points "down" to the center of the Earth. (By convention, the axis which points down is the $\hat{Z}$ direction.) We can define the $\hat{X}$ direction to be along the local horizontal in the general direction of motion. The $\hat{Y}$ direction completes the right hand rule. However, because we'll assume all motion takes place in plane, we won't worry about the y direction. Figure 10-3 shows the entry coordinate system.

We'll also define the *entry flight-path angle, γ,* to be the angle between the local horizontal and the velocity vector. (Note this angle is the same as the flight-path angle, φ, used earlier; however, entry analysts like to use gamma instead, so we'll play along.) Once again, a flight-path angle below the horizon (diving toward the ground) is negative; a flight-path angle above the horizon (climbing up) is positive.

MOTION ANALYSIS PROCESS (MAP) CHECKLIST

❑ Coordinate System

❑ Equations of Motion

❑ Simplifying Assumptions

❑ Initial Conditions

❑ Error Analysis

❑ Testing the Model

Figure 10-2. Motion Analysis Process (MAP) Checklist.

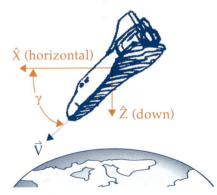

Figure 10-3. Entry Coordinate System. Our entry-coordinate system uses the center of the vehicle at the start of entry as the origin. The orbit plane is the fundamental plane, and the principle direction is down. Here the entry flight-path angle, γ, is shown as the angle between local horizontal and the velocity vector.

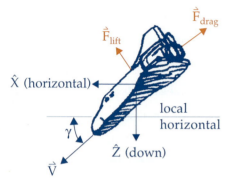

Figure 10-4. Forces on an Entry Vehicle. An entry vehicle could potentially encounter lift, drag, gravity, and other forces.

Figure 10-5. Significant Forces on an Entry Vehicle.

Note: For entry, γ is a negative angle. This has been accounted for in Equation (10-4). i.e. cos (–γ) = cosγ and sin (–γ) = –sinγ.

Next we develop equations of motion. To do this, it's best to start by brainstorming what forces could possibly affect an entering spacecraft. Of course there's gravity (it always seems to get involved) and, because we're travelling through the dense atmosphere just like an airplane, we must also contend with lift and drag. Finally, we'll throw in "other" forces to cover all our bases. A vehicle with all these forces is shown in Figure 10-4. Summing all of these forces, we get

$$\sum \vec{F}_{external} = \vec{F}_{gravity} + \vec{F}_{drag} + \vec{F}_{lift} + \vec{F}_{other} \qquad (10\text{-}2)$$

We can now apply Newton's Second Law, which states

$$\sum \vec{F}_{external} = m\vec{a} \qquad (10\text{-}3)$$

Once again we have a rather complicated equation to solve. Time for some assumptions to bail us out. To make our lives easier, let's assume

- The entry vehicle is a point mass.
- Drag and lift are the dominant forces—all other forces, including gravity, are insignificant. (We'll see why this is a good assumption later.)

With these assumptions, we're left with only two forces—drag and lift. Because we're worried about the size and direction of these forces, we must take a closer look at our force diagram of the vehicle and apply a little trigonometry.

Looking at Figure 10-5, you can see that drag acts in the direction opposite the vehicle's motion, and lift acts perpendicular to this direction. The total inertial forces on the vehicle can then be expressed in the entry coordinate system

$$\sum \vec{F}_{external} = (-F_{drag} \cos\gamma - F_{lift} \sin\gamma)\,\hat{x} + (F_{drag} \sin\gamma - F_{lift} \cos\gamma)\,\hat{z} \quad (10\text{-}4)$$

Next we make some assumptions about lift and drag. Both forces are related to the *dynamic pressure, $\bar{q}$* ("q-bar"), on the vehicle. $\bar{q}$ describes the affect of traveling through an atmosphere of some density, ρ, with some velocity, V.

$$\bar{q} = \frac{\rho V^2}{2} \qquad (10\text{-}5)$$

where
$\bar{q}$ = dynamic pressure (N/m²)
ρ = density (kg/m³)
V = velocity (m/s)

We can then describe lift and drag using two unique properties of vehicle shape—*coefficient of lift, C_L*, and *coefficient of drag, C_D*. Both of these values are computed and then validated using wind tunnels. Combining these quantities with the dynamic pressure and the cross-sectional area of the vehicle, A, we can describe the lift and drag forces as

$$F_{lift} = \bar{q} C_L \, A = \frac{1}{2} \rho V^2 \, C_L \, A \qquad (10\text{-}6)$$

where

F_{lift} = force of lift (N)
C_L = lift coefficient (unitless)
A = cross-sectional area (m^2)
ρ = density (kg/m^2)
V = velocity (m/s)

$$F_{drag} = \bar{q} \, C_D \, A = \frac{1}{2} \rho \, V^2 \, C_D \, A \qquad (10\text{-}7)$$

where

F_{drag} = force of drag (N)
C_D = drag coefficient (unitless)

Our equation of motion now is

$$\sum \vec{F}_{external} = (-\bar{q} C_D A \cos\gamma - \bar{q} C_L A \sin\gamma)\,\hat{x} \qquad (10\text{-}8)$$
$$+ \, (\bar{q} C_D A \sin\gamma - \bar{q} C_L A \cos\gamma)\,\hat{z}$$

For meteors entering the atmosphere, the lift force is almost zero. Even for the Space Shuttle, lift is relatively small when compared to drag. For these reasons, we can assume for now that our vehicle produces no lift, thus, $C_L = 0$. (Actually, the lift generated by the Space Shuttle is enough to significantly change its trajectory, as we'll see in Section 10-4. However, this assumption will greatly simplify our analysis and allow us to demonstrate the trends on entry design.) We can now simplify Equation (10-8) even more to get

$$\sum \vec{F}_{external} = (-\bar{q} C_D A \cos\gamma)\,\hat{x} + (\bar{q} C_D A \sin\gamma)\,\hat{z} \qquad (10\text{-}9)$$

If we divide through by the mass of the vehicle, notice we end up with $C_D A/m$ in both terms

$$\vec{a} = \left(-\bar{q}\frac{C_D A}{m}\cos\gamma\right)\hat{x} + \left(\bar{q}\frac{C_D A}{m}\sin\gamma\right)\hat{z} \qquad (10\text{-}10)$$

where

m = mass (kg)
γ = flight-path angle (deg)

Ever since engineers first began to analyze the trajectories of cannon balls, this product has had a special significance in describing how an object moves through the atmosphere. By convention, engineers flip this term over and call it the *ballistic coefficient, BC.*

$$BC = \frac{m}{C_D A} \qquad (10\text{-}11)$$

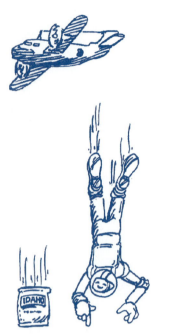

Figure 10-6. Comparing Ballistic Coefficients. A sack of potatoes and an astronaut have about the same ballistic coefficient (BC).

Figure 10-7. Changing BC. With parachute open, the astronaut greatly increases his area, A, and drag coefficient, C_D, thus decreasing his ballistic coefficient, BC, and slowing down much faster than the potatoes.

where
BC = ballistic coefficient (kg/m^2)
m = mass (kg)
C_D = drag coefficient (unitless)
A = cross sectional area (m^2)

From Equations (10-10) and (10-11), we can see the magnitude of our deceleration from drag $|\vec{a}|$ is inversely related to BC

$$|\vec{a}| = \frac{\bar{q}}{BC} \tag{10-12}$$

which means as BC goes up, deceleration goes down.

Let's take a moment to see what BC really represents. Suppose our 150-lb. astronaut and a 150-lb. sack of potatoes are thrown out of an airplane at the same time (same mass, same initial velocity). If the astronaut and the potatoes have about the same mass, m; cross sectional area, A; and drag coefficient, C_D, they have the same BC. Thus, the drag force on each is equal, and they fall at the same rate, as shown in Figure 10-6. What happens if the astronaut opens his parachute? He now slows down significantly faster than the sack of potatoes. But what happens to his BC? His mass stays the same, but when his chute opens, his cross-sectional area and C_D increase dramatically. When C_D and area increase, his BC goes down compared to the sack of potatoes, slowing his descent rate as shown in Figure 10-7. From this example, we see that *an object with a low BC slows down much quicker than an object with a high BC.* In everyday terms, we would say a light, blunt vehicle (low BC) slows down much more rapidly than a heavy, streamlined (high BC) one, as shown in Figure 10-8.

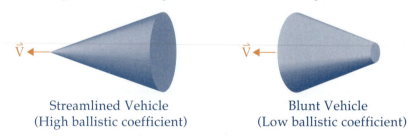

Streamlined Vehicle
(High ballistic coefficient)

Blunt Vehicle
(Low ballistic coefficient)

Figure 10-8. Blunt Versus Streamlined Vehicles. A light, blunt vehicle (low BC) slows down much more rapidly due to drag than a heavy, streamlined (high BC) one.

With all this discussion out of the way, we can turn our attention to the next item on the MAP checklist—Initial Conditions (ICs). These are especially important for entry. As we'll see, the initial entry velocity, V_{Entry}, and the initial flight-path angle, γ, will determine most of the conditions experienced during entry. Determining what these Initial Conditions should be involves many trade-offs for mission and vehicle designers. For entry analysis, we will concentrate on the effects of these ICs and not spend any time on Error Analysis or Testing the Model.

Entry-Motion Analysis in Action

Because the equation of motion we developed for entry in Equation (10-10) is still quite complicated, let's take some time to see how we can use it. We need to understand how the acceleration equation affects a vehicle's velocity and, in turn, its position during entry.

If we give an object a constant acceleration, we can determine its velocity after some time, t, from

$$\vec{V}_{final} = \vec{V}_{initial} + \vec{a}t \qquad (10\text{-}13)$$

where
$\vec{V}_{final}$ = final velocity (m/s)
$\vec{V}_{initial}$ = initial velocity (m/s)
$\vec{a}$ = acceleration (m/s^2)
t   = time (s)

The final position of the object is

$$\vec{R}_{final} = \vec{R}_{initial} + \vec{V}_{initial}\, t + \frac{1}{2}\vec{a}t^2 \qquad (10\text{-}14)$$

Unfortunately, an entry vehicle's acceleration isn't constant. Notice in Equation (10-10) that drag deceleration is a function of velocity, and the velocity will be changing due to drag! This is another example of a transcendental function like the one we discussed in Chapter 8.

Astro Fun Fact
Dinosaurs and Meteors

Every day, 400 tons of micrometeorite dust hit the Earth in the form of minute cosmic particles. Yet, this did not explain what geologist Walter Alvarez discovered in Italy in the late 1970s. He unearthed a half-inch layer of clay deposited 65 million years ago. He named this layer the K-T layer, as the clay lay between the Cretaceous and Tertiary Time periods. Later, a technique called neutron activation found this deposit contained thirty times the normal amount of iridium, an element rare on Earth but abundant in meteors. This evidence led to the theory that a massive meteor collision with the Earth caused the extinction of dinosaurs. The theory, officially called the K-T theory of extinction, appears viable. Possible sites for the meteor's impact include a 190-mile-wide crater off the coast of South America as well as an unknown-sized crater 3500 feet below ground on the Yucatan peninsula (indicated by geographical surface features). Why then, you may ask, did other species survive such an enormous catastrophe? While no scientific explanation can yet answer this question, many scientists believe it may simply have been another event in the natural selection process—survival of the fittest!

Evans, Barry. _The Wrong Way Comet and Other Mysteries of Our Solar System_. Blue Ridge Summit: Tab Books, 1992.

Contributed by Troy Kitch, United States Air Force Academy

Transcendental functions can't be solved directly. How do we deal with it? We use a method first developed by Isaac Newton himself—numerical integration or iteration. Sound complicated? Actually it's not that bad. We'll assume that over some small time interval, Δt, the acceleration *is* constant (a good assumption if Δt is small enough). This will allow us to use the velocity and position equations for constant acceleration during that time interval. By adding together the effects of the acceleration during each time interval, we can determine the cumulative effect on velocity and position. (Of course this means lots of calculations, so it's best to use a computer. We could either write a new computer program or use the built-in flexibility of a spreadsheet. All the analysis you'll see in this chapter was done using a spreadsheet.)

Before diving into a full-scale numerical analysis of entry characteristics, let's see how we can use this basic approach to take a closer look at something in the night sky: a meteor. Imagine one of Earth's many celestial companions wandering through space until it encounters Earth's atmosphere at more than 8 km/s, screaming in at a steep angle. Initially, in the upper reaches of the atmosphere, there is very little drag to slow down the massive chunk of rock. But as the meteor penetrates deeper and deeper, the drag force builds up rapidly, causing it to slow down dramatically. This slowing is like the quick initial deceleration experienced by a rock hitting the surface of a pond. At this point in the meteor's trajectory, its heating rate is also highest, so it begins to glow with temperatures hot enough to melt the iron and nickel it's composed of. If anything is left of the meteor at this point, it will continue to slow down but at a more leisurely pace. Of course, most meteors burn up completely before reaching a planet's surface.

Let's investigate this meteor entry using the numerical procedure we'll use throughout the chapter. The results of the numerical integration are shown in Figure 10-9 where you can see the meteor's velocity profile along with its initial conditions.

We can see from the graph what we expected to find from our discussion of the meteor scenario. Notice in Figure 10-9 the velocity stays nearly constant through the first ten seconds when the meteor is still above most of the atmosphere. But things change rapidly over the next ten seconds. The meteor loses almost 90% of its velocity—almost like hitting a wall. With most of its velocity lost, the deceleration is much lower—it takes 20 seconds more to slow down by another 1000 m/s.

We now have a precise mathematical tool to analyze entry characteristics. We can use this tool to balance all the competing mission requirements by approaching them on two broad fronts:

- Trajectory design, which includes changes to
  - Entry velocity, V_{entry}
  - Entry flight-path angle, γ
- Vehicle design, which includes changes to
  - Vehicle size and shape (BC)
  - Thermal protection systems (TPS)

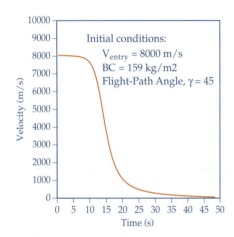

Initial conditions:

V_{entry} = 8000 m/s
BC = 159 kg/m2
Flight-Path Angle, γ = 45

Figure 10-9. Meteor Entering the Atmosphere. This graph shows how abruptly an entering meteor slows down—like a rock hitting the surface of a pond.

Trajectory design involves changing the initial conditions of entry defined by the vehicle's velocity as it enters the effective atmosphere. These initial conditions are the *entry velocity, V_{entry}* and entry flight-path angle, γ. Vehicle design, as we'll see, includes changing the vehicle's shape to alter the BC or designing a thermal-protection system (TPS) to deal with entry heating.

As seen in Figure 10-1, entry design requires iteration. Mission requirements affect the vehicle design. The design drives deceleration, heating, accuracy, and limits on the entry corridor. These, in turn, affect trajectory options, which may change the vehicle design, and so on. In practice, we must continually trade between trajectory and design until we reach some compromise. In the next few sections, we'll explore trajectory options and vehicle design in greater detail.

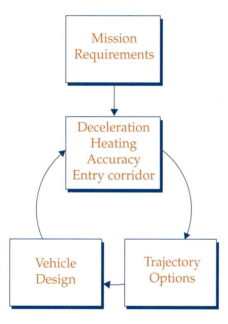

Figure 10-1. Entry Design. Entry design begins with mission requirements. Then engineers must work the trade-offs between vehicle design, deceleration, heating, accuracy, entry corridor, and trajectory options.

▬ Section Review

Key Terms

ballistic coefficient, BC
coefficient of drag, C_D
coefficient of lift, C_L
dynamic pressure, $\bar{q}$
entry coordinate system
entry corridor
entry flight path angle, γ
entry velocity, V_{entry}

Key Equations

$$F_{lift} = \bar{q}C_L \, A = \frac{1}{2}\rho V^2 \, C_L \, A$$

$$F_{drag} = \bar{q} \, C_D \, A = \frac{1}{2}\rho \, V^2 \, C_D \, A$$

$$BC = \frac{m}{C_D \, A}$$

Key Concepts

➤ We must balance four competing requirements for entry design:
  • Deceleration
  • Heating
  • Accuracy
  • Size of the entry corridor

➤ The entry coordinate system is based on
  • Origin—center of vehicle
  • Fundamental plane—vehicle's orbit plane
  • Principle direction—down

➤ To analyze entry trajectories, we must use numerical integration based on the following assumptions
  • Entry vehicle is a point mass
  • Drag is the dominant force—all other forces, including gravity, are insignificant
  • Vehicle produces no lift

➤ Ballistic coefficient, BC, quantifies an object's mass, drag coefficient, and cross-sectional area and predicts how drag will affect it:
  • Light, blunt vehicle—low BC—slows down quickly
  • Heavy, streamlined vehicle—high BC—doesn't slow down quickly

➤ To balance competing requirements, we tackle the entry-design problem on two fronts:
  • Trajectory design—changes to entry velocity, V_{entry}, and entry flight-path angle, γ
  • Vehicle design—changes to vehicle size and shape (BC) and thermal-protection system (TPS)

10.2 Trajectory Options

▬ In This Section You'll Learn to...

☞ Explain how changing the entry velocity and flight-path angle affects deceleration and heating rates

☞ Determine the maximum deceleration and the altitude at which this deceleration occurs for a given set of entry conditions

☞ Determine the maximum heating rate and the altitude at which this rate occurs for a given set of entry conditions

☞ Explain how changing the entry velocity and flight-path angle affects accuracy and size of the entry corridor

Depending on the mission and vehicle characteristics, planners can do only so much with the entry trajectory. For example, the amount of propellent the Space Shuttle can carry for the engines in its Orbital Maneuvering System (OMS) limits how much we can alter its velocity and flight-path angle at entry. Entry conditions for ICBM reentry vehicles, depend on the velocity and flight-path angle of the booster at burnout. In either case, we must know how the entry trajectory affects an entry vehicle's maximum deceleration, heating, accuracy, and the entry corridor's size.

Trajectory and Deceleration

As we saw with our meteor example in the last section, a vehicle entering from space takes time to make its way down into the denser atmosphere. Deceleration builds gradually up to some maximum value, a_{max}, and then begins to taper off. To see how varying the entry velocity and angle affects this maximum deceleration, let's apply our numerical tool to the entry equation of motion we developed in the last section. We'll begin by keeping all other variables constant and change only the initial entry velocity to see its effect on a_{max}. If we set the following initial conditions, we can plot the deceleration versus altitude for various entry velocities.

Vehicle mass = 1000 kg
Nose radius = 2 m
Cross sectional area = 50.3 m^2
C_D = 1.0
BC = 19.9 kg/m^2
Entry flight-path angle, γ = 45°

The graph in Figure 10-11 shows that a higher entry velocity means greater maximum deceleration. This should make sense if you think again

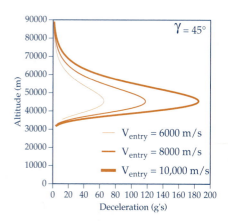

Figure   10-11. Deceleration   Profiles   for Various Entry Velocities. The higher the entry velocity, the greater the maximum deceleration.

about skipping rocks. The harder you throw a rock at the water (the higher the V_{entry}), the bigger the splash it will make (greater a_{max}). Without going into a lengthy derivation, we can find the vehicle's maximum deceleration, and the altitude at which it occurs, from

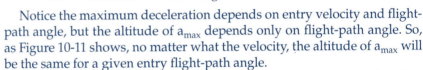

$$a_{max} = \frac{V^2_{entry}\,\beta\,\sin\gamma}{2e} \qquad (10\text{-}15)$$

$$Altitude_{a_{max}} = \frac{1}{\beta}\ln\left(\frac{\rho_o}{BC\ \beta\ \sin\gamma}\right) \qquad (10\text{-}16)$$

where
a_{max} = maximum deceleration (m/s^2)
β = atmospheric scale height, a parameter used to describe the density profile of the atmosphere
  = 0.000139 m^{-1} for Earth
γ = flight-path angle (deg or rad)
e = base of natural logarithm = 2.7182
$\ln$ = natural logarithm of quantity in parenthesis
ρ_o = atmospheric density at sea level = 1.225 kg/m^3
BC = ballistic coefficient (kg/m^2)

Notice the maximum deceleration depends on entry velocity and flight-path angle, but the altitude of a_{max} depends only on flight-path angle. So, as Figure 10-11 shows, no matter what the velocity, the altitude of a_{max} will be the same for a given entry flight-path angle.

Now that we know how V_{entry} affects deceleration, let's look at the other trajectory parameter—flight-path angle, γ. Keeping the same initial conditions as before and fixing the entry velocity at 8 km/s, we can plot the deceleration versus altitude profiles for various flight-path angles.

In Figure 10-12, we see that the steeper the entry angle the more severe the peak deceleration. Once again, this should make sense from the rock-skipping example, in which a steeper angle causes a bigger splash. In addition, we can see that the steeper entry angle dives deeper into the atmosphere before reaching the maximum deceleration.

Now take a look at the amount of maximum deceleration (in g's) for varying entry velocities and flight-path angles. Notice the maximum deceleration can be as much as 200 g's! Because the acceleration from gravity is never more than 1 g, we can conclude the dominant force on a vehicle during entry is drag. This justifies our earlier decision to ignore gravity.

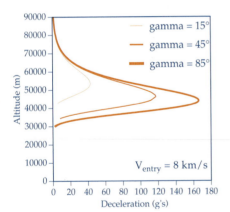

Figure 10-12. Deceleration Profile for Various Entry Flight-Path Angles. The higher the entry flight-path angle (steeper the entry) the greater the maximum deceleration experienced.

Trajectory and Heating

Earlier, we discovered *why* an entry vehicle gets hot—all the energy it starts with must go somewhere (conservation of energy). Now let's look at *how* the vehicle gets hot. Heat can transfer from one place to another by radiation, conduction, and convection. *Radiation* or *radiative heat transfer*, discussed in Chapter 3, involves the transfer of energy from one point to another through electromagnetic waves. If you've ever held your hand in front of a glowing space heater, you've felt radiative heat transfer.

Conduction or *conductive heat transfer* moves heat energy from one point to another through some physical medium. For example, try holding on to one end of a metal rod and sticking the other end in a hot fire. Before too long the end you're holding will get HOT (ouch!). The heat is "conducted" along the metal rod.

Finally, *convection* or *convective heat transfer* takes place when a fluid flows past an object and transfers energy to it or absorbs energy (depending on which is hotter). This is where we get the concept of "wind chill." As a breeze flows past you, heat transfers from your body to the air, cooling you down.

So what's all this have to do with an entering vehicle? If you've ever been on a ski boat, plowing at high speeds through the water, you may have noticed how the water appears to bend around the hull. At the front of the boat, where the hull first meets the water, a bow wave forms so the moving boat never appears to run into the still water. This bow wave continues around both sides of the boat, forming the wake of turbulent water that's so much fun to ski through.

A spacecraft entering the atmosphere at high speeds must plow into the fluid air much like the boat. At extremely high entry speeds, even the wispy upper atmosphere creates a profound effect. In front of the entering spacecraft, a bow wave of sorts forms. This *shock wave* results when air molecules bounce off the front of the vehicle and then collide with the incoming air. The shock wave then bends the flow of air around the vehicle. Depending on the shape of the vehicle, the shock wave can either be attached or detached. If the vehicle is streamlined (high BC, like a cone), the shock wave may attach to the tip. If the vehicle is blunt (low BC, like a rock), the shock wave will detach and curve out in front of the vehicle. Both types of shock waves are shown in Figure 10-13.

So how does the vehicle get hot? As the shock wave slams into the air molecules in front of the entering vehicle, they get hot. They go from a cool, dormant state to an excited state, acquiring heat energy. (To see why, clap your hands together for a few minutes and notice how hot your hands get.) These hot air molecules then transfer some of their heat to the vehicle by convection. Convection is the primary means of heat transfer to a vehicle entering the Earth's atmosphere at speeds under about 15,000 m/s. (For an entry to Mars or some other planet with a different type of atmosphere, this speed will vary.) Above this speed, the air molecules get so hot they begin to transfer much of their energy to the vehicle by radiation.

Without going into all the details of aerodynamics and thermodynamics, we can quantify the *heating rate, $\dot{q}$* ("q dot" or rate of change of heat energy) an entry vehicle experiences. We express this quantity in watts per square centimeter, which is heat energy per unit area per unit time. It's a function of the vehicle's nose radius, its velocity, and the density of the atmosphere. Empirically, for Earth's atmosphere, this becomes approximately

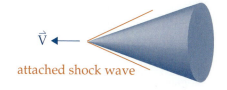

attached shock wave

detached shock wave

Figure 10-13. Attached and Detached Shock Waves. As a vehicle plows into the atmosphere from space a shock wave forms out in front. This shock wave attaches to streamlined vehicles (high BC) but detaches from blunt vehicles (low BC).

$$\dot{q} \cong 1.83 \times 10^{-4} \ V^3 \sqrt{\frac{\rho}{r_{nose}}} \qquad (10\text{-}17)$$

where

$\dot{q}$ = heating rate (W/m²)
V = velocity (m/s)
ρ = air density (kg/m³)
r_{nose} = nose radius of the vehicle (m)

Returning to our numerical analysis of a generic entry vehicle with the same initial conditions as before, we can plot heating rate, $\dot{q}$, versus altitude for various entry velocities. In Figure 10-14 we see that the maximum heating rate increases as the entry velocity goes up. We can find the altitude and velocity where the maximum heating rate occurs using

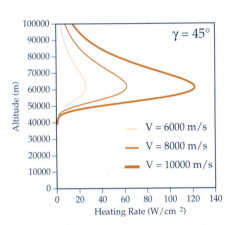

$$\text{Altitude}_{\dot{q}_{max}} = \frac{1}{\beta} \ln\left(\frac{\rho_o}{3BC \ \beta \ \sin\gamma}\right) \qquad (10\text{-}18)$$

where

β = atmospheric scale height = 0.000139 m⁻¹ for Earth
ρ_o = atmospheric density at sea level = 1.225 kg/m³
BC = ballistic coefficient (kg/m²)
γ = flight-path angle (degrees or radians)

Figure 10-14. Variation in Heating Rate for Three Entry Velocities. As the entry velocity increases, the heating rate also increases.

and

$$V_{\dot{q}_{max}} \approx 0.846 \ V_{entry} \qquad (10\text{-}19)$$

where

$V_{\dot{q}_{max}}$ = velocity when maximum heating rate is reached (m/s)
V_{entry} = entry velocity (m/s)

We can also vary the entry flight-path angle, γ, to see how it affects the maximum heating rate. Let's use an entry velocity of 8 km/s again. Keeping all other initial conditions the same and varying γ, we can plot $\dot{q}$ versus altitude for various entry flight-path angles in Figure 10-15.

Notice the correlation between steepness of entry and the severity of the peak heating rate. Recall from our earlier discussion that the steeper the entry the deeper into the atmosphere the vehicle will travel before reaching maximum deceleration. This means the steeper the entry angle the more quickly the vehicle will reach the ground, creating an interesting dilemma for the entry designer:

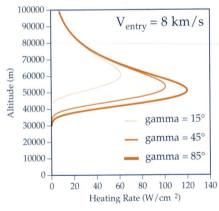

Figure 10-15. Variation in Heating Rate at Different Entry Flight-Path Angles. The steeper the entry angle the higher the peak heating rate.

- Steep entry causes high maximum heating rates but for a short time
- Shallow entry causes low maximum heating rates but for a long time

A steep entry causes very high heating rate but for a brief time, so the overall effect on the vehicle may be small. On the other hand, shallow entries lead to much lower heating rates. However, because heating continues longer, the vehicle is more likely to "soak up" heat and be damaged.

To understand this difference, imagine boiling two pots of water. For the first pot you'll build a fire using large, thick logs. They'll build up a low, steady heating rate lasting for quite some time. Under the second pot you'll place an equal mass of wood but in the form of sawdust. The sawdust will heat up much faster than the logs but will also burn out much more quickly. Which option will better boil the water? Because the logs put out a lower heat rate but for much longer, the water is more likely to soak up this heat and begin to boil. The sawdust burns so quickly that most of its heat simply escapes to the air, so the water in the pot absorbs very little.

This example underscores the importance of considering the heating rate, $\dot{q}$, along with the total heat load, Q. *Total heat load, Q,* is the total amount of thermal energy (m J/m²) the vehicle is exposed to. We find Q by integrating or summing up all the $\dot{q}$'s over the entire entry time. As we've already seen, $\dot{q}$ varies with entry velocity. Q also varies with velocity but *not* with flight-path angle. This makes sense when you consider the heat results from mechanical energy dissipating during entry, which is independent of entry angle. This means, the higher the entry velocity, the higher the total heat load, as shown in Figure 10-16. Thus, although the peak heating rate will vary with flight-path angle, the total heat load for a given entry velocity will be constant.

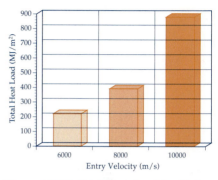

Figure 10-16. Total Heat Load for Various Entry Velocities. The higher the entry velocity, the greater the total heat load.

Again, we face an acute engineering dilemma for manned entry vehicles. We'd like a shallow entry to keep the maximum deceleration low (you don't want to crush the crew), but this means a greater risk of soaking up the heat of entry. Fortunately (for the crew), we have ways to deal with this heat energy, as we'll see in the next section.

Trajectory and Accuracy

Next, we can look at how trajectory affects accuracy. Consider what the atmosphere does to an entering vehicle. Both drag and lift forces will perturb its trajectory from the path it would follow under gravity alone. When we modeled these effects, we used several parameters to quantify how the atmosphere affects the vehicle. Whether we're modeling the density, ρ, or the drag coefficient, C_D, the values we use are, at best, only close to the real values and, at worst, mere approximations. Thus, the actual trajectory path will be somewhat different, so when we try to aim at a particular target we might miss!

To reduce these atmospheric affects, and improve our accuracy, we want a trajectory which spends the least time in the atmosphere. So we choose a high entry velocity and a flight path with a steep entry angle. But as we've just seen, this increases the severity of deceleration and heating. Thus, to achieve highly accurate reentry for ICBMs, these vehicles are built to withstand extremely high g forces and peak heating. Manned vehicles, on the other hand, accept lower accuracy to get lower peak deceleration and heating.

Trajectory and the Entry Corridor

From the definition of entry corridor, we can think of the upper or overshoot boundary as the "skip out" boundary. A vehicle entering the atmosphere above this boundary risks bouncing off the atmosphere and back into space. While hard to quantify exactly, this boundary is set by the minimum deceleration needed to "capture" the vehicle. Changes to entry velocity or flight-path angle don't move this boundary significantly. Therefore, we can change the size of the entry corridor most effectively by tackling the lower or undershoot boundary.

As we've just seen, maximum deceleration and maximum heating rate, the two parameters which fix the undershoot boundary, increase directly with increased entry velocity, V_{entry}, or entry flight-path angle,γ, (steeper entry). Most programs limit maximum deceleration and maximum $\dot{q}$ to certain values. Thus, we could still expand the entry corridor by decreasing V_{entry} or γ. This would give us a larger margin for error in planning the entry trajectory and relieve requirements placed on the control system. Unfortunately, for most missions, V_{entry} and γ are set by the mission orbit and are difficult to change significantly without using rockets to perform large, expensive ΔV. Therefore, as we'll see in the next section, our best options for changing the entry corridor size lie in the vehicle design arena.

Table 10-1 summarizes how trajectory options affect deceleration, heating, accuracy, and entry-corridor size.

Table 10-1.   Trajectory Trade-Offs for Entry Design.

Parameter	Maximum Deceleration	Altitude of Maximum Deceleration	Maximum Heating Rate	Altitude of Maximum Heating Rate	Accuracy	Corridor Width
Entry velocity, V_{entry} (constant γ)						
High	High	Same	High	Same	High	Narrow
Low	Low	Same	Low	Same	Low	Wide
Entry flight-path angle, γ (constant V_{entry})						
Steep	High	Low	High	Low	High	Narrow
Shallow	Low	High	Low	High	Low	Wide

Section Review

Key Terms

conduction
conductive heat transfer
convection
convective heat transfer
heating rate, $\dot{q}$
radiation
radiative heat transfer
shock wave
total heat load, Q

Key Equations

$$a_{max} = \frac{V_{entry}^2 \beta \sin\gamma}{2e}$$

$$\text{Altitude}_{a_{max}} = \frac{1}{\beta}\ln\left(\frac{\rho_o}{BC\ \beta\ \sin\gamma}\right)$$

$$\text{Altitude}_{\dot{q}_{max}} = \frac{1}{\beta}\ln\left(\frac{\rho_o}{3BC\ \beta\ \sin\gamma}\right)$$

$$V_{\dot{q}_{max}} \approx 0.846\ V_{entry}$$

Key Concepts

➤ We can meet mission requirements on the trajectory front by changing
  - Entry velocity, V_{entry}
  - Entry flight-path angle, γ

➤ Increasing entry velocity increases
  - Maximum deceleration, a_{max}
  - Maximum heating rate, $\dot{q}$

➤ Compared to the force of drag, the force of gravity on an entry vehicle is insignificant

➤ Increasing the entry flight-path angle, γ, (steeper entry) increases
  - Maximum deceleration, a_{max}
  - Maximum heating rate, $\dot{q}$

➤ The more time a vehicle spends in the atmosphere, the less accurate it will be. Thus, to increase accuracy, we would use fast, steep entries.

➤ To increase the size of the entry corridor, we would decrease the entry velocity and flight-path angle. However, this is often difficult to do.

➤ Table 10-1 summarizes the trajectory trade-offs for entry design

10.3 Options for Vehicle Design

▬ In This Section You'll Learn to...

☛ Discuss two ways to determine the hypersonic drag coefficient for a given vehicle shape

☛ Discuss the effect of changing the ballistic coefficient on deceleration, heating rate, and entry-corridor width

☛ Discuss three types of thermal-protection systems and how they work

Once we've exhausted all trajectory possibilities, we can turn to options for vehicle design. Here, we have two ways to meet mission requirements:

- Vehicle size and shape
- Thermal-protection systems (TPS)

Vehicle Shape

The entry vehicle's size and shape determine the ballistic coefficient (BC) and the amount of lift it will generate. Because adding lift to the entry problem greatly complicates the analysis, we'll continue to assume we're dealing only with non-lifting vehicles. In the next section, we'll discuss how lift affects the entry problem.

The hardest component of BC to determine for entry vehicles is the drag coefficient, C_D, which depends on the vehicle's shape. At low speeds, we could just stick a model of the vehicle in a wind tunnel and take specific measurements to determine C_D. But at entry speeds approaching 25 times the speed of sound, wind tunnel testing isn't practical because we can't build tunnels that work at those speeds. Instead, we must create mathematical models of this hypersonic flow to find C_D. The most accurate of these models requires us to use high-speed computers to solve the problem. This approach is now a specialized area of aerospace engineering known as computational fluid dynamics.

Fortunately, a simpler but less accurate way will get us close enough for our purpose. We can use an approach introduced more than 300 years ago called Newtonian flow. Yes, you guessed it! Isaac Newton strikes again. Because Newton looked at a fluid as simply a bunch of particles, he assumed his laws of motion must still work. But they didn't at low speeds. Centuries later, however, Newton was vindicated when engineers found his model worked quite well for flow at extremely high speeds. So the grand master of physics was right again—but only for certain situations. Figure 10-17 summarizes these two approaches to analyzing fluid dynamics. Using Newton's approach, we can calculate C_D and thus find BC. Examples using this approach for three simple shapes are in Table 10-2.

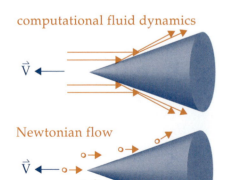

computational fluid dynamics

$\vec{V}$

Newtonian flow

$\vec{V}$

Figure 10-17. Computational Fluid Dynamics (CFD) Versus Newtonian Flow. In CFD, high-speed computers are used to numerically analyze the fluid flow. Newton's approach models the fluid flow as individual particles impacting the vehicle.

Table 10-2. Examples of Estimating BC Using Newton's Approach.

Shape	Example Values	Estimated Ballistic Coefficient
sphere	$D = 2$ m $C_D = 2.0$ $m = 2094$ kg (Assumes density $= 500$ kg/m^3)	$BC \cong 333$ kg/m^2
cone	$\ell = 3.73$ m $\delta_c = 15° =$ cone half angle $r_c = 1$ m = cone radius $C_D \cong 2\,\delta_C^2 = 0.137$ $m = 1954$ kg (Assumes density $= 500$ kg/m^3)	$BC \cong 4543$ kg/m^2
blunted cone	$\ell = 3.04$ m $\delta_c = 15° =$ cone half angle $r_c = 1$ m = cone radius $r_n = 0.304$ m $m = 1932$ kg (Assumes density $= 500$ kg/m^3)	$C_D = (1 - \sin^4\delta_c)\left(\dfrac{r_n}{r_c}\right)^2$ $+ 2\sin^2\delta_c\left[1 - \left(\dfrac{r_n}{r_c}\right)^2\cos^2\delta_c\right]$ $C_D \cong 0.188$ $BC \cong 3266$ kg/m^2

Effects of Vehicle Shape on Deceleration

Now that we have a way to find BC, we can use the numerical tool we used in the last section to see how varying BC changes an entry vehicle's deceleration profile. Let's start by looking at vehicles entering the Earth's atmosphere at an angle of 45° and a velocity of 8000 m/s. Notice something very interesting in Figure 10-18. The maximum deceleration, a_{max}, is the same in all cases! But the altitude of a_{max} varies with BC. This is what we'd expect from Equations (10-15) and (10-16). The higher the BC (the more streamlined the vehicle) the deeper it will dive into the atmosphere before reaching a_{max}. This means a streamlined vehicle will spend less time in the atmosphere and reach the ground long before a blunt vehicle.

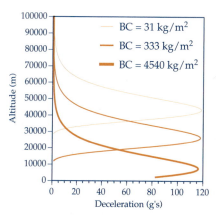

Figure 10-18. Deceleration Profiles for Various Ballistic Coefficient (BC). Note that, regardless of shape, all vehicles experience the same maximum deceleration, but at different altitudes.

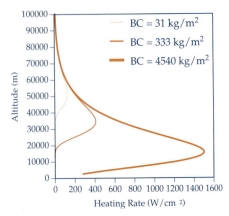

Figure 10-19. Heating Rate Profiles for Various Ballistic Coefficients (BC). Streamlined vehicles have a much higher maximum heating rate, lower in the atmosphere, than blunt vehicles.

Effects of Vehicle Shape on Heating Rate

Now let's see how varying BC affects the maximum heating rate. In Figure 10-19, notice the maximum heating rate is much more severe for the high-BC (streamlined) vehicle and occurs much lower in the atmosphere. The shape of the shock wave surrounding each vehicle causes this difference. Remember the nature of shock waves for blunt and streamlined vehicles shown in Figure 10-13. Blunt vehicles have detached shock waves that spread the heat of entry over a relatively large volume. Furthermore, the air flow near the surface of blunt vehicles tends to inhibit convective heat transfer. Thus, the heating rate for blunt vehicles is relatively low.

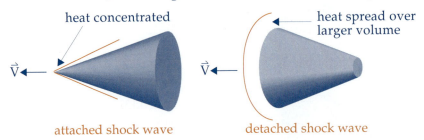

Figure 10-20. Shock Waves and Heating. For streamlined vehicles (high BC), the shock wave is attached, concentrating heat at the tip. For blunt vehicles (low BC), the shock wave is detached, spreading the heat over a larger volume.

Streamlined vehicles, on the other hand, have attached shock waves. This concentrates a large amount of heat near the sharp tip causing it to reach very high temperatures—hot enough to melt most materials. For this reason, "needle-nosed" vehicles are never used. Even streamlined vehicles have a slightly rounded nose to keep the tip from burning off. In addition, the heat around the vehicle is confined to a smaller volume, and the air flow near the surface doesn't inhibit heat transfer as well. As a result, the overall heating rate is higher. See Figure 10-20.

Effects of Vehicle Shape on Accuracy

As we've seen, a more streamlined (high-BC) vehicle gets to maximum deceleration much lower in the atmosphere than a blunt (low-BC) vehicle; thus, it reaches the ground first. We know from earlier discussion that the atmosphere can greatly decrease entry accuracy, so we want to spend as little time in the atmosphere as possible. As a result, we want a streamlined vehicle for better accuracy even though we must accept more severe heating rates. As we'll see, thermal-protection systems can be used to deal with this heating.

Effects of Vehicle Shape on the Entry Corridor

We already know that the entry corridor's upper or overshoot boundary depends on the minimum deceleration for atmospheric capture. Variations in vehicle shape don't affect this end of the corridor significantly. However, we can change the lower or undershoot boundary by changing the limits on deceleration or heating rate. But maximum deceleration is independent of BC, so vehicle shape still doesn't affect this boundary. On the other hand, when the maximum heating rate sets the lower boundary of the corridor, decreasing BC can dramatically decrease the maximum heating rate. This decrease expands the entry corridor and gives us more margin for error.

Table 10-3 summarizes how vehicle shape affects entry parameters.

Table 10-3. Summary of Ballistic Coefficient (BC) Trade-Offs for Entry Design.

Ballistic Coefficient (BC)	Maximum Deceleration	Altitude of Maximum Deceleration	Maximum Heating Rate	Altitude of Maximum Heating Rate	Accuracy	Corridor Width
High (streamlined)	Same	Low	High	Low	High	Narrow
Low (blunt)	Same	High	Low	High	Low	Wider

Thermal-Protection Systems (TPS)

As you know by now, entry can get hot. How do we deal with this massive heat without literally burning up? We'll look at three common ways to deal with entry heating:

- Heat sinks
- Ablation
- Radiative cooling

Heat Sinks

Engineers first started to deal with the prospects of massive entry heating for ICBMs back in the 1950s. At that time, they couldn't get rid of the heat, so they decided to spread it out and store it in the reentry vehicle instead. How does this work? When you put a pan of water over a fire, where does the heat go? Some of it stays in the water and raises its temperature. Now pretend you have one log burning in a fireplace. If you put a five-liter pan and a ten-liter pan of water over the burning log, which pan will boil first? The five-liter pan will because less water is storing the same amount of heat, so the water heats up faster. Thus, when the amount of heat energy we face is fixed (such as for a given set of entry conditions), we can lower the temperature by increasing the material's volume.

This is the basic principle of a *heat sink*—the more material you have to absorb the heat, the less the temperature changes. Entry designers can simply add material to the vehicle to "soak up" the heat and thus prevent the whole thing from melting. In those early days, the heat sink, although heavy, was a simple, effective solution for ICBM entry. For these missions, high entry angles were chosen for better accuracy because the reentry vehicle traveled more quickly through the atmosphere. As a side benefit, the heat sink had to absorb the heat for a relatively short period. Unfortunately, for a given launch vehicle, as a heat sink's mass increased, the available payload mass drastically decreased. Because payload is what they were trying to put on target, this forced designers to consider other options.

Ablation

How do you keep your sodas cold on a hot day at the beach? You put them in a cooler full of ice. At the end of the day, the ice is all gone, and only cold water remains. Why don't you just fill your cooler with cold water to start with? Because ice at 0° C (32° F) is "colder" than water at the same temperature! Huh? When ice goes from a solid at 0° C to a liquid at the same temperature it absorbs a lot of energy. By definition, 1 kilocalorie of heat energy will raise the temperature of 1 liter of water by 1° C. (1 kilocalorie = 1 food calorie, those things we count every day as we eat candy bars.) But to melt 1 kg of ice at 0° C to produce 1 liter of water at the same temperature requires 79.4 kilocalories! This phenomenon, known as the *latent heat of fusion*, explains why your sodas stay colder on ice.

So what does keeping sodas cold have to do with an entry vehicle? Surely we're not going to wrap it in ice? Not exactly, but pretty close! An entry-vehicle designer can take advantage of this concept by coating the vehicle's surface with some material having a very high latent heat of fusion, such as carbon or ceramics. As this material melts or vaporizes, it soaks up large amounts of heat energy, protecting the vehicle. This melting process on an entry vehicle is known as *ablation*. Ablation has been used on the warheads of ICBMs and on all manned entry vehicles until the time of the Space Shuttle. Russian manned vehicles still use this process to protect cosmonauts during entry.

Ablation has one major drawback. By the time you land, part of your vehicle has disappeared! This means you must either build a new vehicle for the next mission or completely refurbish the vehicle. To get around this problem, engineers faced with designing the world's first reusable spaceship came up with a new idea—radiative cooling.

Radiative Cooling

Stick a piece of metal in a very hot fire and, before long, it will begin to glow red hot. Max Planck first explained this process. When you apply heat to an object, it can do one of three things—transmit the heat (like light through a pane of glass), reflect it (like light on a mirror), or absorb it (like a rock in the Sun). As the object absorbs heat, it begins to warm up and, at the same time, radiate some of the heat through *emission*. This emission is what we see when the metal begins to glow.

The amount of energy emitted per square meter, E, is a function of the body's temperature and a surface property called emissivity. *Emissivity, ε,* is a unitless value ($0 < \varepsilon < 1.0$) which measures an object's relative ability to emit energy. A perfect black body would have an emissivity of 1.0. The energy emitted is determined using the Stefan-Boltzmann relationship

$$E = \sigma \varepsilon T^4$$ (10-20)

where

E = emitted energy (W/m^2)

σ = Stefan-Boltzmann constant = 5.67×10^{-8} W/m$^{2\circ}$ K^4

ε = emissivity ($0 < \varepsilon < 1.0$)

T = temperature ($^\circ$K)

As heat energy strikes a body, the body will continue to heat up until the energy emitted balances the energy taken in. At this point the body is said to be in *thermal equilibrium*, where its temperature levels off and stays constant.

If the body being heated has a high emissivity, it will emit almost as much energy as it absorbs. This means it reaches thermal equilibrium sooner, at a relatively low temperature. Reducing equilibrium temperatures by emitting most of the heat energy before it can be absorbed into the vehicle's structure is known as *radiative cooling*. However, even for materials with extremely high emissivities, equilibrium temperatures during entry can still exceed the melting point of aluminum during entry.

These temperatures pose two problems for us. First, we must select a surface-coating material which has a high emissivity and a high melting point, such as a ceramic. Second, if this surface coating were placed directly against the vehicle's aluminum skin, the aluminum would quickly melt. Therefore, we must isolate the hot surface from the vehicle's skin with extremely efficient insulation having a high emissivity.

This artful combination of a surface coating on top of a revolutionary insulator gives us the now famous Shuttle tiles. The insulation in these tiles is made of a highly refined silicate (sand). At the points on the Shuttle's surface where most of the heating takes place, a special coating gives the tiles an emissivity of about 0.8, as well as their characteristic black color, as shown in Figure 10-21.

Figure 10-21. Shuttle Tiles. Space Shuttle tiles combine a surface with high emissivity and an efficient high-temperature insulator. Here we see the workers refurbishing tiles on the bottom of the Shuttle. *(Courtesy of NASA)*

Astro Fun Fact
Shuttle Tiles

(Courtesy of NASA)

We know entry gets hot. For the Space Shuttle, temperatures can exceed 1247°C (2300°F). But the aluminum skin of the Shuttle doesn't reach its maximum temperature of 350° until almost 20 minutes _after_ landing, thanks to perhaps the greatest technical advance of the Shuttle program—tiles. Designed to withstand the aerodynamic loads of ascent and entry, temperature extremes of over 1350°C (2400°F), and repeated usage, they're one of the most unique materials ever invented. To cover the complex contours of the Shuttle surface, over 30,000 individually machined tiles are fit together like a big jigsaw puzzle. Each tile has two pieces—a white silica fiber structure (basically highly refined sand) covered by the characteristic black coating made of reaction-cured glass (RCG). During entry, the RCG dissipates 90% of the heat energy in radiation back to the atmosphere while the white silica fiber structure acts as an insulator to slow the heat transfer from the surface to the inner aluminum skin and bears the brunt of aerodynamic forces.

Refractory Composite Insulation, LI900, LI2200, FRCI, How It Works..., Lockheed Missiles & Space Company, Sunnyvale, CA.

▬ Section Review

Key Terms

ablation
emission
emissivity, ε
heat sink
latent heat of fusion
radiative cooling
thermal equilibrium

Key Equations

$$E = \sigma \varepsilon T^4$$

Key Concepts

➤ We can meet mission requirements on the design front by changing
  • Vehicle size and shape, BC
  • Vehicle thermal protection system (TPS)

➤ Increasing the vehicle's ballistic coefficient, BC,
  • Doesn't change maximum deceleration, a_{max}
  • Increases maximum heating rate, q̇

➤ There are three types of thermal-protection systems:
  • Heat sinks—spread out and store the heat
  • Ablation—melts the vehicle's outer shell, taking heat away
  • Radiative cooling—radiates a large percentage of the heat away from the vehicle before it can be absorbed

Example 10-1

Problem Statement

Long-range sensors determine a re-entry capsule is emitting 45,360 W/m² of energy during entry. If the emissivity of the capsule's surface is 0.8, what is its temperature?

Problem Summary

Given: $E = 45{,}360$ W/m²
$\varepsilon = 0.8$
Find: T

Conceptual Solution

1) Solve Stefan-Boltzmann relationship for T

$$E = \sigma \varepsilon T^4$$

$$T = \sqrt[4]{\frac{E}{\sigma \varepsilon}}$$

Analytical Solution

1) Solve Stefan-Boltzmann equation for T

$$T = \sqrt[4]{\frac{E}{\sigma \varepsilon}}$$

$$T = \sqrt[4]{\frac{45{,}360 \, \dfrac{W}{m^2}}{\left(5.67 \times 10^{-8} \, \dfrac{W}{m^2 K^4}\right)(0.8)}}$$

$T = 1000°$ K

Interpreting the Results

During entry, the capsule's surface reached 1000° K. With the surfaces' emissivity, this means 45,360 W/m² of energy is emitted. Imagine 450 100-watt light bulbs in a 1 m² area!

10.4 Lifting Entry

▬ In This Section You'll Learn to...

☞ Discuss the advantages offered by lifting entry

☞ Explain aerobraking and discuss how it can be used for interplanetary transfer

So far we've assumed the force of lift on our entering vehicles was zero. This allowed us to use a straightforward equation of motion to investigate the trade-offs between entry characteristics. Adding lift to the problem takes it beyond the scope of our simple model. But lift gives us more flexibility. For example, we can use the lifting force to "stretch" the size of the corridor and allow greater margin of error during entry. Controlling lift also improves accuracy over a strictly ballistic entry. We can change the vehicle's angle of attack to improve lift, making the vehicle fly more like an airplane than a rock. This allows the computer or pilot onboard to guide the vehicle directly to the desired landing area.

The Space Shuttle is a great example of a lifting-entry vehicle. About one hour before landing, entry planners send the crew the necessary information to do the deorbit burn. This burn changes the Shuttle's trajectory to enter the atmosphere by establishing a 1° – 2° entry flight-path angle. After this maneuver, the Shuttle is on final approach. Because it has no engines to provide thrust in the atmosphere, it gets only one chance to make a perfect landing, so everything must go okay!

Preparing to hit the atmosphere (just like our skipping stone), the Shuttle assumes a 40° angle of attack. *Angle of attack* describes the angle between the velocity vector and the nose of the vehicle. This high angle of attack exposes the Shuttle's wide, flat bottom to the wind. At about 122,000 m (400,000 ft.), entry interface takes place. This is the point where the atmosphere begins to be dense enough for the true entry phase to begin. From entry interface, more than 6437 km (4000 miles) from the runway, the Shuttle will be on the ground in only about 45 minutes! Figure 10-22 shows the Shuttle's entry profile.

Throughout entry, the Shuttle rolls to change lift direction in a prescribed way, keeping maximum deceleration well below 2 g's. In contrast, earlier manned reentry vehicles, such as Apollo and Gemini, entered much more steeply and so were designed to endure up to 12 g's during entry. Figure 10-23 compares these entry profiles. These roll maneuvers allow the Shuttle to use its lifting ability to steer toward the runway.

Another exciting application of lifting entry is *aerobraking*, which uses aerodynamic forces (drag and lift) to change a vehicle's velocity and, therefore, its trajectory. In Chapter 7 we explored the problem of interplanetary transfer. There we saw that to get from Earth orbit to

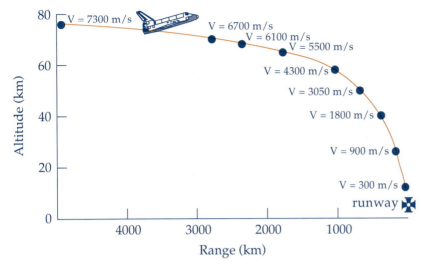

Figure 10-22. Entry Profile for the Space Shuttle. This graph shows the Space Shuttle's altitude and velocity profile for a typical entry.

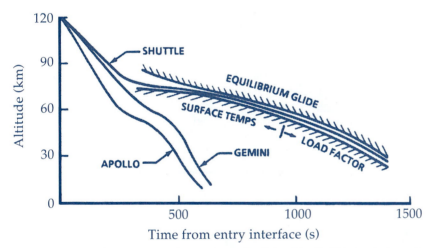

Figure 10-23. Entry Profiles for the Shuttle Versus Gemini and Apollo. This graph shows the difference between entry profiles for Apollo, Gemini, and the Space Shuttle. Notice both Gemini and Apollo entered much more steeply than the Space Shuttle. The Shuttle's entry profile must stay within a tight corridor between equilibrium glide, which insures it will slow down enough to avoid skip out and not over shoot the runway, and surface temps/load factor requirements which determine maximum heating and deceleration.

another planet required us to start our engines twice: ΔV_{boost} to get us on our way and ΔV_{retro} to capture us into orbit around our target planet. But if the target planet has an atmosphere, there's another option. Instead of using engines to slow us down enough to be captured into orbit, we can plan our hyperbolic approach trajectory to take us right into its atmosphere and then use drag to do the equivalent of the ΔV_{retro} burn. We then use our lift to pull us back out of the atmosphere before we crash into the planet! By getting this "free" ΔV, we can save an enormous amount of

Figure 10-24. Aerobraking Concept. This artist's concept shows a heat shield that could be used for aerobraking at Mars or Earth. *(Courtesy of NASA)*

fuel. Calculations show that using aerobraking instead of conventional rocket engines is almost ten times more efficient. This efficiency could mean a tremendous savings in the amount of material that must be put into Earth orbit to mount a mission to Mars. Figure 10-24 shows an artist's conception of an aerobraking vehicle. In his novel *2010: Odyssey Two*, Arthur C. Clarke uses aerobraking to capture a space ship into orbit around Jupiter. Hollywood depicted this in a dramatic scene in the movie made from the novel.

Figure 10-25 shows the aerobraking scenario. On an interplanetary transfer, the vehicle approaches the planet on a hyperbolic trajectory (positive specific mechanical energy with respect to the planet). During the aerobraking maneuver, the vehicle enters the atmosphere at a shallow angle to keep maximum deceleration and heating rate within limits. Once drag has reduced the vehicle's speed enough to capture it into orbit (now negative specific mechanical energy with respect to the planet), the vehicle changes its angle of attack to "pull out" of the atmosphere by using its lift. The atmospheric encounter will place the vehicle on an elliptical orbit around the planet. Because periapsis is now within the atmosphere, the vehicle will reenter with no other changes. Thus, a single burn much smaller than ΔV_{retro} can put us into a circular parking orbit well above the atmosphere.

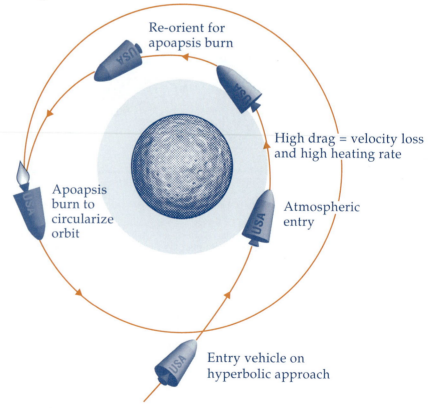

Re-orient for apoapsis burn

High drag = velocity loss and high heating rate

Apoapsis burn to circularize orbit

Atmospheric entry

Entry vehicle on hyperbolic approach

Figure 10-25. Aerobraking. The aerobraking maneuver allows a vehicle to get "free" ΔV by diving into the atmosphere and using drag to slow down.

▆ Section Review

Key Terms

aerobraking
angle of attack

Key Concepts

➤ Applying lift to the entry problem allows us to stretch the size of the entry corridor and improve accuracy by flying the vehicle to the landing site.

➤ The Space Shuttle is a good example of a lifting-entry vehicle. It uses its lifting ability to keep entry deceleration low and fly to a pinpoint runway landing.

➤ Aerobraking can significantly decrease the amount of mass needed for interplanetary transfer. During an aerobraking maneuver, the vehicle dives into the target planet's atmosphere, using drag to slow down enough to be captured into orbit.

References

Chapman, Dean. *An Analysis of the Corridor and Guidance Requirements for Supercircular Entry Into Planetary Atmospheres*, NASA TR R-55, 1960.

Concise Science Dictionary. Oxford: Oxford University Press, U.K. Market House Books, Ltd., 1984.

Eshbach, Suoder, (ed.). *Handbook of Engineering Fundamentals*. 3rd edition. New York, NY: John Wiley & Sons, Inc., 1975.

Entry Guidance Training Manual. ENT GUID 2102, NASA Mission Operations Directorate, Training Division, Flight Training Branch, NASA/Johnson Space Center, Houston, TX, December 1987.

Regan, Frank J. *Reentry Vehicle Dynamics*. AIAA Education Series, J.S. Przemieniecki series ed. in chief. New York, NY: American Institute of Aeronautics and Astronautics, Inc., 1984.

Tauber, Michael E. *A Review of High Speed Convective Heat Transfer Computation Methods*. NASA Technical Paper 2914, 1990.

Tauber, Michael E. *Atmospheric Trajectories*. Chapter for AA213 Atmospheric Entry. NASA/Ames Research Center, Stanford University, 1990.

Tauber, Michael E. *Hypervelocity Flow Fields and Aerodynamics*. Chapter for AA213 Atmospheric Entry. NASA/Ames Research Center, Stanford University, 1990.

Voas, Robert B. *John Glenn's Three Orbits in Friendship 7*. National Geographic, Vol. 121, No. 6. June 1962.

Mission Problems

10.1   Entry Design

1   What mission requirements must entry-design engineers meet?

2   What are the four competing entry requirements?

3   Where does all the heat generated during entry come from?

4   Why would increasing the ability of an entry vehicle to withstand higher g forces not necessarily increase the maximum deceleration requirement for the mission?

5   What is the entry corridor? Define its upper and lower boundaries.

6   Describe the entry coordinate system.

7   What are the potential forces on an entry vehicle? What is the dominant force during entry? Why?

8   Define ballistic coefficient and describe how a blunt versus a streamlined shape affects how a body will slow down due to drag.

9   What two approaches can we use to balance competing entry requirements?

10   Describe entry design.

10.2   Trajectory Options

11 To save fuel, National Aerospace Plane (NASP) engineers want to increase the velocity and the flight-path angle for entry. How will this affect the maximum deceleration and maximum heating rate? The altitudes for maximum deceleration and maximum heating rate?

12 Contact lenses being manufactured in space are returned in an entry capsule to Earth for distribution and sale. If the entry velocity is 7.4 km/s and the entry flight-path angle is 10°, determine the maximum deceleration it'll experience and at what altitude. The capsule's BC is 1000 kg/m^2.

13 For the same capsule in Problem 12, determine the altitude of maximum heating rate and the velocity at which this occurs.

14 What is a shock wave?

10.3   Options for Vehicle Design

15 In what two ways can we determine the hypersonic drag coefficient for a vehicle?

16 Mission planners for a manned Mars spacecraft face two different reentry vehicles. Vehicle A has a high BC; vehicle B has a low BC. Assuming the entry velocity and flight-path angle are the same

for both vehicles, explain any differences in deceleration profile.

17 Compare the advantages and disadvantages the three types of thermal-protection systems.

18 What is latent heat of fusion and how does it relate to ablation?

19 During entry, a meteor reaches a temperature of 1700° K. If its emissivity is 0.25, how much energy is emitted per square meter?

10.4   Lifting Entry

20 What are the advantages offered by a lifting-entry vehicle?

21 How does the Space Shuttle use lift to reach the runway?

22 A vehicle attempting to aerobrake into orbit around Mars needs to achieve an equivalent ΔV_{retro} of 2 km/s. If the entire aerobraking maneuver lasts for 10 minutes, what is the average drag force (in g's) we must attain?

Mission Profile—Space Shuttle

On April 12, 1981, the world's first reusable space ship rocketed into the Florida skies with astronauts John Young and Robert Crippen aboard. The successful flight of STS-1 (Space Transportation System mission 1) heralded a new era which promised to make access to space routine.

Mission Overview

The Space Shuttle, or Space Transportation System as it's sometimes called, is the most complex flying vehicle ever constructed. It has three main parts—the winged orbiter, the external tank (ET), and a pair of solid-rocket boosters (SRBs). The orbiter houses the crew compartment with avionics, payload bay, three Space Shuttle main engines (SSMEs), two orbital maneuvering system (OMS) engines, and 44 reaction control system (RCS) thrusters for attitude control. The ET is a big gas tank holding 790,000 kg (1.58×10^6 lb) of liquid hydrogen and liquid oxygen fed to the three SSMEs on the orbiter through large interconnect valves. The SRBs provide the necessary thrust to get the entire system off the pad at lift-off. A typical Shuttle mission divides into three phases—ascent, on-orbit, and entry.

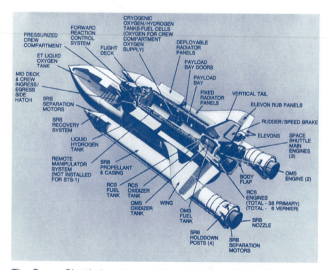

The Space Shuttle has three main parts—the orbiter, external tank, and solid rocket boosters—with various subsystems in each. *(Courtesy of NASA)*

Mission Data

✓ Shuttle Ascent

- T – 8 seconds: The three main engines ignite. As they throttle up to 104% capacity, generating five million newtons of thrust (1.125×10^6 lbf), the entire vehicle pitches forward. If the onboard computers detect any engine problems, all three will be shut down and the mission scrubbed for the day.

- T – 0 seconds: As the vehicle rocks back to upright at T – 0, the mighty SRBs ignite. Each applies a force of more than 11.8 million newtons (2.65×10^6 lbf) to the vehicle, causing it to almost leap off the pad. When the SRBs ignite, there's no way to stop them. At lift-off, the entire Shuttle system has a gross mass of more than two million kg (4.4×10^6 lb).

- T + 60 seconds: The three main engines throttle down to 65% to minimize loads as the vehicle flies through "Max-Q," the region of maximum dynamic pressure. After Max-Q, the engines throttled back up to 104%. [Note: the reason the SSMEs can exceed 100% has to do with engine calibration data established early in the Shuttle's development. 100% is simply a benchmark value which can be safely exceeded by 4%.]

- T + 120 seconds: The two SRBs burn out and are jettisoned to parachute into the ocean where they'll be recovered and refurbished for future missions.

- T + 480 seconds: SSMEs again throttle down to stay below three g's on the vehicle and crew.

- T + 500 seconds: Main engine cut-off (MECO). The ET is jettisoned to burn up in the atmosphere over the Indian Ocean.

- At MECO, the orbiter is not yet in orbit as perigee is well within the atmosphere. The OMS engines must fire at least once to establish a safe orbit.

✓ On-Orbit

- Once on-orbit, the Shuttle uses its OMS engines to change orbits and rendezvous with satellites or to achieve the correct parking orbit and deploy payloads. Typical Shuttle orbits are nearly circular at about 300 km (160 nmi) altitude with an inclination of 28.5° or 57°.

✓ Entry

- De-orbit burn: The entry phase starts with the de-orbit burn of 100 m/s (300 ft/s) about one hour before landing over the Indian Ocean. This burn lowers the Shuttle's orbit for a controlled entry into the atmosphere.

- De-orbit coast: The crew orients the vehicle to present the wide, flat bottom to the atmosphere at a 40° angle of attack.

- Entry interface (EI): EI takes place about 30 minutes before landing at an altitude of 122,000 m (400,000 ft) more than 8300 km (4500 nmi) up range from the landing site. At this altitude, the atmosphere becomes dense enough for aerodynamic forces to begin to be significant. Throughout entry, the guidance, navigation, and control system must manage the Shuttle's energy to guide it through the narrow entry corridor, bleed off enough energy to land safely (but not too much so you overfly the landing site), and maintain acceptable heating levels.

- TACAN acquisition: At about 40,000 m (130,000 ft), the Shuttle's navigation system begins processing data from ground-based Tactical Air Navigation (TACAN) stations—the same ones used by military and commercial aircraft—to update its position and velocity and ensure it's on course for the runway.

- HAC intercept: At about 20,000 m (65,000 ft) the Shuttle intercepts the heading alignment cone (HAC) for a wide (up to 270°) turn to line up on the runway centerline.

- On glide-slope: The Shuttle seems to dive in at the runway, flying a 19° glide slope, much steeper than a convention airliner's 3°.

- Touchdown: The Shuttle touches down at 98 m/s (218 m.p.h.) compared to only 67 m/s (150 m.p.h.) for an airliner.

Mission Impact

As of 1993, the Shuttle has flown over 50 missions. From satellite deployment and retrieval to satellite repair and scientific experiments, the Shuttle has proven its flexibility. However, the emotional impact of the Challenger accident, caused by burn-through in a booster, showed how fragile the complex Shuttle system is and showed that access to space has a long way to go before it's ever routine.

For Discussion

- Has the Space Shuttle been a good investment for the U.S.?

- Given the inherent dangers of space flight, was the public's strong reaction to the Challenger accident justified?

- How would you design a next-generation Shuttle to replace the current one?

References

Space Shuttle System Summary, SSV80-1, Rockwell International Space Systems Group, May 1980.

On Space Shuttle mission STS-51DR, astronauts tried to repair a broken satellite by using a "flyswatter" attached to the end of the Shuttle's robot arm. *(Courtesy of NASA)*

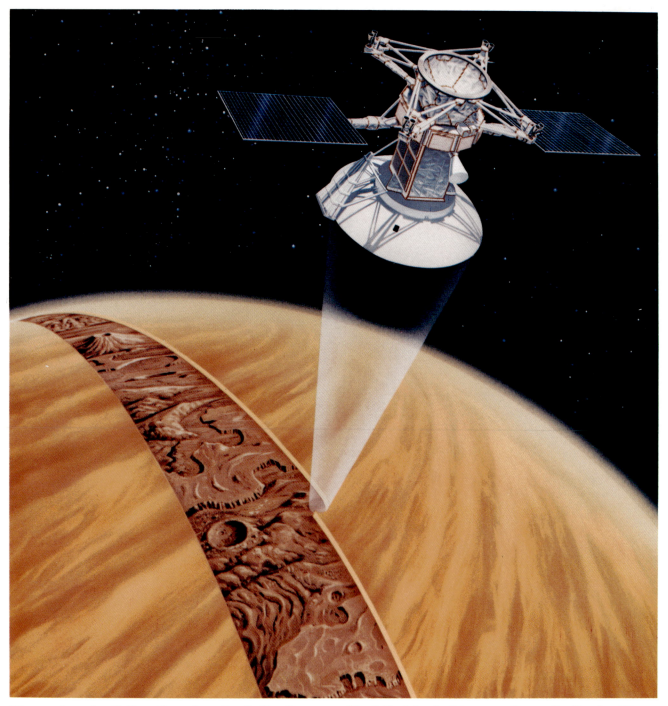

Using an advanced synthetic aperture radar system, the Magellan spacecraft "peels" back the clouds to uncover the secrets of the planet Venus.
(Courtesy of NASA)

Payload and Spacecraft Design

11

In This Chapter You'll Learn to...

☛ Describe the steps in space-mission planning

☛ Describe the physical requirements and limitations of space-based sensors

☛ Identify what the spacecraft bus does

☛ Explain the basic steps in spacecraft design

You Should Already Know...

❏ Elements of space-mission architecture (Chapter 1)

❏ Basic requirements for maneuvering in orbit (Chapter 6)

Man must rise above the Earth—to the top of the atmosphere and beyond—for only then will he fully understand the world in which he lives.

Socrates ca. 450 B.C.

We all use space systems. As you kick back in your easy chair, remote control in hand, the cable television programs you watch arrive through communication satellites. The weather forecasts on the evening news are possible only because weather satellites patiently track the motion of clouds around the globe. In the last several chapters we've focused all our attention on the trajectories these satellites follow through space. You've seen how to select the proper orbit for a mission, launch from a given launch site, and enter the atmosphere for a smooth landing back on Earth. But the orbit is only part of the story.

In this chapter, we'll turn our attention to the satellite itself. We'll discover the importance of doing all our homework during initial planning and how this leads to selecting a payload. This will determine the configuration of the entire spacecraft. Then we'll take a closer look at spacecraft payloads to see how they serve as our eyes and ears in space. Finally, we'll look at the other component of a spacecraft, the bus, to see what it must do to support the payload. When we're finished, we'll appreciate the complexity of designing a vehicle to challenge the final frontier.

11.1 Planning a Space Mission

In This Section You'll Learn to...

- Describe the elements of a space-mission architecture
- Explain the important information contained in the mission statement
- Define the mission subject and its relationship to the payload
- Describe the steps in space-mission planning

All space missions begin with a need, such as the need to economically transmit television shows or to provide world-wide navigation. With this in mind, we can begin to develop space missions to satisfy those needs. Way back in Chapter 1, we introduced the space-mission architecture shown in Figure 11-1. Recall that the unifying theme of this architecture is

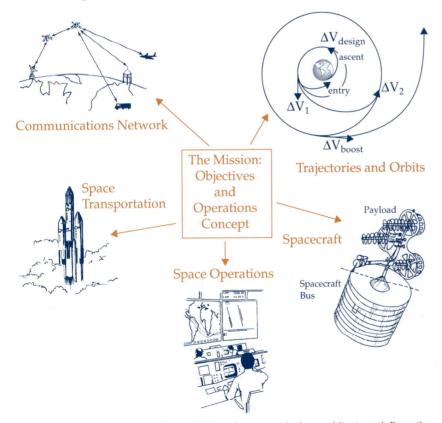

Figure 11-1. Architecture of a Space Mission. A space-mission architecture defines the elements of a space mission and how they interrelate.

the mission itself. The purpose of the mission is in the *mission statement* which tells us

- The mission *objective*—*why* we're doing the mission
- The mission *users*—*who* will benefit from and use the information
- The *operations concept*—*how* all the mission elements fit together

The *mission objective* clearly and concisely states the purpose of the mission and what services or information will be delivered to users. *Users* are the customers who give us the reason for doing a mission. Once we know *why* and *who*, the next question is *how*, which leads us to the operations concept. The *operations concept* describes how people, systems, and all the elements of the mission architecture will interact to satisfy the user.

Forming the mission statement is the first step in space-mission planning, as summarized in Figure 11-2. Only after we fully understand the mission statement can we begin to develop the other elements of the mission architecture, especially the spacecraft. Recall from Chapter 1 that a spacecraft can be divided into two functionally different parts—the payload and the spacecraft bus. The payload consists of the sensors or other instruments that conduct the mission. The bus is a collection of subsystems designed to support the payload. Payload requirements are usually the single biggest driver of the entire spacecraft configuration, so the next step in space-mission planning is understanding what determines an effective payload.

The biggest clue to the payload design comes from the mission statement itself. Somewhere in the description of objectives, users, and operations should be an indication of the mission subject. The *subject* of the mission is what the payload will sense or interact with. The subject can be characterized by such things as its color, size, shape, temperature, chemical composition, or frequency. Once we know what subject we're dealing with, we can select or design a useful payload. With a payload in mind, we can select a practical mission orbit, estimate the overall size of the spacecraft, make a first cut at the booster we'll use, and establish the communication network that ties it all together.

Let's look at an example to demonstrate how mission planning might work. Anyone who's listened to Smokey the Bear knows forest fires are devastating. "Only *you* can prevent forest fires," but once they're started maybe there's still something *we* can do to put them out faster. We want to be able to notify the Forest Service in time for them to contain the fire before it destroys valuable habitat or threatens inhabited areas. Thus, for our example, the mission objective is to detect and locate forest fires in the continental United States, and the user is the Forest Service. One possible operations concept for this mission is to use a spacecraft or number of spacecraft to detect and locate the fires and then communicate this information to Forest Service people at the fire fighting centers. We'll call the spacecraft FireSat.

Now that we've identified the mission objective, the users, and an operations concept, we can determine the subject. The obvious subject for

---

Space-Mission Planning
- Define mission statement
  - Mission objectives
  - Mission users
  - Operations concept
- Identify subject
- Select payload
- Select orbit
- Size spacecraft
- Identify booster
- Identify communications network

Figure 11-2. Planning Process for Space Missions.

this mission is forest fires. But what kind of forest fires? How big or how hot? To investigate this further, we must fully characterize our subject by specifying what the payload will interact with. This may sound trivial, but to the mission designer it's very important. After all, we don't want to send out a forest-fire alert every time someone starts a camp fire. On the other hand we don't want to ignore a multi-acre blaze which could quickly engulf thousands of acres. FireSat must be designed to respond to the right kind of fires. For our mission, the subject is what users define as an appropriately large fire—let's say a fire that covers more than 1 km^2 (about 250 acres).

Now that we know what we're looking for, how do we develop a payload to detect it? A fire generates heat, light, and smoke. Fortunately, we can build sensors to detect each of these attributes of our subject. Imagine you're sitting around a campfire on a clear, cool night. You can feel the heat from the fire. You can see its light, with sparks and ash dancing in the air, and smell the smoke. Electronic devices or sensors can detect some of these same attributes. If we put these sensors on a space vehicle, they become the spacecraft payload. Such a payload might contain a simple camera to detect light or highly sensitive temperature sensors to detect heat. As we'll see, these sensors work in much the same way as human sensors—as they get farther away from the subject they detect less.

Because payloads are so important to mission planning and to the ultimate design of the entire spacecraft, let's continue by looking at the physical limitations on their design in the next section.

— Astro Fun Fact —
Saving Lives From Orbit

(Courtesy of NASA)

In August 1983, cosmonauts aboard the Soviet space station, Salyut-7, noticed a peculiar condition in the Pamir Mountains located in the southern part of what was then the U.S.S.R. A lake was forming from a melting glacier. Realizing that nearby towns could soon face flood waters, the cosmonauts reported the condition to mission control. Radio Moscow reported, "A timely warning from orbit made it possible to evacuate the inhabitants from the threatened areas, and a channel was promptly dug to empty the lake gradually."

Oberg, James A. and Alcestis R. Pioneering Space: Living on the Next Frontier. New York, NY: McGraw-Hill, p. 11-12, 1986.

Contributed by Sloane Englert, United States Air Force Academy

Section Review

Key Terms

mission objectives
mission statement
operations concept
subject
users

Key Concepts

➤ A space-mission architecture consists of
  • The mission itself
  • The trajectory or orbit
  • The spacecraft
  • Space operations
  • Space transportation
  • Communications network

➤ The mission statement tells us
  • The mission objective—clearly, concisely states the mission's purpose and what the system users will need
  • The users—defines the individuals or groups who use the system and its data for their benefit
  • The operations concept—describes how people, systems, and mission elements interact to make the mission successful

➤ The subject of a mission is "what" the spacecraft payload will sense or interact with

➤ Space-mission planning consists of
  • Defining the mission statement
  • Identifying the subject
  • Sizing the payload
  • Selecting an orbit
  • Sizing the spacecraft
  • Identifying a booster
  • Identifying the communication network

11.2 Space Payloads for Remote Sensing

■ In This Section You'll Learn to...

☞ Identify the elements of a remote-sensing system

☞ Describe and compute important aspects of the electromagnetic spectrum

☞ Use Wien's Law and the Stefan-Boltzmann equation to analyze an object's temperature

☞ Identify the two types of remote-sensing payloads and describe their basic functions

As you read this book and listen to the sounds around you, your eyes and ears act as sensors to help you perceive your surroundings. Our eyes see all the colors of the rainbow. Our ears detect tiny disturbances in the pressure of the air around us, which is how we hear. (In space there is no air to transmit sound. All those Hollywood sci-fi movies in which you hear something blowing up in space are bogus!) The payload is the "eyes" and "ears" of a spacecraft. Payloads detect objects on Earth and in space. They also "talk" to other satellites and Earth-based ground systems. Looking and listening from space involves the electromagnetic (EM) spectrum. Most space missions use some part of the EM spectrum. All space missions—communications, navigation, weather, and remote-sensing—rely on the EM spectrum to collect data and interact with other elements of the mission. To illustrate the basic process of payload design, we'll concentrate on *remote sensing*—a broad category of missions designed to monitor the global environment, observe the weather, spy on enemies, search for natural resources, assess agricultural yields, observe space, and even detect forest fires, like the one shown in Figure 11-3.

Sources of EM energy ultimately provide the information necessary to do remote sensing. These sources include the Sun, Earth, and virtually everything else that radiates energy—even you. For example, if a spacecraft needs to track an airplane, the EM radiation for the picture originates at the Sun. This radiation must first travel through space and make it through the atmosphere. It then reflects off the airplane, carrying information back through the atmosphere to be detected by sensors on the spacecraft. The spacecraft sends this information back to users on the ground. Figure 11-4 shows how the elements of a remote-sensing system all work together.

Normally, a group of sensors makes up a space payload. The information they collect is processed and sent to users on the ground (again passing through the atmosphere). The success of a remote-sensing mission depends on the type of energy being sensed and its ability to get

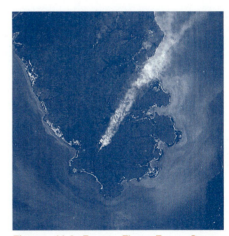

Figure 11-3. Forest Fires From Space. Astronauts on the Space Shuttle captured this view of a forest fire on a tropical island off the North coast of Australia. NASA scientists believe this fire was set on purpose to clear land for grazing. *(Courtesy of NASA)*

through the atmosphere. FireSat, for example, could sense the fire's heat using a part of the spectrum known as infrared. To see how this works, let's take a closer look at EM radiation.

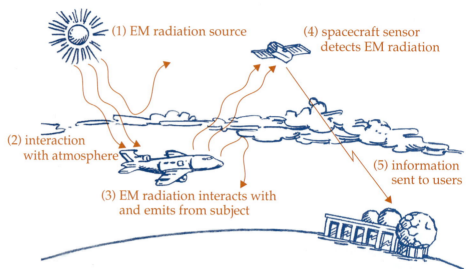

Figure 11-4. Elements of a Remote-Sensing System. A basic remote-sensing system includes an EM radiation source, atmospheric interaction, a subject, a sensor, and an information link to users.

The Electromagnetic Spectrum

EM radiation results from accelerating or "exciting" charged particles—negatively charged particles are electrons and positively charged particles are protons. An *accelerated charge* is one whose speed or direction changes. As a particle heats up, it accelerates, emitting more energy. For example, as metal heats up, the atoms inside get more and more excited, causing electrons to move to higher orbits, from which they eventually drop back down, emitting EM radiation. Thus, the metal becomes red hot and eventually white hot as its temperature increases.

EM radiation is a fickle phenomenon. It can be described as either particle or wave motion. Sometimes it's useful to think about EM radiation as waves, like the ripples on a pond after you've dropped in a rock. EM waves ripple out from the source, traveling at the speed of light: 300,000 km/s (186,000 miles/s). These ripples can be detected by sensors (like our eyes) and interpreted by computers (like our brains). Other times it's more useful to think of radiation as *photons*—tiny, massless bundles of energy emitted from the source at the speed of light. This dual nature of radiation (sometimes called the "wave-icle" theory) can, of course, lead to some confusion. For the most part, except where noted, we'll consider radiation to be waves.

Because payloads can use different parts of the EM spectrum, we need to understand how we describe different types of radiation. Let's look at the EM waves shown in Figure 11-5. The distance from crest to crest is

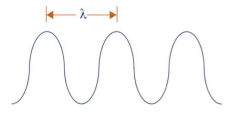

Figure 11-5. Wavelength. We characterize EM radiation in terms of its wavelength or the distance from crest to crest of the waves.

called the *wavelength, λ*. The number of waves that go by in one second is the *frequency, f*.   So, if 10 waves or cycles go by per second, the frequency is 10 cycles/s or 10 Hertz (1 cycle/s = 1 Hz). The time it takes for just one wave to go by is 1/f. That is, if the frequency is 10 Hz, one cycle takes 1/10 second. Now, because we know that distance is just speed multiplied by time, we can relate wavelength, speed, and frequency as

$$\lambda = c\,/f \tag{11-1}$$

where
λ = wavelength (m)
c = speed of light in a vacuum = 3×10^8 m/s
f = frequency (Hz)

Energy in EM radiation is related to the number of waves that hit you over a given time. Thus, ten waves hitting in one second (10 Hz) will deliver twice the energy of five waves (5 Hz). This energy relationship can be expressed as

$$Q = hf \tag{11-2}$$

where
Q = energy (Joules, J)
f = frequency (Hz)
h = Planck's constant = 6.626×10^{-34} J · s

Equation (11-2) shows more energy is available at higher frequencies, so higher frequency waves have more energy than lower waves. That's why we can walk through radio waves all day long, but one large dose of gamma rays, say from a nuclear explosion, can be lethal. Figure 11-6 shows the entire EM spectrum with key parts identified. We use the visible part of the spectrum to observe items of interest or take pictures. We use the infrared part of the spectrum to sense heat. Finally, we use radio wavelengths to transmit voice, television, and radio signals. Most space missions use several different parts of the EM spectrum.

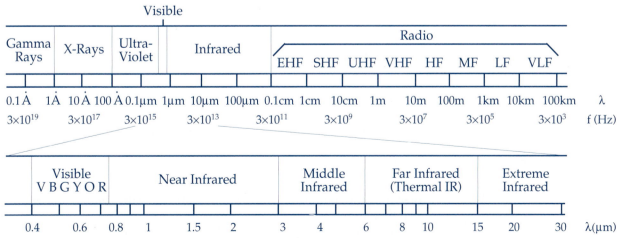

Figure 11-6. EM Spectrum. Here we see the entire EM spectrum. Frequencies are in Hz. Wavelengths are in Angstroms, meters, μm, etc. One Angstrom, Å, is 10^{-10} m, one μm is 10^{-6} m, one cm is 10^{-2} m, and one km is 1000 m. In the visible range, the letters V, B, G, Y, O, and R represent colors of the rainbow.

Sources of Electromagnetic Radiation

Now that we know a little more about what EM radiation is, we can take a closer look at where it comes from. This will allow us to identify a subject and design a payload. Everything above the temperature of absolute 0°K (–273°C or –460°F) emits EM radiation. For sunshine coming from the Sun at 6000° K, this isn't hard to imagine. However, the Earth, at a mere 300° K, emits radiation in the form of "Earth shine" as well. You too, at a robust 310° K (98.6°F), emit EM radiation (and you thought it was your glowing personality). Because everything has *some* temperature, everything emits EM radiation.

All objects will emit energy at different wavelengths depending on their material properties and temperature. The classical explanation for this phenomenon is that thermal radiation begins with accelerated, charged particles near the surface of an object. These charges then emit radiation like tiny little antennas. The thermally excited charges can have different accelerations, which explains why an object emits energy at many different wavelengths.

Max Planck (1858–1947), refined this explanation and helped to usher in the field of quantum physics. He postulated that energy is emitted in tiny bundles or "quanta" called photons. Planck used a black-body as his perfect case. A *black-body* is an object that absorbs and re-emits all of the radiation that strikes it. He was then able to develop a model which related the amount of power given off at specific wavelengths as a function of an object's temperature. If you stick a piece of steel in a hot furnace, it will glow "white hot." Using Planck's relationship, the distribution of wavelength of this glow can then be related directly to the body's temperature. The curves in Figure 11-7 show the wavelengths of black-body radiation for the Sun and Earth.

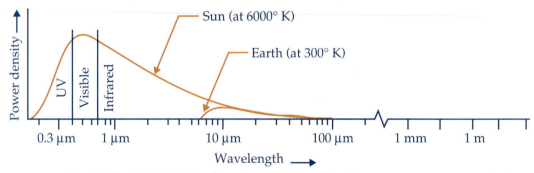

Figure 11-7. Planck's Black-Body Radiation Curve for the Earth and Sun. The hotter the object, such as the Sun, the more EM radiation it will emit at shorter wavelengths (higher frequencies and hence higher energy). UV refers to ultraviolet [1987, Lillesand].

Notice in Figure 11-7 that the peak output for the Sun is in the visible region. Thus, our eyes take advantage of the most abundant type of radiation from the Sun. How convenient! This fact had been noted earlier by Wilhelm Wien (1864–1928). Wien's Displacement Law relates the wavelength (and hence the frequency) of maximum output for a given black-body radiator to its temperature.

$$\lambda_m = \frac{2898°\ K\ \mu m}{T}$$

(11-3)

where
λ_m = wavelength of maximum output (μm)
T = temperature (°K)

For the Sun at 6000° K, λ_m is 0.483 μm, which is in the middle of the range of visible light (0.39–0.74 μm). Just what we'd expect!

Using Wien's law, we can determine the best frequency to use to see a particular subject. For example, if we try to detect the hot plume from a rocket booster, we'd need to know its temperature and then design a sensor tuned to the frequency of maximum output for that temperature.

We can also determine the total power output of black-body radiation. To do so, we use the Stefan-Boltzmann equation discovered by Joseph Stefan (1853–1893) and theoretically derived by Ludwig Boltzmann (1844–1906).

$$E = \varepsilon\ \sigma\ T^4$$

(11-4)

where
E = energy per square meter (W/m^2)
ε = emissivity ($0 < \varepsilon < 1$)
σ = Stefan-Boltzmann constant = 5.67×10^{-8} W/m^2°K^4
T = temperature (°K)

This equation estimates the total amount of power available over all wavelengths for a specified temperature. Note that the power output of a black body goes up as the fourth power of temperature. So, if you double something's temperature, its power output increases 16 times! Recall we used the Stefan-Boltzmann equation in Chapter 10 to estimate the amount of power emitted by a vehicle entering the atmosphere. Later, in Chapter 13, we'll use this same relationship to analyze thermal control for spacecraft.

Seeing Through the Atmosphere

On a crystal clear night, you can see the light from stars thousands of light years away. As far as our eyes are concerned, a clear sky is completely transparent. On a cloudy day, we can't see the Sun, but if you stay out too long you can still get sunburned. How can this be? The atmosphere is selectively transparent to different wavelengths of radiation. On a cloudless day, we see light from the Sun, feel its heat, and get sunburned from its ultraviolet (UV) radiation. On a cloudy day, some of the light and heat are blocked, but the UV still gets through, so you can still get sunburned. Clouds are made of water droplets that block certain wavelengths of radiation. Even on a clear day, different molecules in the Earth's atmosphere block various wavelengths. This protects living things on Earth from the harmful effects of radiation, as discussed in Chapter 3.

For a remote-sensing system, we may want to collect EM radiation reflected or emitted from objects on the Earth's surface while we're above the atmosphere in space. We also need to use radio waves to communicate to and from the surface, again through the atmosphere. Thus, we must be concerned with the wavelengths readily blocked by the atmosphere and those that will pass through. Figure 11-8 shows what percentage of each wavelength gets through the atmosphere.

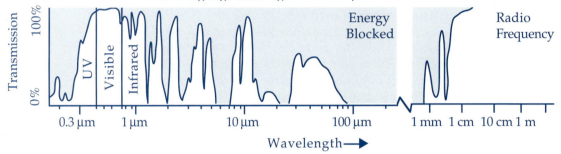

Figure 11-8. Atmospheric Windows. This graph depicts the transmission characteristics of the Earth's atmosphere at various wavelengths. Notice that certain parts of the EM spectrum—visible light, heat, and radio waves—get right through the atmosphere while other wavelengths are blocked. [1987, Lillesand]

Notice in Figure 11-8 that some wavelengths (such as visible light) are completely transmitted while others are almost completely blocked. You can begin to see that we have access to Earth from space through various windows of transmission. *Atmospheric windows* are the wavelengths which give us 80–100% transmission through the atmosphere. Most notable are the visible, infrared, and radio wavelengths. Using the visible and infrared windows, we can peer through the atmosphere to sense properties of objects on Earth from space. We can use the radio-frequency window to pass television and radio signals from studios on the ground through satellites to your living room (and to any aliens that may be listening).

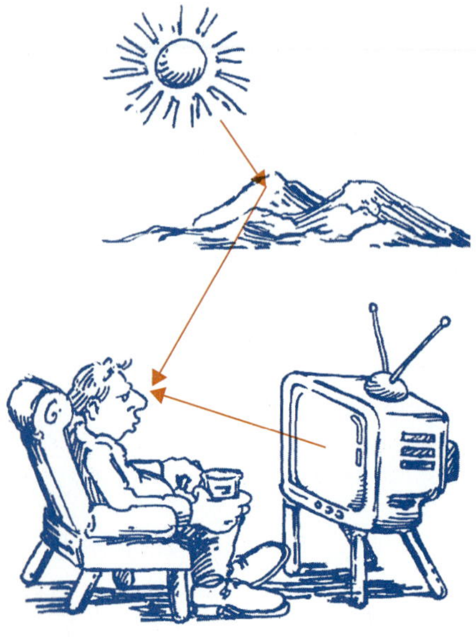

Payload Sensors

Types of payload sensors to collect information about things on Earth or at the edge of the universe literally cover the spectrum. We can classify these sensors into two broad categories:

- Passive sensors
- Active sensors

Passive sensors collect energy radiated by the subject or reflected off the subject by some other radiation source. For example, the subject of the FireSat mission is a forest fire. If we want a passive sensor to detect a fire, we would choose one sensitive to the heat or light a fire emits. Your eyes are a great example of a passive sensor, as Figure 11-9 shows. They detect light that's reflected off or emitted from objects. The source of that light is usually the Sun or the lights in a room. In some cases, it may be an object's temperature (like a red-hot piece of metal) or other energy output (like a television).

Figure 11-9. Passive Sensors. Your eye is a passive sensor. It detects EM radiation other objects reflect or emit.

Passive sensors on spacecraft operate in much the same way as our eyes. They detect EM radiation reflected off objects primarily from sunlight or resulting from the objects' temperature. Like our eyes, passive sensors contain three main parts:

- Radiation collector
- Detector
- Pointing or scanning mechanism

Just like the lenses in our eyes, a lens or antenna collects the radiation reflected or emitted from a target. We usually call the diameter of a lens or antenna its *aperture*. The radiation is then gathered and focused on an image plane containing the detector, similar to our retina. Finally, some pointing or scanning mechanism is needed to gather radiation from different sources just as muscles are used to move our eyes. The image is then focused over some *focal length* onto the detector. Figure 11-10 shows these relationships.

The complexity of the detector depends on the nature of the radiation being sensed. Many passive sensors operate in the visible or infrared (IR) parts of the spectrum. Visual sensors are essentially cameras because they focus visible radiation onto a detector. One type of detector is conventional film. The film contains chemicals which react to the incident radiation, forming an image. The first spy satellites actually took pictures using film and then dropped the film canisters back to Earth, where United States Air Force C-118 and C-130 airplanes caught them in mid-air. Since then, visual

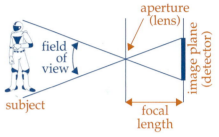

Figure 11-10. Passive Sensor System. This system is similar to a camera. We control the aperture size and focal length to focus the subject on the image plane.

Astro Fun Fact
The Greenhouse Effect

The property of the atmosphere which makes it transparent to some wavelengths but opaque to others causes the greenhouse effect. The greenhouse effect is a result of near-infrared radiation at short wavelengths being allowed into the atmosphere. When this radiation hits the ground and ocean, it's reflected or re-radiated as longer-wave radiation, such as far infrared, which can't penetrate the atmosphere. Thus, energy is "trapped" in the atmosphere as heat. Without the greenhouse effect, the effective temperature of the Earth would be about 0°F (−18°C). Thus, the greenhouse effect is a very useful, natural phenomenon which makes the Earth warm enough to support life. The controversy over the greenhouse effect has to do with man-made things that block Earth's window which allows thermal radiation (heat) to escape. Certain man-made and naturally occurring compounds, such as chlorofluorocarbons (CFCs) and methane, block this window. As a result, the Earth may heat up even more than we want it to. The Mission To Planet Earth spacecraft will observe the greenhouse effect to determine whether human actions, such as pollution and deforestation, or natural phenomena, such as volcanos, may lead to more global warming.

Masters, Prof. Gil. Environmental Science and Technology. Autumn 1989. Course Reader, CE170. Stanford University.

and IR detectors have been produced using semi-conductor materials similar to solar cells. These detectors generate electrical signals proportional to the incident radiation, producing an image. One type in common use is the *charge coupled device (CCD)*, which can also be found in many video cameras. Using CCDs, the visual images can first go into electronic storage and then be sent back to Earth as a data signal like a television broadcast.

Infrared detectors are similar to visible-wavelength detectors, but they sense a different part of the spectrum—heat. In fact, you can buy IR film for your camera. If you close your eyes and hold your hand close to a fire, you can feel the heat and use it to point your hand directly at the fire without peeking. The reason you can detect the fire is that your hand is much cooler than the heat from the fire. To be sensitive enough to detect rather small amounts of infrared energy hundreds of kilometers out in space, infrared detectors often must be very cold—about 77° K to 120° K. To keep them this cold we need liquid nitrogen or some other coolant. Thus, the sensor consists of a telescope, an IR detector, and a refrigeration unit. The Landsat satellites use sensitive IR sensors to image vegetation and other items of interest. A Landsat infrared image is shown in Figure 11-11.

The idea of a telescope as a passive sensor is pretty simple; however, many astronomers use passive sensors in another part of the spectrum—radio frequencies. Scientists are interested in the radio waves emitted by distant stars as well as unique interstellar radio sources such as pulsars. Radio telescopes are really just radio receivers designed to "tune in" to these stellar radio programs. Others try to detect the high-energy gamma and X-rays given off in violent supernovae or that indicate mysterious black holes. Regardless of the frequencies used, all types of astronomy benefit greatly by being in space. Above the atmosphere, many faint signals are not blocked or attenuated, so astronomers can see them more often and with greater resolution.

Scanning mechanisms for passive sensors vary widely in complexity. Some sensors simply stare at the ground and use the satellite's own motion over the Earth to scan the area beneath it. In this case, information collection depends on the *field-of-view (FOV)* of the sensor as shown in Figure 11-12. Our eyes, for example, have an angular field-of-view of about 130°. This means we can detect things to 65° on either side of where we look. The edge of this range is called our *peripheral vision*. There, we can sense an object only if it's moving. To really see something, we use the muscles around our eyes to move them and focus on the object of interest. Similarly, many spacecraft use mirrors to move the image in a sweeping pattern and focus on various objects of interest. The speed at which this motion takes place is called the *scan rate*.

Let's look at a simple example to see all these components at work. Pretend you're driving down the road using a video camera to focus on various objects of interest. The video camera has an optical system made up of a lens with variable-sized aperture so you can vary the focus depending on what you're looking at. The detector is a CCD. It translates

Figure 11-11. Infrared Sensors. This infrared image of the city of Miami clearly shows the contrast between vegetation, coastal water, and metropolitan area. Remote-sensing images of this type are especially useful for city planning. *(Reproduced by permission of Earth Observation Satellite Company, Lanham, MA, U.S.A.)*

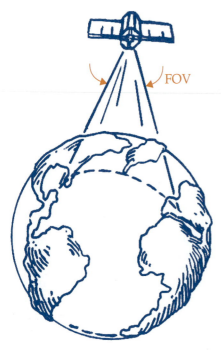

Figure 11-12. Field-of-View (FOV). The FOV of a satellite determines how much of the Earth it can scan at any one time.

the image into a user's signal. You point the camera at various subjects while traveling at some speed relative to them. Thus, in this example, the radiation collector is the camera lens, the detector is the CCD, and your hand and the moving car form the scanning mechanism.

Before you put a camera in space, you must ask "what is the smallest object I can see?" To figure this out we need to introduce the concept of beamwidth, or angular resolution. Imagine that you're standing in a level field in Kansas and far off on the horizon you see two cows in profile. As one cow walks toward the other they appear to overlap. The point at which you can just barely distinguish the two cows separately determines your angular resolution. The *beamwidth* or *angular resolution* is a function of wavelength and aperture diameter.

$$\theta = \frac{1.22\ \lambda}{D} \qquad\qquad (11\text{-}5)$$

where

θ = angular resolution or beamwidth (rads)
λ = wavelength of radiation used (m)
D = aperture diameter (m)

This equation says the smallest angle we can determine is directly proportional to the wavelength we select and indirectly proportional to the size of the lens aperture we use. Aperture describes the diameter of the lens or antenna being used. The smallest physical dimension we can distinguish at a given distance, the *resolution*, is simply the angular resolution times the distance the sensor is from the object we're interested in. Note that the resolution tells us the smallest detectable object, so the smaller the resolution, the better.

$$\text{Res} = \theta\,h = \frac{1.22\ \lambda\ h}{D} \qquad\qquad (11\text{-}6)$$

where
Res = Resolution (m)
θ = angular resolution or beamwidth (rads)
h = distance between the sensor and the viewed object (m)
λ = wavelength of radiation used (m)
D = aperture diameter (m)

Example 11-2 shows how these principles apply to a mission.

Unlike passive sensors, *active sensors* provide their own EM energy. We can use them whenever we need to see through clouds or whenever reflected or emitted EM radiation from the subject is either too faint or blocked by the atmosphere. When you're groping your way across a very dark room, your eyes, which are passive sensors, can't detect enough reflected or emitted EM radiation from the objects in the room. So you hit your toe on that darn coffee table again. To avoid another toe injury, you could use a flashlight to illuminate the objects in the room and allow your eyes to see them. In this way, you are effectively acting as an active-sensor system by shining EM radiation on the object you want to see and then detecting the reflected energy with your sensors (eyes).

Of course, we don't actually put big spot lights in space to shine down on the night side of the planet. But we do "shine" EM radiation from other parts of the spectrum. If you've ever known anyone pulled over for speeding (surely you'd never do such a thing), you're all too familiar with the police radar gun. The radar gun is an active sensor. It shines EM radiation in the radar frequencies onto speeding cars. The radiation reflects off the cars back to a sensor on the gun. The gun then measures this reflected radar energy to determine the car's speed.

Radar can do many other things besides enforce speed limits. Air traffic controllers use radar to keep track of the location and direction of incoming and departing aircraft. Airplanes use onboard radars to "paint" a picture of the terrain below them, even if they can't see the ground through the clouds. This allows the air crew to navigate by comparing terrain features to maps.

Because radar is an active sensor, it works day or night and doesn't depend on sunlight as optical sensors do. How the radar signal is reflected reveals much about the topography of the ground being imaged as well as its composition (such as soil type and presence of subsurface features like ancient river beds). Radar allows us to measure terrain features accurately to construct a 3-D picture of a planet's surface.

Resolution is still an issue with active sensors. Because resolution relates directly to the wavelength of the signal, smaller wavelengths have better resolution than larger wavelengths. Optical sensors use wavelengths on the order of 0.5 μm, while radar systems operate at about 240,000 μm. Thus, for optical and radar systems with the same size aperture, the optical system has almost one-half million times better resolution. For a conventional radar to have the same resolution as an optical system, we must increase the size of the radar's aperture. A conventional radar operating at a wavelength of 240,000 μm would need an aperture of more than 3900 km (6200 miles) to get the same resolution as an optical system with a mere 1 m (3.28 ft.) aperture. Obviously, an aperture this size is impractical. Instead, we've developed signal-processing techniques that can "fool" the electronics into thinking the aperture is much larger than it really is. This is the basis for a *synthetic aperture radar (SAR)*. SARs have been successful in remote sensing of the Earth and Venus. The Magellan spacecraft had a high-resolution (around 150 m or 492 ft.) SAR that gave us a detailed map of more than 98% of Venus, as shown in Figure 11-13.

We've looked at different types of remote-sensing instruments to demonstrate some key concepts of payload design. During space-mission planning, we must carefully specify our objective and characterize the subject before selecting or designing a payload to support a mission. Once we have a payload in mind, we can begin to focus on specific payload requirements such as size, mass, and power, which will drive the spacecraft bus design and determine how to configure the entire system.

Figure 11-13. Radar Images of Venus. The Magellan spacecraft used a synthetic-aperture radar to pierce the dense clouds and return highly accurate images of the surface of Venus. The top view shows portions of the Lada Terra highlands. The large circular feature in the Northwest corner is the 500-km (300-mile) diameter Eithinoha corona formed when hot volcanic magma migrated toward the surface. The bottom view shows a computer-generated 3-D perspective of Venus' surface showing Sapas Mons with an elevation of 4 km (2.5 miles). *(Courtesy of NASA)*

Section Review

Key Terms

accelerated charge
active sensors
angular resolution, θ
aperture
atmospheric windows
beam width
black body
charge coupled device (CCD)
field-of-view (FOV)
focal length
frequency, f
passive sensors
peripheral vision
photons
remote sensing
resolution
scan rate
synthetic aperture radar (SAR)
wavelength, λ

Key Equations

$$\lambda_m = \frac{2898° \text{ K } \mu m}{T}$$

$$E = \varepsilon \, \sigma \, T^4$$

$$\theta = \frac{1.22 \, \lambda}{D}$$

$$Res = \theta \, h = \frac{1.22 \, \lambda \, h}{D}$$

Key Concepts

➤ Sensors are the spacecraft's "eyes" and "ears." Spacecraft "look" and "listen" from space using the EM spectrum

➤ EM radiation can be thought of as waves of energy emanating from some source like ripples in a pond:
  • EM radiation is classified by wavelength—the distance between crests of the waves
  • The shorter the wavelength, the higher the frequency—the number of waves or cycles per second
  • The higher the frequency, the greater the energy of the radiation

➤ Any object with a temperature greater than 0° K emits EM radiation:
  • Planck described how an object's temperature determines its varying radiation frequencies
  • Wien's law tells us the frequency at which the maximum amount of black-body radiation will take place, depending on an object's temperature
  • The Stefan-Boltzmann relationship describes how much energy an object will emit, depending on its temperature

➤ Certain frequencies of EM radiation are blocked by the Earth's atmosphere while other frequencies are allowed to pass through. These are known as atmospheric windows. Engineers must be careful to select frequencies that can penetrate the atmosphere for remote-sensing.

➤ Remote-sensing payloads use passive and active sensors:
  • Passive sensors detect energy reflected or emitted from a subject
  • Active sensors shine EM radiation at the subject and then detect the energy reflected

➤ Sensor resolution refers to the smallest object a sensor can detect. Resolution depends on
  • The sensor's wavelength
  • Distance to the subject
  • Diameter of the sensor's aperture

Example 11-1

Problem Statement

Environmental damage from volcanos can upset many aspects of the global ecology. Select the best frequency for a payload to monitor dormant volcanos in the South Pacific which have an average temperature of 600° K.

Problem Summary

Given: $T = 600°$ K
Find: Frequency of maximum emission

Conceptual Solution

1) Determine wavelength of maximum emission using Equation (11-3).

$$\lambda_m = \frac{2898°\ K\ \mu m}{T}$$

2) Convert wavelength to frequency using Equation (11-1).

$$f = c/\lambda$$

Analytical Solution

1) Determine wavelength of maximum emission

$$\lambda_m = \frac{2898°\ K\ \mu m}{T}$$

$$= \frac{2898°\ K\ \mu m}{600°K} = 4.83\ \mu m$$

2) Convert wavelength to frequency

$$f = \frac{c}{\lambda} = \frac{3 \times 10^8 m/s}{4.83 \times 10^{-6} m}$$

$$f = 6.2 \times 10^{13} hz$$

Interpreting the Results

At a temperature of 600°K, the volcanos will have maximum energy emission at a frequency of 6.2×10^{13} Hz. According to Figure 11-6, this frequently is in the middle of the infrared region.

Example 11-2

Problem Statement

Armed with a small telescope, you want to look for hummingbirds. If the aperture of the primary lens of the telescope is about 5 cm and a hummingbird is about 4 cm across, what is the maximum range at which you could detect one? Assume your eye operates in the middle of the visual range—about 0.5 μm.

Problem Summary

Given: $D = 5$ cm, $\lambda = 0.5$ μm, $Res_{required} = 4$ cm
Find: h_{max}

Conceptual Solution

1) Determine angular resolution
   $\theta = 1.22\,\lambda/D$

2) Determine distance to get required resolution

   $h = \dfrac{Res}{\theta}$

Analytical Solution

1) Determine angular resolution

   $$\theta = \frac{1.22\lambda}{D}$$

   $$\theta = \frac{1.22\,(0.5 \times 10^{-6}\text{m})}{5 \times 10^{-2}\text{m}}$$

   $\theta = 1.22 \times 10^{-5}$ rad

2) Determine distance to get required resolution

   $$h = \frac{Res}{\theta}$$

   $$h = \frac{4 \times 10^{-2}\text{m}}{1.22 \times 10^{-5}}$$

   $h = 3279$ m

Interpreting the Results

In our case, we're looking for hummingbirds during the day, using visible light with a wavelength of about 0.5 μm. We need a resolution of 4 cm, using a telescope with a 5-cm aperture. From Equation (11-5) we determine the angular resolution to be 1.22×10^{-5} radians or 7×10^{-4} degrees. Thus, we should be able to identify our subject at nearly 3300 m (10,827 ft.). When choosing a telescope or other sensor for payloads, we must carefully select the wavelength, aperture size, and distance from the subject to achieve the necessary resolution. In real life, if you're looking for hummingbirds, you probably view an area using your naked eye, looking for motion. Once you see motion, you point the telescope in that direction to determine what caused the motion.

11.3 Spacecraft Design

▬ In This Section You'll Learn to...

- ☞ Describe the basic functions of the spacecraft bus and its key design drivers
- ☞ Describe the preliminary spacecraft-design process

The Spacecraft Bus

As we've seen, the spacecraft bus exists solely to support the payload with all the necessary housekeeping to keep it healthy and "happy." Perhaps the best way to visualize the relationship between the payload and bus is to picture a common, everyday school bus like the one in Figure 11-14. The school bus is designed to take its payload—the kids—to school, and all of its subsystems support this goal.

To design the spacecraft bus, we must understand the specific needs of the payload. For this example, we'd have to know things like

- How far and how fast the kids need to go, so we have a big enough engine and plenty of gas
- How many kids there are so we know how big to make the bus
- How warm to keep the bus so the kids don't freeze or burn up

In addition to knowing these things to directly support the payload, the bus designer must keep in mind other requirements to support the overall mission. This leads to the design of things like the steering system or the horn and radio.

Now that you can see how a school bus works to support its payload, let's extend the analogy to see the specific subsystems that make up a spacecraft bus and the key issues that go into their design.

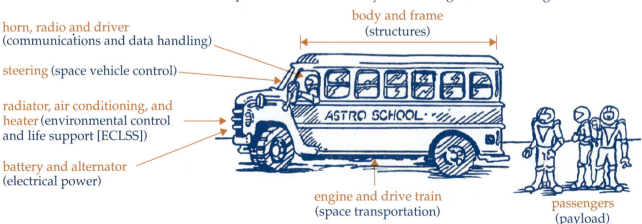

horn, radio and driver
(communications and data handling)

steering (space vehicle control)

radiator, air conditioning, and heater (environmental control and life support [ECLSS])

battery and alternator
(electrical power)

body and frame
(structures)

engine and drive train
(space transportation)

passengers
(payload)

Figure 11-14. A Spacecraft "Bus." The major functions performed on a spacecraft are also performed on a school bus.

Steering—Space Vehicle Control

School bus drivers must know what route they're taking, where they are, and how to steer the bus to get where they need to go. This is all part of controlling their vehicle. On a spacecraft, the space vehicle control system "steers" the vehicle to change

- Attitude—through rotations about the vehicle's center of gravity
- Position and velocity—through translation maneuvers in its orbit

A spacecraft has to have the right *attitude*, or orientation, to point cameras or antennae at targets on the Earth. (Experience has shown there's nothing worse than a satellite with a bad attitude!) The part of the control subsystem that maintains the spacecraft's attitude is the *attitude determination and control subsystem (ADCS)*. In Chapter 12 we'll see how control systems work and how they're used to change a spacecraft's attitude, position, and velocity. The key issues are in the ADCS design and knowing where the spacecraft needs to point and how accurately it needs to point there. In addition, we need to know how the spacecraft determines its attitude and what approach is used to control it.

A spacecraft must also be able to control its position and velocity in space so it can steer into the right orbit or perform the necessary ΔV to rendezvous with another spacecraft. This aspect of controlling the vehicle—called navigation, guidance, and control (NGC)—involves actually moving the vehicle in space:

- *Navigation* is figuring out where the spacecraft is.
- *Guidance* is determining where it's going and how to get there.
- *Control* is using navigation and guidance information to get the spacecraft where it needs to go.

To design the NGC subsystem, we must know the number of mission orbits as well as the number and types of maneuvers planned throughout the mission. To understand how this subsystem operates, we must know how the spacecraft location is determined and how it controls its orbit. Most of today's spacecraft determine and control their attitude using equipment onboard the satellite. However, they determine their position and velocity using ground-based tracking systems and control their orbit using onboard thrusters.

The Horn, Radio, and Driver—Communications and Data Handling

A school district may have hundreds of buses on the road picking up thousands of students. To manage all these buses and people, a complex operations team must oversee which bus goes where and respond to problems. On the bus, drivers have a horn to communicate intentions (or anger and frustration) to other buses and cars. In addition, they have a radio to keep in contact with the dispatcher or call for help in an emergency. A spacecraft also needs to talk to mission controllers and users on the ground and perhaps other spacecraft as well.

For a spacecraft, these two functions are so important that we'll make them two separate elements of the space-mission architecture—operations and the communications network. Space operations involve monitoring the health and status of a spacecraft from a control center. Operators also process vital mission data and ensure its delivery to users. To do all these functions, the operations team relies on a complex communications network tying together spacecraft, tracking sites, operators, and users. In space, the most common means of communication is by radio. Radios on a spacecraft are not that different from the AM/FM radios we listen to every day. In fact, a spacecraft may have several radios to

- Allow controllers to keep track of where the spacecraft is, check how it's doing, and tell it what to do
- Send mission data back to users
- Relay data sent from ground stations

In Chapter 15 we'll see how operations and the communications network tie together to make the mission successful.

Back on the bus, the driver takes in information about the route and the bus's performance and sifts this information through his or her brain to decide how to use it. On a spacecraft, the *communications and data handling subsystem* consists of radios and computers to transmit, receive, and process data. On manned spacecraft, computers also handle routine data, but astronauts can override these systems. A spacecraft may have several computers which work in much the same way as a typical home computer. The size and complexity of onboard systems for data handling depend on the volume of commands and data received, stored, processed, and transmitted, as well as the degree of autonomy built into the vehicle. We'll explore other aspects of data handling in Chapter 13.

Battery and Alternator—Electrical Power

To start the bus in the morning, the driver uses electrical power stored in the battery. Once the engine is running, the alternator keeps the battery charged and provides electrical power to run the lights, the radio, and other electrical components. Just like the school bus, spacecraft depend on electrical power to keep things running. Radios, sensors, mechanisms, and environmental controls all depend on electrical power to operate.

The electrical power used on a spacecraft is no different from the electrical power used to run your television. Unfortunately, in space, there's just no wall to plug into, and an extension cord to Earth would have to be very long! Therefore, a spacecraft must produce its own electrical power from some energy source, just as the school bus uses chemical energy released from burning gasoline to run the engine that turns the alternator. In Chapter 13, we'll see how the *electrical power subsystem (EPS)* converts some energy source, such as solar energy, into usable electrical power and stores it to run the entire spacecraft.

Radiator, Air Conditioner, and Heater— Environmental Control and Life Support

Just as a school bus's engine would overheat without a radiator for thermal control, a spacecraft must also be able to cool components onboard to keep them from overheating. The bus radiator circulates a mixture of water and antifreeze through the engine block, where it heats up and then transfers the heat to the cooler air which rushes through the radiator as you drive along. In Chapter 13 we'll learn how a spacecraft moderates temperatures onboard to keep vital components from being damaged.

If you've ever ridden on a school bus on a cold winter morning or a hot summer day, you know the heater and air conditioner can be a life saver! The heating and cooling systems in a bus or a car are designed to keep the passengers comfortable.

On a manned spacecraft, the *environmental control and life support subsystem (ECLSS)*, protects the crew from the harsh environment of space. It provides a breathable atmosphere at a comfortable temperature and pressure, along with water and food to sustain life. An unmanned satellite doesn't have these tight constraints on passenger comfort, but temperature control is still critical for the payload.

The Body and Frame—Structures

Holding the school bus together are all sorts of beams, struts, and nuts and bolts. These comprise the basic structure of the bus and support the loads it feels as it bounces down the road. The structure of the school bus literally holds it all together. When a driver steps on the gas pedal, you're forced back into your seat as the bus accelerates. When a spacecraft blasts into orbit, it experiences a similar force as the booster rocket accelerates into space. The spacecraft structure must be sturdy enough to handle these high loads. Some spacecraft must also hold together when subjected to spinning up or despinning, as well as all the other thrusting and rotating it will do during the mission. Spacecraft also have many mechanisms which crank, extend, bend, and turn. The structure supports these loads and holds the spacecraft together, along with all the mechanisms that make the spacecraft operate. In Chapter 13 we'll see what kinds of structures hold spacecraft together and learn some of the basic principles of structural analysis.

The Engine and Drive Train—Space Transportation

The bus has an engine and drive train. They supply power to the wheels, which move the bus where the driver wants it to go. The friction of the wheels on the ground pushes the bus along. In space there is nothing to push against, so we must use Newton's 3rd Law of Motion which says that every action has an equal and opposite reaction. The space transportation or propulsion subsystem on a spacecraft produces thrust needed to

- Get into space
- Move in space
- Control spacecraft attitude

Getting into space is so important that it's actually a separate element of a space mission. Once in space, the propulsion subsystem forms part of the spacecraft bus to do maneuvers and control attitude. In Chapter 14 we'll explore how rockets produce smoke and fire to perform all of these functions. To determine the configuration of the spacecraft's propulsion subsystem we need to know the number and size of maneuvers, which translate to the total ΔV required for the mission. We also need to choose an approach to attitude control so we can determine how much propellant we need to carry out the mission.

Figure 11-15 shows many of the subsystems we've just discussed in an exploded diagram of the Magellan spacecraft.

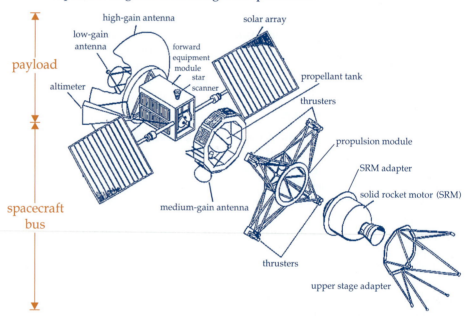

Figure 11-15. Spacecraft Subsystems. This exploded view of the Magellan spacecraft shows the location of many of the subsystems as well as the primary payload. The star scanner is part of the control subsystem. The solar arrays are used to generate electrical power. You can also see the structural elements that hold the spacecraft together, as well as the solid rocket motor and thrusters that make up the space transportation system. The large high-gain antenna is used for communication with Earth as well as to collect payload data as part of the synthetic aperture radar (SAR). The black rectangles on the side of the forward equipment module are louvers used for thermal control. Other elements of the spacecraft bus, such as the data handling subsystem, batteries, and other electronics, are packed safely inside. *(Courtesy of Martin Marietta)*

Designing the Spacecraft

So far in this chapter we've explored space-mission planning as shown at the top of Figure 11-16. We've seen how clearly identifying the mission statement leads to selecting an effective payload. We then took a closer look at how payload sensors use the EM spectrum to physically interact with the subject. In this section, we saw the various subsystems of the spacecraft bus which support the payload. Armed with this information, we can now enter the preliminary spacecraft-design phase shown in the lower part of Figure 11-16.

In this phase, we must assess everything needed to support the payload and conduct the mission with respect to each individual subsystem. For example, to effectively point a camera at a particular location on Earth, the control system must be able to determine the spacecraft's orientation and change it as necessary. The system may also have to maneuver in orbit to change ground-lighting conditions at a particular site.

Figure 11-16 also shows the interdependence of all the subsystems. Notice all subsystems link together in a continuous chain. Continuing with our control-system example, the communications and data handling subsystem must receive and process the command to move the spacecraft and reorient it. Both subsystems need electrical power to carry out the commands. The propulsion subsystem provides the ΔV to change the spacecraft's orbit, but the structure must bear the loads this maneuver will generate. Finally, the environmental control and life support subsystem must dissipate the additional heat that sunlight on the spacecraft will create.

As spacecraft designers, we must be very conscious of this interrelationship between all the bus subsystems. During the design process, seemingly innocent changes to the design of the electrical power subsystem, for example, may have profound effects on the performance of other subsystems. Thus, the design process is, by its very nature, an iterative one. That is, changes to one subsystem lead to changes in another which, in turn, lead to changes in another and... well, you get the idea. We never achieve the perfect design. The best we can hope for is to iterate the process enough times to get a design that satisfies the mission requirements and gets the job done at a reasonable cost.

One of the biggest problems during the design phase is keeping the entire mission in perspective. Too often, people responsible for specific subsystems get so involved in designing their subsystem that they lose sight of how their decisions affect other subsystems and the overall mission performance. Figure 11-17 offers a humorous look at how different subsystem designers often view the overall spacecraft.

Of course, we don't want subsystem managers to look at the spacecraft this way. From a total-quality approach, we want everyone involved with the project to keep one eye clearly focused on the overall mission. Only then can trade-offs between competing requirements be effectively resolved to produce the best result.

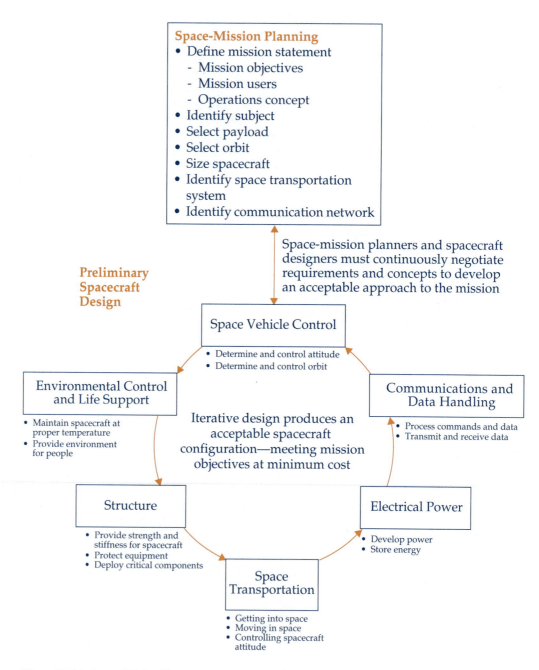

Figure 11-16. Space-Mission Planning and Preliminary Spacecraft Design. Once key design parameters are identified during mission planning, preliminary spacecraft design can begin. In this phase, you must systematically consider trade-offs between competing subsystem requirements.

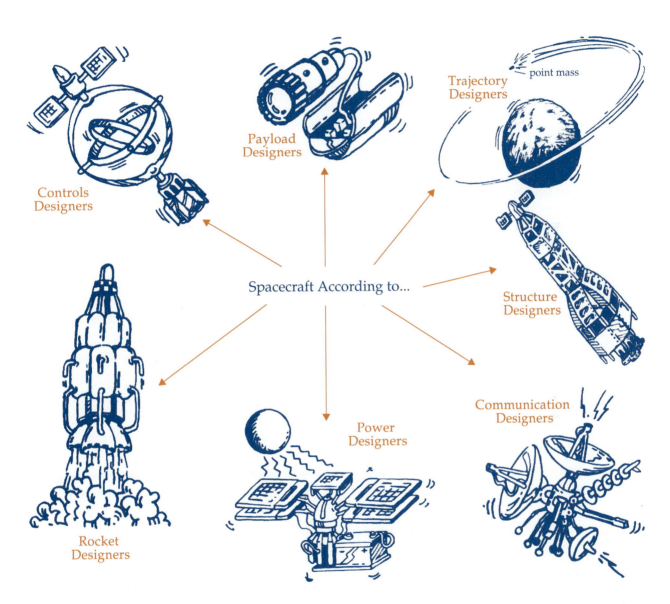

Controls
Designers

Payload
Designers

Trajectory
Designers

point mass

Spacecraft According to...

Structure
Designers

Rocket
Designers

Power
Designers

Communication
Designers

Figure 11-17. The Spacecraft According to the Communication, Controls, Payload, Power, Rocket, Structure, and Trajectory Designers. Sometimes individual subsystem designers get so focused on their subsystem they lose sight of the overall mission.

Section Review

Key Terms

attitude
attitude determination and
   control subsystem (ADCS)
control
communications and data
   handling subsystem
electrical power subsystem (EPS)
environmental control and life
   support subsystem (ECLSS)
guidance
navigation

Key Concepts

➤ The spacecraft bus provides all the housekeeping functions needed to run the payload and get data to users

➤ Space vehicle control
  • Controls the spacecraft's attitude, so it can point in the right direction—through the attitude determination and control subsystem (ADCS)
  • Controls the spacecraft's position and velocity, so it can get to where it needs to go—through the navigation, guidance, and control (NGC) subsystem:
    - Navigation shows you where you are
    - Guidance helps you determine where you need to go
    - Control uses navigation and guidance to get you where you want to be

➤ The data handling subsystem is basically a computer which handles all payload, subsystem, and user data on the spacecraft

➤ The electrical power subsystem (EPS) converts energy from some source into usable electrical power to run the other subsystems and the payload

➤ The environmental control and life support subsystem controls
  • Temperature for all onboard hardware
  • The environment, including air and water, for fragile human payloads

➤ The structure subsystem holds together all the other subsystems and payload and takes launch and mission loads. Mechanisms may also deploy and retract during the mission

➤ The space transportation system consists of a separate space-mission element concerned with getting the spacecraft into space and a spacecraft-bus subsystem involved with maneuvering around in space and, along with the ADCS, helping to control attitude

➤ Space operations involves monitoring and controlling a spacecraft from the ground as well as processing and distributing payload data to users

➤ Communications networks tie operators, spacecraft, remote-tracking sites, and users together, allowing information to flow between them

➤ Preliminary spacecraft design involves iterating on the design over and over, making trade-offs between individual subsystems to reach some compromise design

References

Elachi, Charles. *Scientific American*. December 1982. Vol. 247, No. 6. Radar Images of the Earth From Space.

Halliday, David and Robert Resnick. *Fundamentals of Physics*, 2nd edition. New York, New York: John Wiley and Sons, Inc., 1981.

Masters, Prof. Gil. *Environmental Science and Technology*. Autumn 1989. Course Reader, CE170. Stanford University.

NASA-STD-3000. *Man-Systems Integration Standards*.

Serway, Raymond A. *Physics For Scientists and Engineers*. Vol. II. 1990. Saunders Golden Sunburst Series.

Stanford University. Spring 1990. AA141, Colloquium on Life in Space.

Wertz, James R. and Wiley J. Larson. *Space Mission Analysis and Design*. Dordrecht, Netherlands: Kluwer Academic Publishers, 1991.

Mission Problems

11.1   Planning a Space Mission

1   What three things does the mission statement tell us?

2   What is the subject of a mission and how does it relate to the payload design?

3   Describe the mission objective and operations concept for our proposed FireSat. Why are these two things so important to overall mission design?

4   What is the subject of the proposed FireSat mission?

5   Describe the steps in space-mission planning and relate them to the FireSat mission. How do decisions made in this phase affect what will happen during the rest of the mission?

11.2   Space Payloads For Remote Sensing

6   Describe the primary elements for the FireSat's remote-sensing system.

7   What is the difference between wavelength and frequency?

8   The wavelength of visible light is around 0.5 µm. What is its frequency?

9   Gamma rays have a wavelength of about 10^{-13} m, AM radio waves have a wavelength of about 10^2 m. Which has more energy? Why?

10   What contribution did Max Planck make to our understanding of how objects emit radiation?

11   What is Wien's Law and why is it important?

12 How could you use the Stefan-Boltzmann relationship to determine an object's temperature?

13 Forest fires burn at about 1000°F. What is the wavelength of maximum power output for a forest fire? What is its energy output per square meter?

14 What are atmospheric windows and why are they important to selecting a spacecraft payload?

15 Using Figure 11-8, which frequency would be better to use for a remote-sensing spacecraft— 7.5×10^{13} Hz or 1.03×10^{14} Hz? Why?

16 Describe the differences between passive and active sensors and give examples of each.

17 What are the main parts of a passive sensor?

18 What is sensor resolution?

19 Which sensor has better resolution, one with Res = 5 cm or one with Res = 10 cm?

20 What is the angular resolution of an antenna with a 1 m aperture operating at a wavelength of 1 μm?

21 A remote-sensing payload operates in the visible part of the spectrum ($\lambda = 0.5$ μm). If it has an

aperture of 1 m, what is the highest altitude orbit it should operate in to achieve a resolution of 10 m?

11.3   Spacecraft Design

22 List the various parts of the spacecraft bus and describe functions for each one.

23 Describe what happens during preliminary spacecraft design. What are the key concerns for each subsystem? Why is this an iterative process?

For Discussion

24 What may have been the steps in space-mission planning for the Hubble Space Telescope?

25 What systems did engineers have to plan when designing the satellites for global positioning (navigation)?

Projects

26 The Environmental Protection Agency is interested in detecting the amount of pollution flowing from the mouth of the Mississippi River into the Gulf of Mexico. Plan the space mission for this problem, identify the major trade-offs in selecting a payload, and discuss some of the considerations involved with spacecraft design.

27 Step through mission planning for a manned mission to Mars. Then discuss how the decisions you make during this phase will affect the eventual spacecraft design.

Mission Profile—Landsat

The Earth Resources Technology Satellite (ERTS-1) was designed in the 1960s and launched in 1972. It was the first satellite designed specifically for broad-scale, repetitive observation of the Earth's land areas. The program was renamed Landsat (land satellite) in 1975. Landsat developed as a cooperative, multi-agency government project under NASA's direction. During the 1970's and 80's, Landsat transitioned to a commercial project under the private sector's control. In 1985, the Earth Observation Satellite Company (EOSAT) won a competitive bid to operate Landsat 4 and 5—to collect, archive, distribute, and sell Landsat images and to increase the user base.

Mission Overview

The objective of the Landsat satellite is to conduct remote sensing of Earth's resources, geology, and man-made features and return this data to users on Earth. Landsat employs two primary sensors—the multi-spectral scanner and the thematic mapper. The scanner images in four distinct bands ranging from green to near infrared, with a spatial resolution of 80 m. The mapper images over a larger part of the EM spectrum than the scanner, with seven bands instead of four and a resolution of 30 m.

Mission Data

✓ Landsat was launched in June, 1972. Designed for only a one-year mission, it wasn't retired from service until June 1978, after returning more than 300,000 images of Earth.

✓ Landsat 2 was launched in January, 1975 and remained in service until September 1983. Landsat 3 was launched in March 1978 and retired in September 1983

✓ Landsat 4 was an improved design of 1, 2, and 3, and was launched in July 1982. Because Landsat 4 suffered a power distribution problem in March 1983, it has had to operate at reduced power.

✓ Landsat 5, identical to Landsat 4, was launched in March 1984, with a planned lifetime of three years. It was still in use nine years later in 1993.

✓ All satellites operate in a near-polar sun-synch-ronous orbit crossing the equator at 9:45 A.M. local time.

Mission Impact

For more than 20 years, the Landsat program has produced archived images for most of the Earth's land masses, useful for analyzing long-term and quick-response changes. With the improvement of computer technology in the 1980s to process Landsat images, the applications of this data have exploded. Landsat images have been used for many long-term environmental studies, such as disappearing tropical forests, expanding desert areas, and climatic changes, as well as the Earth's response to natural disasters such as the explosion of Mount St. Helens and fires arising from the Midwestern floods. Landsat imagery has also been used for monitoring oil spills, identifying wildlife habitat, and measuring the growth of urban areas.

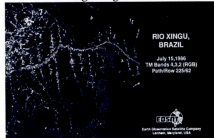

Landsat images can be used to track the destruction of forests. In this image of the Brazilian Rain Forest, clear-cut areas along the roads stand out against the forest background. *(Reproduced by permission of Earth Observation Satellite Company, Lanham, MA, U.S.A.)*

For Discussion

- How have we benefitted from commercial remote-sensing technologies?
- What are the disadvantages of the Landsat sensors? How could the sensors be improved?
- How can we use Landsat data to better manage Earth resources?

Contributors

Steve McGregor and Mark Hatfield, United States Air Force Academy

References

American Society of Remote Sensing. Manual of Remote Sensing. Virginia: The Sheridan Press, 1983.

Campbell, James B. *Introduction to Remote Sensing*. New York, NY: Guilford Press, 1987.

EOSAT Corp., Landsat Data Users Notes, published quarterly.

The manned maneuvering unit (MMU) gives astronauts the freedom to soar through space as a one-person spaceship. *(Courtesy of NASA)*

Space-Vehicle Control Systems

12

In This Chapter You'll Learn to...

- Describe the elements of and uses for control systems
- Apply the control process to attitude control for spacecraft
- Apply the control process to position and velocity control for spacecraft

You Should Already Know...

- Effects of the space environment on spacecraft (Chapter 3)
- Newton's laws of motion (Chapter 4)
- The principle of conservation of momentum (Chapter 4)
- Components and functions of the spacecraft bus (Chapter 11)

The Earth is a cradle of the mind, but we cannot live forever in a cradle.

Konstantine E. Tsiolkovsky
father of Russian astronautics

magine you're a one-person spacecraft, flying the Manned Maneuvering Unit out of the Space Shuttle's payload bay. Your mission is to fly over to a crippled satellite and install a new black box to fix it. You must somehow manipulate the joy sticks in your hands to control your position, velocity, and orientation so you're lined up with the access panel on the spacecraft. How should you rotate? In which direction should you fire your thrusters? Do you speed up or slow down? While this may sound like a fun scenario for a video game, we must answer these questions for nearly all spacecraft. In this chapter we'll begin by examining the basics of any control system and then see how we can apply this process to rotate and move satellites through space.

12.1 Introducing Controls

▬ In This Section You'll Learn to...

☛ Explain the difference between open-loop and closed-loop control systems and give examples of each

☛ Describe the steps in the control process

☛ Use block diagrams to describe the elements of a control system

If you've ever flushed a toilet, driven a car, or turned up the thermostat on a frigid winter night, you've used a control system. A *control system* monitors and changes a system to achieve some desired result.

Let's look at a control problem we're all familiar with—heating a house—to see what we mean. In the old days, we heated our homes with fireplaces. We started with some desired result "It's too cold in here, let's heat things up!" and decided what action to take—"Throw some logs on the fire!" The logs burned, providing heat, and the house warmed up. We can draw a simple diagram for this whole process as shown in Figure 12-1. As you can see, we start with a desired result, decide what to do, do it, and get the desired output.

Or do we? This simple control system has one major drawback. When you go to sleep at night there's no way to ensure that the house will be the desired temperature when you wake up in the morning. This is called an *open-loop control system* because we can't change the inputs to the system based on what's actually happening. Of course, people lived with this kind of heating system for thousands of years (and still do). But eventually we got tired of waking up to a cold house and invented the modern home-heating system. Let's see what makes this modern control system an improvement.

On cold winter nights, you crank up the dial on the thermostat to some desired temperature and wait for things to heat up. After some time, the furnace stokes up, and the temperature starts to rise. At the desired temperature the furnace shuts off. Simple, right? But what's really going on here? As with any control system, you have some desired result you want to achieve—for example, a house at 75°F. This is what you tell the thermostat when you set the dial. For this example, the heating-control system has different jobs. First, it measures or *senses* the current temperature in the house using a thermometer. Next, it *decides* what to do within the "brains" of the thermostat. If the temperature is less than 75°F, it knows things need to heat up. Finally, the furnace carries out the thermostat's decision by putting out heat.

Again, we can draw a simple diagram to illustrate what's going on, as shown in Figure 12-2. We start with a desired output, which we dial into the thermostat. The thermostat compares this temperature with the

Figure 12-1. A Basic Control System. To heat a house we first decide it's too cold then take actions like throwing logs on the fire to warm things up.

current temperature measured by a thermometer. If the current temperature is less than the desired temperature, the furnace is commanded to turn on. The furnace then puts out heat that warms up the house. Because we have a thermometer which constantly measures the current temperature and feeds this information back to the thermostat, we call this a *feedback control system* or *closed-loop control system*.

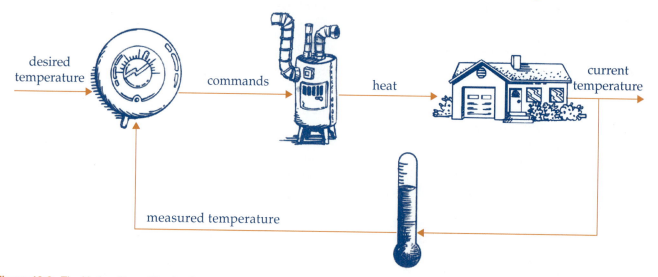

Figure 12-2. The Modern Home-Heating System. A modern home-heating system senses temperature and then decides when to turn on the furnace.

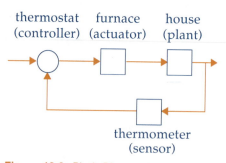

thermostat   furnace   house
(controller) (actuator) (plant)

thermometer
(sensor)

Figure 12-3. Block Diagram for a Heating-Control System. It's easier to represent the elements of a control system using blocks.

Because it's too hard to always draw nice pictures of each element in a control system, we often draw *block diagrams* instead, using a single block to represent each component. Furthermore, we need to establish some basic terms to describe each of the functions. In this example, the deciding function of the thermostat represents the system *controller*, and the thermometer function is the *sensor*. Because the furnace physically changes the thing we're trying to control, we call it the *actuator*. Finally, the house is the thing we're trying to control, which we traditionally call the *plant*. To control it we must understand its behavior (such as adding heat to the house increases its temperature). We can now draw a block diagram for the heating-control system as shown in Figure 12-3.

In this simple example we've identified the four basic tasks all control systems must do

- *Understand* the system's behavior—plant model
- *Observe* the system's current behavior—sensors
- *Decide* what to do—controllers
- *Do* it—actuators

Thousands of different control systems do these basic tasks. Note that open-loop systems do everything except make decisions based on observed behavior. Closed-loop control systems, like the one we described, are in cars, planes, spacecraft, and even the human body. For the most part, we'll focus on two important applications of the control process for spacecraft—controlling

- Attitude—rotating the satellite about its center of mass
- Position and Velocity—translating or moving the satellite

The attitude determination and control system (ADCS) must control attitude—rotating about the center of mass to change where the spacecraft is pointed. The navigation, guidance, and control (NGC) system must control position and velocity to move through space. We'll examine each of these systems in the next two sections.

Section Review

Key Terms

actuator
block diagrams
closed-loop control system
control system
controller
feedback control system
open-loop control system
plant
sensor

Key Concepts

➤ A control system allows us to achieve some desired result or behavior for a system by using four steps
  - Understand the system's behavior—the plant model
  - Observe the system's current behavior—sensors
  - Decide what to do—controllers
  - Do it—actuators

➤ Control systems for a spacecraft control its
  - Attitude (where it's pointing)
  - Position or velocity (where it's and where it's going)

12.2 Attitude Determination and Control Systems

▬ In This Section You'll Learn to...

- ☞ Describe spacecraft attitude in terms of roll, pitch, and yaw
- ☞ Explain what is meant by pointing accuracy
- ☞ Define torque—the way we get something to rotate
- ☞ Explain how Newton's Second Law helps describe attitude dynamics for spacecraft
- ☞ Define precession—an important property of spinning masses
- ☞ Describe the three types of attitude actuators and how each is used
- ☞ Explain the primary disturbance torques on a spacecraft
- ☞ Describe the different types of attitude sensors and explain how they work
- ☞ Compute the change in angular momentum for a momentum-control device given the current and desired angular momentum of the spacecraft
- ☞ Draw the block diagram for a system to control spacecraft attitude and explain what each part does
- ☞ Describe other attitude-control strategies and discuss their advantages and disadvantages

Now let's see how the basic tasks of a home-heating control system apply to the problem of changing and maintaining the spacecraft's attitude. *Attitude* defines a spacecraft's orientation in space. For example, if we want to take pictures of the Earth, we need to align the spacecraft so the payload aims at the subject on the ground. Thus, we would control the spacecraft's attitude so the camera is pointing down. We'll begin by looking at what we mean by attitude in more detail and then explore the system dynamics, actuators, sensors, and controllers we need to monitor and change it.

Having the Right Attitude

Before we go too far, it's important to understand some basic terms used to describe attitude. Recall that when we described the motion of an object we referenced it to some coordinate system. Well, to describe attitude we must again refer to a known coordinate system, but we're now interested in rotation rather than translation. So we define attitude in terms of angles instead of distances. For spacecraft (and aircraft as well), we define attitude in terms of angular rotation with respect to a body-centered coordinate frame, with $\hat{x}$ out the nose, $\hat{y}$ out the left wing, and $\hat{z}$ out the top, as shown in Figure 12-4. We can then describe attitude in terms of roll, pitch, and yaw where *roll* is a rotation about the $\hat{x}$ axis, *pitch* is a rotation about the $\hat{y}$ axis, and *yaw* is a rotation about the $\hat{z}$ axis, as shown in Figure 12-5.

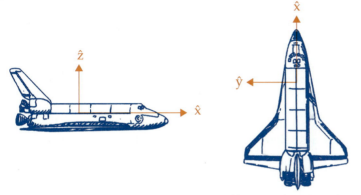

Figure 12-4. Body Frame. Attitude is generally described in terms of rotations about one or more of the body-centered axes.

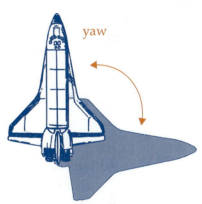

Figure .12-5. Describing Attitude. Spacecraft attitude is generally described in terms of roll, pitch, and yaw about the center of mass.

Now that we know how to describe attitude, how do we determine whether we have the right attitude? We must first know what attitude we need. Typically, we describe attitude-control requirements in terms of accuracy and rate of attitude change. To understand what we mean by attitude or pointing accuracy, pretend you're trying to point a pencil laser beam at a target as in Figure 12-6. Your ability to keep the beam on the target depends on the size of the target and the steadiness of your hand. It should make sense that the smaller the target, the more steady your hand must be to maintain the laser on target. As your hand wavers, the beam will tend to stay within a cone more or less centered on the target. *Pointing* or *attitude accuracy, θ,* is defined in terms of the size of this cone. For a spacecraft trying to point an antenna at a ground station on the Earth, for example, the control system must be accurate enough to keep the radio beam focused on the receiver.

To get a better feel for what we mean by attitude or pointing accuracy, pretend you're trying to keep a laser pointer focused on a target about the size of a dinner plate (25 cm or 10 in). You know the pointing accuracy, ψ, and the distance to the target, r. To find the apparent diameter of the target you can hit, D, as shown in Figure 12-6, we can use

$$D = r\,\psi \qquad (12\text{-}1)$$

where
D = approximate diameter of target (m)
r   = distance to target (m)
ψ = pointing accuracy (rad)

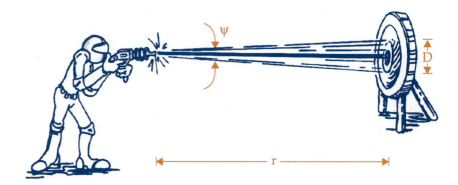

Figure 12-6. Attitude Accuracy. In this example the astronaut is pointing a laser beam at a target. As his hand wavers, the beam will describe a cone more or less centered on the target. Attitude accuracy is defined in terms of the angular size of that cone.

Table 12-1 shows the required pointing accuracy to stay focused on the dinner plate at various distances. Let's put this in space terms. A remote-sensing spacecraft passing directly over you at an altitude of 500 km (310 miles) would need about 0.003° of accuracy to point a laser range finder at your house (D = 30 m). Fortunately, when satellites take pictures of the Earth, the field of view of their camera is wide enough so that actual pointing requirements can be somewhat less—around 0.1° or so.

Table 12-1. Pointing Accuracy.

To Point at a Target the Size of a Dinner Plate at This Distance...	Your Pointing Accuracy Needs to be...
1.4 m (4.7 ft.)	10°
14 m (47 ft.)	1°
140 m (470 ft.)	0.01°
1400 m (0.9 mile)	0.001°

Rate of change of attitude is also an important consideration when designing an attitude-control system. For example, a remote-sensing spacecraft may need to shift its attention between various targets on the ground. This would mean it needs to rapidly change its attitude to focus on these widely separated points of interest. *Slew* is rotation of the pointing axis from one point to another. The speed of attitude change for a spacecraft is defined by its slew rate. *Slew rate* is an angular velocity generally described in terms of radians (or degrees) per second of rotation. Now that we understand more about describing attitude, let's see how we go about controlling it.

Attitude Dynamics for Spacecraft

As we know, all spinning objects—tops, yo-yo's, ice skaters, and even spacecraft—are slaves to Newton's laws of motion. Recall from Chapter 4 that a spinning mass has angular momentum which is a function of its shape and mass distribution, along with its rate of spin. Notice, for example, that a compact object with all the mass concentrated near the center of mass will spin much more easily than an object which has significant mass a long way from the center of mass. As Figure 12-7 shows, this is why figure skaters will bring their arms in to spin faster and extend their arms to slow down. The distribution of mass describes an object's *mass moment of inertia, I*. By knowing both the mass moment of inertia, I, and the object's angular velocity, $\vec{\Omega}$, we can find its angular momentum, $\vec{H}$, in pretty much the same way we found its linear momentum.

$$\boxed{\vec{H} = I\vec{\Omega}} \tag{12-2}$$

where
$\vec{H}$ = angular momentum ($kg \cdot m^2/s$)
I = mass moment of inertia ($kg \cdot m^2$)
$\vec{\Omega}$ = angular velocity (rad/s)

We also know from Chapter 4 that to change an object's momentum we must apply a force. This was explained by Newton's Second Law which we can express as

$$\vec{F} = \frac{d\vec{p}}{dt} = \dot{\vec{p}} \tag{12-3}$$

where
$\vec{F}$ = force (N)
$\vec{p}$ = momentum ($kg \cdot m/s$)
$\dot{\vec{p}}$ = time rate of change of momentum $\left(\dfrac{kg \cdot m/s}{s}\right)$

When you kick a football or serve a volleyball it's not hard to see that applying a force to a mass changes its velocity. But how do you apply force to a rotating mass? If we push on spinning ice skaters they'll start moving in a straight line across the ice while continuing to spin. What if we want to change only their rate or direction of spin but not move them anywhere? Then we must apply a torque. A *torque* is the twisting motion which results when a force is applied over a distance, such as when you use a wrench to turn a bolt. You apply a force some distance away from the bolt which produces a torque as shown in Figure 12-8. A torque in one direction tightens the bolt. A torque in the other direction loosens it.

Mathematically, the direction of torque is defined as the vector cross product of the vector describing the applied force's position with the force vector itself. In other words, you use the good old right-hand rule to turn in the direction of twist, so your thumb will point at the torque vector. In Figure 12-8, you see a force applied to the end of a wrench. The torque vector, $\vec{T}$, points into the bolt. We can compute torque using

mass moment of inertia is low:
skater spins quickly

mass moment of inertia is high:
skater spins more slowly

Figure 12-7. Changing the Mass Moment of Inertia. Ice skaters change their moments of inertia to spin faster or slower by pulling in or extending their arms.

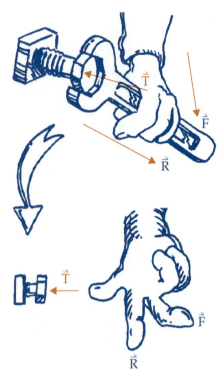

Figure 12-8. Torque. The direction of torque is found using the right hand rule. By wrapping the fingers of your right hand in the direction of spin your thumb will point in the direction of the torque vector. In this case, the torque direction is into the bolt.

$$\boxed{\vec{T} = \vec{R} \times \vec{F}}$$ (12-4)

where
$\vec{T}$ = torque (N · m)
$\vec{R}$ = distance from center of mass to the point where force is applied (m)
$\vec{F}$ = force applied (N)

According to this relationship, you can get more torque with the same force by simply applying the force farther from the center of rotation. Aristotle knew of this effect when he bragged he could move the Earth if given "a fulcrum, a long enough staff, and a place to stand." You don't have to move the Earth to see this effect. All you have to do is push open a door. If you push at the edge of the door, far from the hinges, the door swings right open. If you push on the door right next to the hinges, it's much harder to move.

Returning to Newton's Second Law, we can now see how to relate torque and angular momentum. Just as a force was equal to the time rate of change of linear momentum, torque is the time rate of change of angular momentum. In other words, if you apply a torque to something, its angular momentum will change. We can express this as

$$\vec{T} = \frac{d\vec{H}}{dt} = \dot{\vec{H}}$$ (12-5)

where
$\vec{T}$ = torque (N · m)
$\vec{H}$ = angular momentum (kg · m²/s)
$\dot{\vec{H}}$ = time rate of change of angular momentum $\left(\dfrac{\text{kg} \cdot \text{m}^2/\text{s}}{\text{s}} \right)$

This relationship tells us something we already learned back in Chapter 4. When torque is zero, angular momentum stays constant. Later, we'll see we can use this basic principle to give us accurate attitude sensors, as well as efficient actuators to control attitude. For now let's see how we can use it to analyze how attitude works. Remember that if we apply a force to an object, it will accelerate. Similarly, if we apply a torque to an object, it will start to spin faster and faster. That is, it will experience angular acceleration, $\vec{\alpha}$. Thus,

$$\boxed{\vec{T} = \dot{\vec{H}} = I\vec{\alpha}}$$ (12-6)

where
I = mass moment of inertia (kg · m²)
$\vec{\alpha}$ = angular acceleration (rads/s²)

As we know from our discussion of linear motion, as something accelerates over time, it acquires velocity. If you drop a ball, it accelerates, gains velocity, and falls faster the longer it goes. Similarly, when an object has angular acceleration over time, it gains angular velocity, $\vec{\Omega}$. Thus, to determine a spacecraft's attitude, described by the angle θ, we must look at how long it accelerates and how long it moves at some angular velocity. In other words, by applying a torque to a non-spinning object, we can create angular acceleration, leading to angular velocity and hence a change in angular position. The model for this aspect of spacecraft behavior is the block diagram in Figure 12-9.

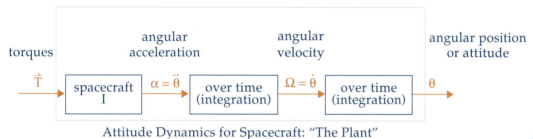

Attitude Dynamics for Spacecraft: "The Plant"

Figure 12-9. Block Diagram of Attitude Dynamics. A torque applied to a spacecraft causes an angular acceleration. Over time this acceleration causes an angular velocity that changes attitude.

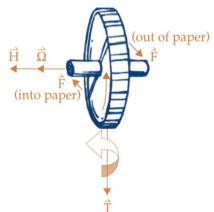

Figure 12-10. Torque Applied to a Spinning Disk. Here, the angular-momentum vector, $\vec{H}$, points to your left. As a force couple is applied to this spinning mass, it creates a torque. In this case the torque axis, $\vec{T}$, is pointed down.

When you apply a torque to a non-spinning object, predictable things happen. For example, when you turn a screw with a screw driver, it rotates in the way you'd expect. But if the object is spinning when the torque is applied, the dynamics get more complicated. As we know, a spinning object has angular momentum. If a torque is applied *parallel* to the angular momentum direction, it causes angular acceleration and velocity. However, if the torque is applied in a direction other than parallel to the angular-momentum vector, something quite different happens.

In Figure 12-10 we have a spinning disk with a force couple (torque) applied to it. You might expect the mass to begin to rotate in the same direction you're torquing it, or clockwise as you look down on it. But that's not what happens! The mass begins to rotate counter-clockwise about an axis that comes out of the page! This phenomenon is known as *precession*.

For the disk shown in Figure 12-11, the $\vec{H}$ vector will begin to move (or precess) toward the $\vec{T}$ vector (things would behave differently if it were a different shape). This precession occurs about a third vector called the precession vector, $\vec{\omega}$, and is at right angles to both $\vec{H}$ and $\vec{T}$. As we'll see, knowing how a spacecraft gains angular velocity and precesses helps us determine how to apply forces in order to adjust the spacecraft's attitude. Note that the direction of precession depends on *how* mass is distributed in the object—its mass moment of inertia. Analyzing *why* is beyond the scope of this book.

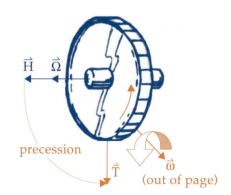

Figure 12-11. Precession of a Spinning Disk. When a torque is applied to this spinning mass, the mass begins to precess by spinning around an axis 90° away from the torque axis.

Spacecraft-Attitude Actuators

Once we've determined what our attitude is, we need to know how to change it. For example, we may want to rotate the spacecraft to point at a new subject. As we've seen, applying a torque will change attitude. That's why we need actuators. Actuators provide "torque on demand" which rotates the spacecraft as needed to take pictures, dump data, or meet other mission requirements. We'll talk about three different types of actuators:

- Thrusters
- Devices to control angular momentum
- Magnetic torquers

Thrusters

Thrusters are perhaps the simplest type of actuator to visualize. *Thrusters* are rockets which rely on "brute force" to rotate the spacecraft. By applying a force with a pair of rockets at some distance away from the spacecraft's center of mass on opposite sides, we get a torque as shown in Figure 12-12. By varying which thruster pair we use and how much force we apply, we can turn the spacecraft in any direction with high accuracy.

Designers must be careful in placing thrusters. Placing thrusters as far from the satellite's center of mass as possible gives them a larger moment arm and allows them to exert the greater torque for a given force. This is evident by looking at Equation (12-4) where you can see that the bigger R is, the more torque we get from the same force. However, because of the principle of precession, when a spacecraft is already spinning, any applied torque in a direction other than the spin axis will cause the spacecraft to rotate at constant velocity about an axis perpendicular to the torque direction.

The biggest advantage of using thrusters is that they work fast, allowing the spacecraft to slew quickly from one attitude to another. But the amount of propellant a spacecraft can carry limits their use. For short missions, such as those flown by the Space Shuttle, this is no problem. For longer missions (months or years), thrusters are used only as a backup and for other purposes we'll discuss later.

Momentum-Control Devices

The most common actuator for controlling attitude is actually a family of systems that all rely on angular momentum. These *momentum-control devices* actively vary the angular momentum of small masses within the spacecraft to change attitude. How can this work? If you stand on a turntable holding a spinning bicycle wheel at arm's length in front of you, you can cause yourself to rotate by moving the bicycle wheel to the left or right. That's because total angular momentum is always conserved. As the bicycle wheel rotates, you must rotate to compensate to keep the total angular momentum constant, as you can see in Figure 12-13. Similarly, a spacecraft's total angular momentum, including that of spinning masses inside it, must be conserved.

From Equation (12-2) we know angular momentum is the product of the mass moment of inertia of a given mass, I, and its angular velocity, $\vec{\Omega}$.

$$\vec{H} = I\vec{\Omega} \qquad (12\text{-}2)$$

Figure 12-12. Thrusters. Thrusters are rockets which apply a force to the satellite at some distance from the center of mass. This force causes a torque that rotates the spacecraft.

Figure 12-13. Bicycle Wheels in Space? You can do a simple experiment to see one way spacecraft control attitude. By standing on a turntable holding a spinning bicycle wheel, shown in the left-hand picture, you can change your attitude by applying a torque to the wheel, as shown in the right-hand picture.

Note we can get the same angular momentum from a large mass (high I) spinning at a relatively slow speed (low $\vec{\Omega}$) as we can from a small mass (low I) spinning at a much higher rate (high $\vec{\Omega}$). So, if we consider a spacecraft and all mass inside it to be one system, we can control where the spacecraft is pointing by changing the angular momentum (rate and direction of spin) of a small mass inside.

There are three types of momentum-control devices:

- Momentum wheels—change spin speed
- Reaction wheels—change spin speed
- Control-moment gyroscopes—change spin direction

A *momentum wheel* is the simplest device for controlling momentum. It consists of a single, constantly spinning wheel within the spacecraft. Because it's always spinning, its angular momentum vector is fixed in inertial space. Thus, the spacecraft's orientation is fixed (because the wheel is attached to the spacecraft). This angular momentum gives the entire system "stiffness," making it resistant to outside torques. Figure 12-14 shows what a momentum wheel looks like. Typical momentum wheels provide 0.1° to 10° of pointing accuracy in two axes.

Figure 12-14. Momentum Wheels. A momentum wheel is a spinning mass inside a spacecraft which can absorb disturbance torques and keep the spacecraft pointing in the right direction. *(Courtesy of Ball Aerospace)*

To maintain the same spacecraft attitude, the momentum wheel will eventually begin to spin faster and faster in response to outside torques. At some point the wheel may be spinning as fast as it can without damaging bearings or other mechanisms. This excess momentum in the wheel must be removed or "dumped." *Momentum dumping* is a way to decrease the angular momentum of spinning masses within the spacecraft

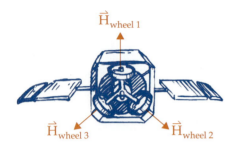

Figure 12-15. Reaction Wheels. Reaction wheels are placed along all three axes of a spacecraft. They are spun up in response to disturbance torques or to rotate the vehicle.

before reaction wheel spins up

$$\vec{H}_{TOT} = \vec{H}_{S/C} = 0$$

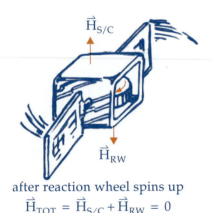

after reaction wheel spins up

$$\vec{H}_{TOT} = \vec{H}_{S/C} + \vec{H}_{RW} = 0$$

Figure 12-16. Reaction Wheels Demonstrate the Conservation of Angular Momentum. The total angular momentum is the sum of the spacecraft's angular momentum plus the momentum of the reaction wheel. In this case, we start with a fixed spacecraft with zero total angular momentum. To rotate the spacecraft in one direction, the reaction wheel is spun-up in the opposite direction such that total angular momentum stays constant.

by applying a torque to the spacecraft. The torque helps to "de-saturate" the spinning mass, reducing its spin rate and thus its angular momentum.

Reaction wheels are also based on momentum, but they don't normally spin until the spacecraft needs to be reoriented or an outside torque is applied. Usually, a spacecraft carries at least three separate reaction wheels oriented at right angles to each other, as seen in Figure 12-15. Often, a fourth wheel is skewed to the other three for redundancy.

When the spacecraft needs to slew to a new location or in response to outside torques, one or more of these wheels are spun up. Let's step through what happens to rotate the satellite. First of all, realize that without any outside torques the total angular momentum of the spacecraft (including the reaction wheels) is conserved. Thus, the angular momentum of the spacecraft plus the angular momentum of the reaction wheels must add up to a constant vector quantity.

Now, imagine one of the reaction wheels being spun up using a motor. As the wheel's rotation rate increases, its angular momentum also increases. But, the total angular momentum of the wheel and spacecraft must always sum to a constant value. So what happens to the spacecraft? Let's look at more specific examples to get a better idea. We can express the total angular momentum of a spacecraft (including reaction wheels) as

$$\vec{H}_{TOT} = \vec{H}_{S/C} + \vec{H}_{RW} \qquad (12\text{-}7)$$

where
$\vec{H}_{TOT}$ = total angular momentum of the spacecraft (kg · m²/s)

$\vec{H}_{S/C}$ = angular momentum of just the spacecraft (kg · m²/s)

$\vec{H}_{RW}$ = angular momentum of reaction wheels (kg · m²/s)

[*Note:* This relationship is vector addition, so $\left|\vec{H}_{TOT}\right| \neq \left|\vec{H}_{S/C}\right| + \left|\vec{H}_{RW}\right|$!]

If a reaction wheel is spun up, its angular momentum increases by an amount $\Delta\vec{H}_{RW}$. Because the total angular momentum must stay constant, the spacecraft's angular momentum *must* automatically decrease to compensate by an amount $\Delta\vec{H}_{S/C}$. The vector increase in the reaction wheel's momentum must exactly equal the decrease in the spacecraft's momentum, or $\Delta\vec{H}_{RW} = \Delta\vec{H}_{S/C}$ to keep constant. Figure 12-16 shows these relationships. To conserve momentum, the spacecraft must either slow its rotation or start rotating in the opposite direction. In either case, the spacecraft's attitude has changed simply by spinning up a small mass inside.

In practice, reaction wheels can be used to get very effective and accurate control of a spacecraft in all three axes. Reaction wheels often can achieve 0.001° [Larson & Wertz, 1992] pointing accuracy. Unfortunately, reaction wheels can also get "saturated," or reach their design limit for rotation speed through repeated or continuous attitude adjustments. When this happens, the wheels must be "de-saturated" with momentum dumping.

The final type of attitude actuator based on momentum is the *control moment gyroscope (CMG)*. They provide pointing accuracies equivalent to reaction wheels but offer much higher slew rates, making them especially useful for tracking other objects. A CMG consists of one or more spinning masses mounted on a framework called a *gimbal* that allows it to rotate freely. Rotating the CMG changes the direction of its angular momentum vector. Again, because total angular momentum is conserved and the CMGs are spinning rapidly, the spacecraft will rotate in a direction to conserve total angular momentum. Because of their expense and complexity, CMGs are used only on systems that require extremely accurate pointing and tracking.

Magnetic Torquers

Magnetic torquers take advantage of the Earth's strong magnetic field which we discussed in Chapter 3. Running an electrical current around a piece of metal onboard creates an electromagnet. This electromagnet will try to align itself to the Earth's magnetic field, dragging the rest of the spacecraft along with it, as seen in Figure 12-17. To see how this works, think about a compass needle that is just a lightweight magnet that can rotate freely. If you've ever played with magnets, you know that one side of a magnet will readily attract and stick to another magnet while the opposite side will repel it. With magnets, opposites attract and likes repel, so the North Pole of a magnet will attract the South Pole of another magnet. The magnet rotating freely in a compass will try to do the same thing. The North end of the compass will tend to point at the Earth's North Pole, and suddenly, you're no longer lost!

Magnetic torquers offer a relatively cheap and simple way to control a spacecraft's attitude. Furthermore, because they run on electrical power which is usually available, they are inexhaustible—unlike thrusters. However, they have two main limitations

- Because their effectiveness depends directly on the strength of the Earth's magnetic field, they become less and less useful in higher orbits

- They're not very accurate; ±30° is the best they can do alone

For these reasons, magnetic torquers are used primarily for momentum dumping in low-Earth orbiting spacecraft. For spacecraft at higher altitudes, such as those in geostationary orbit, thrusters must be used for this purpose.

Disturbance Torques

So why can't we just stick a satellite out in space in the desired attitude and forget about it? As we know from Chapter 3, space can be a nasty place. Over time, if we do nothing, environmental effects called *disturbance torques* will drive the satellite from the desired attitude. These torques are extremely small (in most cases they literally couldn't kill a fly). But just as tiny drops of water can wear away mountains over time, these torques can rotate satellites. We're concerned with four main sources of disturbance torques:

magnetic field lines

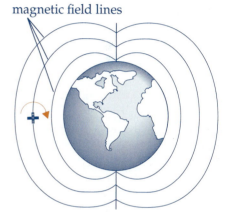

Figure 12-17. Magnetic Torquers. Magnetic torquers take advantage of the disturbance torque from the Earth's magnetic field. Powering electromagnets onboard the spacecraft creates a torque which rotates it.

- Gravity gradient
- Solar-radiation pressure
- Earth's magnetic field
- Atmospheric drag

Let's begin by looking at the *gravity-gradient* torque. Recall from Chapter 4 that Newton said the force of gravity varies inversely with the square of the distance.

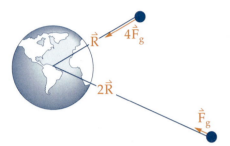

$$\vec{F}_g = \frac{-\mu\, m}{R^2}\hat{R} \tag{12-8}$$

where
$\vec{F}_g$ = Force of gravity (N)
μ = Gravitational parameter of central body (m^3/s^2)
m = Mass of object (kg)
R = Distance from object to central body (m)
$\hat{R}$ = Unit vector in $\vec{R}$ direction (m)

Figure 12-18. Gravitational Force. From Newton's Law of Universal Gravitation we know that gravitational attraction decreases with the square of the distance. Thus, if you double the distance, the gravitational force is only 1/4 as strong.

Thus, as you can see in Figure 12-18, if one object is twice as far from the Earth as a second body, the gravitational force will be one-fourth as large.

This is easy to visualize if the difference in distances from the central body is very great, but it works the same way for very small differences. Imagine we have a dumbbell-shaped spacecraft in Earth orbit. If the dumbbell is hanging vertically, as in Figure 12-19, the lower mass (m_1) will have a slightly greater gravitational force on it than the higher mass (m_2).

If m_2 is directly above m_1, nothing interesting happens. However, if the dumbbell gets displaced slightly off vertical, as in Figure 12-20, this slightly different gravitational force on the two masses will create a torque on the spacecraft that will tend to restore it to vertical. This is fine if you want it to be vertical, but if you don't, your control system must constantly fight against this torque.

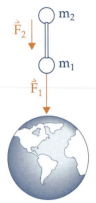

Figure 12-19. Gravity Gradient. In a spacecraft shaped like a "dumb bell" you can see that the gravitational pull on the lower part will be slightly more than on the upper part. $\vec{F}_1 > \vec{F}_2$.

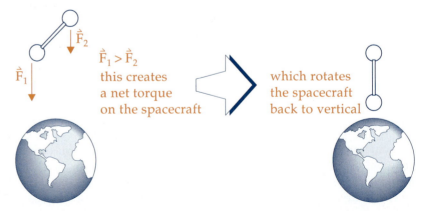

$\vec{F}_1 > \vec{F}_2$
this creates a net torque on the spacecraft

which rotates the spacecraft back to vertical

Figure 12-20. Gravity-Gradient Torque. The slight difference in gravitational pull between the upper and lower part of the spacecraft will tend to rotate the spacecraft to vertical, with its long axis down.

Another disturbance torque for a spacecraft comes from the ever-so-slight force of *solar-radiation pressure* on a surface. We're used to being warmed by the Sun, tanned by the Sun, and even burned by the Sun, but pushed by the Sun? Yes. One way to think about sunlight (or any light for that matter) is as lots of tiny bundles of energy called *photons*. In one of those seeming paradoxes of modern physics, these photons are said to be massless (thus they can travel at the speed of light), but they *do* have momentum. As the photons strike any exposed surface, they transfer this momentum to the surface. Why can't you feel this force when you hold your hand up to the Sun? Because this force is tiny. The force exerted on one square kilometer of surface is only about 5 N (just over one lb_f)! So why worry about it? Over time, even this tiny force acting unevenly over different parts of the spacecraft can exert a torque that the control system must deal with.

The third source of disturbance torque is the Earth's magnetic field. As we've seen, some spacecraft use electromagnets onboard to interact with the Earth's magnetic field, causing them to rotate like a compass needle. As we learned in Chapter 3, because of the impact of charged particles in space a spacecraft can develop a charge on its own. This charge can then cause the spacecraft to be torqued in unwanted directions by the Earth's magnetic field.

Finally, there's drag. As we saw in Chapter 3, in low-Earth orbit the atmosphere applies a drag force to a vehicle, eventually causing it to enter the atmosphere and burn up. Because parts of a spacecraft may have different drag coefficients (solar panels, for example, are like big wings), drag forces on different parts of the spacecraft can also differ. This difference causes a torque. A spacecraft designer can do little to prevent this torque (short of moving the spacecraft to a higher orbit), so again the control system must deal with it.

Spacecraft-Attitude Sensors

As we've seen, an essential element of control through feedback is a device that can see what's happening to the system and report this information to the controller. In other words, we need a sensor. Sensors are the central system's "eyes and ears." They observe the system and transform these observations into signals we can process.

All of us have a built-in attitude-sensor system. As we know from our discussion of the human vestibular system in Chapter 3, we use fluid flowing over tiny hairs in our inner ear along with information from our eyes to detect changes in our physical attitude. For example, are we standing up or falling over? If you suddenly tilt your head to the side, these sensors detect this motion. If your body violently moves or shakes (like when you ride a roller coaster), your eyes and inner ear can get "out of synch," leading to motion sickness. Fortunately, spacecraft don't get sick from all their rotating, but they do need good attitude sensors. So let's take a look at a spacecraft's eyes and ears.

To understand how sensors help us decide our attitude, pretend you're flying the Space Shuttle in low-Earth orbit and need to point the nose at some spot on the surface. You're in the commander's seat facing toward the nose. To point the nose at the surface you must first determine where you're pointed now. How can you do this? The obvious answer is to look out the window at some reference. Let's say you look out and see the sun out the left side window. Would this tell you all you need to know? Unfortunately, no. A single reference point would tell you your current attitude in only two dimensions. In other words, you'd know that the left wing pointed at the Sun and the nose pointed perpendicular to the Sun. But the nose could point in various directions and still be perpendicular to the Sun, so what do you do?

To determine your attitude in three dimensions, you'd need another reference. If you could see some known star out the right-hand window you'd know your orientation with respect to two reference points—the Sun and a star. You could then determine how to change your attitude to point the nose at the Earth.

"Looking out the Window"

When pilots fly along in their airplanes, the easiest way for them to determine attitude is to look out the window (if the weather is good). If the ground is down and the sky is up, they're flying straight and level. Similarly, the simplest way for a spacecraft to determine its attitude is to just "look out the window." Some spacecraft sensors act like your eyes to "see" attitude as well as position. They include

- Earth sensors
- Sun sensors
- Star sensors

Earth sensors are similar to the light sensors used to turn on street lights automatically when it gets dark. For spacecraft in very high orbits, such as those at geostationary altitude where the angular radius of the Earth is about 10°, simply detecting the Earth's location can give accuracies of 0.25° – 0.1° [Larson & Wertz, 1992]. In low-Earth orbit (LEO), however, the Earth fills almost half the sky. For LEO applications, Earth sensors detect only the horizon by focusing on a narrow band of EM radiation emitted by CO_2 in the atmosphere, as shown in Figure 12-21. This gives them accuracies of 0.1°–1° [Larson & Wertz, 1992]. By their nature, Earth sensors can give accurate information about attitude in only two-dimensions. This means the Earth sensor can tell you something about your pitch and roll relative to the horizon, but nothing about yaw.

Sun sensors, the most widely used sensing device, are similar to Earth sensors. As the name implies, the sun sensor is simply a light sensor that measures the spacecraft's attitude with respect to the Sun's location, as seen in Figure 12-22. Sun sensors can achieve accuracies of as much as 0.003° [Larson & Wertz, 1992], but like Earth sensors, can only provide a 2-axis reference.

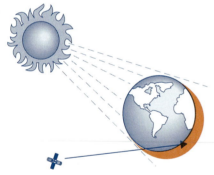

Figure 12-21. Earth Sensors. As their name implies, Earth sensors use the Earth as an attitude reference. One type used for LEO spacecraft detects the Earth's horizon.

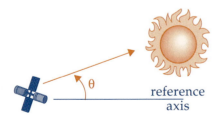

Figure 12-22. Sun Sensor. A sun sensor helps keep track of a satellite's orientation by determining the direction of the Sun.

A more accurate 2-axis reference is a *star sensor*. As Figure 12-23 shows, like the ancient mariners, it measures the spacecraft's attitude with respect to known star locations. These measurements are then compared to accurate maps of the brightest stars stored in the spacecraft's memory. The angle between the known star's position and a reference axis on the spacecraft, θ, then helps us determine the spacecraft's inertial attitude to within 0.0003° [Larson & Wertz, 1992] in some cases.

Each of these sensors provides a 2-D reference onboard. To get a 3-D reference, we often use them together. For example, an Earth sensor may work with a sun sensor to position the satellite in three dimensions. As we'll see, all of these sensors can provide periodic updates to the spacecraft's "ears"—gyroscopes and magnetometers.

Figure 12-23. Star Tracker. A star tracker keeps track of the satellite's orientation by knowing its attitude with respect to the known orientation of the stars.

Gyroscopes

Gyroscopes, like our inner ear, detect changes in attitude directly. One type of gyroscope is basically just a spinning mass. As we know, any spinning mass has angular momentum that is conserved. By using this fundamental principle, we can use the gyroscope to detect an object's angular motion. Two basic principles of gyros make them useful as attitude sensors:

- With no torques, their angular momentum is conserved—they tend to point in the same direction in inertial space
- With torque applied, they precess

Once the angular-momentum vector of a spinning mass is established, it will stay fixed in inertial space unless acted on by an outside torque. For example, let's spin up a gyroscope at 6:00 A.M. with its angular-momentum vector pointed at some convenient inertial reference—say a star just above the Eastern horizon (somewhere off to the right side of the page). We can then observe how conservation of angular momentum works to keep the gyro always pointed in the same inertial direction, as seen in Figure 12-24.

In this case, the angular-momentum vector, $\vec{H}$, will appear to "track" the star because the star is essentially fixed in inertial space. As you hold the gyro, it looks like it's rotating throughout the day. Actually, *you're* moving as the Earth rotates. The gyro is fixed in inertial space. Museums often demonstrate this principal with huge pendulums suspended on long

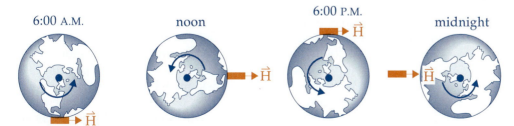

Figure 12-24. Conservation of Angular Momentum. A spinning mass, such as a gyroscope, has angular momentum which is conserved. If we spin up a gyro pointing East at 6 A.M., in this polar view of the Earth, it appears to rotate as the day goes along. Actually, the Earth-bound observer moves as the Earth rotates, but the gyro stays pointed in the same direction in inertial space.

cables. The swinging pendulum is also fixed in inertial space, but as the Earth turns, the pendulum appears to move, knocking over dominos spaced around it.

The second basic principle of gyroscopes relates to their strange motion in response to an applied torque. Earlier, we called this movement precession—rotation with constant angular velocity in a direction 90° away from the direction of the applied torque.

Knowing these two basic principles, you can now see how a gyro can sense attitude. Because its angular-momentum vector stays fixed in inertial space, it will provide a constant reference for inertial direction. One way to do this is to isolate the gyro from torques by mounting it on a gimbal. We then mount the gimbal in a spacecraft and measure the spacecraft's rotation by watching the gimballed gyro's movement. Because the angular-momentum vector is fixed, it will look to us as if the gyro is rotating about the gimbal axis. Actually, the gyro is fixed, and the spacecraft is rotating.

Another way to measure the spacecraft's rotation is to strap the gyro directly to the spacecraft. Then, if the gyro (or the spacecraft which the gyro is in) rotates around an axis perpendicular to the spin vector, a resulting torque will cause the gyro to precess. By measuring this precession rate, we can compute the amount and direction of the spacecraft's rotation and thus determine its new attitude.

Newer types of gyroscopes, called *ring-laser gyroscopes*, don't use these principles of a spinning mass. They use different principles associated with spinning light! A ring-laser gyro consists of a cavity containing a closed path through which two laser beams shine in opposite directions (it's all done with mirrors). As the vehicle rotates, the path lengths travelled by the two beams change, causing a slight change in the frequency of both beams. By measuring this frequency shift, we can compute the vehicle's rate of rotation. By integrating this rate over time, we can then determine the total amount of rotation and hence the vehicle's new orientation. Ring-laser technology promises similar accuracies with greater reliability than the old spinning-mass gyros.

Magnetometers

Another, less accurate means of measuring attitude directly uses knowledge of the Earth's magnetic field. A *magnetometer* is just a fancy magnet which measures not only the direction of the magnetic field like a compass but also its strength. Using an elaborate model of the Earth's magnetic field, the sensor can then determine the orientation of the spacecraft with respect to the Earth. The biggest limitation of magnetometers is the accuracy of the field model, which limits the overall accuracy of the sensor to 0.5°–3° [Larson & Wertz, 1992]. Furthermore, because the magnetic field gets weaker with altitude, magnetometers are useful only on spacecraft in low-Earth orbit.

The Controller

So far, we've looked at the behavior of the system, or plant, to understand how torque affects the spacecraft's angular momentum. We then looked at types of actuators used to generate torques and sources of disturbance torques. Finally, we discussed various sensors used to measure attitude. All that's left is the controller.

The controller's job is to generate commands to the actuators in order to meet the mission designers' requirements for accuracy and slew. To use the information from sensors and continuously adjust commands, the controller must be smart. It has to keep track of what's happening and decide what to do next. To do this right, the controller has to know

- What's happening
- What may happen in the future
- What happened in the past

Knowing what's happening is pretty easy—the controller simply asks the sensor. It then compares this attitude to the desired attitude. The difference between measured and desired attitude is the *error signal*. Based on this error signal, the controller steers in the direction of the proper orientation. That is, if the attitude is 10° off, the controller commands a 10° change. This is known as *proportional control* and is used in some form in virtually all closed-loop control systems.

However, predicting what's going to happen and remembering what's happened can be just as important. For example, if you need to stop at a stop sign, you need to know not only where you are but also how fast you're going, so you can hit the brakes in time. Similarly, to hit the desired attitude, the spacecraft controller must monitor attitude rate as well as current attitude. For you calculus buffs, you may recognize this calculation of rate of change as a derivative. In this case, by knowing the rate of change or "speed" of attitude, the controller can better determine how to command the actuators to achieve better accuracy. This process is called *derivative control*.

Sometimes we can be even more precise by keeping track of how close we're getting to the desired result. One way to do this is for the controller to monitor the difference between the measured and desired attitude. Once the desired attitude is reached, this difference, $\Delta\theta$, will be zero. If the system stopped commanding the actuators at this point, the attitude would immediately begin to drift off due to disturbance torques. A really smart controller, however, would not just look at the instantaneous $\Delta\theta$. Instead, it would keep a running tally, summing the $\Delta\theta$ over time. The result would always be some value other than zero and would tell the controller how much torque to add in a "steady-state" mode to compensate for the disturbance torques. In calculus this process is called integration, so we call this *integral control*. This control algorithm is often used for highly accurate pointing.

Regardless of the exact scheme used, the controller combines its own memory with its current knowledge and an ability to predict future behavior to decide how to command the actuators. We can now complete our block diagram of a spacecraft's attitude-control system in Figure 12-25.

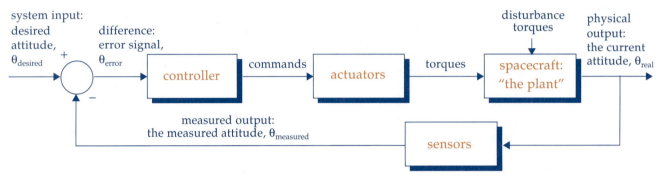

Figure 12-25. System for Controlling Spacecraft Attitude. This system includes a controller, actuators, a spacecraft, and sensors as part of a feedback loop.

Other Attitude-Control Strategies

Besides the active, closed-loop control strategies we just discussed, we can sometimes use other passive, open-loop strategies. Once we start these strategies, they continue with little intervention from the control system. They take advantage of our knowledge of the environment and the system's dynamics. We'll look at

- Spin stabilization
- Gravity-gradient stabilization

Spin Stabilization

Figure 12-26. Fixed Angular-Momentum Vector. Once a spacecraft is spinning, its angular-momentum vector keeps it fixed in inertial space.

Earlier we saw that a spinning mass has unique gyroscopic properties. *Spin-stabilized* spacecraft take advantage of the conservation of angular momentum to maintain the orientation of one axis of the spacecraft. Because the angular-momentum vector, $\vec{H}$, of a spinning mass is fixed in inertial space, the spacecraft will always stay in the same inertial attitude, as shown in Figure 12-26. We can use this fact to achieve inertial pointing accuracies of 0.1°–1° in two axes. For this reason, we often use spin stabilization for deployed satellites. You may have seen a satellite spinning as it leaves the payload bay of the Space Shuttle. This spin gives it enough pointing accuracy to perform the ΔV it needs to transfer to its final orbit. Figure 12-27 shows the direction of the angular-momentum vector.

Perhaps the best example of a spin-stabilized satellite is the Earth itself. The spinning Earth is essentially a giant gyroscope. The Earth's $\vec{H}$ vector points directly out of the North Pole. This $\vec{H}$ stays fixed in inertial space, always pointed at the same place in the sky. When we observe the motion of the stars throughout the night, we see they all appear to rotate about one lone star—The North Star. This is because Earth's $\vec{H}$ vector points at the North Star!

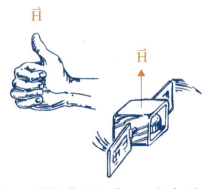

Figure 12-27. Spinning Spacecraft. A spinning spacecraft has angular momentum whose direction is found using the right-hand rule.

While spin-stabilized spacecraft offer a relatively cheap and simple way to control attitude, they are also limited in two important ways.

- We can't use them to point antennas or sensors at the planets they orbit. Imagine if we placed a camera on the spacecraft's spin axis. As you can see in Figure 12-28, during part of the orbit it could point at the Earth, but during the rest of the orbit it will point either tangent to the Earth's surface or directly away from it.

- Because of precession, to change the direction of the spin axis we must apply a torque 90° away from the desired direction of spin.

As with the spin-stabilized system, the *dual-spin system* also uses the fixed angular-momentum vector of a spinning mass. But the dual spin has an actively controlled "de-spun" section which allows it to get around some of the limitations of a conventional spin-stabilized system. Remember that a spin-stabilized system can't keep sensors or antennas on the spacecraft pointed toward the Earth. To solve this problem, dual-spin systems have an inner "de-spun" section. "De-spun" is actually a misnomer. In fact, the "de-spun" section does spin, but at a much slower rate than the outer spinning section.

To allow for pointing of antennas and other sensors, this "de-spun" section spins at exactly the right rate to keep the same axis pointed at Earth. For example, if a spacecraft is in geostationary orbit, the despun platform rotates at "orbit rate" or once every 24 hours, keeping antennas or other sensors focused on the Earth, as shown in Figure 12-29.

The "de-spun" section, of course, makes the spacecraft much more complex. Electrical and other connections must run from the spun to the "de-spun" sections. Highly reliable bearings must allow the two sections to spin at different rates with little friction. Even with these inherent technical challenges, however, dual-spin spacecraft are much more

Figure 12-28. Spin-Stabilized Spacecraft Aren't Good for Earth Pointing. Because spin-stabilized spacecraft stay fixed in inertial space, they're not a good choice for Earth-pointing missions. During part of the orbit they may point toward Earth, but during another part they'll be pointed away.

inner, despun section

outer, spun section

dual-spin spacecraft

spun section provides "stiffness," despun section stays pointed at Earth

Figure 12-29. Dual-Spin Spacecraft. A dual-spin spacecraft uses the inherent "stiffness" of a spinning outer section along with a "de-spun" inner section. The de-spun section actually spins at orbit rate to allow antenna or other sensors to remain pointed at the Earth.

accurate in pointing than passive systems. Accuracies of 1° to 0.1° are common.

Gravity-Gradient Stabilization

Another type of control strategy takes advantage of the gravity-gradient disturbance torque discussed earlier. We can use this torque to keep the spacecraft oriented in a "down" position. A spacecraft doesn't have to be shaped like a dumbbell to take advantage of this effect. Why do we see only one face of the Moon and never the mysterious "dark side"? Because of uneven distribution of mass within the Moon's crust, it's in a gravity-gradient-stabilized attitude with respect to the Earth. To take full advantage of this cheap and reliable way to control attitude, spacecraft will sometimes deploy weighted booms to create a more dumbbell-like shape, as seen in Figure 12-30.

Gravity-gradient control of attitude has drawbacks:

- It controls only two axes—pitch and roll, but not the yaw axis. For some spacecraft, this is no problem, but for others it can seriously affect the mission.

- It isn't very accurate—about ±5° for pointing. That's not enough for optical remote-sensing missions and other missions requiring a steady aim.

gravity-gradient boom

Figure 12-30. Gravity-Gradient Stabilization. Some spacecraft use the gravity-gradient disturbance torque to their advantage. By deploying a long boom they can amplify this effect and keep the spacecraft oriented in a down position.

Astro Fun Fact
The Earth is a Big Gyroscope

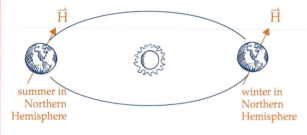

summer in Northern Hemisphere

winter in Northern Hemisphere

The spinning Earth is basically one big gyroscope. How does this affect us? Seasons. As the Earth goes around the Sun, it always stays in the same inertial attitude, which happens to be canted 23.5° with respect to the ecliptic plane. As a result, part of the Earth's orbit has the Northern Hemisphere tilted toward the Sun (summer in the Northern Hemisphere), and the other part of the orbit has the Southern Hemisphere pointed toward the Sun (winter in the Northern Hemisphere). This also accounts for those two obscure lines on a map of the Earth—the Tropic of Cancer at 23.5° N and the Tropic of Capricorn at 23.5° S. These lines represent the Northernmost and Southernmost points reached by the Sun as we travel around in our orbit

▰ Section Review

Key Terms

attitude
attitude accuracy
control moment gyroscope (CMG)
derivative control
disturbance torques
dual-spin system
Earth sensors
error signal
gimbal
gravity-gradient
gyroscopes
integral control
magnetic torquers
magnetometer
mass moment of inertia, I
momentum dumping
momentum wheel
momentum-control devices
parallel
photons
pitch
pointing
precession
proportional control
reaction wheels
ring-laser gyroscopes
roll
slew
slew rate
solar-radiation pressure
spin stabilized
star sensor
Sun sensors
thrusters
torque
yaw

Key Equations

$$\vec{H} = I\vec{\Omega}$$

$$\vec{T} = \vec{R} \times \vec{F}$$

$$\vec{T} = \dot{\vec{H}} = I\vec{\alpha}$$

Key Concepts

➤ To understand a spacecraft's behavior or how it reacts to inputs, we must understand the model of system dynamics based on linear and rotational laws of motion. To rotate a spacecraft we must recognize that
- Angular momentum is always conserved
- A torque describes the direction of a force couple applied to a system
- A torque applied to a non-spinning object (or applied parallel to the direction of spin for a spinning object) will cause angular acceleration, which leads to angular velocity and hence, change in angular orientation.
- A torque applied in a direction other than the direction of spin for a spinning object will cause precession. This means it will begin to rotate at constant angular velocity about an axis perpendicular to the torque direction.

➤ Applying torques to the spacecraft requires spacecraft actuators, including
- Thrusters
- Magnetic torquers
- Momentum devices

➤ A spacecraft experiences many disturbance torques which try to change its attitude, including
- Gravity gradient   • Solar-radiation pressure
- Magnetic   • Atmospheric drag

➤ Sensors help us observe a spacecraft's attitude.
- Sun sensors, horizon sensors, and star trackers are the "eyes" of the spacecraft. They look out the window to determine attitude and position.
- Gyroscopes are the "inner ear" of the spacecraft. They can directly sense changes in attitude because a spinning mass has two important properties:
  - The angular momentum of a spinning mass is constant
  - Torque applied to a spinning mass will cause precession
- Ring-laser gyros use the change in frequency of laser light between two loops of fiber optic material
- Magnetometers measure the direction and magnitude of Earth's magnetic field to determine attitude.

➤ The spacecraft controller decides what to do based on current and historical data from sensors and an understanding of spacecraft behavior.

➤ Other control strategies include
- Spin- or dual-spin stabilization
- Gravity-gradient stabilization

12.3 Navigation, Guidance, and Control Systems

▬ In This Section You'll Learn to...

- ☞ Describe what the navigation, guidance, and control (NGC) system does
- ☞ Explain the basic principles governing vehicle translation
- ☞ Describe the actuators NGC systems use
- ☞ Explain the sensors NGC systems use
- ☞ Draw the block diagram for an inertial-navigation system and explain each component
- ☞ Explain the logic used by the NGC controller to achieve a desired burnout velocity
- ☞ Draw the block diagram for an NGC system and explain each component
- ☞ Determine the additional velocity needed to get a spacecraft from its current velocity to its final velocity

What is Navigation, Guidance, and Control?

The spacecraft's *navigation, guidance, and control (NGC) subsystem* tries to control the spacecraft's position and velocity. It uses the same four steps we've seen for a home-heating system and for attitude control. However, we usually consider the complex subsystems that accomplish these steps separate systems. Recall the four steps for control are to

- *Understand* system behavior—system dynamics
- *Observe* the system's current behavior—sensors
- *Decide* what to do—controllers
- *Do* it—actuators

For NGC in space, the "sensor" is the navigation system. A *navigation system* uses various sensors to determine the spacecraft's current position and velocity. The *guidance system* is the controller, which commands velocity changes to reach the correct position and velocity.

Moving the Spacecraft

The basic principles that explain how to move a spacecraft once again stem from Newton's Laws of Motion. If we apply a force in a certain direction, we accelerate. If we accelerate for some time, we achieve some velocity. If we maintain this velocity for a certain time, we reach some position. That's the basis for modeling the dynamics of the NGC system. The controller knows that to achieve a certain velocity in a certain direction, the vehicle must thrust in the opposite direction. This process is shown in Figure 12-31.

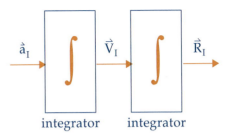

Figure 12-31. Dynamics of an NGC System. If we accelerate for some time, we'll reach some velocity. If we move at some velocity over time, we reach some position. Mathematically, this means we integrate acceleration to get velocity and integrate velocity to get position.

Rockets—The Actuators

The booster system moves to its desired position and velocity using actuators—the rocket engines themselves. We can throttle these engines to change their thrust or use *gimbals* to move them around and turn the direction of thrust. In some cases, boosters use small steering rockets to re-orient themselves or gas injection to turn the thrust direction. But in all cases, the effect is the same—change the amount or direction of thrust.

Navigation—The Sensor

The sensor for a booster-control system is actually a series of sensors working in concert as part of an inertial-navigation system. Remember, we're mainly interested in our velocity vector, $\vec{V}$, and our position vector, $\vec{R}$, with respect to some inertial reference. How does the navigation system find this? It uses sensors to measure acceleration. By applying the basic principles from Newton's laws described above and shown in Figure 12-31, it then "works backward" (integrates) to get velocity and position. Now all we need is some way to measure inertial acceleration. We do this using an accelerometer.

When you stomp on the accelerator in your car, a *contact force* applies to you and the car and throws you back into your seat. Your inner ear detects this contact force as tiny particles of calcium moving past tiny hairs. We use this same action that causes the calcium particles to move in your inner ear to detect acceleration onboard spacecraft and airplanes. A device which can sense acceleration due to contact forces is an *accelerometer.* Basically, an accelerometer is like a mass suspended between two springs. If you suddenly accelerate the box containing the mass and springs by applying a contact force, the springs on one side will compress as the inertia of the mass resists the acceleration. This principle is described in Figure 12-32.

Notice we said that an accelerometer measures only contact forces. But gravity also acts on the vehicle. How do we measure it? We can't. Because gravity acts on all bodies, and we can't "shield" a mass from its effect, we can't construct an instrument to sense gravity directly! Fortunately, armed with Newton's good ol' law of gravitation, shown in Figure 12-33, we can

accelerometer

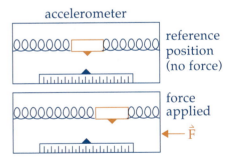

reference position (no force)

force applied

Figure 12-32. Accelerometers. You experience the basic principle of an accelerometer every day. When you push on the accelerator in your car, you feel yourself pushed back into the seat. Accelerometers consist of masses which are likewise displaced when subjected to a contact force. By measuring this displacement, we can find the applied force and hence the acceleration of the vehicle.

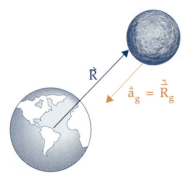

Figure 12-33. Gravitational Acceleration. Gravitational acceleration, $\vec{a}_g = \ddot{\vec{R}}_g$, works "downward," that is, toward the center of the Earth.

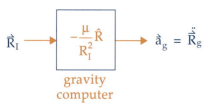

gravity
computer

Figure 12-34. Finding Gravitational Acceleration. We can find the gravitational acceleration, $\vec{a}_g = \ddot{\vec{R}}_g$, by knowing our position, $\vec{R}_1$, and then using Newton's law of universal gravitation as a "gravity computer" to figure it out.

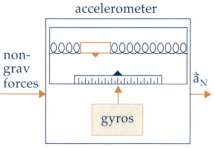

Figure 12-35. Accelerometers and Gyroscopes. To determine the direction as well as the magnitude of the non-gravitational acceleration, $\vec{a}_N$, we use accelerometers with gyroscopes.

compute it. Recall from Chapter 4 that we can find acceleration due to gravity from

$$\vec{a}_g = \ddot{\vec{R}}_g = -\frac{\mu}{R^2}\hat{R}$$

(12-9)

where:

$\vec{a}_g = \ddot{\vec{R}}_g$ = acceleration due to gravity (m/s²)
μ = gravitational parameter of central body (m³/s²)
R = magnitude of position vector (m)
$\hat{R}$ = unit vector

So if we have a "gravity computer," as shown in Figure 12-34, we can compute $\vec{a}_g$ as long as we know the inertial position ($\vec{R}_I$). But wait a minute—that's what we were looking for in the first place! Are we going in circles? Actually, we are, but we'll see how it all works out in just a bit.

We now have the gravitational forces computed from Newton's Law (the "gravs") and the contact, or non-gravitational forces ("grav-nots") measured directly by the accelerometer. Because these constitute all possible accelerations on the vehicle, the total inertial acceleration is the vector sum of the two.

$$\vec{a}_I = \vec{a}_g + \vec{a}_N$$

(12-10)

where:

$\vec{a}_I$ = inertial acceleration (m/s²)
$\vec{a}_g$ = acceleration due to gravity (m/s²)
$\vec{a}_N$ = acceleration due to non-gravitational forces (m/s²) (e.g., lift, drag, thrust, etc.)

Note that this is a vector equation. Thus, the accelerometer needs to tell us the direction and magnitude of the acceleration caused by contact forces. This means we must also know the attitude of the accelerometer which measures the acceleration. To find attitude, we need an attitude sensor. Earlier we discussed using gyroscopes as attitude sensors. Armed with both an accelerometer and a gyroscope, we can then determine $\vec{a}_N$, as shown in Figure 12-35.

Now that we have both $\vec{a}_g$ and $\vec{a}_N$, we can put it all together to get $\vec{a}_I$. We do this by building a little "position and velocity computer," or *inertial navigation system*, which, as we said literally goes in circles as part of a feedback-control system. All we need are the initial conditions of the vehicle (say, at lift-off), and the navigation system does the rest. The elements of an inertial navigation system are shown Figure 12-36.

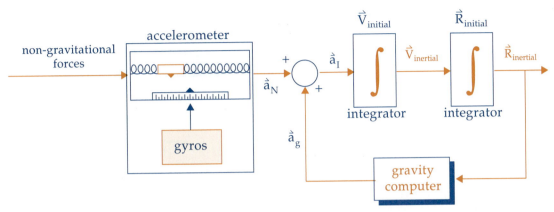

Figure 12-36. Inertial-Navigation System. The inertial-navigation system is the primary "sensor" of the booster-control system. It uses accelerometers, gyros, and a knowledge of Newton's law of gravitation to determine the current inertial position and velocity.

Guidance—The Controller

The NGC system controller knows the velocity it *wants* to achieve—specified by $\vec{V}_{desired}$. Next, it knows its *current* velocity by asking the navigation system to get $\vec{V}_{current}$ (current velocity with respect to an inertial reference frame). The controller must use these two velocities to compute commands that go to the actuators (engines and gimbals) and steer the vehicle in the right direction with the right amount of thrust. Thus, the controller must compute the additional velocity needed $(\Delta \vec{V}_{needed})$ to achieve the desired velocity. Figure 12-37 shows how we can find $\Delta \vec{V}_{needed}$ by subtracting the other two velocity vectors.

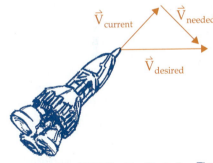

$$\Delta \vec{V}_{needed} = \vec{V}_{desired} - \vec{V}_{current} \qquad (12\text{-}11)$$

where

$\Delta \vec{V}_{needed}$ = velocity change needed to reach desired velocity vector (m/s)

$\vec{V}_{desired}$ = desired velocity vector (m/s)

$\vec{V}_{current}$ = spacecraft's current velocity vector (m/s)

This gives the controller a simple algorithm to use. (More complicated schemes give somewhat more accurate results.) Figure 12-38 shows the entire process for controlling the spacecraft's position and velocity. In Example 12-1, you can see how this whole system works together to determine the velocity the Space Shuttle needs to add during the final phase of powered flight in order to achieve the desired burnout velocity.

Figure 12-37. NGC System Controller. The booster-system controller subtracts the velocity it has from the velocity it wants. It then determines the velocity it needs to get to the desired burnout conditions. Finally, it computes the required steering commands and sends them to the actuators to steer the booster in that direction.

desired output:

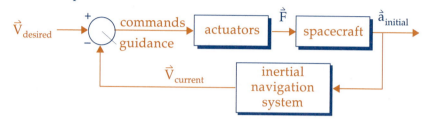

Figure 12-38. Controlling the Vehicle's Position and Velocity. The block diagram shows all elements of the NGC system. Given the desired velocity, $\vec{V}_{desired}$, guidance acts as a controller. It compares this velocity to the inertial velocity from the inertial-navigation system, which acts as a sensor and sends commands to the engines or actuators, which apply a force to the spacecraft.

Astro Fun Fact

Accelerometers and Seat Belts

Most of us use accelerometers every day and we don't even know it! The most common place you come in contact with accelerometers is the seat belt in your automobile. If you've been in an accident, you know that the seat belt tightens up instantly and pulls you into the car seat upon impact. The accelerometer in the seat belt mechanism detects the rapid deceleration and causes the seat belt to tighten. Two types of accelerometers are in most cars. The first is an electromechanical sensor that is a metal ball held in place by a magnet. An impact interrupts the magnetic pull, releases the metal ball, and activates the seat belt. The second is electric. It uses a glass rod that moves horizontally and, when enough deceleration is placed on the system, emits a radio frequency signal which activates the seat belt. Accelerometers in vehicles protect you every time you're in a car.

Adler, U. <u>Automotive Handbook</u>. Stuggart: Robert Bosch GmbH, p. 628–629, 1986.

Contributed by Michael A. Banks, United States Air Force Academy

Section Review

Key Terms

accelerometer
contact force
gimbals
guidance system
inertial navigation system
navigation, guidance, and
    control (NGC) subsystem
navigation system

Key Equations

$$\vec{a}_g = \ddot{\vec{R}}_g = -\frac{\mu}{R^2}\hat{R}$$

$$\vec{a}_I = \vec{a}_g + \vec{a}_N$$

$$\Delta\vec{V}_{needed} = \vec{V}_{desired} - \vec{V}_{current}$$

Key Concepts

➤ The navigation, guidance, and control (NGC) system maintains and changes a vehicle's position and velocity. As with all control systems, it must
  • Understand system behavior—system dynamics
  • Observe the system's current behavior—sensors
  • Decide what to do—controller
  • Do it—actuators

➤ The navigation system is the "sensor." It uses various other sensors to determine current position and velocity

➤ The guidance system is the controller, deciding what actions to take to change position and velocity

➤ To understand the dynamics of a booster system, we must understand Newton's Laws of Motion and know that a force applied over time will cause an acceleration. An acceleration over time causes velocity, and velocity over time changes position.

➤ a booster-system sensor is actually a collection of sensors, including accelerometers and gyroscopes. They combine sensing with Newton's Law of Gravitation to form an inertial-navigation system.

➤ Booster-system actuators change the direction or amount of engine thrust, either by throttling back or physically moving on gimbals

➤ The booster-system controller compares the vehicle's current inertial velocity to the desired velocity in order to compute the velocity it needs to reach burnout conditions. Using this value, it computes and sends commands to the actuators (rocket engines) to achieve the velocity change (magnitude and direction) to get to the desired velocity.

Example 12-1

Problem Statement

The Space Shuttle has launched due East from the Kennedy Space Center (28.5° N, 75° W). Mission designers want the shuttle to achieve a velocity at main engine cut off (MECO) of 7,230 m/s, with a flight-path angle of 10°. The azimuth at MECO should be the same as the launch azimuth. During the last navigation cycle four seconds ago, the navigation system placed the Shuttle at an altitude of 120 km with the following velocity vector in the topocentric-horizon (SEZ) frame—V_Z = 946.1 m/s, V_E = 6950.2 m/s, V_S = –45.7 m/s. During the four seconds since the last navigation update, the inertial measurement units detected the following acceleration: 2.4 g's in the East direction, 1.2 g's in the zenith direction, and –0.4 g's in the South direction. What are the three components of the velocity-to-go vector ($\vec{V}_{desired}$) that the guidance system should compute so the Shuttle can achieve the desired burnout conditions?

Problem Diagram

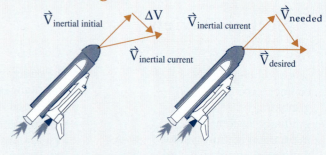

Conceptual Solution

1) Determine desired velocity at burnout, $\vec{V}_{BO}$, using the method from Chapter 9

$$V_{BO_E} = V_{BO}\cos\phi_{BO}\sin\beta_{BO}$$

$$V_{BO_S} = -V_{BO}\cos\phi_{BO}\cos\beta_{BO}$$

$$V_{BO_Z} = V_{BO}\sin\phi_{BO}$$

2) Find local acceleration due to gravity, $\vec{a}_g$

$$\vec{a}_g = -\frac{\mu}{R^2}\hat{R} \equiv \frac{-\mu}{R^2}\hat{z}$$

3) Compute the non-gravitational acceleration, $\vec{a}_n$, over the navigation cycle in m/s²

$$a_{N_E} = (g's\ east)\left(9.798\ \frac{m/s^2}{g}\right)$$

$$a_{N_Z} = (g's\ zenith)\left(9.798\ \frac{m/s^2}{g}\right)$$

$$a_{N_S} = (g's\ south)\left(9.798\ \frac{m/s^2}{g}\right)$$

4) Compute the inertial acceleration, $\vec{a}_I$, during the navigation cycle

$$\vec{a}_I = \vec{a}_g + \vec{a}_N$$

5) Find the change in velocity, $\Delta\vec{V}$, during the navigation cycle

$$\Delta\vec{V} = \vec{a}_I\Delta t$$

Problem Summary

Given: Launch site 28.5° N, 75° W

V_{BO} = 7230 m/s

ϕ_{BO} = 10°

β_{BO} = 90°

Δt = 4 s

Altitude = 120 km

$V_{Z_{initial}}$ = 946.1 m/s

$V_{E_{initial}}$ = 6950.2 m/s

$V_{S_{initial}}$ = –45.7 m/s

Acceleration$_Z$ = 1.2 g's

Acceleration$_E$ = 2.4 g's

Acceleration$_S$ = –0.4 g's

Find: $\vec{V}_{desired}$

Example 12-1 Continued

6) Determine the current inertial-velocity vector, $\vec{V}_{current}$

$$\vec{V}_{current} = \vec{V}_{initial} + \Delta\vec{V}$$

7) Determine $\Delta\vec{V}_{needed}$

$$\Delta\vec{V}_{needed} = \vec{V}_{BO} - \vec{V}_{current}$$

Analytical Solution

1) Determine $\vec{V}_{BO}$

$$V_{BO_E} = V_{BO}\cos\phi_{BO}\sin\beta_{BO}$$
$$= (7230 \text{ m/s})\cos(10°)\sin(90°)$$
$$= 7120.2 \text{ m/s}$$
$$V_{BO_S} = -V_{BO}\cos\phi_{BO}\cos\beta_{BO}$$
$$= -(7230 \text{ m/s})\cos(10°)\cos(90°)$$
$$= 0 \text{ m/s}$$
$$V_{BO_Z} = V_{BO}\sin\phi_{BO}$$
$$= (7230 \text{ m/s})\sin(10°)$$
$$= 1255.5 \text{ m/s}$$

2) Find the local gravitational acceleration

$$\vec{a}_g = -\frac{\mu}{R^2}\hat{z}$$

$$\vec{a}_g = -\frac{\left(3.986\times10^5 \dfrac{\text{km}^3}{\text{s}^2}\right)}{(6378\text{km} + 120\text{km})^2}\hat{z}$$

$$= -9.44\times10^{-3} \frac{\text{km}}{\text{s}^2}\hat{z}$$

$$= -9.44 \text{ m/s}^2 \, \hat{z}$$

3) Compute the non-gravitational acceleration, $\vec{a}_n$, over the navigation cycle

$$a_{n_E} = (2.4 \text{ g's})\left(9.798 \frac{\text{m/s}^2}{\text{g}}\right)$$
$$= 23.52 \text{ m/s}^2$$

$$a_{n_Z} = (1.2 \text{ g's})\left(9.798 \frac{\text{m/s}^2}{\text{g}}\right)$$
$$= 11.76 \text{ m/s}^2$$

$$a_{n_S} = (-0.4 \text{ g's})\left(9.798 \frac{\text{m/s}^2}{\text{g}}\right)$$
$$= -3.92 \text{ m/s}^2$$

4) Compute the inertial acceleration, $\vec{a}_I$, during the navigation cycle

$$\vec{a}_I = \vec{a}_g + \vec{a}_n$$

$$a_{I_E} = 0 + 23.52 \text{m/s}^2$$
$$= 23.52 \text{ m/s}^2$$
$$a_{I_Z} = -9.44 \text{ m/s}^2 + 11.76 \text{ m/s}^2$$
$$= 2.32 \text{ m/s}^2$$
$$a_{I_S} = 0 - 3.92 \text{ m/s}^2$$
$$= -3.92 \text{ m/s}^2$$

5) Find the change in velocity, $\Delta\vec{V}$, during the four-second navigation cycle

$$\Delta\vec{V} = \vec{a}_I\Delta t$$
$$\Delta V_E = (23.52 \text{ m/s}^2)(4 \text{ s})$$
$$= 94.08 \text{ m/s}$$
$$\Delta V_Z = (2.32 \text{ m/s}^2)(4 \text{ s})$$
$$= 9.28 \text{ m/s}$$
$$\Delta V_S = (-3.92 \text{ m/s}^2)(4 \text{ s})$$
$$= -15.68 \text{ m/s}$$

6) Determine the current velocity vector, $\vec{V}_{I\ current}$

$$\vec{V}_{current} = \vec{V}_{I\ initial} + \Delta\vec{V}$$
$$V_{current_E} = (6950.2 \text{ m/s}) + 94.08$$
$$= 7044.28 \text{ m/s}$$
$$V_{current_Z} = (946.1 \text{ m/s}) + (9.28 \text{ m/s})$$
$$= 955.38 \text{ m/s}$$
$$V_{current_S} = (-45.7 \text{ m/s}) + (-15.68 \text{ m/s})$$
$$= -61.38 \text{ m/s}$$

7) Determine $\Delta\vec{V}_{needed}$

$$\Delta\vec{V}_{needed} = \vec{V}_{BO} - \vec{V}_{I\ current}$$
$$\Delta V_{needed_E} = (7120.2 \text{ m/s}) - (7044.28 \text{ m/s})$$
$$= 75.92 \text{ m/s}$$
$$\Delta V_{needed_Z} = (1255.5 \text{ m/s}) - (955.38 \text{ m/s})$$
$$= 300.12 \text{ m/s}$$
$$\Delta V_{needed_S} = (0) - (-61.38 \text{ m/s})$$

Example 12-1 Continued

$$= 61.38 \text{ m/s}^2$$

$$\Delta V_{needed} = \sqrt{V^2_{desired_E} + V^2_{desired_Z} + V^2_{desired_S}}$$

$$= \sqrt{(75.92)^2 + (300.12)^2 + (61.38)^2}$$

$$= 315.60 \text{ m/s}$$

Interpreting the Results

The Shuttle is slightly off course and needs to make some final adjustments to its velocity to achieve the desired burnout conditions. It needs to gain an additional 75.92 m/s in the downrange or East direction and 300.12 m/s in the up or vertical direction to get the correct flight-path angle at MECO. It also needs to correct a slight Northerly motion of 61.38 m/s by steering South.

References

Asimov, Isaac. *Asimov's Biographical Encyclopedia of Science and Technology*. Garden City, NJ: Doubleday and Company, Inc., 1972.

Chetty, P.R.K. *Satellite Power Systems: Energy Conversion, Energy Storage, and Electronic Power Processing*. George Washington University Short Course 1507. October 1991.

Chetty, P.R.K. *Satellite Technology and Its Applications*. THB Professional and Reference Books, New York, NY: McGraw-Hill, Inc., 1991.

Gere, James M., Stephen P. Timoshenko, *Mechanics of Materials*, Boston, MA: PWS Publishers, 1984.

Gonick, Larry and Art Huffman. *The Cartoon Guide to Physics*. New York, NY: Harper Perennial, 1991.

Gordon, J.E., *Structures: Why Things Don't Fall Down*, New York, NY: Da Capp Press, Inc., 1978.

Holman, J.P. *Thermodynamics*. New York, NY: McGraw-Hill Book Co., 1980.

Pitts, Donald R. and Leighton E. Sissom. *Heat Transfer*. Schaum's Outline Series. New York, NY: McGraw-Hill, Inc., 1977.

Wertz, James R. and Wiley J. Larson. *Space Mission Analysis and Design*. Dordrecht, Netherlands: Kluwer Academy Publishers, 1991.

Wertz, James R. [ed.] *Spacecraft Attitude Determination and Control*. Netherlands: D. Reidel Publishing Co., Kluwer Group, 1986.

Mission Problems

12.1 Introducing Controls

1   Define the four steps in control and apply them to some everyday process such as hitting a baseball with a bat.

2   How do spacecraft use control?

3   What are block diagrams and why are they useful?

4   What is a plant model? What is the plant model for the baseball and bat example?

5   What do sensors do in a control system? What sensors do you use to hit a baseball with a bat?

6   Draw a block diagram for the baseball-and-bat system. Assume you are the controller and the bat is the actuator.

12.2 Attitude Determination and Control Systems

7   Give an example of how roll, pitch, and yaw can be used to describe the attitude a remote-sensing spacecraft needs to point a camera at a target on the Earth.

8   What is the difference between attitude accuracy, slew, and slew rate?

9   A spacecraft is not spinning. If a 2.0 N thruster is fired 1.0 m from the spacecraft's center of mass, how much will the angular momentum change? If the spacecraft's moment of inertia is 1000 kg · m², what angular acceleration will the spacecraft have?

10   What are the three main attitude actuators used on spacecraft?

11   How can thrusters be used to torque a spacecraft?

12   How do magnetic torquers work?

13   What are the three momentum-based actuators and how do they work? What are their main differences?

14   Which of the spacecraft's sensors allow it to "look out the window" to determine attitude and position?

15   What unique properties of spinning masses make them useful as gyroscopic attitude sensors?

16   What is a ring-laser gyro and how is it different from a spinning mass?

17   What are disturbance torques and how do they affect a spacecraft?

18   What does a controller do and what types of things does it have to know?

19   Draw the block diagram of the complete system to control a spacecraft's attitude and discuss all of the components.

20   List the advantages and disadvantages of control strategies using gravity gradient, spin, and dual spin.

12.3 Navigation, Guidance, and Control Systems

21   Describe the dynamics of the booster-control system.

22 Describe the purpose and various parts of an inertial-navigation system.

23 How do accelerometers work? What do they measure?

24 The Space Shuttle is at an altitude of 40 km. What gravitational acceleration is it experiencing?

25 What type of actuators are used on booster systems?

26 What is the basic relationship between needed velocity, desired velocity, and current velocity used by the booster controller?

27 A Titan IV rocket has been launched due East from the Kennedy Space Center (28.5° N, 75° W). Mission designers want the booster to achieve a burnout velocity of 8,220 m/s, with a flight-path angle of 5°. The azimuth at engine cut off should be the same as the launch azimuth. During the last navigation cycle five seconds ago, the navigation system placed the booster at an altitude of 60km with the following velocity vector in the topocentric-horizon (SEZ) frame: $V_Z = 544.1$ m/s, $V_E = 436.2$ m/s, $V_S = -5.7$ m/s. During the five seconds since the last navigation update, the inertial measurement units detected the following acceleration

- 6.4 g's in the East direction
- 4.2 g's in the Zenith direction
- –0.6 g's in the South direction

What are the three components of the velocity-to-go vector ($\vec{V}_{desired}$) that the guidance system should compute so the Shuttle can achieve the desired burnout conditions?

Mission Profile—GPS

For centuries, mariners relied on the stars to tell them their location at sea. Today, ships, airplanes, and even spacecraft can still look to the skies to determine where they are. But instead of a crude reckoning based on known star positions, we can now achieve unprecedented accuracy by looking at a man-made constellation of stars better known as NAVSTAR or the global positioning system (GPS).

Mission Overview

The NAVSTAR global positioning system is a space-based radio navigation system which, for any number of users, will provide extremely accurate position and velocity data anywhere on Earth to those equipped with GPS receivers.

Mission Data

✓ The elements of the mission architecture are

- Objective—provide world-wide navigation reference
- User—United States armed forces or anyone with a receiver
- Operations concept—a user picks up signals from four (or more) satellites (either simultaneously or sequentially) to determine three-dimensional position and velocity

✓ The receiver calculates range information by calculating the difference between the current receiver time and the time transmitted in the pulse train, and then multiplying this time by the speed of light.

✓ By using four (or more) signals, the receiver can mathematically eliminate its own clock errors (which are significant), and rely solely on the time kept onboard the satellites by atomic clocks. These clocks would lose or gain only one second every 160,000 years if they were not updated every day.

- Spacecraft—NAVSTAR GPS. The payload consists of two cesium and two rubidium atomic clocks and a communication package to broadcast time and position information to users.
- Space operations—five unmanned stations monitor the GPS block II satellites at Hawaii, Ascension Island, Diego Garcia, Kwajalein, and

Falcon AFB in Colorado. Falcon AFB also serves as the master control station.

- Booster—Delta II
- Communication network—United States Air Force's satellite-tracking and control network

Mission Impact

During Operation Desert Storm, almost every element of the coalition used GPS. Supporting everything from precision air strikes to lights-out delivery of meals to foot soldiers in the field, the GPS proved itself a true "force multiplier." Receivers were in such demand that soldiers bought civilian sets out of their own pockets so they could navigate in the featureless deserts of the Middle East. Because there's no limit to the use of GPS signals, the volume and scope of its applications are also limitless. Cars are now being equipped with "moving maps," which use GPS and digital maps stored on compact disks (CDs). GPS may soon be a household word on par with even radio and television.

For Discussion

- What other applications can you think of for GPS?
- What are its potential limitations?
- Could it be used for interplanetary or interstellar navigation? Why or why not?
- The United States government provides this service to the world basically free of charge. Is this the right thing to do?

Contributor

Kirk Emig, United States Air Force Academy

References

Logsdon, Tom. *The Navstar Global Positioning System.* New York, NY: VanNostrand Reinhold, 1992.

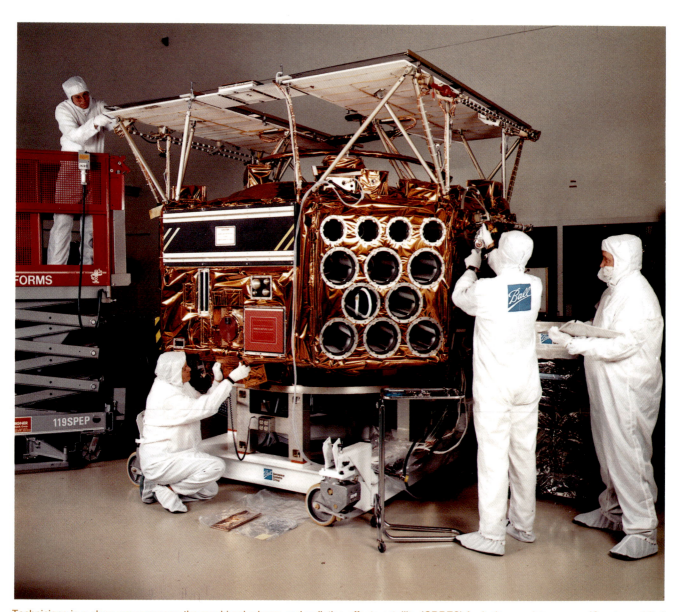

Technicians in a clean room prepare the combined release and radiation effects satellite (CRRES) for its journey into space. *(Courtesy of Ball Aerospace)*

Spacecraft Subsystems

13

▰ In This Chapter You'll Learn to...

☛ Describe the principle functions and requirements of the data handling subsystem

☛ Describe the principle functions and requirements of the electrical power subsystem

☛ Describe the principle functions and requirements of the environmental control and life support subsystem

☛ Describe the principle functions and requirements of the structure

▰ You Should Already Know...

❏ Basic functions of the spacecraft bus's subsystems (Chapter 11)

❏ Effects of the space environment on spacecraft (Chapter 3)

❏ Black-body radiation, Stefan-Boltzmann relationship (Chapter 11)

❏ Basic elements of a control system (Chapter 12)

You would make a ship sail against the winds and currents by lighting a bonfire under her deck ... I have no time for such nonsense.

*Napoleon commenting
on Fulton's steamship*

Being a spacecraft is a lot of work. You get packed into the nose cone of a rocket sitting on tons of explosives. You get blasted into orbit on a bumpy ride that subjects you to many g's. Then you get dumped into the cold vacuum of space to fend for yourself. You spend the rest of your life keeping the payload happy, supplying it with electrical power, keeping it not-too-hot and not-too-cold, pointing its sensors in the right direction, and processing its data.

In this chapter, we'll see how the spacecraft does all this. In Chapter 11, we looked briefly at what the spacecraft bus does and the design process. In the last chapter, we explored spacecraft-control subsystems. In the next chapter, we'll explore space-transportation systems. Here, we'll explore all of the bus's other subsystems in greater detail. We'll begin by seeing how the bus processes data—how computers are used to oversee what goes on in all the subsystems. Next, we'll see where all the electrical power comes from to operate the spacecraft. After that, we'll explore the environmental control and life support subsystem to see how temperatures and other environmental factors are maintained to keep machines and people happy. Finally, we'll look at what holds the whole thing together—the structure—to see what requirements it must meet. So let's look under the hood and see how a spacecraft really works.

13.1 Communications and Data Handling

In This Section You'll Learn to...

☞ Describe what the data handling subsystem does

☞ Describe basically what the data handling subsystem's needs to do its job

Back in Chapter 11 we described the communications and data handling subsystem. This subsystem acquires, stores, and transfers all information for the spacecraft. The part of this subsystem devoted to communications transfers data and commands between the spacecraft, the ground system, and other spacecraft. The part of this subsystem devoted to data handling collects payload or mission data and housekeeping data on all the subsystems and stores it until it can be transmitted to the ground. Some spacecraft combine communications and data handling into one subsystem, as we did in Chapter 11, whereas others separate them. We'll discuss communications in detail in Chapter 15 and talk about data handling here.

Nowadays, computers are everywhere. They're in your car, your microwave, and even in the human body as part of prosthetic devices and pacemakers. Spacecraft are no different. All spacecraft must have some was to process and store data. In this section we'll see that a spacecraft's data handling subsystem works just like a home computer shown in Figure 13-1.

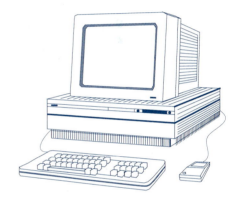

Figure 13-1. A Home Computer. A spacecraft's data handling subsystem works just like an ordinary home computer.

Functions

The data handling subsystem consists of a computer or series of computers programmed to

• Process and store payload data
• Gather and process routine housekeeping data
• Process and carry out instructions from ground controllers

The spacecraft bus exists to serve the payload, so the data-handling subsystem processes and stores payload data. For example, a remote-sensing payload may monitor levels of ozone in the upper atmosphere. In this case, instead of sending all the data directly to Earth, the data handling subsystem could filter the data to search for specific changes in ozone concentration from day to day. It would then store this much smaller amount of data and send it to Earth the next time the spacecraft passes over a ground station.

Along with supporting the payload, the data handling subsystem coordinates all the routine functions of the other subsystems. In Chapter 12, we discussed the elements of a control system. As you recall, the "brain" of a control system is the controller. In a sense, the data handling subsystem serves as the controller for the entire spacecraft. It may control position, velocity, attitude, electrical-power levels, temperature, and antenna pointing.

Finally, the data handling subsystem responds to various commands from ground controllers. These commands can be simple and routine, such as telling the system to close a valve, or they can be quite complex. In the extreme case, flight controllers may have to re-program the entire computer from the ground. When the Voyager spacecraft had trouble pointing the camera after flying by Saturn, ground controllers had to extensively modify Voyager's programs from more than three billion km (two billion miles) away to get it working again. Although these functions may sound complicated and appear to require massive, state-of-the-art computers, they're actually fairly easy to do. In fact, until recently, most spacecraft have had about the same amount of computing power as a modern washing machine! All data handling subsystems share the same basic elements, so let's see how all these elements fit together.

Background

In a basic sense, your body contains all the components of a computer. As you read, your eyes act as sensors, collecting data which is passed on to your brain. Data passing into a computer is known as *input*. Different areas of your brain do different things, such as processing visual images. As hardware, your brain is just a hunk of gray matter; fortunately, it has software that lets you access memory to define words and understand what you read. You then store this information into memory (hopefully long enough to get you through the next test) or bring it out by writing it down or talking to your friends about it. Data leaving a computer, in whatever form, is known as *output*. Let's look at what the data handling subsystem needs to carry out these same functions.

Hardware

As shown in Figure 13-2, a data handling subsystem has four main types of hardware:

- Central processing unit (CPU)
- Sensors and other devices to input data
- Memory storage units
- Actuators and other output devices

The *central processing unit (CPU)* is the "brain" of the data handling subsystem. It processes data, accesses memory, computes, makes decisions, and causes things to happen through actuators and device drivers. If you open up a typical home computer, the CPU looks like a black square box about two inches on a side with all sorts of wires running into it. But inside the CPU is a miniature metropolis of activity. Try to imagine tiny cars driving around an intricate network of streets within the CPU-city. Each car represents a bit of information. Traffic lights at every

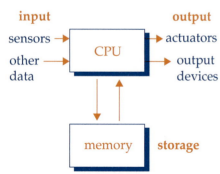

Figure 13-2. Parts of a Data Handling Subsystem. The data handling subsystem consists of a central processing unit (CPU), memory, and input or output devices.

corner control the flow of this information traffic. Each light follows an overriding set of instructions or programs contained in memory and laid down as law by the city planners. Information traffic within the CPU moves at nearly the speed of light. Every year, computer designers are able to cram more and more capability into smaller and smaller CPUs. In 1979, a state-of-the-art CPU contained 68,000 traffic lights or *transistors*. In 1993, 1.2 million transistors commonly fill the same space. CPUs with tens or even hundreds of millions of transistors will soon be available.

On a home computer, data input to the CPU comes through the keyboard and mouse. On spacecraft, data input comes through sensors tied directly to the CPU and to data lines which allow mission controllers to send information from the ground. When additional information is needed, the CPU is able to access and use memory. Memory can come in many different forms. *Read only memory (ROM)* stores fixed programs and data that won't change throughout the mission. This is non-volatile memory that remains constant even if the computer shuts off. Within a home computer, ROM stores basic instructions which allow the computer to turn on and talk to input devices.

For other long-term memory storage, spacecraft use tape drives or solid-state memory. This long-term, non-volatile memory stores large amounts of data such as specific programs or payload data. The problem with tapes is that it takes time for the CPU to access this information. Thus, computers also use *random access memory (RAM)*. Information flows from long-term storage into RAM where the CPU can access it at electronic speeds. Unfortunately, RAM usually holds much less information than tape drives or other devices for long-term memory storage. In addition, RAM is volatile memory: all information stored in RAM vanishes when power shuts off.

Astro Fun Fact
Spacecraft Computers Have Progressed— but not Far

In the early days of the space program, onboard mechanical devices—timers run by small electric motors— controlled the spacecraft. These contained reels of 35 mm mylar tape with holes punched at various intervals. As the tape passed over small brushes, the brushes made contact through the holes and turned devices on or off. The first computer that stored commands and their associated times used long pieces of wire coiled into a box called acoustic delay line devices. A motor twisted the end of the wire clockwise for a "1" and counter clockwise for a "0." The bit then traveled down the wire to a "receiver" at the other end, where it was amplified and fed back to the beginning. Capacity of this digital storage device was limited by the length of the wire and how fast the bit traveled. We then progressed to normal computers on spacecraft, but they have always been behind the state-of-the-art. Current spacecraft contain computers that are about ten years behind technology. This is because people are very reluctant to put risky stuff into orbit, and new technology is risky. It's hard to fix a computer when it's orbiting 300 km above us.

Jackson R. Ferguson, United States Air Force Academy

A home computer puts out information through the monitor or printer, but a spacecraft has several ways to do so. When the CPU acts as a controller, it sends commands to fire rockets or spin reaction wheels. These actuators are carrying out the computer's commands. In other cases, output is information about payload data or subsystem health and status which is transmitted back to Earth using the communication network.

Software

Hardware is useless without specialized software to carry out instructions. *Operating-system software* runs the entire computer and schedules the running of all application software. On home computers, this is the basic set of instructions which allow the computer to start up and run. *Application software* handles specific tasks such as word processing or spreadsheets. On a spacecraft, for example, a separate application program may run soon after deployment to spin-up the spacecraft or extend antennas. Other programs control sensors, collect data, and so on.

Bits, Bytes, and Throughput

Computers think in binary, that is, in terms of 1s and 0s. Each 1 or 0 is called a *bit*. Bits of data or instructions can be combined in groups of eight to form *bytes*. Depending on the computer, 8, 16, 32, etc. bits define a *word*. Thus, for a computer using 32-bit words, one word is four bytes. Within the computer, a clock runs to determine how fast data moves. Each time the computer executes one instruction, it completes one cycle. Thus, we can determine how fast all these bits, bytes, and words are processed. We measure this *throughput* or *processing speed* in cycles per second (or Hertz, Hz) and sometimes call it "clock speed" on home computers. Computers also store data. Their storage capacity depends on how many bits must be remembered for later.

To see all this in action, let's take a simple task like typing a letter and convert all the required actions into terms a computer could understand. If you could type 100 words per minute, that would be about 600 characters per minute (average of six letters per word). On a typical keyboard, 256 unique characters are possible (capital, lower case, &, $, etc.). This means we need 1 byte (8 bits) to represent all possible characters (11111111 binary = 256). Thus, you must be able to process 600 characters per minute times 8 bits per character or 4800 bits per minute.

Now let's assume that for each character you type, your brain needs ten instructions (to tell you where the key is, how to move your fingers, etc.) and for each word you type you need to execute 100 more instructions (to tell you what the word means, how to spell it, etc.). We can then find your processing rate by first multiplying the characters per minute by the instructions per character to get 6000 instructions per minute. We then add this number to the additional 10,000 instructions per minute needed to process each word (100 words per minute times 100 instructions per word) to get a total of 16,000 instructions per minute. If your brain goes though six cycles for each instruction (to remember it and process it), we need

96,000 cycles per minute or 1600 cycles per second. Thus, your brain needs a throughput of 1.6 kHz (1600 cycles/s).

Now, what about storing all this data? If you had to memorize everything you typed and you typed for one hour, you'd have to remember 4800 bits per minute times 60 minutes or 288,000 bits—that's 36,000 (36 k) bytes (your brain will get full).

Design Considerations

When buying a new home computer you're automatically faced with balancing two competing requirements—cost and ability. If you're given a blank check to work with, you could buy a super-computer which could do billions of operations per second and serve hundreds of users simultaneously. Of course, this would be a big waste of money if all you need is something to type your term paper on. When designing a spacecraft's data handling subsystem we must trade off the same two requirements.

Assessing the cost of a particular system is pretty easy. We simply measure it in terms of dollars. However, to assess the system's ability we must consider not only its processing speed, memory capacity, and other performance parameters relatively easy to measure but also its *autonomy* and ability to survive in the space environment. The smarter a system, the more it can work independently (autonomously) from ground controllers. A spacecraft computer on a probe to the outer planets, for example, needs a lot of autonomy because of the time lag in communicating back to ground controllers. Autonomy also means a spacecraft can selectively send back only the most relevant payload data. This decreases requirements for the communication system and makes it easier for users on the ground. Autonomy, of course, means a more complex system, which means more money.

Data handling subsystems must also survive in space. Recall from Chapter 3 that, in the harsh space environment, computers are bombarded by charged particles which cause single-event phenomena (SEP). SEPs can severely disrupt electrical processes within the CPU itself or change software stored in memory. Because of this demanding environment, we're typically limited to "space-hardened" systems or those with a proven track record in space. Unfortunately, testing to certify space-hardened systems is expensive, and the potential market is relatively small. So these systems cost far more than "off-the-shelf" systems and are typically several years behind the state-of-the-art in terms of processing speed, memory, capacity, and so on.

▰ Section Review

Key Terms

application software
autonomy
bit
bytes
central processing unit (CPU)
input
operating-system software
output
processing speed
random access memory (RAM)
read only memory (ROM)
throughput
transistors
word

Key Concepts

➤ The communication and data handling subsystem acquires, stores, and transfers all information for the spacecraft. (See Chapter 15)
  • Communications transfers data and commands between the spacecraft and the ground system
  • Data handling collects and stores payload and housekeeping data on the spacecraft.

➤ Data handling onboard a spacecraft requires computers which are the "brains" of the satellite. The system moves and stores data and carries out operations pre-programmed or sent from the ground.

➤ The spacecraft's data handling subsystem
  • Processes and stores payload data
  • Gathers and processes routine housekeeping data
  • Processes and carries out instructions from ground controllers

➤ Data handling subsystems have four kinds of hardware:
  • Central processing unit (CPU)
  • Sensors and other devices for inputting data
  • Units for memory storage
  • Actuators and other output devices

➤ Data handling subsystems use
  • Operating system software
  • Applications software

➤ Computers think in a binary language of 1's and 0's:
  • Each 1 or 0 is a bit
  • Eight bits combine to make a byte
  • A collection of bits makes up a word

➤ Throughput is the rate at which bits, bytes, and words are processed

➤ Mission designers must trade off a data handling subsystem's
  • Cost
  • Capability—performance parameters, autonomy, and survivability

13.2   Electrical Power

▬ In This Section You'll Learn to...

- ☞ Describe what the electrical power subsystem (EPS) does
- ☞ Define and determine basic concepts and parameters of electrical power
- ☞ Identify the main energy sources used by the electrical power subsystems
- ☞ Discuss how solar energy is used to produce electrical power
- ☞ Determine how long a spacecraft eclipse lasts
- ☞ Describe how batteries and fuel cells are used on spacecraft and discuss some of their limitations
- ☞ Discuss how and why nuclear energy is used in space
- ☞ Discuss the requirements for power regulation, distribution, and control

We need electrical power to run the payload and most spacecraft subsystems. Unfortunately, space doesn't have wall outlets, and an extension cord would be way too long! So we need an *electrical power subsystem (EPS)* to convert some energy source into usable electrical power for the entire spacecraft. The EPS

- Converts energy from some source into electrical power
- Stores energy
- Regulates, distributes, and controls power throughout the spacecraft

The EPS must convert raw energy from some convenient source, usually the Sun, into electrical power that other equipment on the spacecraft can use. Once this energy is produced, it often must be stored, especially in the case of solar electric systems, for times when the satellite is in the Earth's shadow or when some unique mission requirement calls for a power boost. Finally, all this power must be doled out to the various subsystems, which sometimes have different requirements. Furthermore, all subsystems must be protected from unpredicted power surges or spikes, much as your home has a fuse or breaker box to keep appliances from drawing too much power and destroying themselves or causing fires. Before diving into a discussion of the EPS, which will involve looking at watts, volts, and things like that, we need to review some basic principles of electricity.

Basic Principles

Six terms and concepts will help us explain electrical power:

- Charge
- Coulomb's Law
- Voltage
- Current
- Resistance
- Power

To understand all of these electrical concepts, it's useful to look at physical analogies that are easier to get our arms around. To begin with, *charge* is the basic unit of electricity. It's similar in concept to mass. Recall from Chapter 4 that mass is a basic property of matter which describes its behavior in terms of how much stuff it has or how much gravitational force it generates. We know that applying a force, $\vec{F}$, to a mass will cause it to accelerate an amount, $\vec{a}$, according to Newton's famous second law, $\vec{F} = m\vec{a}$.

Similarly, charge is a basic property of matter, like mass, for which we can develop laws of behavior. Charge comes in two flavors—positive (+) and negative (–). If something lacks charge or has an equal number of positive and negative charges, we say it is *neutral*. Opposite charges attract, and like charges repel. For mass, the basic unit is the gram. For charge, the basic unit is the *Coulomb (C)*. One electron has 1.6×10^{-19} Coulombs of negative charge. One proton has exactly the same amount of positive charge.

The force of attraction or repulsion between charges is called an *electrostatic force*. This force is shown in Figure 13-3 and can be quantified using Coulomb's Law.

Coulomb's Law. The force of attraction (or repulsion) between two charges is directly proportional to the amount of each charge and inversely proportional to the square of the distance between them.

Note that this sounds a lot like Newton's Universal Law of Gravitation. We can write an equally similar equation

$$\vec{F} = K \frac{Qq}{R^2} \hat{R} \tag{13-1}$$

where
$\vec{F}$ = electrostatic force on charge 1 (N)
K = constant ($9 \times 10^9 \, \text{N} \cdot \text{m}^2/\text{C}^2$)
Q = value of charge 1 (Coulombs, C)
q = value of charge 2 (Coulombs, C)
R = distance between charges (m)
$\hat{R}$ = unit vector in $\vec{R}$ direction

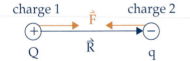

charge 1   charge 2

Figure 13-3. Electrostatic Force. Coulomb's Law of electrostatic force is very similar to Newton's Universal Law of Gravitation.

As with mass, when you apply a force to a charge it starts to move. Again, a physical analogy is useful. Imagine that a charge is like a water droplet. Under pressure, the water droplet flows through a hose. Similarly, we define *current, i*, to be the rate at which charges are flowing through some given area, like the cross section of wire. The unit for current is *amperes (amps)*, which is equal to the charges flowing per second.

$$\text{amps} = \text{charges/time (Coulombs/s)}$$

$$i = \frac{dQ}{dt} \tag{13-2}$$

where
i   = current (amps)
dQ = flow of charges (Coulomb)
dt = time (s)

In electricity, *potential* or *voltage* is similar to the concept of potential energy from mechanics. Remember, you must expend energy to move an object against a gravitational field. For example, when you climb a flight of stairs, you expend energy which is stored as potential energy. If you were to jump out of a window at the top of the stairs, this potential energy would be transformed into kinetic energy by the gravitational field as you plummet to the ground. Similarly, we describe electrical potential as the energy an electrical field can transmit to a unit charge. Returning to our garden hose analogy, it's like water pressure: the greater the pressure, the faster water flows. We use the unit of *volts, v*, to describe this relationship.

$$\text{energy per charge} = \text{volts} = \frac{\text{Joules}}{\text{Coulomb}}$$

For example, a 12-volt battery in a car can deliver 12 Joules of energy to each coulomb of charge. A battery with higher voltage would have more potential and could deliver more energy per coulomb of charge. The terms *potential* and *voltage* can be used interchangeably.

In electricity, *resistance, R*, tries to prevent charges from flowing. Just as a kink in a garden hose slows down the flow of water, a resistor slows down the current. The unit of resistance is the *ohm, Ω*. Ohm's Law relates the current to the voltage pushing the charge and the amount of resistance to that push.

$$i = \frac{V}{R} \tag{13-3}$$

where
i   = current (amps)
V = voltage (volts)
R = resistance (ohms)

This equation tells us that current is directly proportional to voltage and inversely proportional to resistance. Does this make sense? For a given resistance, the higher the voltage, the greater the energy delivered to a unit charge; thus, more charges flow, and current increases. It's as if you increase the pressure in a garden hose. The more pressure on the water (higher the voltage), the more water flows (the higher the current). Furthermore, the higher the resistance, the lower the flow of charges and the lower the current (put a kink in the garden hose and less water flows).

Power is the amount of energy delivered per unit time. The unit of power is the *watt, W*, and is defined as one joule of energy per second.

$$1\,W = 1\,J/s$$

Power is also equal to the product of the voltage and the current. This makes sense if you look at what voltage and current represent

$$V = \frac{energy}{charge}, \text{ and } i = \frac{charge}{time} \text{ so,}$$

$$P = \frac{energy}{\cancel{charge}} \times \frac{\cancel{charge}}{time} = \frac{energy}{time}$$

Thus,

$$\boxed{P = iV} \tag{13-4}$$

where
P = power (W)
i   = current (amps)
V = voltage (volts)

Finally, we'll define a loop through which current flows as a *circuit*. For example, when you turn on a lamp, you are closing a switch and establishing a circuit of current. Thus, electrons can begin to flow through the light bulb, causing it to heat up, glow, and give off light. The flow of current in a circuit is generally defined to be from positive to negative, as seen in Figure 13-4. Current can either be direct, DC, or alternating, AC. With DC, the current always flows in one direction. For AC, the direction of current flow switches back and forth at some cycle rate. Spacecraft most often use DC in space because energy sources tend to be DC. Thus, by using DC circuits, we can avoid the need for complicated converters to go from DC to AC. On Earth, the standard current in the U.S. is 60 Hz AC (meaning 60 cycles per second). AC is more efficient for very high power demands and allows for transporting power over long distances through high-voltage transmission lines with lower losses. Now that we understand some of the basic terms of electricity, let's explore the energy sources available to spacecraft and see how they're used.

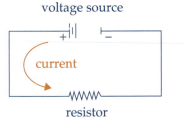

Figure 13-4. Current. Current is defined as the flow of charge around a closed path. Current can either be direct, DC, or alternating, AC.

Energy Sources

Solar Energy

Remember that a power system must convert energy from some source into usable electrical power. Because the Sun is the energy source that drives life on Earth, it's an obvious choice for spacecraft as well. We'll look at two ways to harness solar energy for use in space—solar cells and dynamic solar power systems.

By far the most common way to convert energy into electricity on a spacecraft is the *solar cell* or *photovoltaic, PV, cell*. We shine light on a solar cell, and current flows out. But how? A solar cell is basically a thin wafer of silicon, gallium arsenide, or other semi-conductor crystal. As photons from sunlight strike the cell, they transmit their energy to the atoms in the cell, thus freeing electrons. These electrons then move across tiny junctions of silicon and phosphorous, or similar materials, within the cell. This decreases the resistance in the cell, and the freed electrons are able to flow. Recall, a flow of electrons (charges) is defined as a current. Thus, we have electricity!

Solar cells were first developed at Bell Labs in the early 1950s. As the Space Age dawned in the late 1950s, their ability to produce electrical power from abundant solar energy quickly offered spacecraft designers a way to produce reliable electrical power without lugging lots of heavy batteries into orbit.

Unfortunately, solar cells now on the market aren't very efficient. We define the efficiency of a solar cell in terms of its *energy conversion efficiency,* η, (Greek letter eta), where $\eta = 100\%$ means all the solar energy striking the surface is converted to electrical energy. Although laboratory specimens have converted energy at efficiencies of nearly 30%, typical production cells provide around 15%. Silicon cells average around 14% while, newer, more expensive gallium-arsenide cells exceed 18%. This means that only about 15% of the solar energy that strikes the surface converts to electrical energy. The rest either reflects or ends up as heat.

Practically speaking, only the component of solar energy hitting the solar cell perpendicular to the surface is transformed into electrical energy. For this reason, we try to keep solar cells constantly facing the Sun. The total power output of a cell depends on its overall efficiency, the intensity of the solar power, and the Sun angle. The *angle of incidence, θ,* describes the angle between a line perpendicular to the cell and the Sun line. Figure 13-5 shows solar energy striking a solar cell at some angle of incidence. With this information, we can now express the output power density (P_{OUT}) as

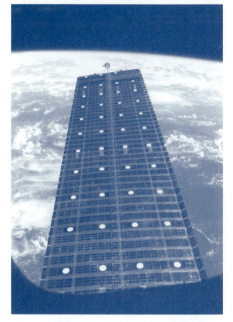

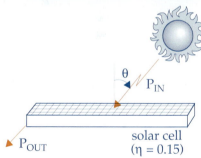

$$\boxed{P_{OUT} = P_{IN}\ \eta\ \cos\theta} \qquad (13\text{-}5)$$

where
P_{OUT} = the solar cell's output power density (W/m²)
P_{IN} = solar input power density (W/m²)
η = the solar cell's energy-conversion efficiency
    (typically < 0.15)
θ = incidence angle (deg or rad)
    perpendicular to surface = 0°

Figure 13-5. Photovoltaic or Solar Cells. A photovoltaic cell, like these being deployed from the Space Shuttle, takes incident solar energy and produces electrical power at some efficiency (usually around 15%). *(Photo courtesy of NASA)*

The input density of solar power decreases with the square of the distance from the Sun. Near Earth, the density of this incident solar power is about 1358 W/m². In comparison, near Venus, it increases to 2596 W/m², and out at Mars, it decreases to only 585 W/m². As you go farther from the Sun, out near Jupiter, the solar energy is a mere 50 W/m². For this reason, solar cells become impractical for spacecraft going to Jupiter or any of the other outer planets, because the solar arrays you'd need would be too big.

For satellites needing hundreds of watts of power, one small solar cell just doesn't hack it. For this reason, hundreds of cells are wired together to combine their output. Solar cells can be mounted in one of several ways. The simplest way is to attach all the cells to the outside of the spacecraft. This is called a *body-mounted solar array*.

Body-mounted arrays are simpler and don't require complicated wiring options. However, in most cases, body-mounted arrays are used on spacecraft with spin or dual-spin stability, such as the IntelSat spacecraft, shown in Figure 13-6. For this reason, less than half of the cells "see" the Sun at any one time. The rest are completely shaded by the spacecraft's body or have an angle of incidence so high that they produce very little effective power. For a cylindrical spacecraft, less than one-third (actually it's $1/\pi$) of the spacecraft's total surface can generate electricity at any one time. Thus, designers give a spacecraft enough total surface area to generate the required power.

Another disadvantage of body-mounted arrays, and all types of arrays that don't actively track the Sun, is the angle of incidence due to the Earth's tilt with respect to the ecliptic plane. Because the Earth tilts 23.5° with respect to the ecliptic, the apparent Sun angle changes as the Earth moves around its orbit. At the winter solstice (December 21), the Northern Hemisphere tilts away from the Sun, so the Sun appears to strike directly on 23.5° South latitude (the tropic of Capricorn). At the summer solstice (June 21), the Northern Hemisphere tilts toward the Sun, so the Sun's rays appear to strike directly on 23.5° North latitude (the tropic of Cancer). At the vernal (March 21) and autumnal (September 21) equinoxes, the Earth's tilt is perpendicular to the Sun, so the Sun's rays appear to strike directly on the equator as shown in Figure 13-7.

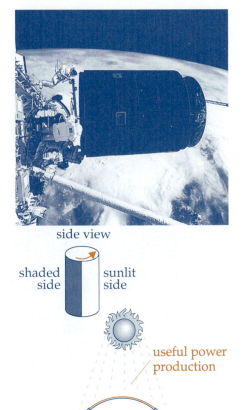

side view

shaded side   sunlit side

useful power production

angle of incidence too high   top view   angle of incidence too high

completely shaded

Figure 13-6. Body-Mounted Arrays. The IntelSat spacecraft, shown here being repaired by astronauts, uses body-mounted solar arrays covering the entire cylindrical surface. As the diagram shows, approximately 1/3 of the solar cells produce power at any one time as the spacecraft spins, due to shading and high incidence angle. *(Courtesy of NASA)*

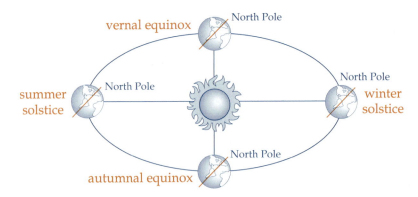

vernal equinox   North Pole

summer solstice   North Pole

North Pole   winter solstice

autumnal equinox   North Pole

Figure 13-7. Changing Sun Angle. The Earth tilts 23.5° with respect to the ecliptic plane. Because the Earth's angular momentum is constant as the Earth moves around the Sun, its orientation changes with respect to the Sun. This gives us seasons on Earth and causes the Sun angle on solar arrays to change for spacecraft in Earth orbits.

What does this have to do with solar power? Any spacecraft in Earth orbit will experience this same change in Sun angle during the year. Thus, for spacecraft in equatorial orbits with body-mounted arrays that don't track the Sun, the incidence angle will change by ± 23.5° during the year. We must plan for the worst case of $\theta = 23.5°$ when designing for body-mounted solar cells.

We can overcome this problem with Sun angle by actively tracking the Sun with our solar arrays and thus keeping the angle of incidence near zero. To do this, the solar cells must be mounted on movable arrays. A complex control system must then track the Sun and move the arrays accordingly. Thus, while solar-tracking arrays do provide the best power output, their increased complexity is a drawback for some applications. In addition, separately movable arrays won't work on spin- or dual-spin stabilized spacecraft. Figure 13-8 shows a typical sun-tracking array as used on the Hubble Telescope.

Besides Sun angle, several other environmental factors can degrade the performance of solar cells:

- Temperature
- Radiation and charged particles
- Eclipses

Solar cells are very sensitive to temperature, being most efficient at low temperatures and losing efficiency at higher temperatures. Solar cells lose from 0.025% to 0.075% of their efficiency per °C as the temperature increases above 28°C. For example, an array that's 15% efficient at 28° C would be only about 14.75% efficient at 38° C. This means that thermal control for solar arrays is very important.

Solar cells and their cover glass are also extremely sensitive to the radiation and charged particles in space. As radiation and particles hit the solar arrays, the arrays degrade. Depending on their orbit, solar arrays can lose up to 40% of their effectiveness over ten years. For this reason, spacecraft designers deliberately over-design the size of the solar arrays for the start of the mission (called *beginning-of-life (BOL)*), knowing that over the lifetime of the mission the power levels will degrade. Thus, beginning-of-life power must be high enough so that by the end of the mission (called *end-of-life (EOL)*), enough power is still available to run all spacecraft systems.

One other difficulty in using solar power is the eclipse experienced as an orbiting spacecraft periodically passes into Earth's shadow, as shown in Figure 13-9. When this happens, the incident solar energy goes to zero, and the solar cells stop producing power. The length of orbital eclipses depends mostly on the spacecraft's altitude, which determines the *Earth's angular radius, ρ*, as shown in Figure 13-10 (we can ignore inclination to get a good worst-case approximation).

$$\rho = \sin^{-1}\left(\frac{R_{Earth}}{h + R_{Earth}}\right) \qquad (13\text{-}6)$$

Figure 13-8. Sun-Tracking Arrays. Sun-tracking solar arrays, like the ones shown on the Hubble Telescope, get higher efficiency by maintaining a zero angle of incidence—that is, the sunlight is perpendicular to the surface. *(Courtesy of NASA)*

Figure 13-9. Earth's Shadow. Most spacecraft, pass into Earth's shadow, once each orbit, causing an eclipse and blocking input to solar cells.

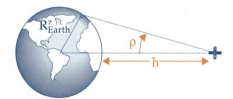

Figure 13-10. Earth's Angular Radius. As viewed from a spacecraft in orbit, the Earth has an angular radius, ρ, which depends on the spacecraft's altitude, h.

where

ρ   = Earth's angular radius viewed from space (deg)
h   = orbit altitude (km)
R_{Earth} = radius of Earth = 6378 km

We can then find the total time in eclipse during one orbit by finding the fraction of the orbit period spent in the Earth's shadow. Note: this will give us an approximate maximum eclipse time. Actual eclipse times will be shorter due to orbit inclination and other factors.

$$TE = \frac{2\,\rho}{360°} \times P \qquad (13\text{-}7)$$

where
TE = maximum time of eclipse (min)
ρ   = angular radius of the Earth (deg)
P   = orbit period (min)

For example, a spacecraft in low-Earth orbit (say at an altitude of 550 km or 343 miles) will experience 15 eclipses per day (one per orbit) or about 5500 per year. Each will last up to 36 minutes (depending on the inclination). In contrast, a geostationary orbit (at an altitude of 35,000 km [21,834 miles] and an inclination of 0°) experiences only 90 eclipses per year at a maximum of 72 minutes.

Another way to convert solar energy into electrical energy is through heat. *Dynamic solar power systems* focus the Sun's energy to heat some working fluid to high temperatures. This hot fluid then drives an electric generator. This is the same process used today in virtually all electrical-power plants. We burn a fossil fuel to produce heat used to boil water. The steam is then used to turn a turbine and produce electricity.

Dynamic solar systems use the same kind of generators to produce electricity. All we need is a way to first turn solar energy into mechanical energy. How? By focusing incoming solar energy into a concentrated area and thus producing extremely high temperatures. We then use this heat to expand some working fluid. The fluid drives a turbine to produce a form of mechanical energy which, in turn, drives the generator. While the system is exposed to sunlight, excess heat is stored by melting some type of salt. During eclipse times, any heat is removed from the salt as it solidifies to continue to drive the system. In this way we don't need batteries. No dynamic solar systems are now in space, but they could produce large amounts of power for future industrial applications.

Chemical Energy

Because solar-powered systems periodically pass into the Earth's shadow, we need some means of storing electrical power. We'll look at two types of chemical-energy systems for storing power—batteries and fuel cells.

Rechargeable batteries allow the spacecraft to store electrical power while in sunlight for use during eclipses. Batteries store electrical energy in the form of chemical energy. If you stick two different types of metal nails into a lemon and connect the two nails to a voltmeter, you'll find that an electrical potential exists. Thus, the lemon produces a current. In this case, the citric acid of the lemon acts as an *electrolyte*—a fluid containing ions. The two nails become the *electrodes*—conductors which emit or collect electrons. One nail collects positive ions (called the *anode*) the other collects negative ions (the *cathode*).

All types of batteries work by this same basic principle. As electrical energy goes into the battery (charging), a chemical reaction takes place in the electrolyte, effectively storing the electrical energy as chemical energy. As the electrical energy goes back out (discharging), the chemical reaction within the electrolyte reverses. Depending on the type of battery, the battery can be charged, discharged, and then recharged over and over again.

Spacecraft can use either primary or secondary batteries, similar to the ones shown in Figure 13-11. *Primary batteries*, like those in portable stereos, provide the sole source of electrical power. They're designed for a single, short use and can't be recharged. For very short missions (less than one month), primary batteries work pretty well. It's easier to load enough primary batteries to last throughout the brief mission than to provide some way to recharge them. One common type of primary battery for spacecraft is made of silver-zinc.

Most space missions last several years, however, so secondary batteries are much more common. *Secondary batteries* provide a backup source of electrical power and can be discharged and recharged repeatedly like your car battery, so they're an obvious choice to store electrical power during eclipses. The most common type of secondary batteries used onboard spacecraft are just like the rechargeable batteries found in video camcorders—nickel-cadmium, NiCd. NiCd batteries can be discharged and recharged thousands of times. NiCd batteries also have high energy density. This means they can store a relatively high amount of electrical energy for their weight—about 25 – 30 W · hr/kg. Nickel-hydrogen, NiH, batteries, recently developed for space applications, have an even better energy density. Figure 13-12 shows examples of NiH batteries.

A battery's life depends on the *depth-of-discharge (DOD)* and the number of discharge and recharge cycles it undergoes. DOD is the percentage of total stored energy removed during any discharge period. The smaller the DOD, the more times a battery can be cycled before it eventually dies. For example, a NiCd battery can be cycled as many as 21,000 times at a 25% DOD, but only 800 times at 75% DOD. You can see this same effect with car batteries. If you repeatedly leave your lights on all day while the car is parked (causing a DOD of nearly 100%), your battery will wear out much faster than the manufacturer's advertised lifetime. Eventually, no matter how long you try to charge it, it just won't hold a charge.

The total number of charge or discharge cycles batteries undergo depends on the size of the orbit. Lower orbits have shorter periods and experience more eclipses during their lives than spacecraft in higher orbits. Because of this, more total battery capacity (meaning more

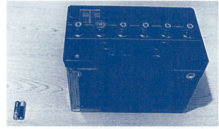

Figure 13-11. Primary and Secondary Batteries. Primary batteries are like the ones used in portable stereos—they can't be recharged. Secondary batteries, like car batteries, can be recharged and discharged many times.

Figure 13-12. NiH Batteries. The top view shows a single nickel-hydrogen (NiH) battery cell. The bottom view shows a set of NiH cells wired to provide total power storage requirements. *(Courtesy of NASA)*

batteries) are needed for low-Earth orbit spacecraft (with nearly 5500 cycles per year) versus those in geostationary orbit (with only 90 cycles per year). Figure 13-13 shows the relationship between DOD and the number of cycles for NiCd and NiH batteries.

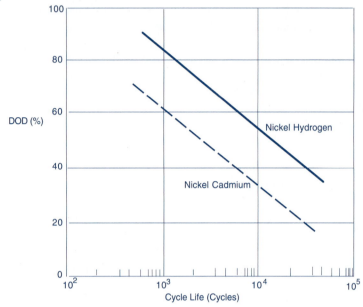

Figure 13-13. Depth of Discharge Versus Number of Eclipse Cycles for Typical NiCd and NiH Batteries. [Larson & Wertz, 1992]

Space-mission operators must also be concerned with battery memory. This is a problem you may have with NiCd batteries even in your own video camera. If the batteries are repeatedly discharged a small amount (say 25%) and then recharged, they may develop *battery memory* so they can't be discharged beyond 25% when needed. To avoid this, operators will periodically "condition" the batteries by taking them down 80% or more before recharging.

Batteries also have stringent thermal requirements. Because of the nature of the chemical reactions in the electrolyte, extremely high or extremely low temperatures can greatly affect their ability to hold a charge. Again, car batteries illustrate this point. If you've ever experienced the frustration of a dead car battery on a bitter, cold winter morning, you know how battery charge can decline when exposed to extreme cold. Operational temperature ranges for NiCd batteries are 10° to 30°C.

Batteries can be sized based on the total amount of energy they can hold. This is usually expressed in terms of W · hrs. Remember that W = energy/time, so W · hr is really energy/time × time = energy. Example 13-3 shows how you'd go about determining the battery capacity needed for a typical mission.

Fuel cells are the second type of chemical-energy systems found onboard spacecraft. When two highly reactive compounds are brought together, their chemical reaction can produce a current. Fuel cells have appeared on all United States manned spacecraft since Gemini (except for Skylab), because these missions have lasted two weeks or less and required power of 1000 W or more. Typically, fuel cells use gaseous hydrogen, H_2, and oxygen, O_2, as the reactants. From elementary chemistry, you know that these two compounds combine to produce H_2O (water). Thus, using fuel cells for manned missions gives us an added bonus—drinking water as a by-product of electrical power production. (Drink it and you could get a real jolt!) Figure 13-14 shows a typical fuel cell.

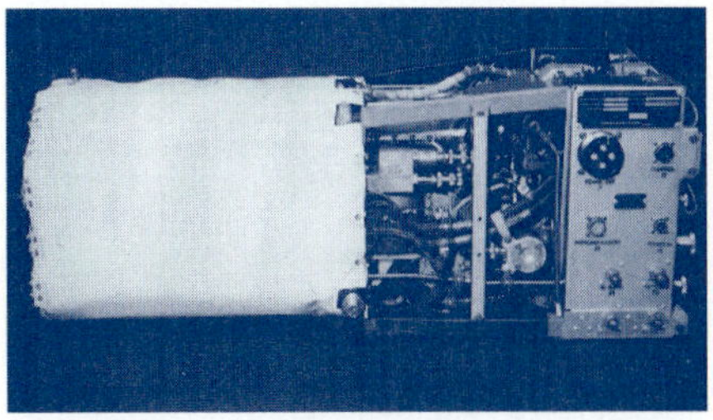

The main drawback of fuel cells is the limited amount of H_2 and O_2 that can be practically brought along on a mission. As missions go beyond a month, solar cells or other power options are much more efficient.

Figure 13-14. Fuel Cells. Fuel cells, like these shown for the Space Shuttle, combine gaseous hydrogen (H_2) and oxygen (O_2) to get electrical power, plus drinking water as a by-product. *(Courtesy of NASA)*

Nuclear Energy

We get nuclear energy from two sources—radioisotope thermoelectric generators and thermal-cycle fission systems. *Radioisotope thermoelectric generators (RTGs)* use the heat generated by the natural decay of radioactive isotopes to produce electricity. Radioactive isotopes of uranium or plutonium naturally decay into non-radioactive elements over time. The time it takes for one half of the material to decay is called the *half-life*. Half-lives of some radioactive isotopes can last tens, hundreds, or even thousands of years. As the radioactive material decays, it gives off tremendous amounts of heat. In an enclosed container, the temperature can reach many hundreds of degrees.

Astro Fun Fact
How Safe are RTGs?

Because RTGs contain radioactive material such as uranium or plutonium, their launch poses obvious public health concerns. To ensure maximum safety, RTGs are designed to withstand every conceivable launch failure and then undergo an elaborate testing program aimed at precisely quantifying the level of risk. During the testing program, hundreds of thousands of computer simulations look at how the RTG could be damaged due to booster explosions, accidental reentry, or impact with the ground. In addition to the computer simulations, RTG containment cases are shot at by bullets traveling 684 m/s (1530 m.p.h.), burned by solid rocket propellant, hurled at high velocity at concrete and steel, and shot at by SRB fragments launched by gas guns and rocket sleds. Fuel elements, clad in iridium, are also thrown with high velocity at concrete, steel, and sand to ensure their integrity in an accident. The results from all these tests indicate the probability of exposing even one person to 6.37×10^{-3} REM is 16/10,000 (more than one hundred times less than simply living one year at sea level). When exposure to significant dosages is considered, the probabilities enter the realm of the incredible. In comparison, the probability of 1000 fatalities from a dam failure during the same time period of the launch is only 1/100. So, are there risks to launching RTGs? Yes, but the risks are extremely small and well understood. It's much more likely someone will be injured driving a car to protest a launch than in an RTG accident.

Executive Summary of the Final Safety Analysis Report for the Ulysses Mission, Prepared for U.S. Department of Energy Office of Special Applications, General Purpose Heat Source Radioisotope Thermoelectric Generator Program, ULS-FSAR-006, NUS Corporation, March 1990.

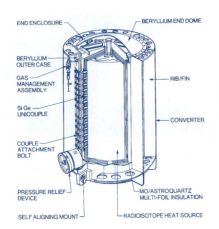

END ENCLOSURE

BERYLLIUM END DOME

BERYLLIUM OUTER CASE

GAS MANAGEMENT ASSEMBLY

SiGe UNICOUPLE

COUPLE ATTACHMENT BOLT

RIB/FIN

CONVERTER

PRESSURE RELIEF DEVICE

SELF ALIGNING MOUNT

MO/ASTROQUARTZ MULTI-FOIL INSULATION

RADIOISOTOPE HEAT SOURCE

Figure 13-15. Radioisotope Thermoelectric Generator (RTG). RTGs like this one used on the Voyager spacecraft produce electrical power by converting the heat from radioactive decay into electricity through thermocouples. *(Courtesy of NASA)*

So where does the electricity come from? Figure 13-15 shows a cut-away view of the RTG used for the Voyager spacecraft. RTGs use thermocouples to transform heat into electricity. *Thermocouples* are bi-metallic strips formed from two unlike metals. As one end of the thermocouple heats up, the difference in resistance between the two metals causes free electrons to begin to flow, thus producing a current. Thermocouples are used extensively in industry—often as thermometers—because we can compute temperature by measuring the current produced.

To construct an RTG, we place thermocouples against a contained heat source—the decaying radioisotopes. In this way, an RTG can put out sustained electrical power for many years. For example, RTGs on the Voyager spacecraft are still going after nearly twenty years in space.

RTGs have two major drawbacks over other power options: expense and political concerns. Radioactive material is, of course, expensive by itself. To design, construct, and test a self-contained power system housed in a containment shell which can survive launch disasters is even more expensive. (Typically, RTGs cost $15,000/W versus solar cells at $3,000/W.) Furthermore, any use of radioactive material brings public scrutiny. Concerned activists tried to block the launch of the Galileo spacecraft, which was powered by an RTG. Fortunately, mission planners were able to convince the courts that RTGs could easily contain the radioactive material during a launch disaster. For these reasons, RTGs are used only for missions which preclude any other type of power system. Typically, this means missions to the outer planets (beyond the orbit of Mars), where solar energy is so weak that solar cells become impractical.

Thermal-cycle fission systems produce electrical power in much the same way nuclear power plants do on Earth. Nuclear fission is the process of splitting apart an atom, which can be done easily only with extremely heavy atoms like uranium or plutonium. Rather than the slow decay witnessed with RTGs, fission is a much more dramatic process which releases tremendous amounts of heat in a relatively short time. With fission, a single neutron collides with the nucleus of an atom causing it to split apart. This produces two new atoms of lower atomic number. The kinetic energy of this fission product produces heat, gamma rays, charged particles, and neutrons. The neutrons then collide with other atoms to continue the fission in a controlled chain reaction.

On Earth, heat from nuclear fission is used in power plants to boil water and produce steam. As the steam expands, it drives turbines which turn generators. Thus, nuclear energy produces thermal energy, which produces mechanical energy, which produces electrical energy. A spacecraft with a fission reactor as the heat source could use a closed thermal cycle like the one discussed with dynamic solar systems to do the same thing. Another simple option for thermal-cycle fission power is called a thermionic system. *Thermionic systems* use the heat produced by fission to produce a flow of electrons (current directly by heating a metallic emitter.

Thermal-cycle fission systems would be best for applications needing extremely high power, such as those proposed for orbital missile defensive systems or for manned outposts on the Moon or Mars. But these systems have the same drawbacks as RTGs—high cost and political fallout from launching a fully fueled nuclear reactor. The Soviet Union used a small-scale thermionic nuclear reactor in their RORSAT spacecraft which created ecological and political problems when one crashed in Canada. Research is underway to find other applications for thermionic systems.

Power Distribution and Control

Regardless of the electrical-power source—solar cells, RTG, etc.—the power must somehow get to the spacecraft's subsystems and payloads that run on electrical power. In addition, if batteries are present, we need more power to charge them. A spacecraft's electrical power is normally distributed over a "power grid" which other parts of the spacecraft plug into. In some cases, the wiring and interfaces needed to connect all parts of the spacecraft can account for as much as 10% of the entire mass of the EPS.

Furthermore, power levels (and hence current or voltage levels) can vary throughout a spacecraft's orbit and lifetime. These large changes in current or voltage can severely damage a spacecraft's subsystems and scientific equipment. For this reason, we need some means of regulating the power throughout the spacecraft. In your home, you have a "fuse box" which may or may not actually contain fuses. Most newer homes have breakers instead of fuses, but they do the same thing: protect circuits from excessive current. The EPS provides the same protection to other subsystems. If power surges, due to discharging caused by buildup of charged particles, for example, breakers or fuses prevent damage to equipment.

▬ Section Review

Key Terms

amperes (amps)
angle of incidence, θ
anode
battery memory
beginning-of-life (BOL)
body-mounted solar array
cathode
charge
circuit
Coulomb (C)
current, i
depth-of-discharge (DOD)
dynamic solar power systems
Earth's angular radius, ρ
electrical power subsystem (EPS)
electrodes
electrolyte

Key Concepts

➤ The spacecraft's electrical power subsystem (EPS) converts some energy source, such as solar energy, into usable electric power to run the spacecraft. The EPS

• Converts energy from some source into electrical power

• Stores energy

• Regulates, distributes, and controls power throughout the spacecraft

➤ Charge, current, voltage, and power are the basic principles of electrical power

➤ Solar-energy systems use one of two methods to get electrical power:

• Photovoltaic cells convert sunlight directly into electricity. This is by far the most common system onboard spacecraft.

Continued on next page

Key Terms (Continued)

electrostatic force
end-of-life (EOL)
energy conversion efficiency, η
fuel cells
half-life
neutral
ohm, Ω
photovoltaic, PV, cell
potential
power
primary batteries
radioisotope thermoelectric
    generators (RTGs)
resistance, R
secondary batteries
solar cell
thermal-cycle fission systems
thermionic systems
thermocouples
voltage
volts, v
watt, W

Key Equations

$$i = \frac{V}{R}$$

$$P = iV$$

$$P_{OUT} = P_{IN}\ \eta\ \cos\theta$$

$$\rho = \sin^{-1}\left(\frac{R_{Earth}}{h + R_{Earth}}\right)$$

$$TE = \frac{2\ \rho}{360°} \times P$$

Key Concepts (Continued)

- Dynamic solar-power systems use heat from the Sun to produce electricity through mechanical means

➤ Chemical-energy systems use either batteries or fuel cells
  - Batteries are either primary or secondary
    - Primary batteries are the sole source of power. They can't be recharged and are used for short, one-shot missions.
    - Secondary batteries provide backup power and can be recharged many times
  - The lifetime of secondary batteries is limited by the depth of discharge (DOD) and the number of cycles
  - Fuel cells produce power by combining two reactants such as oxygen and hydrogen. They have been used on most United States manned missions but have limited lifetimes.

➤ Spacecraft can use nuclear energy to produce electrical power in two ways—radioisotope thermoelectric generators (RTGs) and thermal-cycle fission systems. RTGs are the most common and have been used for various interplanetary missions.

➤ The EPS distributes and controls electrical power to all onboard "users." The EPS must supply power in the right form to all other subsystems and protect the entire spacecraft from voltage and power spikes.

Example 13-1

Problem Statement

Engineers planning a remote-sensing payload for a new small satellite have determined the sensor has 10 ohms of resistance. If the power-system operates at 10 V, what current will go through the payload?

Conceptual Solution

1) Find i using

$$i = \frac{V}{R}$$

Problem Summary

Given: R = 10 Ω, V = 10 V
Find: i

Analytical Solution

$$i = \frac{V}{R}$$

$$i = \frac{10V}{10\Omega} = 1 \text{ amp}$$

Problem Diagram

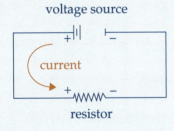

voltage source

current

resistor

Interpreting the Results

The payload with 10 ohms of resistance will draw 1 amp of current from an electrical-power bus with 10 V potential.

438

Example 13-2

Problem Statement

A remote-sensing spacecraft designed to monitor crop production is in a circular orbit at 300 km altitude. What is the maximum time the spacecraft will be in shadow during any orbit? If the mission is designed to last ten years, how many eclipses will it experience?

Problem Summary

Given: $h = 300$ km
Find: Max TE, number of eclipses in ten years

Problem Diagram

Conceptual Solution

1) Find Earth's angular radius

$$\rho = \sin^{-1}\left(\frac{R_{Earth}}{h + R_{Earth}}\right)$$

2) Find orbit period

$$P = 2\pi\sqrt{\frac{a^3}{\mu}}$$

3) Find max TE

$$TE = \frac{2\rho}{360°} \times P$$

4) Find number of eclipses in ten years

$$\frac{1 \text{ eclipse}}{\text{orbit}} \times \frac{\#\text{orbits}}{\text{day}} \times \frac{\#\text{days}}{10\text{yr}}$$

Analytical Solution

1) Find Earth's angular radius

$$\rho = \sin^{-1}\left(\frac{R_{Earth}}{h + R_{Earth}}\right) = \sin^{-1}\left(\frac{6378 \text{ km}}{300 \text{ km} + 6378 \text{ km}}\right)$$

$$= 72.76°$$

2) Find orbit period

$$P = 2\pi\sqrt{\frac{a^3}{\mu}} = 2\pi\sqrt{\frac{(6678\text{km})^3}{3.986005\times10^5\frac{\text{km}^3}{\text{s}^2}}} = 90.52 \text{ min}$$

3) Find max TE

$$TE = \frac{2\rho}{360°} \times P = \frac{2(72.76°)}{360°}(90.52 \text{ min})$$

$$= 36.58 \text{ min}$$

4) Find number of eclipses in ten years

$$\frac{1 \text{ eclipse}}{\text{orbit}} \times \frac{\#\text{orbits}}{\text{day}} \times \frac{\#\text{days}}{10\text{yr}}$$

$$\frac{\#\text{orbits}}{\text{day}} = \frac{(24 \text{ hr})(60 \text{ min/hr})}{P(\text{min})} = \frac{1440}{90.52} = 15.9$$

$$\frac{\#\text{days}}{10 \text{ yr}} = (365 \text{ days})(10 \text{ yr}) = 3650 \text{ days}$$

number of eclipses $= (15.9)(3650) = 58,065$

Interpreting the Results

A remote-sensing satellite in low-Earth orbit will experience a maximum eclipse time of 36.58 minutes per orbit. This value is important for planning battery capacity. Over its ten-year lifetime, the spacecraft will experience more than 58,000 eclipses.

Example 13-3

A satellite researching the atmosphere in low-Earth orbit requires 1000 W of continuous power. The solar cells can meet all power needs except during eclipses. If the maximum eclipse time is 25 minutes and the maximum DOD is 25%, what battery capacity (in $W \cdot hr$) is needed?

Problem Summary

Given: $P_{required}$ = 1000 W
TE = 25 min.
DOD = 25%
Find: Battery Capacity

Conceptual Solution

1) Find energy needed during eclipse
   Energy Required = $P_{required} \times$ TE

2) Find battery capacity
   Energy in batteries = energy required/DOD

Analytical Solution

1) Find energy needed during eclipse
   Energy required = Power × Time, so

   $$\text{Energy required} = 1000 \text{ W} \times \frac{25}{60}\text{hr}$$

   Energy required = 417 W · hr

2) Find battery capacity

   $$\text{Energy in batteries} = \text{Energy required} \times \frac{100\%}{\text{DOD}}$$

   $$\text{Energy in batteries} = 417 \text{ W hr} \times \frac{100\%}{25\%} = 1667 \text{ W·hr}$$

Interpreting the Results

For this mission, we would select enough batteries to give us a capacity of 1667 W · hr.

13.3 Environmental Control and Life Support

▬ In This Section You'll Learn to...

☞ Explain what the environmental control and life support subsystem does

☞ List the primary sources of heat for a spacecraft

☞ Describe the concept of thermal equilibrium

☞ Describe the three means of heat transfer—conduction, convection, and radiation—explain which a spacecraft uses most

☞ Describe the various ways to control heat outside and inside a spacecraft

☞ Describe what life support subsystems do

As we know from Chapter 3, space is a rough place—for humans and machines. Thus, we need some way to keep the payload and all the subsystems onboard (including the crew) healthy and happy. That's what the *environmental control and life support subsystem (ECLSS)* does. We'll focus on how it keeps subsystems from overheating or getting too cold on an unmanned spacecraft and then introduce some of the complications caused by placing fragile humans onboard.

Heat Sources in Space

A spacecraft orbiting happily along through space is perhaps the ultimate example of an isolated system. We can quite easily analyze everything that goes in and out. One of those things is heat. If more heat goes into a spacecraft than leaves it, what remains increases the spacecraft's temperature. If more heat leaves than goes in, the spacecraft begins to cool off. Because sensitive equipment and payloads (including humans) can't survive wide temperature swings, we like to keep the heat flow *in* equal to the heat flow *out*. Thus, the spacecraft's average temperature should stay constant, or in *thermal equilibrium*

To regulate and control the amount of heat that gets in and out of the spacecraft, we must have a *thermal control subsystem*. Just as the furnace and air conditioner do in our homes, the thermal control subsystem regulates and moderates the spacecraft's overall temperature. Typically, the biggest problem for the thermal control system is getting rid of heat. For the most part, temperatures inside an unmanned spacecraft need to be maintained at normal room temperature (20°C or about 70°F). In some cases, specific payloads may have more demanding requirements.

Infrared sensors, for example, often require temperatures of about 80° K (–193° C or –316° F).

As Figure 13-16 shows, heat in space comes from three main sources:

- The Sun
- The Earth
- Inside the spacecraft

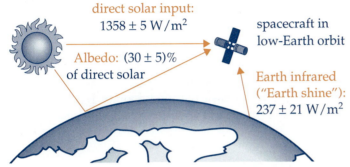

direct solar input:
1358 ± 5 W/m²

spacecraft in low-Earth orbit

Albedo: (30 ± 5)% of direct solar

Earth infrared ("Earth shine"): 237 ± 21 W/m²

Figure 13-16. Sources of Heat for Spacecraft. A spacecraft gets heat from the Sun, from the Earth (both reflected and emitted energy), and from within itself.

Near Earth, the biggest source of heat for a spacecraft is the Sun—about 1358 W/m². We all know how hot we get standing out in the Sun on a summer day. For a satellite in space, the heat from the Sun is much more intense because no atmosphere moderates the temperature. On the side facing the Sun, the surface of a spacecraft can reach many hundreds of degrees Kelvin. On the side away from the Sun, the temperature can plunge to only a few degrees Kelvin.

So what is the temperature in space? The Earth is about 300° K, while temperatures in space range from 900°–1300° K. Sounds hot, but is it? On Earth we measure temperature using a thermometer. The fluid in the thermometer expands when heated by molecules striking it. Temperature is proportional to the velocity of the molecules. In space, the molecules are traveling faster but there aren't very many of them. So while the temperature appears higher in space, the effect on people and materials is much less than the equivalent temperature on Earth.

For satellites in low-Earth orbit, the Earth itself is also an important source of heat based on two effects. The first results from sunlight reflecting off Earth—called *albedo*. It accounts for as much as 400 W/m², or 30% of the total incident energy on a spacecraft. Another important source is "Earth shine," or the infrared energy the Earth emits directly as a result of its temperature. This accounts for another 237 W/m² or about 12% of the incident energy on a spacecraft.

Internal sources also build up heat. Electrical components running onboard and power sources such as RTGs produce waste heat. If you've ever placed your hand on the top of your television after it's been on a while, you know how hot it can get. In your living room, the heat from the television quickly distributes throughout the room. Otherwise, your television would overheat and be damaged. Unfortunately, as we'll see, spacecraft can't so easily dump their excess heat. Thus, the thermal control subsystem must cool the spacecraft enough to prevent damage to equipment.

To maintain thermal equilibrium, the thermal control subsystem must balance all these inputs and keep an acceptable, constant overall temperature. This means the heat coming in plus the heat produced internally must equal the total heat ejected.

$$\text{Heat In} + \text{Internal Heat} = \text{Heat Out}$$

$$(\text{For thermal equilibrium})$$

Our biggest challenge is to control external and internal heat flow. Later we'll see exactly what we have in our bag of tricks to handle these two aspects of the problem. But first, because any discussion of thermal control requires an understanding of basic thermodynamics, let's take a step back to review some of the principles of heat transfer.

Heat Transfer

Recall from Chapter 3 that heat can be transferred from one point to another through

- Conduction

- Convection

- Radiation

If you hold one end of a long metal rod and put the other end into a fire, as in Figure 13-17, what happens? You get burned! The heat from the fire somehow manages to flow right up the metal rod. When heat flows from hot to cold through some physical medium (in this case the rod), we call it *conduction*. Heat conduction is something we experience every day. It's the reason we put insulation in the walls of our home to prevent heat from the inside being conducted outside (and vice versa in the summer time). We can describe the amount of heat transfer through conduction using the Fourier Law (developed by J.B.J. Fourier [1768-1830]).

$$q = -KA\frac{\Delta T}{\Delta X} \tag{13-8}$$

where

q	= heat energy conducted per unit time (W)
K	= thermal conductivity of the material (W/°K m)
A	= cross-sectional area of the material (m²)
ΔT	= temperature difference between two sides of the material (°K)
ΔX	= distance between "hot" and "cold" locations in the material (m)

This relationship indicates heat will flow faster if the material is a better conductor of heat (high K), such as metal rather than wood. It will also

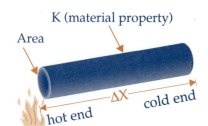

Figure 13-17. Conduction. Heat conduction occurs when heat flows through some solid medium from a hot point to a cooler point.

flow faster if a larger area is available, if the temperature difference is great, or if the distance is small. We use this principle to insulate the walls in our homes by making thick walls (large ΔX) and using poor heat conductors (insulators) which have low K. As we'll see, conduction works well for moving heat around inside a spacecraft. However, in the vacuum of space, there's nothing for a spacecraft to conduct, so we can't use it to get heat out of a spacecraft.

If you've ever boiled a pot of water on the stove, you've seen convection at work. Unlike conduction, which relies on heat flow through a solid medium, *convection* transfers heat to a fluid medium flowing past the heat source. Convection relies on gravity or some other force to push the liquid past the heat source. Look at how water boils in a pot on the stove as in Figure 13-18. Water on the bottom of the pot nearest the heat source gets hot first through conduction directly from the source. As the water gets hot, it expands slightly, making it a bit less dense than the water above it. At the same time, contact forces caused by gravity pull everything to the bottom of the pot. Thus, the cooler, denser water at the top of the pot displaces the warmer, less dense water at the bottom. Once on the bottom, this cooler water also heats up, expands, and rises. A convection current then continues as water begins to flow past the heat source, driven by the force of gravity.

The free-fall environment of space has no contact forces to cause cooler water to replace the warmer water. For convection to work in space, we must supply the force needed to move the fluid. This can be done within the spacecraft but the vacuum outside doesn't allow convection to take place, so we can't use convection to rid the spacecraft of heat.

Thus, as we said back in Chapter 3, we're left with only one real option for ejecting heat—radiation. If you've ever basked in the warm glow of an electric space heater, you've felt the power of heat transfer by radiation. *Radiation* is the means of transferring energy (such as heat) through space. More specifically, radiative heat transfer occurs through EM radiation. Recall from Chapter 11, EM radiation can be described in terms of waves (or particles) emitted from some energy source. Recall that a red-hot piece of metal can be thought of as a black-body radiator. The intense heat energy in the hot metal causes it to emit EM radiation. In this case, the frequency of the EM radiation is in the visible, red portion of the EM spectrum. We can use the Stefan-Boltzmann Law to describe the heat-power transfer by radiation.

heated water rises

cooler water sinks

heat source

Figure 13-18. Convection. Convection occurs when some driving force, such as gravity, moves the medium (usually a liquid or gas) past a heat source.

$$q = \sigma \varepsilon \, A T^4 \qquad (13\text{-}9)$$

where
q = heat-power transfer per unit time (W)
σ = Stefan-Boltzmann's constant (5.67×10^{-8} W/m^2 K^4)
ε = emissivity ($0 \le \varepsilon \le 1$)
A = area of black body (m^2)
T = temperature of black body (°K)

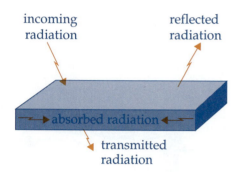

incoming
radiation

reflected
radiation

absorbed radiation

transmitted
radiation

Figure 13-19. Radiation. Radiation striking a surface is either reflected, absorbed, or transmitted.

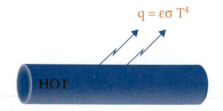

$$q = \varepsilon \sigma \, T^4$$

HOT

Figure 13-20. Emissivity. Any object with a temperature above 0° K (meaning basically everything in the universe) emits (EM) radiation per the Stephan-Boltzmann relationship. The greater the emissivity of the material the more energy it emits at a given temperature.

This relationship tells us that, as the temperature of a black body increases, the amount of heat power it emits increases as the fourth power of the temperature. Thus, if we double the temperature, the amount of energy emitted will increase sixteen times.

As Figure 13-19 shows, when radiation strikes a surface, it can be reflected, absorbed, or transmitted. Reflected radiation is just like reflected light from a mirror. This is radiation that basically "bounces" off the surface. We use the symbol ρ (Greek rho) to quantify the *reflectivity* of a surface (not to be confused with Earth's angular radius from Section 13-2). Reflectivity can be thought of as a percentage; that is, $\rho = 0.3$ means 30% of the radiation is reflected.

Absorbed radiation is energy the surface captures, just as a sponge soaks up water. Absorbed radiation will eventually cause the surface to heat up. We use the symbol α (Greek alpha) to quantify *absorptivity* Absorptivity can also be thought of as a percentage; that is, $\alpha = 0.5$ means 50% of the radiation is absorbed.

Transmitted radiation is energy that passes right through (like visible light through a pane of glass). We use the symbol τ to quantify *transmissivity*. Transmissivity can be thought of as a percentage; that is, $\tau = 0.2$ means 20% of the radiation is transmitted.

Because of conservation of energy, all of the radiation must be accounted for. So the sum of the reflected, absorbed, and transmitted radiation energy must equal the incoming energy. Another way of looking at this is:

$$\tau + \alpha + \rho = 1 \tag{13-10}$$

where
τ = transmissivity $(0 \leq \tau \leq 1)$
α = absorptivity $(0 \leq \alpha \leq 1)$
ρ = reflectivity $(0 \leq \rho \leq 1)$

As an object absorbs energy, the kinetic energy of individual molecules increases and the objects get hotter. As Figure 13-20 shows, all objects above absolute zero (0° K) will emit radiation. But not all materials emit heat with the same efficiency. We call a material's ability to emit heat its *emissivity, ε*. A pure black body has an emissivity of 1.0. The black tiles used on the Space Shuttle are designed with a very high emissivity ($\varepsilon \cong 0.8$).

Now that we've explored the basic options for heat transfer, let's see how they are used in a spacecraft. As we said earlier, we must manage heat coming into and out of the spacecraft, as well as heat generated inside it.

Methods for Thermal Control

External Thermal Control

Even with the physical limitations on heat transfer in space, various systems can eject heat from spacecraft and balance the heat internally between various subsystems. We've already seen how heat gets into the spacecraft from the Sun or Earth. Our first line of defense is to keep heat from being absorbed in the first place. One of the simplest, but most innovative ways to do this is to slowly rotate the spacecraft about an axis perpendicular to the Sun. In this "barbeque" mode, the spacecraft is alternately heated by the Sun and then exposed to the cold of space, maintaining a moderate surface temperature without hot spots. The Apollo spacecraft used this method all the way to the Moon and back.

Another relatively easy way to control heat input involves the material on the spacecraft's surface. As we've seen, radiation is either absorbed, reflected, or transmitted. By changing the surface material of the spacecraft, we can vary its total absorptivity. To do this, we use special coatings on the spacecraft's surface, as well as insulation around thermally sensitive areas. These coatings reflect much of the heat and protect instruments from the heat that does come in. We use similar "special" coatings all the time. If you were going to play golf on a sunny day, chances are you'd wear light-colored clothing that reflects and doesn't absorb as much heat. If you were playing during the cold of winter, you'd probably wear a dark color to reflect less and absorb more heat. The same principle applies to a spacecraft. By changing the absorptivity, we can regulate how much heat is absorbed. We can similarly vary emissivity to regulate how much of the absorbed heat is re-radiated to space. We can change this ratio of heat absorbed to emitted (α/ε) by carefully selecting materials to keep the surface temperature at the right level.

Gold is one of the best surfaces for retaining the heat it absorbs. If you compare it to black paint, you'll find that gold surfaces retain thirteen times more heat (have higher α/ε) than ordinary black paint. Obviously, a solid-gold spacecraft would be far too heavy (not to mention, expensive). So we vacuum-deposit "gold plating" for very special applications with particularly severe thermal demands. Normally we use multiple layers of insulation composed of alternating sheets of an aluminized, plastic-like material called *kapton* with a mesh-like material as a spacer. Kapton often looks like gold foil and is similar to the "space blankets" sold in sporting goods stores to keep you warm during emergencies (another great spin-off from space technology!) We can meet nearly 85% of a spacecraft's thermal-control demands simply by choosing the right surface coatings and insulation. Figure 13-21 shows reflective foil used for thermal control on the outside of the upper atmospheric research satellite (UARS).

Inevitably, some heat does get absorbed, so how can a spacecraft transfer this heat? In space, surrounded by a vacuum, a spacecraft can't conduct or convect heat away. Of course, we could transfer the heat to some fluid, such as water, and then dump it overboard. The Space Shuttle sometimes uses this method to remove excess heat with a device called a *flash evaporator*. Water is pumped around hot subsystems, cooling them through convection, and then vented overboard. Unfortunately, this method works only as long as you have extra water onboard. For long missions, this is impractical.

Often, the most effective means of dumping the excess heat collected or generated by a spacecraft is to design special areas on the spacecraft's surface with low absorptivity and very high emissivity (low α/ε). Heat

Figure 13-21. Thermal Coatings. We can meet nearly 85% of a spacecraft's thermal-control demands by choosing the right coatings and insulation. Here we see foil wrapping used on the upper atmospheric research satellite (UARS). *(Courtesy of NASA)*

concentrated in these areas can then be readily emitted into space. These surfaces are called *radiators*. Radiators are like windows which allow hot components on the inside of a spacecraft to radiate their heat out into the cold of space. Often a radiator is simply a section of glass coating over a particularly hot section of the spacecraft. This greatly increases the emissivity of that part of the spacecraft so more heat will radiate away. By design, radiators have very low absorptivity and high reflectivity, so less heat comes in to begin with. The radiators on the Space Shuttle are quite evident on the inside of the payload bay doors, as shown in Figure 13-22.

Radiators can be simple or complex. Simple radiators rely on *second-surface mirrors* or *optical-surface reflectors* which reflect incoming heat and efficiently emit the rest. More complex radiators have "venetian blinds" called *louvers* over them that can open or close to better regulate the amount of heat radiated.

Internal Thermal Control

Inside the spacecraft we have different problems. Each subsystem has different thermal requirements, and we must keep them all happy. To begin with, some systems like to stay toasty. For example, propellant lines need to stay warm so they don't freeze. Otherwise, they could burst, spewing propellant everywhere, causing *big* problems. For this reason, they usually have heaters. *Heaters* are just like little electric blankets wrapped around components to keep them warm. They're connected to thermostats (again, just like your electric blanket), which turn the heaters on and off to maintain the required temperature. Of course, because heaters need electrical power, they demand more of the electrical power subsystem.

Most subsystems must stay cool, so getting rid of heat and moving heat around are the most important problems inside the spacecraft. How we solve this problem depends on two things—how fast we need to move the heat and how much heat we need to move. To remove modest amounts of heat from spacecraft components when we're in no big hurry, the easiest way is simply to establish a heat-conduction path from the hot component to a passive external radiator. This can be as simple as a piece of metal connecting the two.

As the amount of heat and the speed with which we must move it increases, we need more complex, active thermal-control methods such as *heat pipes*. Heat pipes are hollow tubes closed on both ends filled with some fluid like ammonia, as shown in Figure 13-23. As one end of the pipe is exposed to a heat source, the ammonia absorbs this heat and vaporizes. The ammonia vapor then flows through the pipe and carries the heat away to the cold end, usually connected to a radiator. At the radiator, the ammonia gas loses its heat and re-condenses as a liquid. It then flows back to the other end along a wick—just as liquid would flow through a candle wick. This is a form of convection in which the work is being done not by gravity but by the wicking effect.

A heat pipe takes advantage of the latent heat absorbed when liquids vaporize. What do we mean by this? If you heat water on the stove, how

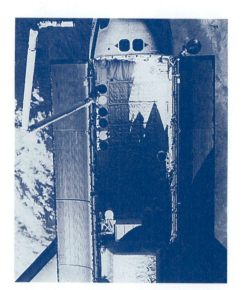

Figure 13-22. Radiators. Radiators, like the ones on the inside of the Space Shuttle's payload bay doors are areas of low absorptivity and high emissivity which can eject waste heat transferred to them. *(Courtesy of NASA)*

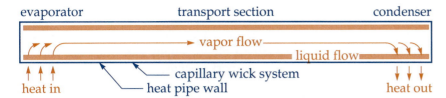

Figure 13-23. Heat Pipes. Heat pipes employ some liquid with a low boiling point inside a hollow tube. As the liquid absorbs heat at the hot end, it vaporizes and carries the heat to the cool end. There it re-condenses and is "wicked" back to the hot end.

hot can it get? Only about 100° C (212° F). No matter how long it's heated, it can reach only this temperature—the boiling point of H_2O. As the water boils and adds heat energy, it changes phase (is vaporized) from a liquid to a gas (steam). Steam is not limited to 100° C; it can get much, much hotter and thus can store more heat. *Latent heat of vaporization* is the principal of storing additional heat in a liquid as it changes phase. If you look at the graph of energy input versus temperature for water (or almost any substance for that matter) in Figure 13-24, you'll see where this latent heat comes in. As the ammonia in a heat pipe vaporizes, it absorbs a large amount of heat due to this phenomenon.

In some cases, heat pipes aren't enough to remove large amounts of heat quickly, so we must resort to other, more complex cooling devices. One option is to use some type of cryogenic cooling, such as liquid nitrogen (70°K), or refrigeration units to keep subsystems cool. In this case the super-cold nitrogen flows by the hot components, transferring heat to the coolant by convection. This method quickly removes vast amounts of heat but has the disadvantage of requiring bulky and unreliable refrigeration units or tanks of liquid nitrogen which eventually get used up.

Another innovative system uses paraffin or some other *phase-change material* with a relatively low melting point to remove heat from a component during times of peak thermal demand. As the paraffin absorbs heat, it melts. When the component is no longer in use and stops producing heat, the melted paraffin conducts or radiates this heat to other parts of the spacecraft to be ejected by radiation. As the paraffin cools, it solidifies for use during the next peak demand cycle. This thermal control method tends to be very reliable because it has no moving parts and the paraffin essentially never wears out. What makes this so efficient is the same principle which makes ice a good thing to put in your cooler—latent heat of fusion. *Latent heat of fusion* is the same basic idea as latent heat of vaporization, but with melting instead of boiling. As the ice melts in your cooler, it takes heat out of your sodas. We applied this principle back in Chapter 10 to remove heat through ablation from spacecraft entering the atmosphere.

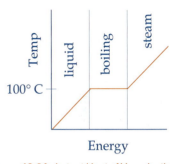

Figure 13-24. Latent Heat of Vaporization. As heat is added to a liquid, its temperature increases linearly until it reaches the boiling point. Then, the temperature stays constant as more heat is added. This additional heat needed to change the phase of the substance from a liquid to a gas is known as latent heat of vaporization.

Life Support Subsystems

As we learned in Chapter 3, space is a hostile place. Charged particles, radiation, vacuum, and free-fall are all potentially harmful, even fatal, to unprepared humans. The life support system provides a "home away from home" for space travelers. To understand its requirements, we can look at humans as just another system (self-loading baggage), like the payload or the data handling subsystem. As such, humans have various needs which the life support subsystem must provide.

In Chapter 3 we focused on the physiological dangers of charged particles and radiation. There we discussed the free-fall environment and the potential problems it poses. Because these inherent dangers tend to have long-term effects, the life support subsystem can't change them. Thus, assuming the thermal control subsystem is maintaining a livable temperature, we'll focus here on the basic necessities to keep humans alive for even a brief time in space:

- Oxygen—at the right pressure
- Water—for drinking, hygiene, and humidity
- Food
- Waste management

From a systems perspective, we can look at human requirements in terms of inputs and outputs. Figure 13-25 shows the relative amounts of oxygen, water, and food an average person needs for minimum life support and the amounts of waste they produce. Let's take a look at each of these requirements to better see what the life support subsystem must deliver.

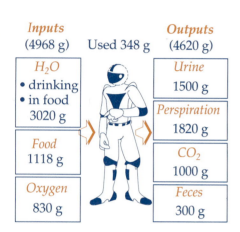

Inputs
(4968 g)   Used 348 g   **Outputs**
(4620 g)

H_2O • drinking • in food 3020 g		*Urine* 1500 g
Food 1118 g		*Perspiration* 1820 g
Oxygen 830 g		CO_2 1000 g
		Feces 300 g

Figure 13-25. Humans as a System. Like any other system, humans take some amount of input, process it, and produce output. Here we see the approximate daily food, water, and oxygen requirements for an astronaut and the corresponding urine, perspiration, CO_2, and feces produced. (Adapted from Nicogossian, et al and Chang, et al)

Astro Fun Fact

Cosmonauts Rescue Crippled Space Station

Power out, life support systems failing, temperature dropping. . .sounds like an episode of Star Trek. But that's the situation faced by cosmonauts Vladimir Dzhanibekov and Viktor Savinykh when they blasted off from the Baikonur Cosmodrome on June 16, 1985, on a mission to rescue the crippled Salyut 7 space station. With the station's automated rendezvous equipment down, the crew had to make a tricky manual docking. Once attached to the station, they cautiously opened the valve into the station to sample its atmosphere. Luckily, the station still held pressure, but it was cold! "Ice was everywhere," Dzhanibekov later commented, "on the instruments, control panels, windows. Mold from past occupations was frozen to the walls." Because their suit thermometers only went down to 0° C, the crew ingeniously decided to spit on the walls and time how long it took to freeze. Using this crude estimate, they determined the station was at –10° C (14° F). The crew's first order of business was to recharge the station batteries so heat, light, and ventilation could be restored. Even bundled up "like babies in a Moscow winter," they found they could stand the cold for only about 40 minutes at a time before retreating into the refuge of their Soyuz capsule. Other problems plagued them as well. Without ventilators to circulate the air, carbon dioxide from their breath hung around their heads, causing headaches and sluggishness. After nearly 24 hours of constant work, they switched on the power and "suddenly the lights turned on and ventilators started whirring. . .the station was saved."

Canby, Thomas Y. <u>Are the Soviets Ahead in Space?</u> National Geographics Vol. 170, No. 4. Washington, D.C.: National Geographics Society, October 1986.

Oxygen

At sea level, we breathe air at a pressure of 101 kPa (14.7 psi). Of this, 20.9% is oxygen (O_2), 78.0% is nitrogen (N_2), 0.04% is carbon dioxide (CO_2) and the rest consists of various trace gasses like argon. During respiration, our lungs take in all of these gasses but only the oxygen gets used. Our bodies use it to "burn" other chemicals as part of our metabolism. Within the lungs, O_2 is transferred to the blood in exchange for a metabolic by-product, CO_2. Exhaling, we dump CO_2 back into the air around us. On Earth, this waste CO_2 is eventually taken up by plants which exchange it back for O_2, and the process continues. In space, it's not that simple.

To provide a breathable atmosphere in space, the life support subsystem must provide O_2 at a high enough partial pressure to allow for comfortable breathing. The *partial pressure* refers to the amount of the total pressure accounted for by a particular gas. Thus, at sea level, the partial pressure of O_2 (PPO$_2$) is 20.9% of 101 kPa (14.7 psi) or around 21 kPa (3.07 psi). After becoming acclimated, people living at high altitudes (above 2000 m or about 6000 ft) show little discomfort with PPO$_2$ of 13.8 kPa (2.0 psi) or less. On the Space Shuttle, PPO$_2$ is maintained at 22 ± 1.7 kPa (3.2 ± 0.25 psi).

Besides keeping the PPO$_2$ high enough, we must also be concerned about getting it too high. Breathing oxygen at too high a pressure is literally toxic. This is a problem which scuba divers must also worry about during deep dives. A PPO$_2$ of less than 48 kPa (7 psi) is usually safe.

Besides providing adequate oxygen to breathe, we must consider other trade-offs. We want the PPO$_2$ to be low enough so it doesn't create a fire hazard in the crew cabin. This was the problem during the Apollo 1 fire. At that time, the cabin atmosphere was pure oxygen, which led to the untimely deaths of three astronauts when a wiring problem caused a fire during a routine ground test. The pure oxygen atmosphere was blamed for causing the fire to spread much more rapidly than it would have in a normal O_2/N_2 atmosphere. Since then, cabin atmospheres have contained a mixture of oxygen and nitrogen to decrease this fire hazard. The correct mixture and pressure of gasses is also important for thermal control. Convective heat transferred to the cabin atmosphere is also used to cool electronic components, so atmospheric composition and circulation must support that function.

A final concern for cabin air is that it creates a problem for astronauts leaving the spacecraft for extravehicular activity (EVAs or space walks). The Space Shuttle maintains a sea-level pressure of 101 kPa (14.7 psi). Because of design limitations, space suits operate at 29.6 kPa (4.3 psi). To avoid potential decompression problems, the Shuttle is taken down to 70.3 kPa (10.2 psi) 12 hours before a planned EVA. Even then, astronauts must breathe pure oxygen for 3 – 4 hours before the EVA to purge nitrogen from their bodies. Otherwise, the nitrogen would form bubbles in the blood, causing a potentially deadly problem known as "the bends," which is all too familiar to scuba divers. Given the relative infrequency of EVAs, these procedures are considered a minor inconvenience. On space stations or Moon bases, daily EVAs could require new space suits or a different cabin atmosphere.

Where does the air for the life support subsystem come from? For Space Shuttle missions, tanks hold liquid oxygen and nitrogen. As liquids, they take up much less volume than as gasses. The liquid is then warmed and evaporated into a gas to maintain the correct PPO_2 and to replenish atmosphere which is vented or lost through small leaks. Part of the life support subsystem is feedback control, with sensors to constantly monitor the pressure and composition of cabin air and alert the crew and ground controllers to any problems before they can become a health hazard.

Water and Food

With air taken care of, we can now turn our attention to one of the simpler pleasures of life—eating and drinking. We normally go about our day eating and drinking without much concern about the total mass we consume. For space missions, every gram taken to orbit represents a huge cost, so we want very little waste. On the other hand, because you can't call out for pizza, it's important to carry enough water and food for any contingency. Thus, we must fully understand the crew's needs to design the life support subsystem.

Water is needed onboard for many reasons. Minimum human needs for drinking water to stay alive are about two liters per day (about 2 kg or 4.4 lbs.). Another liter or so of water is needed for food preparation and rehydration. Besides this minimal amount of water to maintain life, astronauts need water for personal hygiene (washing, shaving, etc.) as well as doing the dishes and washing clothes. All told, this can add up to more than 20 liters per person per day.

We also need food. The average human needs about 13 calories per pound of body weight per day to maintain their present weight. This means a typical 68 kg (150 lb.) astronaut needs at least 1950 calories per day—more for strenuous EVAs. Food in space has come a long way since the tubes of peanut butter and Tang used during the Gemini missions. Nowadays, the food astronauts eat isn't all that different from what we're use to on Earth. Figure 13-26 shows an astronaut ready to attack floating tortillas. Planners pretty much defer to the recommended daily allowances (RDA) of carbohydrates, protein, fat, vitamins, and minerals we're all use to reading about on the labels of everything we eat. To conserve weight and volume, much of the food is dehydrated or freeze-dried and then rehydrated on orbit. For short-term missions, fresh fruit and other perishable items are also taken along—space and weight permitting. Depending on how it's packaged and the total calories needed, we plan as much as 2 kg of food per person per day.

Figure 13-26. Food in Space. Food astronauts eat isn't all that different from Earth food. However, to save space, more dehydrated and freeze-dried foods are often used. Above you can see a typical meal tray used on the Shuttle. Below, astronaut James Wetherbee prepares his lunch while a package of tortillas floats in front of him. *(Courtesy of NASA)*

Where does all this food and water come from? For U.S. manned flights during Apollo and on the Space Shuttle, ample water has been available as a by-product of the fuel cells used to produce electrical power. Thus, astronauts have had the luxury of using as much water as they want. Waste water is simply dumped overboard into the vacuum of space. All the necessary food has also been carried along. For extended missions lasting several months or more, such as those flown by the Soviet/Russian Mir space station, unmanned re-supply spacecraft are sent up every few months with more groceries.

For missions to the Moon or Mars, it's no longer practical to take along all the supplies or rely on re-supply missions. Instead, we need to establish a closed system which can reclaim and recycle water and other waste. Such closed-loop systems could recycle urine, feces, and CO_2 to provide water and food to the crew, as shown in Figure 13-27. While this may not sound appetizing, it promises to greatly reduce the amount of mass needed for very long missions. Scientists are investigating life support subsystems which can effectively reclaim and recycle water, the heaviest item to take along. One limited approach to this idea is to reclaim so called "gray" water (used for washing and rinsing) and reuse it for purposes other than drinking. Other scientists are looking beyond such limited systems to ones that would fully recycle nearly everything onboard and provide all oxygen, water, and food crew members would need for years at a time. However, such systems are still far in the future.

Waste Management

Humans produce waste in the form of urine, feces, and CO_2 simply as a by-product of living. Collecting and disposing of this waste in an effective and healthy manner is one of the biggest demands on the life support subsystem. Urine and feces pose health risks as well as odor problems. CO_2 poses a more subtle problem. Unless it is removed from the air, its concentration will build up, eventually causing increased heart and respiratory rates, a change in the acidity of the body, and other health complications.

Collecting urine and feces in a free-fall environment brings up one of the most commonly asked questions of the entire space program "How do you go to the bathroom in space?" In the early days of the space program, those dashing young astronauts with the "right stuff" were subjected to a most humbling experience. They collected urine and feces using inconvenient and messy methods euphemistically called *intimate-contact devices*. Because they were all males at that time, they collected urine using a roll-on cuff placed over the penis and connected to a bag. They collected feces using a simple diaper, or, even more messy, a colostomy-type bag taped or placed over the buttocks. Urine and feces could then either be dumped overboard or returned to Earth for analysis and disposal.

The Skylab program ushered in a new era in free-fall toilets. For the first time, intimate contact devices were no longer necessary. An advanced version of this system is now on the Space Shuttle, as shown in Figure 13-28. However, the free-fall toilet created (and still creates) considerable challenge to engineers. We tend to take for granted all the work that gravity does for us every time we go to the bathroom. On Earth, urine and feces fall away from our bodies; in orbit, it's a different story. In free-fall, you, your urine, and feces are all falling at the same rate. As a result, waste isn't compelled to move away from you, so it tends to just float next to your body in a smelly blob. To get around this problem, forced air is used to create a suction effect, pulling both urine and feces into a waste-collection system where they can be contained and disposed of.

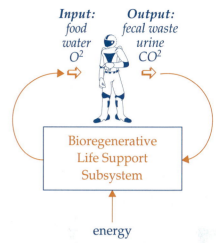

Input: food water O^2

Output: fecal waste urine CO^2

Bioregenerative Life Support Subsystem

energy

Figure 13-27. Bioregenerative Life Support Subsystems. For long space missions to Mars or beyond, bioregenerative life support subsystems may be needed. These systems allow us to "close the loop" and recycle all human waste into food, water, and oxygen.

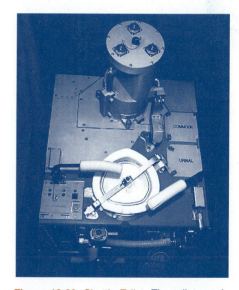

Figure 13-28. Shuttle Toilet. The toilet used by astronauts on the Space Shuttle is designed to compensate for the free-fall environment. On Earth, gravity does all the work; in free-fall, it's not so easy. *(Courtesy of NASA)*

Unfortunately, this method doesn't work nearly as well as good ol' gravity, but at least it's a vast improvement over the older methods.

In comparison, removing CO_2 from the air is much simpler and far less messy. On the Space Shuttle, air is filtered through canisters containing lithium hydroxide and charcoal. The lithium hydroxide chemically reacts with the CO_2, trapping it in the filter. The charcoal absorbs odors and other contaminants. The crew must change these canisters periodically during the flight as one astronaut is shown doing in Figure 13-29. On Skylab, for missions lasting up to 84 days, CO_2 filtration was done using a molecular sieve which was then "baked out" and re-used.

Someday, closed-loop life support subsystems, like the ones mentioned earlier, may make it easier for astronauts to eat, drink, breathe, and go to the bathroom. Until then, these pioneers in the high frontier must accept some hardship.

Figure 13-29. "Scrubbing CO_2." Lithium hydroxide canisters, like the ones being changed out here, are used to remove carbon dioxide (CO_2) from the air on the Space Shuttle. *(Courtesy of NASA)*

■ Section Review

Key Terms

absorptivity, α
albedo
conduction
convection
emissivity, ε
environmental control and life
    support subsystem (ECLSS)
flash evaporator
heat pipes
heaters
intimate-contact devices
kapton
latent heat of fusion
latent heat of vaporization
louvers
optical-surface reflectors
partial pressure
phase-change material
radiation
radiators
reflectivity
second-surface mirrors
thermal equilibrium
thermal control subsystem
transmissivity

Key Equations

$$q = \sigma \varepsilon \, AT^4$$

Key Concepts

➤ The thermal control subsystem on a spacecraft regulates and controls the temperature of sensitive components

➤ A spacecraft heats up from three main sources:
- The Sun
- The Earth (for spacecraft in low-Earth orbits)—from albedo and "Earth shine"
- Electrical components within the spacecraft

➤ Heat can transfer in one of three ways:
- Conduction
- Convection
- Radiation

➤ Various systems control heat on spacecraft:
- External heat transfer can use only radiation in the vacuum of space
  - Engineers use special coatings and several layers of insulation to change the absorptivity and emissivity of the spacecraft's surface, which prevents heat from getting in
  - Radiators, sometimes with venetian-blind attachments called louvers, help eject heat
- Internal heat transfer uses increasingly complex methods depending on the amount of heat that must be moved and how fast you want to move it
  - Heaters keep some systems, such as propellant lines, warm
  - Simple conducting paths can remove modest amounts of heat over long periods
  - Heat pipes use the latent heat of vaporization of a liquid like ammonia to quickly move heat from one point to another
  - Refrigerators or cryogenic cooling using liquid nitrogen can quickly remove large amounts of heat. However, their complexity make them less reliable for longer missions.
  - Phase-change materials such as paraffin, which use the latent heat of fusion (like ice in a cooler), can efficiently remove heat from individual components

➤ Life support subsystems keep humans alive in space:
- They provide a pressurized environment with all the comforts of home—air, water, food, and shelter

Example 13-4

Problem Statement

A cube, 1 m on a side, is in orbit around the Earth, well beyond the orbit of the Moon (we can ignore Earth IR and albedo). It's painted with black paint ($\alpha/\varepsilon = 1.11$). Assuming no internal heat source, the solar input (1358 W/m^2) is on one side only, and thermal energy is emitted from six sides. Find the cube's equilibrium temperature.

Problem Summary

Given: cube, 1 m on a side
$\alpha/\varepsilon = 1.11$
no internal heat
$S = 1358$ W/m^2

Find: $T_{equilibrium}$

Problem Diagram

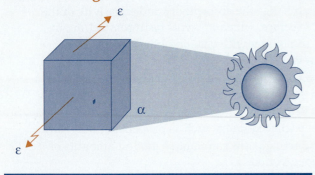

Conceptual Solution

1) Balance heat$_{in}$ = heat$_{out}$, solve for T

$$q_{in} = q_{out}$$

$$A_{in} S\alpha = \sigma \varepsilon A_{out} T^4$$

$$T^4 = \frac{A_{in} S\alpha}{\sigma \varepsilon A_{out}}$$

$$T = \sqrt[4]{\frac{A_{in} S\alpha}{\sigma \varepsilon A_{out}}} = \sqrt[4]{\frac{A S}{\sigma A_{out}}\left(\frac{\alpha}{\varepsilon}\right)}$$

Analytical Solution

1) Solve for T

$$T = \sqrt[4]{\frac{A_{in} S\alpha}{\sigma \varepsilon A_{out}}} = \sqrt[4]{\frac{A S}{\sigma A_{out}}\left(\frac{\alpha}{\varepsilon}\right)}$$

$$T = \sqrt[4]{\frac{(1 \text{ m}^2)(1358 \text{ W/m}^2)(1.11)}{\left(5.67 \times 10^{-8} \frac{\text{W}}{\text{m}^2}\,{}^\circ\text{K}^4\right)(6 \text{ m}^2)}}$$

$$T = 258^\circ \text{ K}$$

Interpreting the Results

A 1 m^2 cube covered with black paint would reach an equilibrium temperature of 258° K (–15.15° C or 4.73° F) when exposed to sunlight in space.

13.4 Structures

▬ In This Section You'll Learn to...

☞ Discuss the main functions of the spacecraft's structures and mechanisms

☞ Discuss basic terms and concepts of structural mechanics

☞ Describe the spacecraft's structural components

It may seem strange to refer to the basic structure of the spacecraft as a system. It doesn't turn on and off, fire rockets, or talk to other spacecraft, but without the structure you wouldn't *have* a spacecraft. The structure holds all the other subsystems together and withstands the stresses, strains, and vibrations of lift-off and operations. In addition, the structure supports all the mechanical components that push, pull, extend, or pivot to do various tasks. In many cases, the success of the entire mission may depend on one mechanism working correctly. For example, solar arrays normally fold up during launch and then deploy by mechanisms once the spacecraft reaches orbit. If the solar panels don't extend, the spacecraft won't be able to get any power. The mission will fail!

Without a strong enough structure, the entire spacecraft would collapse before it got into space. In this section we'll see what the structure does, review some basic mechanics to see how it works, and then look at some of the types of structures commonly used.

Functions

All together, the structure usually accounts for about 20% of the spacecraft's total dry weight (total weight minus propellant). We can think of it as having two parts. The *primary structure* carries most of the loads the spacecraft must withstand. A *load* is any force pushing, pulling, or vibrating the structure. The *secondary structure* holds together all the lightweight wires, pipes, doors, and brackets inside and outside the spacecraft. We'll focus our attention on the primary structure because it contributes most to the weight of the spacecraft.

The primary structure is designed to handle various loads during all phases of the mission, including

- Prelaunch loads
  - Assembly
  - Transporting
  - Testing
  - Stacking

- Launch loads
  - G forces
  - Vibrations
- Operations loads
  - Repetitive loads on orbit—mechanical the thermal cycling
  - Upperstage acceleration
  - Deployment
  - On orbit vibrations
- Entry and Landing
  - Deceleration
  - Heating
  - Landing

Of these, ascent loads are typically the most severe, so we must make sure the spacecraft structure can withstand its violent ride into space. Loads during launch can be quite high, up to nine g's in some cases, depending on the booster rocket. Remember, a load of one g is your weight at the surface of the Earth, so a nine-g load is nine times the weight of the spacecraft. For example, it's the force you'd feel if you could accelerate your car at 88 m/s^2 (288 ft/s^2). That's 0 to 100 km/hr (60 m.p.h.) in 0.3 seconds! But even more severe than the launch loads are vibrations during ascent. If not properly designed, the spacecraft could literally shake apart during launch.

Because structural design requires an understanding of how specific loads affect a particular type of structure, we need to review basic structural mechanics.

Mechanics of Structures

Whether you're building a bridge, a car, or a spacecraft, structures work pretty much the same way. Basically, a structure is anything that carries a load. For example, your skeleton, muscles, and connective tissue form the structure of your body. This structure allows you to stand up under the pull of Earth's powerful gravitational field. And it has enough built-in safety factors to allow you to run and jump—sometimes causing loads many times those due to gravity.

What kinds of things must a structure be designed to deal with? Basically, you can

- Push, pull, twist, or bend it—causing loads, stress and strain
- Shake it—causing vibrations
- Change its temperature—causing thermal stress

Any or all of these things put demands on the structure in some way and can deform or break it if it's not built strong enough. To understand this in more detail, let's take a look at how all of these things affect a structure.

Types of Loads

Loads applied to any structure can either axial, lateral, or torsional. *Axial loads* are those applied parallel to the longitudinal axis of the structure. *Lateral loads* those are applied parallel to the lateral axis of the structure. *Torsional loads*, on the other hand, are produced by twisting or applying a torque to the structure.

To see all of these loads in action, pretend you're playing with a soda can as shown in Figure 13-30. If you push in the can on both ends you're applying an axial load. In this case, the load is said to be one of *compression* or a *compressive load*. If you pull out on both ends you're again applying an axial load but in this case it said to be one of *tension* or a *tensile load*. If you now secure one end of the can to a table top and push or pull on the side of the can you're applying a lateral load. Notice that in this case, one side of the can is in compression while the other is in tension. Finally, if you twist the can as though you're wringing out a wet towel, you're applying a torsional load.

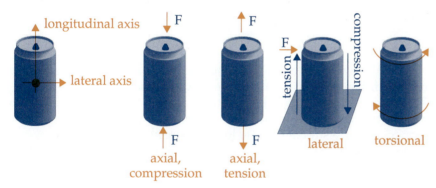

Figure 13-30. Types of Loads. Loads on any structure can be classified as axial (compression or tension), lateral, or torsional.

When axial or lateral loads are applied to a structure, bending can occur. Its easiest to visualize bending for the case of lateral loads. Notice in Figure 13-31 that the lateral load causes a bending moment about the attachment point. A *bending moment, M*, is the result of a load applied some distance away from an attachment point.

$$M = F\,d \qquad\qquad (13\text{-}11)$$

where
M = bending moment (N · m)
F = force (N)
d = distance between load and attachment point (m)

Bending moments can be especially dangerous because even a very small load, when applied over a great enough distance, can cause significant bending.

Throughout a spacecraft's life it will be subjected to all of these types of loads. During launch, compressive loads are most severe. Therefore,

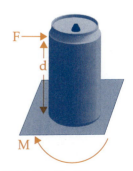

Figure 13-31. Bending Moment. The lateral load applied to the fixed soda can cause a bending moment, M, about the attachment point.

boosters and spacecraft are designed to handle up to 10 g's of axial, compressive load. The attitude control system on the booster works to ensure these loads are directed through the long axis where the vehicle is strongest, otherwise, lateral loads could easily tear the vehicle apart. To achieve the greatest strength with the minimum weight, boosters are designed a lot like soda cans. They are strong in the axial direction—the direction of thrust—but weak in the lateral direction. To illustrate this, have a friend stand on top of a an empty soda can. If your friend is careful, she can do this without crushing it. While she balances on top, you can apply a very slight lateral load to the can and watch your friend come crushing down on it, making the can ready for the recycle bin. Any time a structure, such as a can, fails under a compressive load, it's said to *buckle*.

Once a spacecraft reaches orbit, other types of loads will become important to its structural design. During deployment, the spacecraft may be subjected to a momentary shock causing severe compressive load. Later, as solar cells are deployed or the spacecraft is spun up, load bearing members within the spacecraft will be subjected to tensile and torsional loads.

Stress, Strain, and Shear

Now that we know a little bit about the types of loads that a structure must withstand, we can determine the effect of those loads on the structure. The fundamental question of structural analysis is "will it break?" To answer this question we can look at three parameters which quantify changes within a structure due to loading: stress, strain, and shear.

When you hear someone talk about stress, you probably think about that late night before the big project is due. But for structures, stress has an entirely different meaning. When you apply a axial or lateral load to a structure, that load is essentially spread out over its entire cross-sectional area. Thus, *stress* is the applied load per area. We can compute it using

$$\sigma = \frac{F}{A}$$

(13-12)

where
σ = stress (N/m^2)
F = applied load (N)
A = cross sectional area (m^2)

Notice the units on stress are the same as pressure, force per unit area. One way to visualize the affect of stress on a structure is to imagine inflating a car tire. The more air you pump in, the greater the pressure in the tire. Similarly, the greater the load applied to a given structure, the greater the stress. Depending on the direction of the applied load, stress is either compressive (pushing in) or tensile (pulling out).

As a structure undergoes stress, it begins to deform.   Imagine pulling on a hunk of Play-doh. As you pull (applying an axial load), it stretches out. This change length, or deformation, due to an applied load is known as *strain*. We determine the strain in a structure by looking at the ratio of a structure's change in length compared to its original length as shown in Figure 13-32.

$$\varepsilon = \frac{\Delta L}{L}$$

(13-13)

where
ε = strain (m/m)
ΔL = change in length (m)
L = original length (m)

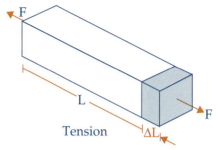

Figure 13-32. Strain. Strain describes how an object changes in length due to stress.

All types of loads—axial, lateral, and torsional—can cause a third effect within a structure known as shear. When transverse loads are applied to our soda can, as shown in Figure 13-33, the magnitude of the internal force within the structure that results from these transverse loads, or due to a torsional load, is known as *shear*. The resulting stress generated by the shearing force is known as *shear stress*.

Whether or not a structure will actually break as a result of too much stress, strain or shear stress depends on its material properties. In other words, the same amount of strain that will break apart a steel girder wouldn't even phase a hunk of Play-doh. We'll look briefly at how we quantify these material properties later in this section.

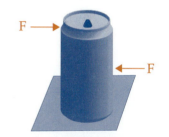

Figure 13-33. Shear. When a transverse load, as shown above, is applied to a structure, the resulting shear will cause shear stress. Depending on how they're applied, torsional or axial loads may also cause shear.

Vibrations

Getting into space can get you all shook up. If you've ever ridden on an old wooden roller coaster or in the back of a pickup going down a bumpy road, you've experienced vibrations similar to launch. How an object responds to vibration depends on its natural frequency. An object's *natural frequency* is the frequency at which it will vibrate if given one sudden impulse and then left to itself. When you vibrate an object at any frequency *other* than its natural frequency nothing interesting happens. The applied or *forcing vibration* gets damped out so no great harm is done. This is because as the forcing vibration is applied, sometimes it works to increase the amplitude at which the object vibrates and sometimes it works to decrease it depending on the timing. However, when you vibrate something at a frequency *equal* to its natural frequency, something very interesting happens—it begins to *resonate*.

To understand resonance, think back to when you used to play on a swing, as shown in Figure 13-34. After giving yourself a little push to get started, you pumped your legs at just the right time, gradually building up the amplitude of your swing. What you were doing was timing the pump of your legs to coincide with the frequency at which you were swinging. In other words, you provided a cyclic input equal to the current natural frequency. This allowed you to work *with* the swing instead of *against* it every time.

Resonance is powerful. Because the forcing vibration amplifies the oscillation of the structure, even a very small input, over time, can cause

Figure 13-34. Resonance. When you play on a swing you pump your legs at just the right time to increase the amplitude at which you're swinging. In this way, the forcing vibration is in resonance with the natural frequency of the swing, so you go higher.

an undamped structure to build up large-amplitude vibrations. One of the classic examples of resonance is an opera singer who is able to shatter a glass with just the sound of her voice. Setting the pitch of her note to coincide with the natural frequency of the glass causes large-amplitude vibrations, which eventually cause the glass to break.

A more dangerous resonance occurs during earthquakes. During the 1985 earthquake in Mexico City, scientists recorded the primary frequency of the forcing vibration to be around 0.5 Hz. After the quake, they found some of the most severe damage occurred in buildings between 10 and 14 stories tall while taller and shorter buildings were relatively unaffected. It turns out that the natural frequency of 10–14 story buildings is about 0.5 Hz!

One way to estimate the natural frequency of a structure is to assume it's basically a spring. The frequency then depends on its mass and something called a spring constant. The *spring constant, k*, is a measure of how much force it takes to compress a spring, or any structure, by a certain amount. Knowing these values, we can compute the natural frequency using

$$f_{natural} = \frac{1}{2\pi}\sqrt{\frac{k}{m}} \qquad (13\text{-}14)$$

where

$f_{natural}$ = natural frequency (Hz)
k   = spring constant (N/m)
m   = mass (kg)

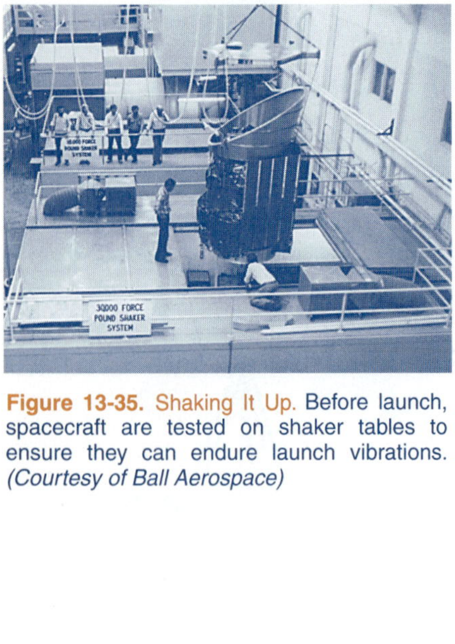

During launch, the booster will vibrate at some known rate (published by the booster manufacturer). To avoid resonances that could damage any part of the spacecraft, we must design the spacecraft so its natural frequencies are different from the booster's frequencies. Prior to launch, spacecraft are tested on shaker tables, as shown in Figure 13-35, to ensure they'll survive the trip. Once in space, there are other vibrations to consider. For example, thruster firings, astronaut movements, and thermal shock as the spacecraft moves in and out of sunlight can all induce enough vibrations to cause problems for sensitive instruments trying to stay locked onto a target.

Figure 13-35. Shaking It Up. Before launch, spacecraft are tested on shaker tables to ensure they can endure launch vibrations. *(Courtesy of Ball Aerospace)*

Thermal Stress

Have you ever wondered where all those cracks in the sidewalk come from? As a material heats up, it expands. Similarly, as it cools, it contracts. This is no problem if it has plenty of space to expand into or isn't constrained from contracting. But in the case of the sidewalk, which is locked in on both sides, this expansion and contraction produces stresses which eventually cause it to crack. Of course a similar problem can occur in a space structure. As it goes in and out of Earth's shadow, it almost literally goes from fire to ice. These frequent thermal cycles can cause stresses in the structure which, if not planned for, can degrade the mission. For example, once the Hubble Space Telescope was put into orbit, engineers found that thermal expansion of the beam holding the solar array caused vibration in the entire structure. This made it very difficult to keep its huge telescope focused on distant objects.

Thermal expansion and contraction can be quantified by knowing an object's length and a material property called its *coefficient of thermal expansion, α*. We can then find the change in length, δ, from

$$\delta = \alpha \, (\Delta T) \, L \qquad\qquad (13\text{–}15)$$

where
δ = expansion or contraction (m)
α = coefficient of thermal expansion (1/°C)
ΔT = temperature change (°C)
L = original length (m)

Material Properties

The ultimate effect of stress, strain, vibration, bending, and thermal stress depends on what the structure is made of—its material properties. When choosing what material to use to build a structure, we must consider its

- Thermal properties—how much will it expand and contract?
- Strength and stiffness—how well will it take the loads?
- Density—how much will it weigh?
- Cost
- Ease of manufacturing—how easy is it to bend, cut, and weld?

We've already seen how different materials react to thermal stress. Materials react differently to loads as well. For example, you could easily pull apart a hunk of Play-doh, but only Superman could do the same thing with a hunk of steel. To describe this basic property of materials, we can once again treat any structure like a spring. This allows us to apply Hooke's Law, named for English mathematician Robert Hooke (1635 – 1703), which describes the stress in a structure in terms of its strain and a new parameter—modulus of elasticity, E.

$$\sigma = E \, \varepsilon \qquad\qquad (13\text{-}16)$$

where
σ = stress (N/m^2)
E = modulus of elasticity or Young's modulus (N/m^2)
ε = strain (m/m)

Modulus of elasticity, E, also called *Young's modulus* after English scientist Thomas Young (1773 – 1829), is a basic property of any material describing the amount of deformation it will undergo for a given amount of stress. For example, steel has a modulus of elasticity of around 200×10^9 N/m^2, but wood is only about 2×10^6 N/m^2, telling us what we already knew— steal is stiffer than wood.

Young's modulus, along with other measures of a material's behavior, can then be used for selecting which material to use (aluminum or steel or Play-doh).

As different materials are subjected to stresses and strain, they behave in predictable ways. As a load is applied to most metals, they first begin to deform elastically. That is, they stretch but then return to their original size and shape when the load is removed. In this *elastic region*, Young's modulus is constant, so the relationship between stress and strain is linear. This elastic region will be maintained up to a point called the *proportional limit*. At this point, Young's modulus is no longer constant, and the material will be permanently deformed. If you continue to apply a load after the material reaches its proportional limit, you'll reach a *yield point*, where a residual strain of 0.2% will remain after the load is removed. Applying load after the yield point will eventually lead to the material failing or breaking. This is known as the *ultimate failure* point.

Any material that can withstand significant stress before failing is called *ductile*. For example, Play-doh would be considered ductile because you can stretch it pretty far before it breaks. A material which breaks after only a small amount of strain is called *brittle*. (That's why it's called peanut brittle, not peanut ductile). The stress-strain curve in Figure 13-36 illustrates all of these relationships.

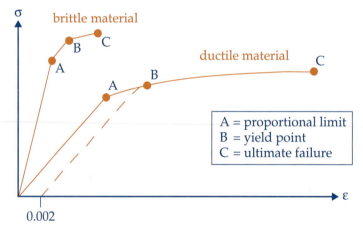

Figure 13-36. Stress-Strain Curves. Typical stress-strain curves show the difference between ductile and brittle materials.

In most cases, we assume a material is *homogeneous*; that is, it has the same properties throughout and in all directions. This is a pretty good assumption for common spacecraft materials, such as aluminum, but not all materials behave this way. Wood, the oldest known building material, for example, is generally not homogeneous. It has a grain and tends to be much stronger with the grain than across it. Ironically, spacecraft engineers are re-discovering this ancient material, or at least its basic properties, in the form of composites.

Composites are basically "designer" materials meant to carry a specific load in a specific way. Like wood, composites can be made to be very strong in specific directions. What makes them so attractive for aerospace applications (as well as tennis rackets, etc.) is that they can achieve the same strength at a fraction of the weight of aluminum or other traditional materials. After all, if a structure needs to be strong only in *one* direction, why not build it to be strong only in that direction and save lots of weight?

A third consideration in material selection is density. For a given strength, we'd like to choose the least dense material to reduce the overall weight of a specific structural element. Of course, cost is also a big driver in deciding what type of material to use. Even if your analysis indicates it would be the best material, it's doubtful you'll build a spacecraft entirely out of diamond! Thus, we must always try to balance the best material with reasonable costs. Another important indication of cost is ease of manufacture. Again, analysis may show one material can handle the loads better than another, but if it's too difficult or dangerous to manufacture, it's not practical to use. A good example is beryllium, which has many advantageous material properties but is toxic to work with.

Types of Structures

To handle all loads (and especially the launch load) we use different types of structural components. Among these are beams, trusses, panels, monocoque structures, and semi-monocoque structures.

Beams are long, straight components which support loads either lengthwise (axial) or crosswise (lateral). If you hold a book out at arm's length, your arm is a beam, attached at the shoulder, with a lateral load at the end point (your hand). An example of a beam is shown in Figure 13-37.

Trusses are essentially a collection of beams, all connected at various angles to achieve a higher degree of strength than you could get from an individual beam. Take a look at any large bridge and you're likely to see trusses. Figure 13-38 shows astronauts constructing a truss structure in the Space Shuttle's payload bay.

Panels are thin walls designed to enclose a structure and, in some cases, handle loads. Often panels are constructed from two thin panels separated by a filler. This makes a "panel sandwich" as shown in Figure 13-39 with thin plates of aluminum as the bread and a lightweight honeycomb material of aluminum or other material in between. The advantage of the honeycomb is that, at a fraction of the weight, it offers the same or even better strength and stiffness as an equal-sized panel of solid metal. You can see this honeycomb concept in an average cardboard box. If you tear apart one of the larger boxes, you'll see thin paper outer layers separated by corrugated paper to achieve high strength with low weight. Honeycomb panels are in the sides of the Magellan spacecraft.

A *monocoque* structure is just a fancy name for a thing-walled tube, but the tube doesn't have to be perfectly round. Monocoque structures can be almost any shape as long as they consist of one unbroken "cell" of material. A soda can is a good example of a monocoque structure and

Figure 13-37. Beams. Beams are straight components which support loads.

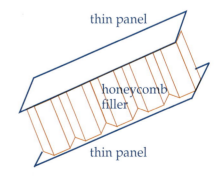

Figure 13-39. Honeycomb Panels. Panels support loads in a variety of ways. Honeycomb panels achieve great strength with light weight.

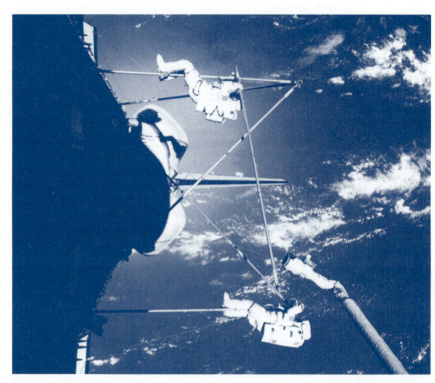

Figure 13-38. Astronauts Assemble Truss. Truss structures, like the one being assembled by astronauts, may one day be used to form large structures in space. *(Courtesy of NASA)*

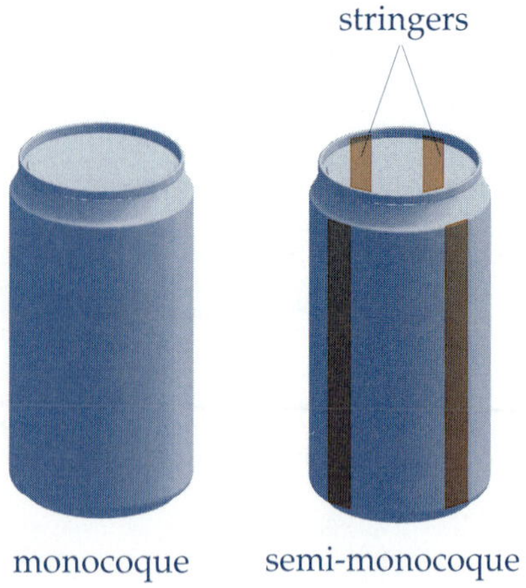

stringers

monocoque   semi-monocoque

Figure 13-40. Monocoque and Semi-Monocoque Structures. Monocoque structures are basically thin-walled tubes carrying a load. Semi-monocoque structures achieve greater strength through the use of stringers as stiffeners.

demonstrates their usefulness. Soda cans are designed to take tremendous loads (try standing on one sometime!) and yet be extremely light. Monocoque structures are used onboard spacecraft for the same reason. The Magellan spacecraft also shows this structure. *Semi-monocoque* is simply a monocoque structure with added reinforcements called stringers. *Stringers* are designed to increase strength without significantly increasing weight. An example of each is shown in Figure 13-40.

As you can see, designing a structure, like all design efforts, is iterative. Once we reach a preliminary design, we must go back to square one to verify that the design will work. The results from this step may require us to select new materials or to redesign. Eventually, a compromise design is reached which satisfies mission requirements and provides the necessary support to hold the spacecraft together.

While few people notice the engineering beauty of a bridge they're driving over, everyone notices if something breaks. Depending on how you look at it, the spacecraft structure can be one of the most interesting subsystems onboard. After all, every other system literally leans on it for support.

Section Review

Key Terms

axial loads
beams
bending moment, M
brittle
buckle
coefficient of thermal expansion, α
composites
compression
compressive load
ductile
elastic region
forcing vibration
homogeneous
lateral loads
modulus of elasticity, E
monocoque
natural frequency
panels
primary structure
proportional limit
resonate
secondary structure
semi-monocoque
shear
shear stress
spring constant, k
strain
stress
stringers
tensile load
tension
torsional loads
trusses
ultimate failure
yield point
Young's modulus

Key Concepts

➤ The spacecraft structure accounts for up to 20% of the dry weight of the entire spacecraft:
  - The primary structure carries all the major loads. The most stressful of these are the vibrations and other loads associated with launch.
  - The secondary structure holds all the individual subsystems together and attaches them to the primary structure

➤ The loads a structure must be designed to take fall into four types:
  - Shaking—this causes vibrations. Any structure has a natural frequency of vibration depending on its mass and inherent stiffness. Designers must ensure the natural frequency of the spacecraft structure is different from the natural frequency of the booster; otherwise, dangerous resonance can occur, which could tear the structure apart.
  - Pushing (compression) or pulling (tension)—this causes stress and strain
  - Bending—bending can lead to more stress
  - Changes in temperature—heating causes expansion; cooling causes contraction. Both can lead to stress in the structure.

➤ Designers choose material for spacecraft structures based on various qualities:
  - Thermal properties—quantified by coefficient of thermal expansion
  - Stiffness—quantified by Young's modulus
  - Density
  - Cost
  - Ease of manufacturing

➤ Several structural components can be used in any spacecraft:
  - Beams
  - Trusses
  - Panels
  - Monocoque and semi-monocoque

Continued on next page

Key Equations

$$M = F\,d$$

$$\sigma = \frac{F}{A}$$

$$\varepsilon = \frac{\Delta L}{L}$$

$$f_{natural} = \frac{1}{2\pi}\sqrt{\frac{k}{m}}$$

$$\delta = \alpha\,(\Delta T)\,L$$

$$\sigma = E\,\varepsilon$$

Key Concepts (Continued)

➤ Designing a spacecraft structure requires several steps:
- Classify loads environment
- Determine spacecraft mass characteristics
- Select structural materials to minimize weight, temperatures, cost, and other requirements
- Do a preliminary design of the primary structure to take primary loads
- Do a preliminary design of the secondary structure to support subsystems
- Analyze the structure design to verify it will meet requirements with enough margin

Example 13-5

Problem Statement

A 2 m-long beam, 0.1 m in diameter, is part of a truss structure supporting a new space telescope. During docking maneuvers with the Space Shuttle, the beam is subjected to a 1000 N axial, compressive load. Is the stress in the beam under the proportional limit of 10^7 N/m²?

Problem Summary

Given: L = 2 m
       D = 0.1 m
       F = 1000 N
Find: σ

Problem Diagram

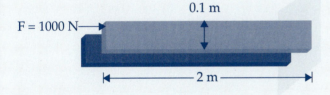

0.1 m

F = 1000 N→

2 m

Conceptual Solution

1) Solve for area, A
   $$A = \pi\, r^2$$

2) Solve for stress, σ
   $$\sigma = F/A$$

3) Compare to proportional limit

Analytical Solution

1) Solve for area, A
   $$A = \pi\, r^2$$
   $$A = \pi(0.05 \text{ m})^2 = 7.854 \times 10^{-3} \text{ m}^2$$

2) Solve for stress, σ
   $$\sigma = \frac{F}{A} = \frac{1000 \text{ N}}{7.854 \times 10^{-3} \text{m}^2} = 1.273 \times 10^5 \text{N/m}^2$$

3) Compare to proportional limit

4) 1.273×10^5 N/m² $< 10^7$ N/m²

Interpreting the Results

The 1000 N load on this beam will cause a 1.273×10^5 N/m² stress, well below the proportional limit.

Example 13-6

Problem Statement

At a temperature of 25° C a beam forming part of a space truss structure is 10.1 m in length. As the beam enters full sunlight, it reaches a temperature of 100° C. If the coefficient of thermal expansion is $3 \times 10^{-5}/°$ C, how much strain is added to the beam?

Problem Summary

Given: $L_{initial} = 10.1$ m
$T_{initial} = 25°$ C
$\alpha = 3 \times 10^{-5}/°$ C
$T_{final} = 100°$ C

Find: ε

Conceptual Solution

1) Solve for expansion, δ, ΔL
$$\delta = \alpha \, (\Delta T) \, L_{initial}$$
$$\Delta L = \delta$$

2) Solve for strain, ε
$$\varepsilon = \Delta L / L_{initial}$$

Analytical Solution

1) Solve for expansion, δ
$$\delta = \alpha \, (\Delta T) \, L_{initial}$$
$$\delta = (3 \times 10^{-5}/° \text{ C}) \, (100° \text{ C} - 25° \text{ C}) \, (10.1 \text{ m})$$
$$\delta = 0.0227 \text{ m}$$
$$\Delta L = 0.0227 \text{ m}$$

2) Solve for strain, ε
$$\varepsilon = \frac{\Delta L}{L_{initial}} = \frac{0.0227 \text{ m}}{10.1 \text{ m}} = 2.248 \times 10^{-3} \text{m/m}$$

Interpreting the Results

After entering the sunlight the beam will expand by 0.0227 m (almost a full inch). This is a strain of 2.248×10^{-3} m/m.

References

Asimov, Isaac. *Asimov's Biographical Encyclopedia of Science and Technology.* Garden City, NJ: Doubleday and Company, Inc., 1972.

Beer, Ferdinand P. and Russel E. Johnson, Jr. *Statics and Mechanics of Materials.* New York, NY: McGraw-Hill Inc., 1992.

Chang, Prof. I. Dee (Stanford University), Dr. John Billingham (NASA Ames), Dr. Alan Hargen (NASA Ames). "Colloquium on Life in Space". Spring, 1990.

Chetty, P.R.K. *Satellite Power Systems: Energy Conversion, Energy Storage, and Electronic Power Processing.* George Washington University Short Course 1507. October 1991.

Chetty, P.R.K. *Satellite Technology and Its Applications.* New York, NY: THB Professional and Reference Books, McGraw-Hill, Inc., 1991.

Doherty, Paul. "Catch a Wave." *Exploring.* Vol. 16, No. 4, (Winter 1992): 18–22.

Gere, James M. and Stephen P. Timoshenko. *Mechanics of Materials.* Boston, MA: PWS Publishers, 1984.

Gonick, Larry and Art Huffman. *The Cartoon Guide to Physics.* New York, NY: Harper Perennial, 1991.

Gordon, J.E. *Structures: Why Things Don't Fall Down,* New York, NY: Da Capp Press, Inc., 1978.

Gunston, Bill. *Jane's Aerospace Dictionary.* New Edition. London, U.K.: Jane's Publishing Co., Ltd., 1986.

Holman, J.P. *Thermodynamics.* New York, NY: McGraw-Hill Book Company. 1980.

Pitts, Donald R. and Leighton E. Sissom. *Heat Transfer.* Schaum's Outline Series. New York, NY: McGraw-Hill, Inc., 1977.

Wertz, James R. and Wiley J. Larson. *Space Mission Analysis and Design.* Dordrecht, the Netherlands: Kluwer Academy Publishers, 1991.

Nicogossian, Arnauld E., Huntoon, Carolyn Leach, Sam L. Pool. *Space Physiology and Medicine.* 2nd Edition, Philadelphia, PA: Lea & Febiger, 1989.

MacElroy, Robert D. Course Notes from AA 129, Life in Space, Stanford University, 1990.

Poole, Lon. "Inside the Processor", *MacWorld*, October 1992, pp. 136–143.

Mission Problems

13.1   Data Handling

1   List what the spacecraft's data handling subsystem does and oversees.

2   What are the parts of the data-handling hardware?

3   What are the two types of software and what do they do?

4   Define bit, byte, and throughput.

5   Designers are looking at the data handling subsystems for two different spacecraft. One will orbit the planet Pluto to take high-resolution photographs of this cold, mysterious planet. The other will be in low-Earth orbit to detect forest fires. Discuss the trade-offs in complexity for onboard data-handling for each of these two spacecraft.

13.2   Electrical Power

6   A 30-amp circuit has a 10 Ω resistor in it. What is the voltage drop across the resistor?

7   A spacecraft designer wants to supply 500 W of power at 28 V. What will be the current?

8   A spacecraft in Earth orbit has its solar panels at a 15° angle to the Sun. If the efficiency of the solar cells is 12%, what is the power output? (Assume average incident solar power density for Earth = 1358 W/m^2).

9   What is the basic operating principle of a dynamic solar power system?

10   What types of chemical power systems are used onboard spacecraft?

11   What is the difference between primary and secondary batteries?

12   A spacecraft is in a low-Earth orbit at an altitude of 350 km. What is the maximum time of eclipse for this orbit?

13   A spacecraft with a requirement for 500 W of continuous power is in an orbit with a maximum eclipse time of 32 minutes. If the maximum depth of discharge (DOD) for the batteries is 30%, what battery capacity do we need? (Give answer in W·hr).

14   What does the electrical power subsystem do?

15   Give an example of a mission which would need an RTG.

13.3   Environmental Control and Life Support

16   What does the spacecraft's environmental control and life support subsystem do?

17   Describe the potential sources of heat for a spacecraft.

18   Describe the three mechanisms of heat transfer. Give examples of each from your everyday life.

19   You're holding a 0.1 m metal rod 0.01 m in diameter into a pot of boiling water (100°C). If the thermal conductivity of the rod is 100 W/°K· m, and your hand is at normal body temperature (37°C), what is the rate of energy transfer along the rod?

20   A section of Space Shuttle tile is exposed to 1200° K on entry. If the tile is 0.1 m thick with an area of 0.01 m^2, what is the rate of heat transfer through the tile? The thermal conductivity of shuttle tile is about 0.108 W/°K · m.

21   A 1 m^2 spacecraft radiator with an emissivity of 0.85 is to be used to eject 100 Joules of heat per second. What temperature must it be?

22   A cube 1 m on a side is in interplanetary space the same distance from the Sun as the Earth. If the absorptivity, α, of the surface is 0.3 and the emissivity, ε, is 0.7, what's the thermal-equilibrium temperature of the cube? Assume no internal heat.

23   List the various ways a spacecraft can transfer heat internally.

24   Define latent heat of fusion and describe how this principle can be used onboard a spacecraft.

25 Discuss ways a spacecraft can eject heat.

26 Describe the basic operating principal of a spacecraft radiator. How is it different from a louver?

27 Your car is said to have a "radiator," but does it really radiate? Explain.

13.4 Structures

28 Approximately what percentage of a typical spacecraft's mass is structure?

29 What is the difference between primary and secondary structure?

30 List the loads that a spacecraft structure may be subjected to. Which of these is typically most severe?

31 What are the four basic types of structural members? Give examples of their use in everyday life.

32 What is natural frequency? Why must engineers be concerned with it?

33 What is the natural frequency, in Hz, of a 10 kg spring with a spring constant of 0.1 N/m?

34 An Earth-observation satellite will be launched on a booster with high-amplitude ascent vibrations at a frequency of 25 Hz. If the mass of the spacecraft is 1000 kg, what spring constant must the structure avoid to prevent resonance?

35 A cylindrical rod 10 m long with a diameter of 0.1 m is subjected to a 1000 N axial tensile load. What is the stress in the rod?

36 If the rod in Problem 35 deforms by 0.001 m, what is the strain?

37 If the rod in Problem 35 is bolted to the side of the spacecraft on one end and subjected to a 200 N lateral force on the other end, what is the bending moment?

38 The rod in Problem 35 has a coefficient of thermal expansion of 0.01/° C. If the 10 m length was measured at room temperature (21° C), what will the strain be if the temperature is 41° C? If the temperature is 1° C?

39 What factors do engineers consider in choosing a material for a structure?

40 A 5 m beam is subjected to a $100 \, \text{N/m}^2$ compressive stress. If the Young's modulus of the material is $50 \times 10^9 \, \text{N/m}^2$, what is the strain in the beam?

Mission Profile—HST

The Earth's atmosphere has always been a problem for astronomers—it distorts and attenuates incoming light. The Hubble Space Telescope (HST) was designed to provide a remarkable new view of the galaxy and beyond. Placed above the Earth's atmosphere, Hubble can detect objects 25 times fainter than any visible from Earth's surface. Thus, astronomers see a universe almost 250 times larger than what is visible on Earth.

Mission Overview

Hubble's mission is to provide an orbiting platform for space-based astronomy, avoiding the interference of the Earth's atmosphere. HST was designed with three main abilities:

✓ High angular resolution to provide fine image detail

✓ Ultraviolet performance to provide ultraviolet images and spectra

✓ High sensitivity to detect very faint objects

Mission Data

✓ Circular orbit at 607 km (377 miles) altitude and 28.5° inclination

✓ Pointing accuracy of 0.00000278° (0.01 ± 0.007 arcsec) for up to 24 hours. This means it could focus on a penny at a distance of more than 200 km.

✓ Large aperture mirror built using honeycomb-sandwich, reducing weight by a factor of four over solid glass

✓ Internal structure holds optical components aligned within 2.54×10^{-4} cm (1/10,000 in.) through extreme temperature changes

Mission Impact

At the start of this 15-year NASA mission, Hubble's thousands of observations have lent substantial credibility to its creators' claims. It has made many outstanding new discoveries and will continue to make observations that place the HST program at the forefront of astronomy. With future servicing missions Hubble's usefulness will continue to improve. It will remain on the cutting edge of science and technology for years to come.

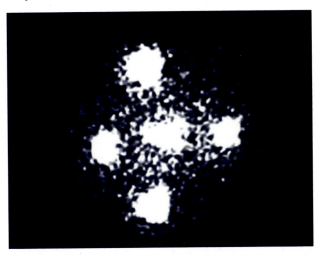

Hubble's faint-object camera captured this image of a rare cosmic sight—gravitational lens G2237 + 0305—sometimes called the "Einstein Cross." This photograph shows four images of a single quasar approximately eight billion light years away. The multiple images are caused by an aspect of the theory of relativity. The mass of galaxy between us and the quasar (the fuzzy image in the middle) distorts space, causing the light to be refracted as if it passed through a big lens. *(Courtesy of NASA)*

For Discussion

• HST's mirror has a flaw reducing its effectiveness. What lesson does this teach us about the design and management of space projects?

• How do we justify the cost of purely scientific missions like HST?

Contributors

Mari D. Brenneman and John D. Slezak, United States Air Force Academy

References

National Aeronautics and Space Administration. *Hubble Space Telescope Update: 18 Months in Orbit.* Washington: Government Printing Office, 1992.

National Aeronautics and Space Administration. *Hubble Space Telescope, Media Reference Guide.* Sunnyvale, CA: Lockheed Missiles & Space Co., Inc.

The Space Shuttle Atlantis rockets into the sky, powered by its two mighty, solid-rocket boosters and three main engines. *(Courtesy of NASA)*

Space Transportation Systems

14

In This Chapter You'll Learn to...

☛ Explain basic rocket principles

☛ Discuss the basic types of rocket systems and their operating principles

☛ Discuss the principles of rocket staging and how to determine the velocity change from a staged booster

You Should Already Know...

❏ Newton's laws of Motion and Conservation of Linear Momentum (Chapter 4)

❏ ΔV requirements for on-orbit maneuvers (Chapter 6)

❏ ΔV required to reach orbit (Chapter 9)

❏ Components and functions of control systems (Chapter 12)

❏ Disturbance torque from solar pressure (Chapter 12)

Mankind will not remain on Earth forever, but in its quest for light and space will at first timidly penetrate beyond the confines of the atmosphere, and later will conquer for itself all the space near the Sun.

Konstantine E. Tsiolkovsky
father of Russian astronautics

Few parts of a space mission get the adrenaline flowing like the launch. As a fully-fueled booster sits on the pad, it contains tons of highly-flammable propellant just waiting for the ignition signal to start the controlled explosion that will hurl it into orbit. All space-transportation systems use rockets. In this chapter we'll see how rockets work and look at some different types of rocket systems. We'll then look at boosters to understand the problems they face in getting payloads into orbit, including staging and control issues.

Rockets do three things for a spacecraft:

- Get it into space
- Move it around in space
- Change its attitude

In this chapter, we'll concentrate on getting the spacecraft into space by using the massive engines on a booster rocket. In Chapter 9 we saw how much ΔV is needed to get a spacecraft from the surface of the Earth into space. The booster must produce enough velocity change to overcome losses due to gravity and drag and, in the case of very high-inclination or retrograde orbits, overcome the Earth's rotational velocity.

Once the spacecraft is in orbit, the space-transportation system must provide the ΔV necessary to make orbit corrections and maneuvers. In Chapters 6 and 7 we explored how a spacecraft maneuvers during the life of the mission, doing Hohmann transfers, plane changes, rendezvous, and even interplanetary transfers. Depending on the size of these maneuvers (how much ΔV is required), the propulsion system may be integrated into the spacecraft, like the small rockets on the Space Shuttle for orbital maneuvering and attitude control, or be a separate, upperstage rocket, like the inertial upperstage (IUS) used to boost large satellites into geostationary orbits or toward the planets.

The spacecraft's propulsion system also contributes to attitude control. In Chapter 12 we explained how thrusters can be used as actuators for attitude control. These thrusters are either the sole method of attitude control (like the Shuttle's reaction-control system), or they supplement the primary system and provide for momentum dumping. All of these functions for space transportation are summarized in Figure 14-1.

Before we look at the options for getting into space, let's step back to see how rockets really work. After all, you can't be a *real* rocket scientist until you know some basic rocket theory!

controls
attitude

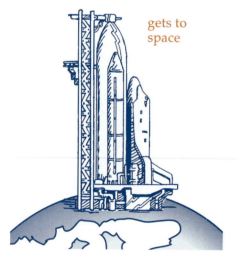

maneuvers

gets to
space

Figure 14-1. What the Space Transportation System Does. The space transportation system gets us into space, moves us around when we get there, and helps control our attitude.

14.1 Rocket Basics

▬ In This Section You'll Learn to...

☛ Explain the basic physical principles of rockets

☛ Define and determine important parameters describing rocket performance—thrust, specific impulse, and ΔV

Thrust and Velocity Change

A rocket is basically something that spews out hot gasses. In that way, it's really no different from a balloon. All of us have blown up a balloon and let go of the stem to watch it fly wildly around the room as shown in Figure 14-2. What makes the balloon go? Recall from Chapter 4 we talked about

Newton's Third Law. For every action there is an equal but opposite reaction.

The skin of the balloon is stretched tight, squeezing the air inside so it has a higher pressure than the air in the room. To equalize this pressure, the air is forced out the stem. Because of Newton's Third Law, as the air is forced out in one direction (the action), an equal force pushes the balloon in the opposite direction (the reaction). The balloon flies so wildly because it's unstable. If you attached some light paper fins to the base of the balloon, it would fly fairly straight and you'd have a mini rocket!

Now that you understand the basic concept of a rocket, let's look at this action/reaction situation a little differently to better see how a rocket produces thrust. Consider an astronaut perched in a wagon armed with a load of rocks as shown in Figure 14-3. If he's initially at rest and begins to throw the rocks in one direction, because of Newton's Third law, an equal but opposite force will move him (and the wagon load of rocks) in the other direction. To throw the rocks, the astronaut had to apply a force to them. This force is identical in magnitude but opposite in direction to the force applied to the wagon. However, remember the concept of conservation of linear momentum we discussed in Chapter 4. It tells us the change in speed of the rock (because it has less mass) will be greater than the change in speed of the wagon. This is basically how a rocket works. It expends energy to eject mass out the back at high speeds, pushing the rocket in the opposite direction. The mass is being ejected at a rate we'll call the *mass flow rate*, $\Delta m / \Delta t = \dot{m}$ (kg/s).

A simple rocket engine has three basic parts—the combustion chamber, a throat, and a nozzle—as shown in Figure 14-4. The *combustion chamber*, as the name implies, is where the propellant burns, producing hot gasses. The *throat* is designed to constrict the flow of hot gasses which build up in

Figure 14-2. A Simple Rocket. A balloon is the simplest form of a rocket. The pressure exerted on the air in the balloon by the stretched balloon skin causes the air to escape out the stem at high speeds. This gives the air molecules momentum in one direction, causing the balloon to go in the opposite direction.

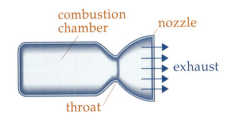

Figure 14-3. A One-Person Rocket. An astronaut throwing rocks out the back of a wagon is a simple rocket. He uses his muscles to accelerate the rocks in one direction, leading to an equal but opposite force on the wagon, pushing it in the opposite direction.

Figure 14-4. Parts of a Rocket. A rocket has three basic parts—a combustion chamber, a throat, and a nozzle.

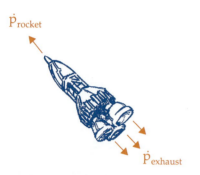

Figure 14-5. Conservation of Momentum. As the hot gasses from the rocket engine gain momentum in our direction, the rocket gains momentum in the other direction.

the combustion chamber, controlling the chamber pressure and mass flow rate. The hot gasses reach the speed of sound in the throat. The *nozzle* is designed to efficiently direct the flow of gasses in the desired direction. The hot gasses have mass and move out the back at some *exit velocity, V_{exit}*. For typical chemical rockets like those used by the Space Shuttle, V_{exit} can be as high as 3 km/s.

Recall from Chapter 4 that linear momentum was defined as mass times velocity. This means the ejected mass acquires momentum at a rate of $\dot{m}V_{exit}$. But remember, momentum is always conserved! So as the momentum of the ejected mass increases in one direction, the momentum of the rocket must likewise be increasing in the other direction, as shown in Figure 14-5. Dropping the vector notation to look at magnitudes only, we can say

$$\dot{P}_{rocket} = \dot{P}_{exhaust}$$

$$\boxed{\dot{P}_{rocket} = \dot{m}V_{exit}} \qquad (14\text{-}1)$$

where

$\dot{P}_{rocket}$ = time rate of change of momentum of the rocket (N)
$\dot{m}$ = mass flow rate of exhaust products (kg/s)
V_{exit} = exit velocity of exhaust (m/s)

Notice this momentum change has the same units as force. Typically, we like to consider this momentum change to be equivalent to a force on the rocket that we'll call the *momentum thrust*.

$$F_{momentum\ thrust} = \dot{m}V_{exit}$$

where

$F_{momentum\ thrust}$ = effective thrust on the rocket from momentum change (N)
$\dot{m}$ = mass flow rate of exhaust products (kg/s)
V_{exit} = exit velocity of exhaust products (m/s)

But momentum change isn't the only consideration when determining the total thrust on a rocket. To see this, consider an imaginary "control volume" drawn around a rocket as shown in Figure 14-6. Acting on the boundaries of this volume, we've drawn the atmospheric pressure and $P_{atmosphere}$ acting on all sides except at the nozzle exit. At the nozzle exit, the pressure is P_{exit}, drawn inward for consistency. Notice that due to symmetry, $P_{atmosphere}$ cancels out everywhere except in the direction parallel to momentum thrust over an area equal to nozzle exit area, A_{exit}. The net force exerted on the rocket from this pressure differential is called *pressure thrust*. It's equal to the difference between exit pressure, P_{exit}, and atmospheric pressure, $P_{atmosphere}$, times the exit area, A_{exit}. Its magnitude is

$$F_{pressure\ thrust} = A_{exit}(P_{exit} - P_{atmosphere})$$

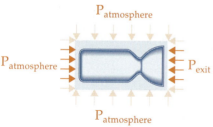

Figure 14-6. Pressure Thrust. Pressure thrust on a rocket results from the difference between exit pressure and atmospheric pressure.

where
$F_{pressure\ thrust}$ = pressure thrust (N)
A_{exit} = exit area of nozzle (m^2)
P_{exit} = exit pressure (N/m^2)
$P_{atmosphere}$ = atmospheric pressure (N/m^2)

The magnitude of the total thrust on the rocket can then be expressed using the total rocket thrust equation

$$F_{thrust} = \dot{m}V_{exit} + A_{exit}(P_{exit} - P_{atmosphere}) \qquad (14\text{-}2)$$

where
F_{thrust} = total rocket thrust (N)
$\dot{m}$ = mass flow rate (kg/s)
V_{exit} = exit velocity of exhaust products (m/s)
A_{exit} = exit area of nozzle (m^2)
P_{exit} = exit pressure (N/m^2)
$P_{atmosphere}$ = atmospheric pressure (N/m^2)

To further simplify this equation, we can define the *effective exhaust velocity, c*, as

$$c \equiv V_{exit} + \frac{A_{exit}}{\dot{m}}(P_{exit} - P_{atmosphere}) \qquad (14\text{-}3)$$

where
c = effective exhaust velocity (m/s)
V_{exit} = exit velocity of exhaust products (m/s)
A_{exit} = exit area of nozzle (m^2)
$\dot{m}$ = mass flow rate (kg/s)
P_{exit} = exit pressure (N/m^2)
$P_{atmosphere}$ = atmospheric pressure (N/m^2)

Which means we can write the total rocket thrust equation as

$$F_{thrust} = \dot{m}c \qquad (14\text{-}4)$$

where
F_{thrust} = total rocket thrust (N)
c = effective exhaust velocity (m/s)
$\dot{m}$ = mass flow rate (kg/s)

This relationship should make sense from our astronaut example. He can increase the thrust on the wagon by either increasing the rate at which he throws out the bricks (higher $\dot{m}$) or throwing the bricks faster (higher c). Or he could do both.

To get a rocket off the ground, the total thrust must be greater than the weight of the entire vehicle. We refer to the ratio of the thrust produced to the vehicle's weight as the *thrust-to-weight ratio*. Thus, to get a rocket off the ground, the thrust-to-weight ratio must be greater than 1. For example, the thrust-to-weight ratio for the Atlas booster is around 1.2, and the Space Shuttle's around 1.6.

An astronaut strapped to the top of a rocket feels this thrust force as an acceleration or g-load. Thus, for a rocket with constant thrust, an astronaut will feel a g-load or acceleration which depends on the rocket's mass. But this mass will decrease as propellant burns and is ejected out the back. This means the acceleration will tend to increase over time. To keep the overall g-load on the Space Shuttle under 3 g's, the main engines are throttled down about six minutes into the launch to decrease their thrust as propellant burns.

A casual glance at the total rocket thrust in Equation (14-2) may lead you to conclude you'd want to make $P_{exit} \gg P_{atmosphere}$ to maximize total thrust. Although this would increase the amount of thrust generated, it would do so at the expense of efficiency. To understand this, we must return to the basic components of the rocket engine and see how they all work together. Within the combustion chamber we're burning propellant to produce gasses at high temperatures and pressures. This gas is getting crowded in there and it wants to go somewhere. The only escape route is through the throat. Gas can escape through the throat only as fast as the speed of sound at that temperature and pressure, so the flow is known as *sonic flow* or *choked flow*.

Once past the throat, this sonic flow expands as it enters the nozzle, accelerating to supersonic speeds. Where does this energy to increase the speed of the gasses come from? Basically, the nozzle changes the random, thermal energy in the hot gasses into organized kinetic energy. This means the gasses are cooled as they increase in speed.

To get the most out of the energy in the hot exhaust gasses, we'd like to convert as much of that thermal energy to kinetic energy as possible without incurring losses due to other considerations. The purpose of the nozzle is to expand the flow of gasses from sonic speeds in the throat to supersonic speeds at the exit. For supersonic flow, as the gasses expand they *increase* in velocity while decreasing in pressure. Thus, the higher the V_{exit}, the lower the P_{exit}. For the ideal case, it turns out the pressure thrust should be zero ($P_{exit} - P_{atmosphere} = 0$), which means the exit pressure equals the atmospheric pressure ($P_{exit} = P_{atmosphere}$). In this case, the exit velocity and thus the momentum thrust is maximized.

$$F_{thrust\ ideal} = \dot{m}V_e$$

What happens when $P_{exit} \neq P_{atmosphere}$? When this happens, we have a rocket that's not as efficient as it could be. We must consider two possible situations:

- Overexpansion: $P_{exit} < P_{atmosphere}$. This is often the case for a rocket at lift-off. Because the launch pad is normally near sea level, the atmospheric pressure is at a maximum. This atmospheric pressure can cause shock waves to form just inside the nozzle. These shock waves represent areas where kinetic energy turns back into thermal energy. In other words, we've robbed kinetic energy from the flow, lowering the exhaust velocity and thus decreasing the overall thrust.

- Underexpansion: $P_{exit} > P_{atmosphere}$. In this case, the exhaust gasses have not expanded as much as they could have within the nozzle and thus there's a "loss" in the sense that we've not used all the energy we potentially could have. This is the normal case for a rocket operating in a vacuum because P_{exit} is always higher than $P_{atmosphere}$ ($P_{atmosphere}$ = 0 in vacuum). Unfortunately, you'd need an infinitely long nozzle to expand the flow to zero pressure, so in practice we must accept some loss in efficiency. All cases of expansion are illustrated in Figure 14-7.

In real life, it's nearly impossible to have a perfectly expanded nozzle. Thus, rocket scientists must put up with some inherent inefficiencies in all rockets. For example, when we design a booster rocket to carry a payload into orbit, we must deal with the changing atmospheric pressure as the rocket gains altitude. It would be nice if the nozzle could change shape throughout the trajectory to change the exit pressure as atmospheric pressure changes. Unfortunately, with current technology, the weight penalty to do this is too great. Instead, we design the nozzle to achieve ideal expansion at some design altitude about 2/3 of the way from the altitude of engine ignition to the altitude of engine cut-off. For example, if a rocket is designed to take you from sea level to 60,000 meters, a good choice for the desired exit pressure would be the atmospheric pressure at about 40,000 meters altitude. As a result, our rocket would (by design) be overexpanded below 40,000 meters and underexpanded above 40,000 meters. As you can see in Figure 14-8, a nozzle designed in this way offers better overall performance than one designed to be ideally expanded only at sea level.

ideally expanded nozzle

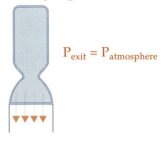

$P_{exit} = P_{atmosphere}$

overexpanded nozzle

$P_{exit} < P_{atmosphere}$

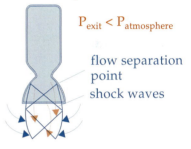

flow separation point

shock waves

underexpanded nozzle

$P_{exit} > P_{atmosphere}$

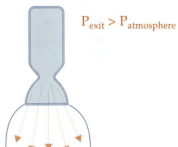

Figure 14-7. Nozzle Expansion. To take advantage of all the energy in the exhaust flow, we'd like P_{exit} to equal $P_{atmosphere}$ in the ideal case. When $P_{exit} < P_{atmosphere}$, the nozzle is overexpanded, causing a loss of energy in the flow. When $P_{exit} > P_{atmosphere}$, the nozzle is underexpanded and not all available energy has been used.

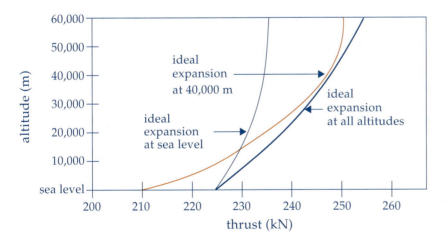

Figure 14-8. Thrust Versus Altitude for Different Nozzle Designs. Because we don't yet have a rocket nozzle that is ideally expanded at all possible altitudes, we typically design the nozzle for ideal expansion 2/3 of the way up, in this case at 40,000 m. This offers better overall performance than one designed to be ideally expanded only at sea level.

The Ideal Rocket Equation

When you take a long trip in your car, one of your main concerns is having enough gas in your tank. This concern is equally important for a trip into space. But how do we determine how much "gas" we need for a trip in space? First of all, in rocket terms, the mass expelled out the back, whatever its original form, is *propellant*. To figure how much propellant is needed for a given trip, we must have some relationship between the change in velocity required and amount of propellant. We can find this relationship by setting the thrust produced by the rocket engine equal to the momentum change of the entire vehicle.

$$F_{thrust} = \dot{m}c = \frac{\Delta p_{rocket}}{\Delta t}$$

where

F_{thrust} = effective thrust on the rocket (N)

$\dot{m}$ = mass flow rate (kg/s)

c = effective exhaust velocity (m/s)

$\dfrac{\Delta p_{rocket}}{\Delta t}$ = time rate of change of momentum for the rocket (N)

From this relationship we can come up with the *ideal rocket equation*. (See Appendix C for the complete derivation.) It tells us how much propellant we need to get a certain ΔV.

$$\Delta V = c \ln\left(\frac{m_{initial}}{m_{final}}\right) \qquad (14\text{-}5)$$

where

ΔV = velocity change (m/s)

c = effective exhaust velocity (m/s)

$\ln$ = natural logarithm of quantity in parenthesis

$m_{initial}$ = initial mass of vehicle before firing engine (kg)

m_{final} = final mass of vehicle after firing engine (kg)

Equation (14-5) is one of the single most useful relationships of rocket propulsion. Armed with this relationship, we can determine how much propellant we need to do anything from stopping the spin of a spacecraft in orbit to launching a satellite to another solar system. Notice that we're taking the natural logarithm of the ratio of initial to final mass. *The difference between initial and final mass represents the amount of propellant used.* ΔV is also a function of the effective exhaust velocity. This should make sense because, as the propellant moves out the back faster, momentum changes more, and the rocket goes faster. Example 14-1 shows one application for this very useful equation.

Rocket Efficiency

Some rockets are more efficient than others. For example, one rocket may need 100 kg of propellant to change velocity by 100 m/s while another needs only 50 kg. How do we measure this efficiency? To answer this question we must first introduce the concept of impulse. Returning to our astronaut throwing bricks out of the wagon, realize that, to give the brick velocity, he must apply a force to it over some length of time. This is explained by Newton's Second Law, which we can express as

$$F = \frac{\Delta p}{\Delta t}$$

If you multiply both sides by Δt, you end up with

$$F\Delta t = \Delta p$$

The left side of this equation represents a force, F, such as the force you would generate if you threw a brick, acting over some time, Δt. The result of that force acting over time is on the right side of the equation, where Δp represents the change in momentum as a result of the force. Thus, to achieve some change in momentum, we can use a large force acting over a short time or a smaller force acting over a longer time.

We define a force acting over some period as the *total impulse, I*.

$$\boxed{I \equiv F\ \Delta t = \Delta p} \qquad (14\text{-}6)$$

where
I = total impulse (N · s)
F = force (N)
Δt = time (s)
Δp = momentum change (N · s)

Impulse works the same way for rockets as it does for a brick. We want to change the velocity and hence the momentum of our rocket by some amount, so we must apply some impulse. But as we've seen, we can achieve the same impulse for a rocket by applying a small thrust over a long time or a large thrust over a short time. So, although total impulse is useful for sizing the engine and determining the amount of propellant needed, it doesn't tell us much about the engine itself. To compare the performance of different types of rockets, we need a new parameter we'll call specific impulse. *Specific impulse, I_{sp},* is the ratio of the total impulse to the rocket's change in weight as propellant is ejected. Recall in Chapter 4, when we defined specific mechanical energy and angular momentum, we divided by mass rather than weight. Using weight rather than mass to calculate I_{sp} is simply a convention established by the founders of rocket science and, as a result, I_{sp} has the unusual units of seconds.

$$\boxed{I_{sp} \equiv \frac{I}{\Delta W}} \qquad (14\text{-}7)$$

where

I_{sp} = specific impulse (s)

I = total impulse (N · S)

ΔW = change in weight (N)

Substituting for total impulse

$$I_{sp} = \frac{F_{thrust}\,\Delta t}{\Delta W}$$

$$\boxed{I_{sp} = \frac{F_{thrust}}{\dot{W}}}\qquad(14\text{-}8)$$

where

I_{sp} = specific impulse (s)

F_{thrust} = force of thrust (N)

$\dot{W}$ = weight flow rate (N/s)

Basically, I_{sp} represents the ratio of what you get (momentum change) to what you spend (propellant). So the bigger the I_{sp}, the more efficient the rocket. To get a more useful expression for I_{sp}, we can replace the weight flow rate with the mass flow rate, $\dot{m}$, times the gravitational acceleration at sea level (a constant).

$$I_{sp} = \frac{F_{thrust}}{\dot{m}g_o}\qquad(14\text{-}9)$$

where

$\dot{m}$ = mass flow rate

g_o = gravitational acceleration constant = 9.81 m/s^2 (sea level)

Earlier, we found the force of thrust in terms of the mass flow rate and the effective exhaust velocity; by substituting Equation (14-4) into Equation (14-9), we get a different expression for I_{sp}.

$$\boxed{I_{sp} = \frac{c}{g_o}}\qquad(14\text{-}10)$$

where

I_{sp} = specific impulse (s)

c = effective exhaust velocity (m/s)

g_o = gravitational acceleration at sea level (9.81 m/s^2)

Notice g_o is a constant value representing the acceleration due to gravity at sea level, which we use to calibrate the equation. This means *no matter where we go in the universe, we humans will use the same value of g_o to measure engine performance.*

We can substitute the definition of I_{sp} into the rocket Equation (14-5) to compute the ΔV for a rocket if we know the I_{sp}, as well as the initial and final rocket mass.

$$\Delta V = I_{sp} \; g_o \; \ln\left(\frac{m_{initial}}{m_{final}}\right)$$

(14-11)

where

ΔV	= velocity change (m/s)
I_{sp}	= specific impulse (s)
g_o	= gravitational acceleration at sea level (9.81 m/s^2)
ln	= natural logarithm of quantity in parenthesis
$m_{initial}$	= initial mass of vehicle before firing engine (kg)
m_{final}	= final mass of vehicle after firing engine (kg)

As a measure of engine performance, I_{sp} is like the miles per gallon (MPG) rating given for new cars. The higher the I_{sp} is for a rocket, the more "bang for your buck" you get, which means you get more ΔV for a given amount of propellant. Another way to think about I_{sp} is that the faster a propulsion system can expel propellant, the more efficient it is. Thus, for a given required ΔV, a rocket with a higher I_{sp} can do the job with less propellant. This is demonstrated in Example 14-2. However, as we'll see in the next section, when we look at the I_{sp} for specific rocket systems, I_{sp} doesn't tell the whole story.

Section Review

Key Terms

choked flow
combustion chamber
effective exhaust velocity, c
exit velocity, V_{exit}
ideal rocket equation
mass flow rate
momentum thrust
nozzle
pressure thrust
propellant
specific impulse, I_{sp}
sonic flow
throat
thrust-to-weight ratio
total impulse, I

Key Equations

$$\dot{P}_{rocket} = \dot{m}V_{exit}$$

$$F_{thrust} = \dot{m}V_{exit} + A_{exit}(P_{exit} - P_{atmosphere})$$

$$c \equiv V_{exit} + \frac{A_{exit}}{\dot{m}}(P_{exit} - P_{atmosphere})$$

$$F_{thrust} = \dot{m}c$$

$$\Delta V = c \ln\left(\frac{m_{initial}}{m_{final}}\right)$$

$$I \equiv F \Delta t = \Delta p$$

$$I_{sp} \equiv \frac{I}{\Delta W}$$

$$I_{sp} = \frac{F_{thrust}}{\dot{W}}$$

$$I_{sp} = \frac{c}{g_o}$$

$$\Delta V = I_{sp} g_o \ln\left(\frac{m_{initial}}{m_{final}}\right)$$

Key Concepts

➤ A rocket's ability to produce thrust is explained by Newton's laws of motion.

➤ A rocket has three main parts:
  • The combustion chamber—place where propellant burns to produce hot gases
  • The throat—small area which constricts the flow of hot gases accelerating it to sonic speeds and controls chamber pressure and mass flow rate
  • The nozzle—accelerates the flow to supersonic speeds and directs it in the direction opposite to the rocket's motion

➤ A rocket's total thrust comes from two different sources:
  • Momentum thrust results from Newton's Third Law and conservation of momentum. As the momentum of the rocket exhaust increases in one direction, the rocket's momentum increases the same amount in the opposite direction.
  • Pressure thrust results from the difference in pressure between the exhaust gases and the outside atmospheric pressure. An ideally expanded nozzle produces no pressure thrust.

➤ To determine the amount of velocity change a rocket gets from a given amount of propellant, we need to know the effective exhaust velocity, c, and use the ideal rocket equation.

➤ Specific impulse, I_{sp}, measures a rocket's efficiency; the higher the I_{sp}, the more efficiently the rocket uses propellent to generate thrust. It represents the ratio of the force of thrust to the weight flow rate and has units of seconds.

Example 14-1

Problem Statement

An advanced inertial upperstage (IUS) is being designed to boost a new cable TV satellite from a low-altitude parking orbit to geosynchronous orbit. The ΔV for the first burn of the Hohmann transfer is 3.34 km/s, and the effective exhaust velocity of the IUS is 3000 m/s. If the mass of just the structure of the IUS, without propellant, is 100 kg and the satellite mass is 1000 kg, what mass of propellant should be loaded into the upperstage?

Problem Summary

Given: $\Delta V = 3.34$ km/s
$c = 3000$ m/s
$m_{structure} = 100$ kg
$m_{satellite} = 1000$ kg

Find: $m_{propellant}$

Problem Diagram

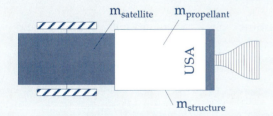

$m_{satellite}$ $m_{propellant}$

USA

$m_{structure}$

Conceptual Solution

1) Determine the final mass of the IUS plus payload at the end of the burn

$$m_{final} = m_{structure} + m_{satellite}$$

2) Determine the initial mass of the IUS using the rocket equation

$$\Delta V = c \ln\left(\frac{m_{initial}}{m_{final}}\right)$$

3) Solve for the mass of the propellant needed by subtracting the final mass

$$m_{propellant} = m_{initial} - m_{final}$$

Analytical Solution

1) Determine the final mass of the IUS plus payload at the end of the burn

$$m_{final} = m_{structure} + m_{satellite}$$

$$m_{final} = 100 \text{ kg} + 1000 \text{ kg} = 1100 \text{ kg}$$

2) Determine the initial mass of the IUS using the rocket equation

$$\Delta V = c \ln\left(\frac{m_{initial}}{m_{final}}\right)$$

$$m_{initial} = m_{final}\, e^{\frac{\Delta V}{c}}$$

$$m_{initial} = (1100 \text{ kg})\, e^{\frac{3340 \text{m/s}}{3000 \text{m/s}}}$$

$$m_{initial} = 3349 \text{ kg}$$

3) Solve for the mass of the propellant needed by subtracting the final mass

$$m_{propellant} = m_{initial} - m_{final}$$

$$m_{propellant} = 3349 \text{ kg} - 1100 \text{ kg} = 2249 \text{ kg}$$

Interpreting the Results

In this case, 2249 kg of propellant must be loaded on the IUS to achieve the desired ΔV. This means more than 67% of the initial vehicle mass is propellant! And this is for a relatively efficient rocket motor with an I_{sp} of over 300 seconds.

Example 14-2

Problem Statement

Two rockets are candidates for a space mission. Rocket 1 has an I_{sp} of 300 seconds, and rocket 2 has an I_{sp} of 350 seconds. If the total ΔV needed for the mission is 1000 m/s, how much more propellant will rocket 1 need over the life of the mission? Assume the dry mass of the spacecraft is 1000 kg.

Problem Summary

Given: Rocket 1 I_{sp} = 300 seconds
           Rocket 2 I_{sp} = 350 seconds
           M_{final} = 1000 kg
           ΔV = 1000 m/s
Find: Difference in $M_{propellant}$ for rockets 1 and 2

Conceptual Solution

1) Solve rocket equation for $M_{initial}$ of rocket 1

$$\Delta V = I_{sp}\, g_o\, \ln\left(\frac{m_{initial}}{m_{final}}\right)$$

$$m_{initial-rocket\,1} = m_{final-rocket\,1}\, e^{\frac{\Delta V}{I_{sp}g_o}}$$

2) Solve rocket equation for $M_{initial}$ of rocket 2

$$\Delta V = I_{sp}\, g_o\, \ln\left(\frac{m_{initial}}{m_{final}}\right)$$

$$m_{initial-rocket\,2} = m_{final-rocket\,2}\, e^{\frac{\Delta V}{I_{sp}g_o}}$$

3) Determine $M_{propellant}$ of rockets 1 and 2 and subtract

$$m_{propellant\,rocket\,1} = m_{initial\,rocket\,1} - m_{final\,rocket\,1}$$

$$m_{propellant\,rocket\,2} = m_{initial\,rocket\,2} - m_{final\,rocket\,2}$$

Analytical Solution

1) Solve rocket equation for $m_{initial}$ of rocket 1

$$\Delta V = I_{sp}\, g_o\, \ln\left(\frac{m_{initial}}{m_{final}}\right)$$

$$m_{initial-rocket\,1} = m_{final-rocket\,1}\, e^{\frac{\Delta V}{I_{sp}g_o}}$$

$$= (1000\ kg)\, e^{\frac{1000\ m/s}{(300\ s)\,(9.81\ m/s^2)}}$$

$$m_{initial\,rocket\,1} = (1000\ kg)\, e^{0.3398}$$

$$m_{initial\,rocket\,1} = 1405\ kg$$

2) Solve rocket equation for $m_{initial}$ of rocket 2

$$\Delta V = I_{sp}\, g_o\, \ln\left(\frac{m_{initial}}{m_{final}}\right)$$

$$m_{initial-rocket\,2} = m_{final-rocket\,2}\, e^{\frac{\Delta V}{I_{sp}g_o}}$$

$$= (1000\ kg)\, e^{\frac{1000\ m/s}{(350\ s)\,(9.81\ m/s)}}$$

$$m_{initial\,rocket\,2} = 1338\ kg$$

3) Determine $m_{propellant}$ of rockets 1 and 2 and subtract

$$m_{propellant\,rocket\,1} = m_{initial\,rocket\,1} - m_{final\,rocket\,1}$$

$$m_{propellant\,rocket\,1} = (1405\ kg) - (1000\ kg) = 405\ kg$$

$$m_{propellant\,rocket\,2} = m_{initial\,rocket\,2} - m_{final\,rocket\,2}$$

$$m_{propellant\,rocket\,2} = (1338\ kg) - (1000\ kg) = 338\ kg$$

Propellant difference =

$$m_{propellant\,rocket\,1} - m_{propellant\,rocket\,2}$$

Propellant difference =

$$(405\ kg) - (338\ kg) = 67\ kg$$

Interpreting the Results

Rocket 1 is about 14% (50/350) less efficient than rocket 2. As a result, it needs nearly 20% (67/338) more propellant than rocket 2 to do the same mission.

14.2 Rocket Systems

▰ In This Section You'll Learn to...

> ☛ Describe the basic types of rocket systems and their operating principles

Now that we know all about thrust and specific impulse, we can look at some real-life rocket engines to see how they're put together. Because all rockets operate by the same basic principles, they all have the same basic elements, as shown in Figure 14-9. They all use an energy source to accelerate some mass of propellant to high velocity and expel it out the back, producing thrust. To do this, some energy-conversion process takes the raw energy and converts it into a form the accelerator can use. For example, burning hydrogen and oxygen converts chemical energy to thermal energy. At the same time, a propellant-feed system moves propellant from the storage tanks to the accelerator. For liquid propellants, this is done using pumps and other plumbing. The accelerator can be as simple as a nozzle or as complex as a high-voltage electric grid.

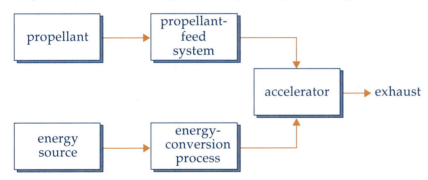

Figure 14-9. Components of a Rocket System. All rocket systems have the same basic components. They use a propellant-feed system and convert energy to move propellant into a mass accelerator producing thrust.

While all rockets have these same basic components, we can further subdivide types of rockets into three broad categories:

- *Thermodynamic systems*—create thrust by converting thermal energy (heat) into kinetic energy (mass moving at high velocity)

- *Electrodynamic systems*—create thrust by using electric or magnetic fields to accelerate charged particles and give them kinetic energy.

- *Exotic systems*—create thrust through some means other than thermodynamic and electrodynamic impulse. Exotic systems are mostly still on the drawing board or in the theoretical stage.

Let's explore these systems to see what makes them tick.

Thermodynamic Propulsion

Thermodynamic systems produce thrust by changing thermal energy into kinetic energy. They usually burn something (a chemical reaction) to produce exhaust products with high thermal energy. A nozzle then converts the random thermal energy in the exhaust products into organized kinetic energy in one direction to propel the rocket. Not too surprisingly, the efficiency of a thermodynamic system, measured by I_{sp}, depends on the combustion temperature and the propellant's molecular weight. *Molecular weight* is a measure of the weight per molecule of propellant. Hydrogen is the lightest element, whereas lead, for example, is much, much heavier. To improve efficiency for thermodynamic systems, we increase the combustion temperature or decrease the molecular weight. This can be summarized by

$$I_{sp} \propto \sqrt{\frac{T_{combustion}}{M}}$$

(14-12)

where
I_{sp} = specific impulse (s)
$T_{combustion}$ = Combustion temperature
M   = molecular weight
[*Note:* the symbol "$\propto$" means proportional to]

As a result, the most efficient thermodynamic systems operate at the highest temperature with propellant having the lowest molecular weight. Hydrogen is often used as propellant because it meets both requirements. Thermodynamic systems include

- Cold-gas
- Chemical
- Nuclear thermal
- Arc-jet
- Resisto-jet

Cold-gas systems are the simplest of all. They work just like the balloon example discussed in the last section by merely exhausting a pressurized gas out of a nozzle. No combustion chamber is needed. These systems are very reliable and can be turned on and off repeatedly, producing small pulses at precise thrust levels—a desirable characteristic for attitude-control systems. A good example of a cold-gas system is the manned maneuvering unit (MMU) used by Shuttle astronauts. The MMU uses compressed nitrogen to give astronauts complete freedom to maneuver. Unfortunately, due to their low thrust and I_{sp}, cold-gas systems are typically work only for attitude control or special applications such as the MMU. To get greater thrust and efficiency, we must look to other types of systems.

Chemical rockets are another type of thermodynamic system and are either liquid, solid, or hybrid in form. All chemical systems use a chemical reaction to achieve combustion (rapid burning), which produces a hot exhaust product.

491

Let's begin with liquid propulsion systems. The simplest type of liquid rocket uses a single type of propellant called a *mono-propellant*. A catalyst is used to decompose the mono-propellant, such as hydrazine and expel the by-products of decomposition out a nozzle. More complex liquid rockets use two propellants and are thus called *bi-propellant* systems. They combine a fuel, such as liquid hydrogen (LH_2) and an oxidizer, such as liquid oxygen (LOX), in a combustion chamber. There, the two compounds chemically react to form heat plus reactant products.

To accomplish all this, the fuel and oxidizer must be stored in tanks and then moved to the combustion chamber and injected in just the right proportion. For small rockets, we can force the propellant into the combustion chamber using an inert pressurized gas, such as helium or nitrogen. For very large rocket engines, such as the Space Shuttle's main engines, shown in Figure 14-10, vast quantities of fuel and oxidizer must burn each second. Simple pressure-fed systems can't deliver this high volume of propellant. Instead, we use pumps, not that much different from the ones used to empty swimming pools. However, these pumps are vastly more powerful. In fact, the fuel pump on the shuttle could empty an average-size swimming pool in only 25 seconds! To drive these massive pumps, we use large turbines fueled by burning the same oxidizer and fuel used by the engines. Figure 14-11 shows a schematic for a basic pump-fed, bi-propellant liquid rocket.

Depending on the propellant used, the combustion process may need something to ignite the process (a lot like the spark plug in your car). LH_2/LOX systems, for example, need an ignitor to ensure complete combustion starts completely. Other systems use *hypergolic propellant*—the fuel and oxidizer react immediately upon contact, so no ignition is required. Hydrazine and nitrogen tetroxide are examples of a hypergolic fuel and oxidizer. This propellant also is storable at near room temperature for long periods. The Titan, an early intercontinental ballistic missile (ICBM), used these propellants because they could be stored in the missile until needed.

In contrast, *cryogenic propellants*, such as liquid hydrogen and oxygen, must be super-cooled to hundreds of degrees below zero. Because they require such cold temperatures and are hard to store for long periods, cryogenic propellants are used only in the first part of a mission. However, engineers are working on ways to store cryogenics for longer missions, such as those to Mars. *Space-storable propellants* are used whenever the propellant needs to stretch over days or years. The Space Shuttle uses both cryogenic and space-storable propellants. Hydrogen and oxygen burn during the first eight minutes of the mission to deliver the vehicle into orbit. Space storables fuel the orbital-maneuvering engines and reaction-control thrusters throughout the rest of the mission. Because of their lower molecular weight, cryogenic propellants offer a higher I_{sp} than space-storable ones. For example, the Space Shuttle's main engines use LH_2/LOX for an I_{sp} of about 455 seconds, whereas the orbital-maneuvering engines using storable hydrazine/nitrogen tetroxide have an I_{sp} of about 313 seconds. As an aside, we usually refer to liquid propulsion systems as

Figure 14-10. Shuttle's Main Engines. The Space Shuttle's main engines are a good example of a liquid rocket. They burn liquid hydrogen and liquid oxygen in tremendous quantities. *(Courtesy of NASA)*

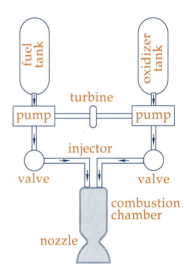

Figure 14-11. Pump-Fed Liquid Rocket System. A typical pump-fed liquid rocket system combines a fuel and an oxidizer using turbine driven pumps to force the propellant into a combustion chamber where it is burned. This hot exhaust is then expanded out a nozzle producing thrust.

"engines" because, like your car engine, they rely on fuel being fed from an external source.

If you've ever watched fireworks on the 4th of July, you've seen solid propulsion systems. The Chinese first developed these rockets thousands of years ago. A solid rocket is called a "motor" because it's self-contained and consists of some propellant loaded into a container fitted with a nozzle and an ignitor. Because of their simplicity, solid rocket motors are inherently more reliable and cheaper to produce than liquid rocket engines. Unfortunately, they tend to have significantly lower I_{sp}.

Solid rocket motors consist of a mixture of fuel and oxidizer blended together in the right proportion and solidified. A typical propellant mixture contains an asphalt (just like the kind used to pave roads) fuel and an ammonium perchlorate ($NH_4 ClO_4$ used in fertilizer) oxidizer. The oxidizer is about 75% of the propellant's weight. The entire mixture is held together by a binder such as hydroxyl terminated polybutadiene (HTPB or basically rubber).

Because the propellant is solid, the thrust depends on the burning rate and the burning surface area. The larger the surface area burning, the faster the propellant will be used up and the higher the thrust. Burning rates depend on the type of fuel and oxidizer, their mixture ratio, and how the solid is designed to burn. The simplest type of burn profile is end-burning—like a cigarette—producing a flat thrust profile. For greater flexibility in tailoring the thrust profile, internal burning motors are normally used. During casting, designers can shape the hollow inner core of the solid motor to adjust the surface area, so they can control burning rate and thrust. Both types of solids and their thrust profiles are shown in Figure 14-12. The Space Shuttle's solid rocket motors, for example, have a core shape designed so the thrust will decrease about 55 seconds into the flight to lower the aerodynamic forces on the vehicle.

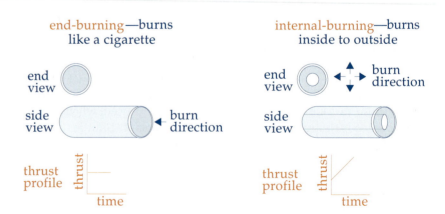

Figure 14-12. Designs and Thrust Profiles for a Solid Rocket Motor. Engineers can tailor the thrust profile of a solid rocket motor by changing its burn profile. An end-burning profile has a neutral thrust profile (constant thrust over time). An internal burning tube, on the other hand, has a progressive profile, meaning the thrust increases over time.

The I_{sp} of a solid rocket motor depends on the fuel and oxidizer. Once the motor is cast, the I_{sp} and thrust level are fixed. It's very difficult, if not impossible, to vary the thrust as you could with a liquid engine. The I_{sp} for solid rocket motors typically ranges from 200 – 300 seconds, with the Shuttle's solid rocket boosters (SRBs) having an I_{sp} of 261 seconds. Stopping the thrust may also be a problem. With a liquid rocket, we simply turn off the flow of propellant to shut the engine down. But we must literally blow a solid rocket motor up to stop the thrust.

Hybrid propulsion systems combine aspects of liquid and solid systems. Typically, a liquid or gaseous oxidizer is injected into a thrust chamber lined with a solid fuel as seen in Figure 14-13. Once ignited, the oxidizer combines with the fuel and burns. This process is like burning a log in the fireplace. Oxygen from the air combines with the log (fuel) in a fast oxidation process and burns. If you take away the air (throttle the oxidizer), the fire dies out. If you use a bellows or blow on the fire, you increase the flow of air, and the fire heats up. A properly designed hybrid rocket can offer the best of both worlds. They are safe to handle and store, like a solid, but can be throttled and restarted like a liquid engine. Their efficiencies and thrust levels are comparable to solids. Unfortunately, at this time hybrid rocket research and experience lags far behind research into both liquid and solid systems.

Nuclear thermal rockets don't need a fuel and oxidizer to create heat. As you can see in Figure 14-14, they use a nuclear-fission reactor to heat up some propellant, usually liquid hydrogen, to very high temperatures and then expand it out a nozzle. They can achieve very high temperatures using a propellant with a low molecular weight and obtain specific impulses of more than 1000 seconds. The United States Air Force has sponsored an emerging technology called a *particle-bed reactor*. It increases the surface area of the fuel, so the working fluid is allowed to reach even higher temperatures and hence, higher power densities and efficiencies, than other fission reactor concepts envisioned now. Particle-bed systems offer the added advantage of higher thrust-to-weight ratio and the potential for non-radioactive exhaust products.

Because of their relatively high thrust and better efficiencies, nuclear thermal rockets offer a distinct advantage over chemical systems for manned planetary missions. These missions must minimize transit time to decrease the detrimental effects of free-fall and solar or cosmic radiation on the human body. Ironically, future astronauts may escape the danger of solar and cosmic radiation by using the radiation source of a nuclear reactor to get them there faster.

Arc-jet thrusters offer a higher I_{sp} than nuclear thermal systems, but at much lower thrust. An arc-jet rocket produces thrust by passing propellant through an electric arc. This interaction heats the propellant to a high temperature, after which it can expand through a nozzle. Arc-jet systems can reach a theoretical I_{sp} of more than 2500 seconds. However, they do this at a thrust level measured in only tens of newtons (a few pounds). Thus, they are good only for limited space applications. A schematic for an arc-jet system is in Figure 14-15.

Figure 14-13. Hybrid Rocket System. A hybrid rocket system consists of a liquid or gaseous oxidizer which then combines with a solid fuel to burn it "nature's way."

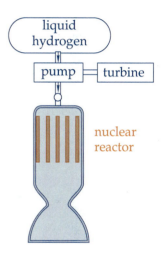

Figure 14-14. Nuclear Thermal Rocket. A nuclear thermal rocket uses a nuclear reactor to heat up some working propellant, such as liquid hydrogen. When the hydrogen reaches high temperatures, it expands out a nozzle to produce thrust.

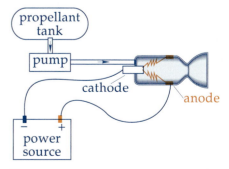

Figure 14-15. Arc-jet Thruster. An arc-jet thruster works by passing a propellant through an electric arc, after which the heated propellant expands through a nozzle to produce thrust.

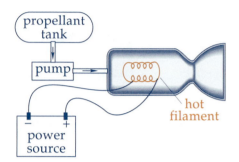

Figure 14-16. Resisto-jet Thruster. A resisto-jet system works by heating a filament (like the ones in a light bulb or toaster) inside the combustion chamber. As propellant passes over the filament, it heats up and expands out a nozzle.

Resisto-jet propulsion systems work much like a toaster, as you can see in Figure 14-16. Electrical energy heats up a filament inside a combustion chamber, heating a propellant. While resisto-jets are less efficient than arc-jets, with I_{sp} of only about 250 – 900 seconds, they offer a simple, well understood technology with longer mission lifetimes. Like arc-jets, they have low thrust levels—in the several-newton range.

Electrodynamic Propulsion

Thermodynamic rocket motors are designed to produce high thrust for short periods—minutes for chemical and hours for nuclear or arc-jet systems. But each is limited by I_{sp} and the amount of propellant it can carry. Electrodynamic systems, on the other hand, produce very small thrust for long periods—months or even years. By continually thrusting for so long, even a very small thrust can achieve velocities much higher than those from thermodynamic systems. Electrodynamic systems use:

- Ion rockets—electrostatic systems
- Plasma rockets—electromagnetic systems

Ion Rockets

Ion thrusters accelerate a fully ionized stream of atoms through a charged grid by means of an electric field, as shown in Figure 14-17. Remember, an *ion* is an atom with excess charge. The propellant enters the motor at an ionization station where one or more electrons are stripped from the atoms, creating positively charged ions. The ions accelerate throughout the motor by interacting with an electric field called an acceleration grid and exit the motor at velocities that theoretically can be as high as 1,000,000 m/s, yielding an I_{sp} of over 100,000 seconds. Unfortunately, the thrust levels are extremely low—only one newton or less (< 0.25 lbs.).

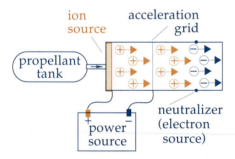

Figure 14-17. Ion Thruster. An ion thruster works by ionizing a propellant and then accelerating it using an electric field.

Ion motors have been tested in Earth orbit with promising results. One such test was the space electric-rocket test (SERT II) which demonstrated long-term storage in space (more than ten years) and ability to restart more than 300 times. Of course, ion motors require large amounts of electrical power, which the spacecraft's electrical-power system must provide.

Plasma Rockets

Plasma rocket motors use an electric arc and magnetic field to accelerate a neutral plasma of electrons, neutrons, and neutrally charged atoms to high velocity. An electric arc passes through a plasma, which acts as a conductor. As the arc forms, a magnetic field acts on it to accelerate the arc and the plasma along a channel and out the motor, as you can see in Figure 14-18.

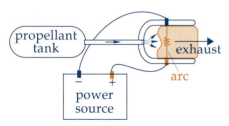

Figure 14-18. Plasma Rocket System. A plasma system accelerates a plasma of propellant using an electric arc and a magnetic field.

With a strong electric arc and magnetic field, the theoretical exit velocity for a plasma motor is very high—on the order of 50,000 m/s. This gives a theoretical I_{sp} of more than 5000 seconds. Unfortunately, to accelerate the entire plasma and not just the ions, a very dense plasma must exist. This density increases molecular collisions with the motor casing and can cause structural failure. Furthermore, the power needed to create the plasma by ionizing the propellant is very high. Despite these difficulties, experimental pulsed-plasma motors did small station-keeping maneuvers on the U.S. Navy's Nova series spacecraft launched in the early 1980's. Using nitrogen as propellant, a magnetoplasmadynamic (MPD) thruster has been ground tested with an I_{sp} of 4000 seconds. Plasma thrusters promise to be extremely efficient for small but frequent maneuvers. They could also be useful for long missions to other planets.

Table 14-1 compares the performance of various thermodynamic and electrodynamic propulsion systems.

Exotic Propulsion Systems

Exotic propulsion systems are those "far out" ideas which are still on the drawing board. One which may soon work is the solar sail. In Chapter 12 we mentioned solar pressure as one source of disturbance torque for a spacecraft. In Chapter 11 we discussed the nature of radiation as being both waves and particles called photons. If you think about EM radiation as photons, you can visualize the momentum imparted to a vehicle as it's struck by radiation. As a result, electro-magnetic radiation from the sun exerts pressure on an exposed surface. Why don't you feel this solar pressure "blowing" on your face on a sunny day? Because it's *very* small. However, if you could construct (and control!) a *very* large *solar sail*, you could catch this solar pressure and literally sail around the solar system. Of course, the farther you go from the sun the less solar pressure you feel, so a solar sail would work best inside the orbit of Mars. How big a sail would you need? To produce five newtons of thrust (about one pound), you'd need one square kilometer of sail! That's a sail 1 kilometer on a side! To achieve escape velocity from a low-Earth orbit (assuming a total spacecraft mass of only 10 kg), this force would have to be applied for more than 17 years! But the solar sail uses no propellant, so the thrust you get is basically "free."

Instead of using the momentum imparted by striking photons, another way to use this same concept is to turn it around. Based on Newton's Third Law of Action and Reaction, if the photons exert momentum when they strike something, a force must be imparted to it whenever photons are ejected. A simple flashlight is a good example of a *photon rocket* motor. Energy stored in the batteries is used to heat up a filament in the bulb, thus producing light or photons. Because those photons have momentum, they impart an equal but opposite impulse to the flashlight. Of course, in a typical 9-volt flashlight, the energies involved are so tiny that you never notice the flashlight pushing against your hand. However, if we could build a *very* large power source to produce the photons, practical photon rockets would be possible. Because the exit velocities are at the speed of light, the I_{sp} of such a system would approach 30,000,000 seconds! That's 10,000 times more efficient than the Space Shuttle's main engines! The thrust for such a motor would be very low, but by thrusting for several years, a spacecraft could begin to approach some significant fraction of the speed of light. When such a system becomes possible, interstellar travel may be in our grasp.

Table 14-1. Comparisons Between Various Propulsion Systems.

Type	Typical Propellants	Thrust (N)	I_{sp} (s)	Advantages	Disadvantages
Cold Gas	N_2 He H_2	$0.2 - 2700$	$60 - 225$	• Very simple • Ideal for attitude control • "Unlimited" cycling	• Low I_{sp} • Redundancy required
Liquid	H_2/LOX Kerosene/LOX N_2H_4 N_2H_4/N_2O_4	$10 - 6 \times 10^6$	$270 - 530$	• Can throttle • Very high thrust	• Complex • Expensive • Explosive hazard
Solid	$Al/NH_4 ClO_4$ Asphalt/$NH_4 ClO_4$ HTPB* binder	$0.1 - 1.2 \times 10^7$	$200 - 300$	• Very simple • Very high thrust • Inexpensive • Storable	• Can't throttle • Can't stop/restart • Difficult to vector thrust • Significant explosive hazard
Hybrid	Plexiglass/LOX HTPB*/LOX	$10 - 3.0 \times 10^{5**}$	$250 - 350$	• Can throttle • Simple • Non-explosive • Environmentally safe	• Difficult to build very high-thrust motors • Lower I_{sp} than liquid systems
Nuclear Thermal	H_2	$3000 - 6 \times 10^5$	$700 - 1100$	• High I_{sp} • High thrust • Can cycle • Can throttle	• Nuclear contamination • Shielding required • Lower thrust-to-weight than chemical systems
Arcjet	H_2 NH_3	$0.05 - 40$	$500 - 2500$	• High thrust (electric) • High I_{sp}	• Limited life-span • High power requirement
Resistojet	NH_3 H_2	$5 \times 10^{-4} - 10$	$250 - 900$	• Simple • Long life	• High power requirement • Lowest I_{sp} electric system
Ion	Cs Hg Xe C_{60}	$0.02 - 2.0$	$5,000 - 10,000$	• Very high I_{sp} • Very long useful life • "Unlimited" cycling	• Very low thrust • High power requirement
Plasma	H_2 Ar Xe	$0.2 - 200$	$2,000 - 10,000$	• Relatively high thrust • Very high I_{sp}	• Limited life-span • High power requirement • Limited ability to cycle

* Demonstrated HTPB = "rubber"
** Proposed Shuttle SRB replacement was designed to yield 5×10^6 Newtons thrust

Astro Fun Fact

F = –ma and Other Far Out Rocket Ideas

To zip around the solar system in a few days or establish colonies around other stars we'll need even more exotic propulsion systems. One promising, relatively near-term concept will be familiar to Star Trek fans—anti-matter. When matter and anti-matter come in contact, both are annihilated and energy is released. By harnessing this energy, we can theoretically build a booster to carry 2.2 tons of payload into low-Earth orbit using only ten milligrams of anti-matter [Forward, 1993]. An even more exotic concept proposed by Dr. Robert Forward involves the use of negative matter. While anti-matter still obeys the normal physical laws we're used to, negative matter does the opposite. This means if you push on negative matter it comes toward you (F = -ma)! If negative matter existed and if you could manipulate it into the density of a black hole (two very big "if's"!), then you could put a chunk of it inside a spacecraft and push on it to move around. Other concepts rocket scientists are exploring include the use of large, rotating space tethers, beamed laser propulsion, interstellar ramjets, and even pulsed nuclear rockets which would use the shock from exploding nuclear bombs one at a time under the ship to propel you forward.

Forward, Robert L. Advanced Propulsion Systems. Space Propulsion Analysis and Design, advanced copy, 1993.

Forward, Robert L. 21st Century Space Propulsion. The Journal of Practical Applications in Space. Vol. 2, Issue No. 2. Arlington, VA: High Frontier, Inc., Winter, 1991.

As we approach the speed of light, however, physics as we know it begins to take on a weird twist. One by-product of Albert Einstein's theory of relativity says that, as your velocity approaches the speed of light, your perception of time begins to change relative to a "fixed" observer. This leads to the so-called "twin paradox," illustrated in Figure 14-19. Suppose we have a set of twins, a sister and a brother. If the sister leaves the brother behind to set off on a space mission and travels near the speed of light, when she returns she'll find her brother will have aged much more than

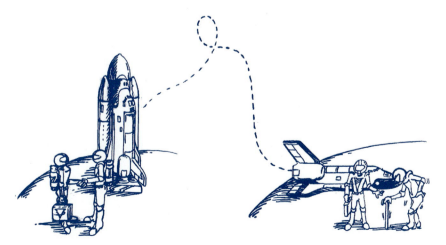

Figure 14-19. Twin Paradox. Einstein's theory of relativity tells us that if one twin leaves Earth and travels at speeds near the speed of light, when she returns she'll find her twin left on Earth will have aged more than she.

she has! In other words, while the mission will have seemed to last only a few years for the sister, tens or even hundreds of years may have passed for her brother. This *time dilation effect*, sometimes called a "Tau factor", τ, can be expressed using the Lorentz transformation

$$\tau = \frac{t_{starship}}{t_{Earth}} = \sqrt{1 - \frac{V^2}{c^2}}$$ (14-13)

where

$t_{starship}$ = time measured on a starship
t_{Earth} = time measured on the Earth
V = starship velocity
c = speed of light = 300,000 km/s

The tau factor tells us the ratio of how time passes aboard a speeding starship versus how time passes back on Earth as demonstrated in Example 14-3. As the spacecraft's velocity approaches the speed of light, τ gets very small, meaning that time on the ship is passing much more slowly than it does back on Earth. While this may seem convenient for readers thinking about a weekend journey to Alpha Centauri, Einstein's theory also places a severe "speed limit" on would-be space travelers. As the velocity of a spacecraft increases, its effective mass also increases. Thus, as the ship's velocity approaches the speed of light, we need more and more thrust to get a given change in velocity. To make the ship go at the speed of light, we'd need an infinite amount of energy. For this reason alone, travel at near the speed of light is well beyond current technology.

But travel at light speeds and beyond is common using the propulsion systems of science fiction. Perhaps the best known exotic system is the Warp Drive of the Star Trek television series. Each week, this exotic system drives the crew of the Enterprise well beyond the speed of light, enabling them to visit various solar systems in a single episode. Who knows? For years, scientists and engineers said travel beyond the speed of sound, the so-called "sound barrier," was impossible. But in October 1947, Chuck Yeager proved them all wrong while piloting the Bell X-1. Today, jet planes routinely travel at speeds well beyond the speed of sound. Perhaps by the 23rd century some future Chuck Yeager will break another speed barrier and take a spacecraft beyond the speed of light.

▬ Section Review

Key Terms

arc-jet thrusters
bi-propellant
chemical rockets
cold-gas systems
cryogenic propellants
electrodynamic systems
exotic systems
hybrid propulsion systems
hypergolic propellant
ion
molecular weight
mono-propellant
nuclear thermal rockets
particle-bed reactor
photon rocket
resisto-jet propulsion systems
solar sail
solid rocket motors
space-storable propellants
thermodynamic systems
time dilation effect

Key Equations

$$I_{sp} \propto \sqrt{\frac{T_{combustion}}{M}}$$

$$\tau \doteq \frac{t_{starship}}{t_{Earth}} = \sqrt{1 - \frac{V^2}{c^2}}$$

Key Concepts

➤ All rocket systems have the same elements:
  • Propellant
  • Energy Source
  • Propellant-feed system
  • Energy-conversion system
  • Accelerator

➤ All types of rockets fit into three broad categories:
  • Thermodynamic systems—convert thermal energy (heat) into kinetic energy
  • Electrodynamic systems—convert electrical energy into kinetic energy
  • Exotic systems—create thrust through some other means

➤ Thermodynamic systems include
  • Cold gas
  • Chemical
  • Nuclear Thermal
  • Arc-jet
  • Resisto-jet

➤ Thermodynamic systems can increase specific impulse by
  • Increasing combustion temperature
  • Decreasing the propellant's molecular weight

➤ Chemical rockets are most common. Three basic types are
  • Liquid
  • Solid
  • Hybrid

➤ Table 14-1 compares the types of thermodynamic systems

➤ Electrodynamic systems include
  • Ion Thrusters
  • Plasma Thrusters

➤ Table 14-1 compares the types of electrodynamic systems

➤ Exotic propulsion methods include
  • Solar sails
  • Photon rockets

➤ Exotic systems may one day propel spacecraft at speeds near the speed of light. When this happens, we'll have to worry about the time dilation effect predicted by Einstein's theory of relativity.

Example 14-3

Problem Statement

In the 22nd century the starship *Sting Ray* is traveling at 250,000 km/s near the star Tau Ceti—11.8 light years from Earth. If one month goes by on the ship, how much time passes on Earth? Assume Earth is relatively motionless compared to the speed of the ship.

Problem Summary

Given: $V=250,000$ km/s

$t_{starship} = 1$ month

Find: time passage on Earth

Conceptual Solution

1) Determine the Tau-factor

$$\tau = \frac{t_{starship}}{t_{Earth}} = \sqrt{1 - \frac{V^2}{c^2}}$$

2) Divide one month by τ to get time passage on Earth

Analytical Solution

1) Determine the Tau-factor

$$\tau = \frac{t_{starship}}{t_{Earth}} = \sqrt{1 - \frac{V^2}{c^2}}$$

$$\tau = \sqrt{1 - \frac{(250,000 \text{km/s})^2}{(300,000 \text{km/s})^2}} = 0.553$$

2) Divide one month by τ to get time passage on Earth

$$t_{Earth} = \frac{t_{strship}}{\tau} = \frac{1 \text{ month}}{0.553}$$

$$= 1.81 \text{ months} \cong 56 \text{ days}$$

Interpreting the Results

Because they are traveling near the speed of light (about 83% of c), time is dilated, so while one month goes by on the ship, nearly two months pass on Earth.

14.3 Booster Staging

▬ In This Section You'll Learn to...

☞ Discuss the advantages and disadvantages of booster staging

☞ Determine the velocity change staging can achieve

Now that we've seen all the types of rocket systems available, let's see how we use these systems to solve perhaps the most important problem of astronautics: getting into space. Appendix E summarizes key parameters about the booster systems most often used for putting spacecraft into orbit. Boosters come in all different shapes and sizes, from the mighty Space Shuttle to the tiny Pegasus. However, in all cases, boosters in use have two common characteristics:

- Chemical rocket systems
- Multiple stages

Even though chemical rockets aren't as efficient as some options discussed in the last section, they offer very high thrust and, more importantly, very high thrust-to-weight ratios. Only chemical rockets can now produce thrust-to-weight ratios greater than one. To understand why they all use multiple stages, we must explore the advantages of staging.

Staging

When you look at a typical booster, most of what you see is the propellant tank. Why carry all that extra tank weight along once part of the propellant has been used up? Instead, why not split the propellant into smaller tanks and then drop the tanks as they empty out? During World War II (before fuel tanks became too expensive), fighter planes flying long distances used this idea in the form of "drop tanks." These tanks provide extra fuel for long flights and were dropped off as soon as they were empty, thus lightening and streamlining the plane. This is the basic concept of *staging*. Figure 14-20 shows a Saturn 1B during staging.

Stages consist of propellant tanks and rocket engines. As each stage burns out, it's dropped off, and the engines of the next stage start up (hopefully) to continue the trip into space. As expended stages drop off, the booster's mass decreases. This means a smaller engine can keep the booster on track into orbit.

Table 14-2 gives an example of how staging can increase the amount of payload delivered to orbit. As you can see, for this simple example the staged vehicle can deliver twice the payload to orbit as a like-sized, single-staged vehicle with the same total propellant mass—even after adding 10% to the structure's overall mass to account for the extra engines and

Figure 14-20. Saturn 1B Staging. This long-range photo shows a Saturn 1B rocket staging. Notice the first stage drops off as the second stage ignites. Staging increases the efficiency of a booster by dropping off deadweight. *(Courtesy of NASA)*

plumbing needed for staging. That's why all boosters use staging. (Engineers are working on single-stage to orbit [SSTO] designs. If the booster can be completely reusable, we may be willing to accept the loss in payload.)

Table 14-2. Comparing a Single-Stage and Two-Stage Booster.

Booster	Parameters	Payload to Orbit
Single Stage $m_{payload} = 84$ kg $m_{structure} = 250$ kg $m_{propellant} = 1500$ kg engine $I_{sp} = 480$ s	$\Delta V_{design} = 8000$ m/s $I_{sp} = 480$ s $m_{structure} = 250$ kg $m_{propellant} = 1500$ kg	$m_{payload} = 84$ kg
Two Stage $m_{payload} = 175$ kg $m_{propellant} = 750$ kg $m_{structure} = 140$ kg engine $I_{sp} = 480$ s $m_{propellant} = 750$ kg $m_{structure} = 140$ kg engine $I_{sp} = 480$ s	$\Delta V_{design} = 8000$ m/s **Stage 2** $I_{sp} = 480$ s $m_{structure} = 140$ kg $m_{propellant} = 750$ kg **Stage 1** $I_{sp} = 480$ s $m_{structure} = 140$ kg $m_{propellant} = 750$ kg	$m_{payload} = 175$ kg

As you can see in Table 14-2, for both cases the size of the payload we can deliver to orbit compared to the weight of the entire booster is pretty small—5% or less. About 80% or more of a typical booster is propellant. The other 15% or so is structure, avionics, and other "deadweight." Obviously, we could get more payload into space if only our engines were more efficient. However, with engines operating at or near state-of-the-art, the only other option, as the examples show, is to get rid of deadweight on the way into orbit.

Now let's see how we can use our rocket equation to analyze the total ΔV we get from a staged vehicle. We'll start with

$$\Delta V = I_{sp}g_o \ln\left(\frac{m_{intial}}{m_{final}}\right)$$

We must recognize that, for a staged vehicle, each stage has an initial and a final mass. Also, the I_{sp} may be different for the engine(s) in different stages. To get the total ΔV of the staged vehicle, we must add the ΔV for each stage. This gives us the following relationship for the ΔV of a staged vehicle with n stages.

$$\Delta V_{total} = \Delta V_{stage\ 1} + \Delta V_{stage\ 2} + ... + \Delta V_{stage\ n} \qquad (14\text{-}14)$$

$$
\begin{aligned}
\Delta V_{total} = {} & I_{sp\ stage\ 1}\ g_o \ln\left(\frac{m_{initial\ stage\ 1}}{m_{final\ stage\ 1}}\right) \\[2mm]
& + I_{sp\ stage\ 2}\ g_o \ln\left(\frac{m_{initial\ stage\ 2}}{m_{final\ stage\ 2}}\right) + ... \\[2mm]
& + I_{sp\ stage\ n}\ g_o \ln\left(\frac{m_{initial\ stage\ n}}{m_{final\ stage\ n}}\right)
\end{aligned}
\qquad (14\text{-}15)
$$

where
ΔV_{total} = total ΔV from all stages (m/s)
$I_{sp\ stage\ n}$ = specific impulse of stage n (s)
g_o = gravitational acceleration at sea level (9.81 m/s^2)
$m_{initial\ stage\ n}$ = initial mass of stage n (kg)
$m_{final\ stage\ n}$ = final mass of stage n (kg)

What is the initial and final mass of stage 1? The initial mass is easy; it's just the mass of the entire vehicle at lift-off. But what about the final mass of stage 1? Here we have to go back to our definition of final mass when we developed the rocket equation. Final mass of any stage is the initial mass of that stage (including the mass of subsequent stages) less the mass of propellant burned in that stage. So for stage 1

$$m_{final\ stage\ 1} = m_{initial\ vehicle} - m_{propellant\ stage\ 1}$$

Similarly, we can develop a relationship for the initial and final mass of stage 2, stage 3, and so on.

$$m_{initial\ stage\ 2} = m_{final\ stage\ 1} - m_{structure\ stage\ 1}$$

$$m_{final\ stage\ 2} = m_{initial\ stage\ 2} - m_{propellant\ stage\ 2}$$

Example 14-4 shows how to compute the total ΔV for a staged vehicle.

Overall, staging has several unique advantages over a single-stage vehicle. It

- Reduces the vehicle's total weight for a given payload and ΔV requirement

- Increases the total payload mass we can deliver to space for the same-sized vehicle
- Increases the total velocity we can achieve for the same-sized vehicle
- Decreases the engine efficiency (I_{sp}) required to deliver a given-sized payload to orbit

But, as the old saying goes, "There ain't no such thing as a free lunch" (or launch)! In other words, all of these advantages of staging come with drawbacks. These include

- Increased complexity because of the extra set of engines and plumbing
- Decreased reliability because we're depending on extra sets of engines and the plumbing to work
- Increased total cost because more complex vehicles cost more to build

Another interesting limitation of staging has to do with the law of diminishing returns. So far, you may be ready to conclude that if two stages are good, four stages must be twice as good. But this isn't necessarily the case. Although the first added stage significantly improves performance, each additional stage enhances it less. By the time we add a fourth or fifth stage, the increased complexity offsets the small extra gain in performance. That's why most boosters have only two or three stages.

Astro Fun Fact
Single Stage to Orbit

Most boosters in use are high-performance, multi-stage systems with more than 15% of the total mass taken up by structure. Employing them is similar to using an Indy race car to pull freight on the interstate and then abandoning it when you get to your destination. However, new advances in Space Age materials—such as composites made of carbon, epoxy, and graphite—may make relatively simple, singe-stage to orbit (SSTO) boosters practical. Using these new materials, perhaps only 8% of the booster would be structure. Such a system would use one stage to lift a small payload to low-Earth orbit and then return to Earth, using the same engine to slow down for landing. One SSTO concept being developed by the McDonnell-Douglas Company is the Delta Clipper. It would take off and land vertically—more like the Lunar lander than the Space Shuttle. The full-scale vehicle, dubbed the DC-Y, is designed to deliver 5000 kg to LEO. The National Aerospace Plane (NASP) is an example of an SSTO vehicle using horizontal takeoff and landing. NASP would use atmospheric oxygen along with liquid hydrogen in a supersonic combustion ram-jet (or SCRAMJET) to get most of the way into orbit and then use a small rocket to achieve orbit once outside of the atmosphere. Someday you may be booking a flight on the Delta Clipper or NASP for a free-fall vacation in low-Earth orbit!

(Courtesy of Pratt and Whitney)

Edward H. Kolcum. <u>Aviation Week and Space Technology</u>. "Delta Clipper Partners Set Goal for Single-Stage-to-Orbit Vehicle." Feb. 3, 1992, p. 55–56.

Mike Lydon, United States Air Force Academy

Section Review

Key Terms

staging

Key Equations

$$\Delta V_{total} = I_{sp\ stage\ 1}\ g_o \ln\left(\frac{m_{initial\ stage\ 1}}{m_{final\ stage\ 1}}\right)$$

$$+\ I_{sp\ stage\ 2}\ g_o \ln\left(\frac{m_{initial\ stage\ 2}}{m_{final\ stage\ 2}}\right) + ...$$

$$+\ I_{sp\ stage\ n}\ g_o \ln\left(\frac{m_{initial\ stage\ n}}{m_{final\ stage\ n}}\right)$$

Key Concepts

➤ Appendix E summarizes major characteristics of some common booster systems.

➤ All booster systems in use rely on chemical rockets so they can achieve a thrust-to-weight ratio greater than 1

➤ By using staging on booster systems, we can
  • Reduce the total vehicle weight for a given payload and ΔV requirement
  • Increase the total mass of payload we can deliver to space for the same-sized vehicle
  • Increase the total velocity we can achieve for the same-sized vehicle
  • Decrease the engine efficiency (I_{sp}) required to deliver a given-sized payload to orbit

➤ But staging also has several disadvantages:
  • Increased complexity, because we need extra engines and plumbing
  • Decreased reliability, because we depend on an extra set of engines and plumbing
  • Increased total cost, because building a more complex vehicle costs more

Example 14-4

Problem Statement

An experimental two-stage booster is preparing to launch from the Kennedy Space Center. The booster must deliver a total ΔV (ΔV_{design}) of 10,000 m/s. The total mass of the second stage, including structure and propellant, is 12,000 kg, 9000 kg of which is propellant. The payload mass is 2000 kg. The I_{sp} of the first stage is 350 seconds and of the second stage is 400 seconds. The structural mass of the first stage is 8000 kg. What mass of propellant must be loaded on the first stage to achieve the required ΔV_{design}? What is the vehicle's total mass at lift-off?

Problem Summary

Given: 2 stages

$m_{payload} = 2000$ kg

$m_{structure\text{-}2} + m_{propellant\text{-}2} = 12{,}000$ kg

$m_{propellant\text{-}2} = 9000$ kg

$m_{structure\text{-}1} = 8000$ kg

$I_{sp\text{-}1} = 350$ s

$I_{sp\text{-}2} = 400$ s

$\Delta V_{design} = 10{,}000$ m/s

Find: $m_{propellant\text{-}1}$

$m_{initial}$

Problem Diagram

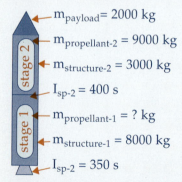

$m_{payload} = 2000$ kg

$m_{propellant\text{-}2} = 9000$ kg

stage 2

$m_{structure\text{-}2} = 3000$ kg

$I_{sp\text{-}2} = 400$ s

stage 1

$m_{propellant\text{-}1} = ?$ kg

$m_{structure\text{-}1} = 8000$ kg

$I_{sp\text{-}2} = 350$ s

Conceptual Solution

1) Determine $\Delta V_{stage\ 2}$

$$\Delta V_{stage\text{-}2} = I_{sp\text{-}2} g_o \times$$

$$\ln\left(\frac{m_{structure\text{-}2} + m_{propellant\text{-}2} + m_{payload}}{m_{structure\text{-}2} + m_{payload}}\right)$$

2) Determine required ΔV of stage 1

$$\Delta V_{stage\text{-}1} = \Delta V_{design} - \Delta V_{stage\text{-}2}$$

3) Determine initial mass of stage 1

$$\Delta V_{stage\text{-}1} = I_{sp\text{-}1} g_o \times$$

$$\ln\left(\frac{m_{initial}}{m_{structure\text{-}2} + m_{propellant\text{-}2} + m_{payload} + m_{structure\text{-}1}}\right)$$

4) Determine mass of propellant in stage 1

$$m_{propellant\text{-}1} = m_{initial} -$$

$$(m_{structure\text{-}1} + m_{structure\text{-}2} + m_{propellant\text{-}2} + m_{payload})$$

Analytical Solution

1) Determine $\Delta V_{stage\ 2}$

$$\Delta V_{stage\text{-}2} = I_{sp\text{-}2} g_o \times$$

$$\ln\left(\frac{m_{structure\text{-}2} + m_{propellant\text{-}2} + m_{payload}}{m_{structure\text{-}2} + m_{payload}}\right)$$

$$= (400\ s)(9.81\,m/s^2)$$

$$\ln\left(\frac{14{,}000\ kg}{12{,}000\ kg + 2000\ kg - 9000\ kg}\right)$$

$$\Delta V_{stage\ 2} = 4{,}040\ m/s$$

2) Determine required ΔV of first stage

$$\Delta V_{stage\ 1} = \Delta V_{design} - \Delta V_{stage\ 2}$$

$$= 10{,}000\ m/s - 4040\ m/s$$

$$\Delta V_{stage\ 1} = 5960\ m/s$$

Example 14-4 (Continued)

3) Determine initial mass of stage 1

$$\Delta V_{stage\text{-}1} = I_{sp\text{-}1} g_o \times$$

$$\ln \left(\frac{m_{initial}}{m_{structure\text{-}1} + m_{structure\text{-}2} + m_{propellant\text{-}2} + m_{payload}} \right)$$

$$5960 \text{ m/s} = (350 \text{ s}) (9.81 \text{ m/s}^2) \ln$$

$$\left(\frac{m_{initial}}{8000 \text{ kg} + 12{,}000 \text{ kg} + 2000 \text{ kg}} \right)$$

$$m_{initital} = 22{,}000 \text{ kg } e^{\frac{5960 \text{ m/s}}{(350 \text{ s}) (9.81 \text{ m/s}^2)}}$$

$$m_{initial} = 124{,}821 \text{ kg}$$

4) Determine mass of propellant in stage 1

$$m_{propellant\text{-}1} = m_{initial} -$$

$$(m_{structure\text{-}1} + m_{structure\text{-}2} + m_{propellant\text{-}2} + m_{payload})$$

$$= 124{,}821 - (8000 \text{ kg} + 3000 \text{ kg} + 9000 \text{ kg} + 2000 \text{kg})$$

$$m_{propellant\text{-}1} = 102{,}821 \text{ kg}$$

Interpreting the Results

The total mass of this booster vehicle at lift-off is 124,821 kg (113 tons). About 82% of this mass is propellant in the first stage alone (102,821 kg/124,821 kg). Less than 2% of the total lift-off mass is actually payload (2000 kg/124,821 kg).

References

Einstein, Albert. *Relativity: The Special and the General Theory*. New York, NY: Bonanza Books, 1961. Distributed by Crown Publishers, Inc. for the estate of Albert Einstein.

Isakowitz, Steven J. *International Reference Guide to Space Launch Systems*. Washington, D.C.: American Institute of Aeronautics and Astronautics (AIAA), 1991.

Wertz, James R. and Wiley J. Larson. *Space Mission Analysis and Design*. Dordrecht, Netherlands: Kluwer Academy Publishers, 1991.

Mission Problems

14.1   Rocket Basics

1   What are Newton's Laws of Motion?

2   What are the basic parts of a rocket engine and what does each do?

3   Describe how a rocket produces thrust from both momentum change and pressure differences.

4   What is the rocket-thrust equation? What does each term represent?

5   What is effective exhaust velocity?

6   Describe the difference between overexpanded, underexpanded, and ideally expanded rocket nozzles.   How do engineers try to get the best nozzle design by picking a design altitude?

7   What is the ideal rocket equation? Define each term.

8   While on its way into orbit, the Space Shuttle, with an initial mass of 100,000 kg, burns 1000 kg of propellant through its orbital maneuvering system's engines (c = 2000 m/s). How much ΔV is achieved?

9   A remote-sensing spacecraft needs to correct its orbit by 10 m/s. If the effective exhaust velocity of the orbital-maneuvering thruster is 1000 m/s, and the spacecraft's initial mass including propellant is 1000 kg, how much propellant will the maneuver require?

10   What is impulse? Give an example from your daily life.

11 How is specific impulse defined?

12 What are the units of I_{sp}? What does an I_{sp} of 300 seconds mean?

13 Engineers are evaluating two different thrusters for a new communication satellite. System 1 has an I_{sp} of 100 seconds and system 2 an I_{sp} of 150 seconds. If the total ΔV the system must deliver over the life of the spacecraft is 500 m/s, how much propellant will you save by using system 2 instead of system 1? Assume the initial mass in both cases is 1000 kg.

14.2   Rocket Systems

14 What are the basic elements of any rocket system?

15 What are the three basic categories of rocket systems?

16 List the five types of thermodynamic propulsion systems and briefly describe how each works. Compare their relative performances.

17 What two qualities of a thermodynamic rocket engine or motor affect the I_{sp} it can produce? How

do we choose these qualities to produce the highest I_{sp}?

18 How does a thermodynamic-propulsion system produce thrust? How does an electrodynamic-propulsion system produce thrust?

19 What type of propellant system is best suited for a fast-reaction, silo-based ICBM?

20 What is an advantage of a solid-fueled chemical engine over a liquid-fueled one?

21 List the two main types of electrodynamic rocket thrusters, describe their basic operating principles, and compare their performance.

22 How does a solar sail work?

23 The starship *Endeavour* travels at 80% of the speed of light for one year (relative to the crew.) How much time will pass relative to those of us on Earth?

14.3   Booster Staging

24 What are the advantages and disadvantages of staging?

25 A new three-stage rocket for delivering small payloads to low-Earth orbit is being analyzed. It has these characteristics:

- I_{sp} stage 1 = 300 s

- I_{sp} stage 2 = 350 s

- I_{sp} stage 3 = 400 s

- Payload mass = 1500 kg

- Structure mass stage 1 = 10,000 kg

- Structure mass stages 2 and 3 = 7500 kg each

- Propellant mass stage 1 = 50,000 kg

- Propellant mass stage 2 = 40,000 kg

- Propellant mass stage 3 = 35,000 kg

a) What is the initial mass of the entire vehicle?

b) What is the final mass of stage 1?

c) What is the ΔV of stage 1?

d) What is the initial mass of stage 2?

e) What is the final mass of stage 2?

f) What is the ΔV of stage 2?

g) What is the initial mass of stage 3?

h) What is the final mass of stage 3?

i) What is the ΔV of stage 3?

j) What is the total ΔV of the booster?

26 A two-stage rocket with the following characteristics must produce a total ΔV of 8000 m/s.

- I_{sp} stage 1 = 400 s

- I_{sp} stage 2 = 450 s

- Payload mass = 100 kg

- Structure mass stage 1 = 8000 kg

- Structure mass stage 2 = 6000 kg

If the ΔV for stage 2 is 3000 m/s, what is the vehicle's total mass at lift-off?

27 NASA is working on a new two-stage rocket. The I_{sp} of stage 1 is 300 seconds. The I_{sp} of stage 2 is 400 seconds. The structural mass of stage 1 is 1000 kg and of stage 2 is 800 kg. Propellant loading is 75,000 kg for stage 1 and 50,000 kg for stage 2. Engineers want to place a 1200-kg payload into a circular orbit at an altitude of 200 km. The proposed launch site is Kennedy Space Center (L_o = 28.5°) with an inclination of 28.5°. Assume ΔV_{losses} are 800 m/s. Can this rocket do the job? [Hint: you may need to review ΔV_{design} calculations from Chapter 9.]

28 A small solid-rocket motor has an I_{sp} of 70 seconds. If the propellant weighs 0.0015 kg and is burned in 0.2 seconds, how much thrust does the motor generate? (Assume thrust and propellant mass flow rate are constant throughout the burn.)

Mission Profile—Pegasus

April 5, 1990, marked a significant addition to launch vehicles. A new three-stage booster named Pegasus was air-launched from a NASA B-52 aircraft over the Dryden Flight Research Center in California. Other launchers use a vertical launch profile from a fixed site. After dropping from a transport aircraft, this new launch vehicle ignited its engines to boost the payload into orbit. In a link to space heritage, the NASA B-52 aircraft used for the launch was also used to launch X-15s in the 1960s.

Mission Overview

Developed by Orbital Sciences Corporation, Pegasus was developed with support from the Defense Advanced Research Project Agency (DARPA) and the United States Air Force. The three-stage, solid-propellant booster provides flexible launching of small satellites into low-Earth orbit. The mobile launch platform allows the vendor to better meet the customer's needs.

Mission Data

✓ The Pegasus vehicle is carried to an altitude of about 12,000 m (40,000 ft) and dropped at a speed of approximately Mach 0.8

✓ The booster is 15.5 m (50.9 ft) long, with a launch weight of about 19,000 kg (42,000 lb.)

✓ The three stages use solid propellant:

- Stage 1
  I_{sp} = 295.3 s (vac)
  Thrust = 486,700 N (109,420 lb.)

- Stage 2
  I_{sp} = 295.5 s (vac)
  Thrust = 122,800 N (27,600 lb.)

- Stage 3
  I_{sp} = 291.1 s (vac)
  Thrust = 34,560 N (7770 lb.)

✓ Pegasus can launch to any azimuth (i.e., any inclination) with the following payloads:

- 455 kg (1000 lb.) into a 185 km (100 nm) circular orbit at a 28° inclination

- 365 kg (800 lb.) into a 185 km (100 nm) circular orbit at a 90° inclination

Mission Impact

Pegasus will reduce problems in range safety because it can fly from any point on Earth. Pegasus also can launch into many more inclinations. Finally, Pegasus promises a faster turnaround time between launches because it has no launch pad to refurbish.

Artists conception of the Pegasus air-launched space booster in lifting ascent to orbit. *(Courtesy of Orbital Sciences Corporation)*

For Discussion

- What other benefits can we derive from a flexible launch system like Pegasus?

- What dangers arise with a mobile launch system?

- Will systems like Pegasus encourage a greater shift from government launch services to those from the private sector?

Contributor

Steve Crumpton, United States Air Force Academy

References

Isakowitz, Steven J. *International Reference Guide to Space Launch Systems*. AIAA, 1991.

Wilson, Andrew. *Interavia Space Director*. Alexandria, VA: Jane's Information Group, 1990.

Space operations experts work "behind the scenes" at the Mission Control Center of NASA's Jet Propulsion Laboratory, which supports spacecraft around the solar system. *(Courtesy of NASA)*

Space Operations and Communication Networks

15

▬ In This Chapter You'll Learn to...

☞ Describe the major functions of space operations

☞ Describe the main parts of a space mission's communication network

☞ Describe basic communication principles and determine key parameters of system design

▬ You Should Already Know...

❏ Elements of space-mission architecture (Chapter 1)

❏ Definition of an operations concept for space missions (Chapter 11)

❏ Electromagnetic spectrum and black-body radiation (Chapter 11)

Astronaut Swigert: "Okay Houston, we've had a problem here."
Mission Control Capcom: "This is Houston, say again please."

Fateful words that began
the dramatic Apollo 13 ordeal

P eople are the most important part of a space mission. So far we've focused on trajectories and hardware, but people plan, support, direct, and even risk their lives to complete a mission. With the seven simple words quoted at the beginning of this chapter, the crew of Apollo 13 set in motion a heroic effort by the ground-control team at the Mission Control Center in Houston, Texas. Hundreds of engineers and operators worked around the clock to devise a plan to save three astronauts facing death 250,000 miles from home. The resourcefulness of these dedicated men and women and the cool reactions of the crew ensured their safe return to Earth.

Throughout most of the space age, Houston has been the hub of U.S. manned space programs. On TV you see a bunch of people seated at consoles staring at video screens. But what are all these people doing? Why does it seem to take so many of them to pull off just one mission?

Throughout this text we've referred to the operations concept of the mission—the unifying principle that describes the relationship between hardware systems, mission data and services, and their ultimate users—people. Now we'll turn our attention to what this concept involves. We'll explore the various aspects of mission operations, from mission design, to launch, to collecting data on orbit. At every step we'll see how people make it happen. Finally, we'll focus on an integral part of the operations concept, the most important tool operators have at their disposal—the communications network that ties spacecraft, astronauts, boosters, operators, and users into an efficient team.

15.1 Space Operations

In This Section You'll Learn to...

☞ Identify the three phases of space operations

☞ Discuss what space operations people do during each phase

In Chapter 11, we learned how important understanding the operations concept is to preliminary mission design. A space mission we can't operate is like a car we can't drive. During all phases of a mission, engineers and users must be acutely aware of the operator's needs and limits. Typically, we consider operations important only during and after launch. However, the role of the operator is equally critical years earlier—during the initial mission planning. We'll explore how operators carry out critical functions during the three phases of a mission—mission planning and design, launch and orbit checkout, and normal operations.

Mission Planning and Design

During mission planning and design, operators are involved in

- Refining the operations concept
- Developing the mission timeline
- Developing training plans and simulator requirements
- Assembling the operations team
- Simulating the mission
- Reviewing mission readiness

The *operations concept* drives the entire system design. This concept describes how users will get their data or use the mission services—what they need and when. Operators help by bringing a "real-world" perspective to the operations concept. They can help limit the scope of the design and select what is feasible from competing objectives. For example, in Chapter 11 we explained design using a hypothetical mission called FireSat to detect forest fires. During mission planning for FireSat, operators refine how the spacecraft, ground controllers, and Forest Service work together. Their suggestions also determine how the communication network is designed and configured to support the mission.

The *mission timeline* is a detailed script clearly defining all events that take place during the mission and when they must occur. From roll-out of the booster to mission end, the sequence of events must be carefully organized to ensure that one action doesn't get ahead of another. For example, we'd like to keep the spacecraft running on ground power until the last possible minute to save onboard batteries. The mission timeline would lay out exactly when in the countdown sequence to cut off external power. Because operators help execute the timeline, they must help develop it.

Operators spend lots of time training. When it's time for launch they need to be intimately familiar with every aspect of the mission and what to do when something goes wrong. The only way they can do that is the same way a musician gets to Carnegie Hall—practice, practice, practice. For operators, practice involves complex dress rehearsals known as *simulations*. During mission planning, operators must decide how to train for the actual mission and define simulator requirements to support that training. For new systems, this process can take months or even years to complete. Operators working with other support people must define hardware and software requirements to do simulations and plan entire training programs.

Next, the operations team must be assembled. Members of the operations team are called *flight controllers*. To effectively use expensive simulator time, mission managers must identify team members as early as possible to cover four basic areas. First there's a team leader, called a *flight director* for Space Shuttle missions, who coordinates the input from other team members. The flight director sits in the "hot seat." It's up to him or her to make the final decisions on what to do throughout the mission.

Under the flight director, the team can be divided into three other areas. *Subsystem specialists* are experts on individual parts of the spacecraft and the booster. For example, one person may devote all his or her attention to monitoring the electrical-power system while another watches the attitude-control system. *Ground-system specialists* link the operations team and spacecraft together. They maintain the computers in the control center and ensure the complex communication network which links ground to spacecraft is up and running. They gather the data from the spacecraft and tracking sites, deliver it to the control center, and relay commands from operators back to the spacecraft. The final members of the team are the *payload specialists*. They're responsible for the payload—its health, status, and operation. It's up to them to point cameras or antennas to collect valuable mission data. They process and deliver this data to users quickly and efficiently. Figure 15-1 shows an attentive flight controller at his console.

Figure 15-1. The Shuttle's Mission Control Center (MCC). Within the MCC, flight controllers are at consoles monitoring every aspect of a mission. *(Courtesy of NASA)*

With the team assembled, it's time to start rehearsing for the mission by doing simulations. During a simulation (or "sim" for short), devious trainers feed simulated mission data and disaster scenarios to operators at their consoles. Over months or even years, the operators should see almost every problem that could conceivably pop up during the mission. By learning to deal calmly and efficiently with "worst on worst" cases, operators develop the confidence to deal with the routine anomalies that will inevitably occur. Figure 15-2 shows astronauts practicing for an upcoming flight.

Finally, everything is in place. The spacecraft and booster are waiting on the pad, and all that remains is to give the "go" for launch. At this point, operators, spacecraft manufacturers, and *mission managers*—the spacecraft "owners" who have final say about what goes on with the mission—sit down to a comprehensive review of flight readiness. They review all aspects of the flight—hardware, software, and the status of operator

Figure 15-2. Shuttle Mission Simulator (SMS). Months or even years before any mission, astronauts practice every aspect of the flight in the SMS. *(Courtesy of NASA)*

training. When the mission managers are completely satisfied that all risks have been minimized (they can't eliminate risks), they give the go-ahead for launch.

Launch and Orbit Checkout

On launch day all the months and years of analyzing, planning, building, testing, and training come to an end and real-time operations begin. *Real time* refers to reacting second-by-second to events as they happen during the mission. Operators use this term to distinguish real-time operations from simulated or post-flight operations. The ground-operations team at the launch site pulls out all the stops as the last few hours tick away toward launch. During this phase, launch operators must

- Support the launch team
- Monitor booster and spacecraft telemetry
- Track the booster trajectory
- Send commands to the booster and spacecraft
- Plan and monitor spacecraft maneuvers
- Validate system performance

Hours before lift-off, the team begins loading millions of gallons of propellant into the booster. Meanwhile, weather forecasters at the launch site carefully watch the skies to ensure conditions are right for launch. Weather balloons are sent high above the launch area to measure upper-level winds. If these winds are too high, the booster may be damaged as it accelerates into orbit.

In support of the launch team, operators in the mission-control center carefully check the status of all the spacecraft subsystems displayed to them on their console screens. Such information is known as telemetry. *Telemetry* literally means "far measurement." Telemetry data includes information on the health and status of the spacecraft and its payload, as

---

Astro Fun Fact

Global Positioning System in Action

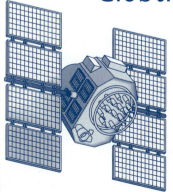

The global positioning system (GPS) was used extensively during the Gulf War. Even with an incomplete GPS constellation, ground troops found Navstar indispensable for navigation in the featureless desert. Because only a limited number of satellites were available for GPS operations, the military couldn't afford to lose any of them to malfunctions. Lieutenants working at Falcon Air Force Base, in Colorado Springs, Colorado, were able to salvage one older satellite whose attitude-control system was broken. They spun the spacecraft up using gas jets so that, in an inertially stable mode, it pointed toward the Earth whenever it was over the war zone. Air Force ingenuity does it again.

Dr. Jackson R. Ferguson, Jr., United States Air Force Academy.

Contributed by Carlee Bishop, United States Air Force Academy

well as the communications network which transmits the telemetry to the control center. After launch, telemetry monitoring moves into high gear. Thousands of separate measurements are taken onboard—engine temperature and pressure, battery voltage, attitude, and many, many more. All of these are displayed to the operators who are poised to react should something go wrong.

To receive this data, antennae on the ground must stay pointed at the booster during launch and at the spacecraft after deployment. To do this, controllers must know where the spacecraft is at all times, so they rely on a world-wide tracking network. In Chapter 8 we discussed tracking and predicting a vehicle's trajectory. This process relies on ground- or space-based parts of the communication network taking measurements in one of three ways:

- Receiving and analyzing the spacecraft's telemetry signal
- Using radar similar to those air traffic controllers use to keep track of inbound and outbound airplanes at busy airports
- Sending the spacecraft a signal which an onboard transponder receives and retransmits to reveal position and velocity

The flight-control team uses this tracking data to determine position and velocity vectors for the spacecraft. From there, they can derive the spacecraft's classical orbital elements, as we demonstrated in Chapter 5, and predict its path as we showed in Chapter 8. (So that's what all that was for!)

Equally attentive to the tracking data throughout the launch is another person with a thankless job—the *range safety officer (RSO)*. The RSO sits with a finger poised over a button which will destroy the booster on command. Why would anyone do such a thing? Recall in Chapter 9 we discussed the inclinations physically attainable from a given launch site. In reality, other constraints further limit the orbits we can achieve from a certain site. Chief among these is safety. Because most boosters drop off stages on their way into orbit, it wouldn't be safe to launch these boosters over populated areas. That's why most launch sites are on the coast, so they can launch over the water and avoid population centers. Figure 15-3 shows the range of available inclinations for launches from Kennedy Space Center in Florida and Vandenberg Air Force Base in California. The RSO must carefully monitor the booster's trajectory into orbit and send commands to destroy it if it veers off course and threatens humans.

The RSO isn't the only one who can send commands up to the booster. Throughout every phase of the launch and on orbit, controllers are prepared to send commands to the booster as well as the spacecraft. *Commands* are instruction sets telling the onboard computers to take some specific action or update some critical part of the software. This commanding function again takes place using the communication network. Commands can tell the spacecraft to charge batteries, open or close sun shades, or reorient a sensor to point at a new target.

Commands can either be real-time or stored. *Real-time commands* are implemented immediately on receipt. *Stored commands* are sent with a time tag to be carried out some time in the future or tied to some other mission

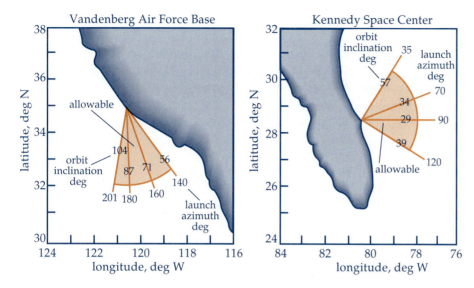

Figure 15-3. Available Inclinations. While physical limits constrain the range of available inclinations (see Chapter 9), politics and safety also play a part. Here we see the range of available inclinations and corresponding launch azimuths for launches from Kennedy Space Center and the U.S. Air Force's launch facility at Vandenberg, AFB in California.

event which hasn't happened yet, such as antenna deployment. Most important commands are sent in two stages. This means the command goes up to the spacecraft and is then echoed back to ground controllers for verification. Once controllers confirm that the command received by the spacecraft is the one sent, they send a second command to enable it in the onboard software. Two-stage commands ensure that important information doesn't get garbled during transmission and better secure the spacecraft from outside interference. In addition, certain commands are always stored on the spacecraft in case something goes wrong with the communications link or with the spacecraft itself. For example, if the spacecraft should become disoriented for some period while it's out of contact with the ground, it will use these stored commands to fly safely until it can be recovered.

Once the spacecraft moves into its parking orbit, the flight controllers must monitor the transfer out to the final mission orbit if required. For spacecraft bound for geostationary orbit, a large upperstage starts the maneuver which raises apogee to geostationary altitude. At apogee another ΔV moves the spacecraft to its final orbit. Again, ground controllers must track the spacecraft continuously to get it into the proper orbit.

With the spacecraft safely in its final orbit, the operators must verify that all subsystems are working normally. They carefully check each subsystem to determine if it can support the payload. Years before, operators and mission engineers worked out detailed procedures, called *flight rules*, telling them what to do during any possible contingency to save the mission. If everything goes well, operators will turn the spacecraft over to another team to head up normal operations as the spacecraft begins fulfilling mission objectives.

Normal Operations

Finally, after years, or even decades, of planning, designing, building, running simulations, and enduring the dramatic events of launch, the spacecraft is ready to do the mission. Normal operations is as close as space ever gets to "routine." Ground controllers must

- Monitor spacecraft telemetry
- Track and maneuver the spacecraft
- Send commands and resolve anomalies
- Process and distribute payload data
- Maintain the communications network
- Continue training

Just as during launch and early orbit checkout, three of the operator's primary tasks are to monitor telemetry on the spacecraft's health and status, track and maneuver the spacecraft, and send commands for routine housekeeping or to resolve anomalies. For Space Shuttle missions, operators work around the clock to ensure mission performance and safety. Commands are sent directly to the Space Shuttle's computers or, sometimes, in the form of the Shuttle's teletype instructions to the astronauts, as you can see in Figure 15-4. For less complex and more mature missions, such as the global-positioning system (GPS), controllers may take only a "snap-shot" of telemetry and send commands once a day or so.

Perhaps the most important things operators do during this phase of the mission are collecting, processing, and distributing payload data to users. (After all, the spacecraft was built and launched in the first place to satisfy a user's need.) This involves overseeing antenna pointing and communication traffic on communication satellites or processing large amounts of data sent back from remote-sensing spacecraft, such as the upper atmospheric research satellite (UARS) shown in Figure 15-5. For the first example, operator intervention can be fairly routine and require very little work. For the second example, hundreds of hours may be involved in processing volumes of data which rival that stored in the entire Library of Congress!

Along with baby-sitting the spacecraft, operators must also maintain the complex communication network which keeps them in touch with their space assets. This involves routine maintenance at remote tracking sites, upgrades to control-center hardware and software, and even integrating new relay satellites.

And training never ends. Especially for spacecraft in orbit five years or more, several different operations teams are likely to be involved over the years. As controllers move on to other jobs, they must train their replacements and document procedures. The simulations used before the mission help controllers to train new operators and rehearse contingencies.

As you can see from this overview of mission operations, *lots* of people are involved. Supporting this army of people along with the intricate communications network that ties them together is *expensive*. In Chapter 16, we'll explore the economic aspects of a mission in greater detail, but for now it's easy to see why operations alone often account for one-third of a

Figure 15-4. Teletype Commands. One way commands are sent to the Space Shuttle is through teletype instructions. Here you see astronaut Bo Bobko swimming in a detailed set of procedures for a rendezvous mission. *(Courtesy of NASA)*

Figure 15-5. Upper Atmospheric research Satellite (UARS). Spacecraft such as the UARS used to monitor the Earth's environment generate huge amounts of data which ground controllers and users must collect and process. *(Courtesy of NASA)*

large program's cost. For this reason, engineers and operators are constantly looking for new ways to streamline operations and cut costs. One way to do this is to place more functions onboard the spacecraft itself, eliminating the need for some costly ground functions.

Astro Fun Fact

"Okay, Houston, we've had a problem!"

More than 200,000 miles away from home on April 13, 1970, Apollo 13 was in trouble. Unknown to everyone at the time, one of the O_2 tanks suppling the three fuel cells that power the command module had exploded, tearing a gaping hole in the side of the service module and creating a life-or-death situation for astronauts Jim Lovell, Fred Haise, and Jack Swigert. On the ground, activities in the Mission Operations Control Room in Houston, Texas, immediately shifted into high gear as flight controllers struggled with limited data to resolve the problem and get the crew home safely. As the minutes ticked away, the entire flight-control team pulled together in what controller Glynn Lunney later called, "the crowning achievement" of the program. "We are as close as you get to the edge" he said "and were still able to pull back." To pull back from that edge, the team had to make hundreds of crucial, irrevocable decisions. They developed entirely new procedures to use the Lunar Landing Module (LLM) as a "life boat" to provide for the crew's needs until the LEM's engines, designed to carry the crew to the lunar surface and back, could be used to provide the crucial ΔV to put them safely on course back to Earth. Against all odds, the flight-control team and crew did everything right. As an entire planet watched and prayed, Apollo 13 splashed down safely after the most dramatic space mission ever.

Charles Murray and Catherine Bly Cox. Apollo The Race to the Moon. New York, N.Y.: Simon & Schuster, 1989.

Section Review

Key Terms

commands
flight controllers
flight director
flight rules
ground-system specialists
mission managers
mission timeline
operations concept
payload specialists
range safety officer (RSO)
real time
real-time commands
simulations
stored commands
subsystem specialists
telemetry

Key Concepts

➤ The operations concept is important during every phase of a space mission

➤ During mission planning and design, operators must
  • Refine the operations concept
  • Develop the mission timeline
  • Develop training plans and simulator requirements
  • Assemble the operations team
  • Conduct mission simulations
  • Participate in the mission readiness review

➤ During launch and orbit checkout, operators must
  • Support the launch team
  • Monitor booster and spacecraft telemetry
  • Track the booster's trajectory
  • Send commands to the booster and spacecraft
  • Plan and monitor spacecraft maneuvers
  • Validate system performance

➤ During normal operations, controllers
  • Monitor spacecraft telemetry
  • Track the spacecraft
  • Send commands and resolve anomalies
  • Process and distribute payload data
  • Maintain the communications network
  • Continue training

➤ Space-mission operations are expensive, so we must decrease costs by making operations more efficient and spacecraft more autonomous

15.2 Communication Networks

In This Section You'll Learn to...

☞ Describe the basic elements of a space mission's communication network

☞ Discuss the basic physical principles of communication

☞ Determine some of the basic parameters associated with a communication link

Throughout this section and in other parts of the book we've referred to the communication network that ties together all the other elements of a space mission. At long last, let's take a look at these vital links. The communications network is the glue that holds space missions together. In Chapter 13, we talked about the data-handling subsystem on the spacecraft and how it keeps track of all the information onboard, including housekeeping, payload data, and commands from ground controllers. The communication network passes this information from spacecraft to spacecraft and from spacecraft to the ground.

Communication-Network Architecture

In general, communications is the exchange of messages and information. For space missions, *communication* is the exchange of commands and engineering data between the spacecraft and ground controllers, as well as the processing and transmission of payload data to users. In this section we're going to focus on the different ways spacecraft can communicate—the communication architecture. Figure 15-6 shows an example of a communication architecture. The *communication architecture* is the configuration of satellites and ground stations in a space system and the network that links them together. It has four elements:

- *Spacecraft*—the spaceborne elements of the system
- *Ground stations*—the Earth-based antennas and receivers that talk to the spacecraft
- *Control center*—the command authority which controls the spacecraft and all other elements in the network
- *Relay satellites*—additional spacecraft which link the primary spacecraft with ground stations

Information moves between these elements on various links:

- *Uplink*—data sent from a ground station to the primary spacecraft
- *Downlink*—data sent from the primary spacecraft to a ground station
- *Forward link*—data sent from a ground station to the primary spacecraft through a relay satellite

- *Return link*—data sent from the primary spacecraft to a ground station through a relay satellite

- *Crosslink*—data sent on either the forward link or return link

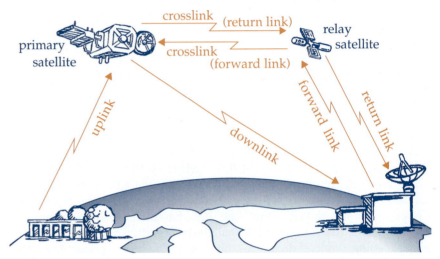

Figure 15-6. Communication Architecture. A communication architecture for space missions consists of ground and space-based elements tied together through different communication paths or links.

We're all used to communicating. We talk to our friends and family all day long. But let's take a moment to dissect the communication process so we can better understand the problems faced in building communications for a spacecraft. Our objectives in this section are to identify what design parameters we can change to build better systems.

Imagine that you go over to a friend's house to talk about your French homework. What conditions are necessary for you and your friend to communicate? We need to consider

- Distance
- Language
- Speed
- Environment

First of all, you should be within shouting distance so your words reach your friend. The farther away you are from your friend, the louder you must talk to be heard. If you're too far away, you'll never communicate. Another issue to consider is language. You and your friend must be able to understand and speak the same language. If she knows French much better than you do, it could be difficult to communicate. Next, there is the speed at which you talk. Have you ever tried to listen to someone who speaks very fast? If you have, you know that it's sometimes hard to catch every word. If they speak too quickly, you can't process the words fast enough to understand the message.

So, if you and your friend are within range and speaking the same language at a reasonable rate, you should be able to communicate, right? Wrong! You also need to consider your environment. Imagine that, as you and your friend are talking, her little brothers and sisters run into the room screaming loudly. Their screaming represents noise. If the kids' screaming is too loud, you'll need to raise your voice so your friend can still hear you above the noise. In other words, your *signal*—the volume and content of the message you're speaking—must be louder than the noise. That is, to be heard, your *signal-to-noise ratio* must be greater than 1. What's important is the ratio of the volume of your speech to the volume of their noise.

Now let's see what all this has to do with spacecraft communications. To communicate effectively from one spacecraft to another or to a ground station, we must consider the distance or range between the speaker—called the *transmitter*—and the listener—called the *receiver*. We must also have a transmitter and receiver that both understand the language or code that's used. Furthermore, the receiver must be able to handle the signal's speed or *data rate*. Finally, the volume or *signal strength* at the receiver must be higher than the overall noise in the system. To see all these concepts in practice, we can now focus on some basics of satellite communication.

Communication Basics

Unfortunately, spacecraft can't just shout down to the ground. Instead, we rely on radio to communicate across space. In Chapter 11 we discussed the basics of electromagnetic radiation. Communication takes place using radio frequencies of EM radiation. Your car stereo illustrates the basic principles of radio, as shown in Figure 15-7. When you turn on your stereo, you receive signals from the radio station through electromagnetic

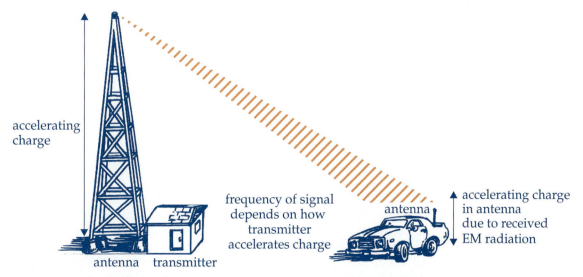

accelerating charge

frequency of signal depends on how transmitter accelerates charge

antenna

accelerating charge in antenna due to received EM radiation

antenna   transmitter

Figure 15-7. Basic Principle of Radio. A radio station produces a signal by accelerating charged particles in the antenna. The signal then travels out as EM radiation until it's received by your car antenna, where charged particles again accelerate to produce the music you hear.

radiation. Remember, EM radiation is produced by an accelerating charge. As this charge accelerates, an electric field forms and induces a magnetic field, which induces an electric field, and so on. James Maxwell (1831–1879) first developed this concept. The frequency at which this charge accelerates determines the frequency of the EM radiation. The faster the charge accelerates, the higher the frequency. The station broadcasts a *carrier signal* at some specific frequency set by the Federal Communications Commission (FCC). The message being sent—music, news, or mission data—is then superimposed on top of the carrier signal using some type of modulation scheme. The schemes we're most familiar with are amplitude or frequency modulation (AM and FM). Spacecraft applications use other schemes as well. This signal travels out from the station's antenna and hits your radio antenna. There, more charges are accelerated. Your receiver detects this charge movement in the antenna and re-translates it to the original signal. The message is demodulated from the carrier signal and, suddenly, you're listening to tunes while cruising down the road.

Now we want to take a closer look at communication systems to understand some of the basic principles and limitations for constructing satellite communications. Let's use a light bulb to demonstrate some of these key principles. Like a radio, a light bulb puts out EM radiation, but at a different frequency. Thus, it's easier to visualize how the radiation gets out. If you put a light bulb in the center of a room, as shown in Figure 15-8, light radiates outward in all directions (assuming it's a perfect bulb with no light blockage). The intensity or brightness of the light at some distance from the bulb is called the *power flux density, F.* Of course, the farther you get from a light bulb, the dimmer it appears. In other words, the power flux density, which you perceive as brightness, decreases as you get farther away. As it turns out, the brightness actually decreases with the square of the distance because all the output is distributed over the surface of a sphere surrounding the source. This can be expressed as

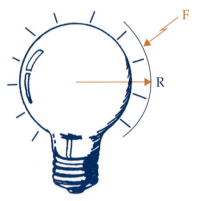

Figure 15-8. Power Flux Density. An ideal light bulb radiates equally in all directions. The brightness, or power flux density, F, at any given distance, R, depends on the bulb's output, P.

$$F = \frac{\text{Power}}{\text{Surface area of a sphere}} = \frac{P}{4\pi R^2} \qquad (15\text{-}1)$$

where
P = power rating of the light bulb (W)
R = distance from the bulb in (m)
F = power flux density (W/m²)

We know that visible light, like that of a light bulb, is simply electromagnetic radiation. Radiation moving equally in all directions, like our light bulb example from Figure 15-8, is called *omni-directional* or *isotropic.* Now, what if we wanted to increase the brightness or power flux density in only one area using the same bulb. As Figure 15-9 shows, that's just what a flashlight does. This time we're still using our ideal light bulb but we've put a parabolic-shaped mirror on one side of it. Thus, much of the light reflects off the mirror in the opposite direction, and we end up with a directed beam of light—a spotlight—rather than an omni-

Figure 15-9. Directed Output From a Light Bulb. A parabolic mirror can direct the bulb output to give us an effective spot light. The mirror allows us to focus the bulb's energy into a smaller area, thus increasing the gain.

directional source. In this way, we've effectively been able to concentrate much of the light energy into a smaller area. As a result, we get a brightness in that one particular area which is much, much greater than it was when the bulb was allowed to radiate in all directions. We've been able to "gain" extra power density by using the parabolic mirror.

This is the basic principle of an antenna. Instead of broadcasting in all directions, wasting all that energy, spacecraft use directional antennas that point at the user. The user employs another directional antenna to receive the signal. Like our mirror, these antennas are often parabolic-shaped to transmit and collect the radio energy efficiently. The *gain* of an antenna is the ratio of the energy it transmits in a certain direction to the energy that would be available from an omni-directional source. In other words, the gain for omni-directional broadcast would be 1, whereas the gain for a broadcast using a directed antenna would be greater than 1. The general expression for gain is

$$G = \frac{\text{Energy on target with directed antenna}}{\text{Energy on target with omni-directional antenna}}$$

We can relate these two values for energy to an antenna's area, its efficiency, and the wavelength of the energy we're using by

$$G = \frac{4\pi A \eta}{\lambda^2} = \frac{4\pi A_e}{\lambda^2} \qquad (15\text{-}2)$$

where
G = gain (unitless)
A = physical area of the antenna (m^2)
η = antenna efficiency (about 0.55 for parabolic antennas)
A_e = effective area of the antenna (= $A\eta$, m^2)
λ = wavelength of signal (m)

This relationship tells us that if we want to increase the gain of an antenna we can either increase its effective area or decrease the wavelength at which we operate. Transmitting and receiving antennas use the same expression for gain.

If we multiply the transmitter's power output by its antenna gain, we get an expression that represents the amount of power an isotropic transmitter would have to put out to get the same amount of power on target. We call this the *effective isotropic radiated power, EIRP*

$$EIRP = P_t \, G_t \tag{15-3}$$

where
EIRP = Effective Isotropic Radiated Power (W)
P_t = power output of the transmitter (W)
G_t = gain of the transmitter antenna

How much of the transmitter power does the receiver collect? Think about collecting rainfall in a bucket. The amount of rain water you collect depends on how hard it's raining—the rain's density—and the bucket's size or cross-sectional area. Similarly, the signal strength at the receiver is a function of the power flux density at the receiver and the area of the receiver antenna. The resulting expression for the signal intercepted by the receiving antenna is then

$$S = \left(\frac{P_t G_t}{4\pi R^2}\right) A_{e_{receiver}} \tag{15-4}$$

where

$\left(\dfrac{P_t G_t}{4\pi R^2}\right)$ = the effective power of the transmitter spread out over a sphere of radius R (W)

$A_{e_{receiver}}$ = effective area of the receiving antenna (m^2)

Solving the right-hand expression in Equation (15-2) for $A_{e_{receiver}}$ and substituting into Equation (15-4) results in

$$\boxed{S = P_t G_t \left(\frac{\lambda}{4\pi R}\right)^2 G_r} \tag{15-5}$$

where
S   = signal received (W)
$\left(\dfrac{\lambda}{4\pi R}\right)^2$ = space loss (0 < space loss < 1.0)
G_r = receiving antenna gain (computed the same way as transmitter gain) (unitless)

Notice we have a term representing space loss. *Space loss* is not a loss in the sense of power being absorbed; it accounts for the way energy spreads out as an electromagnetic wave travels away from the transmitting source. As distance increases, this term becomes smaller, which means space losses get worse. This makes sense because a smaller term multiplied by the receiver and transmitter gains actually *decreases* the effective signal.

So we now have several ways to increase the received signal:

• Increase the transmitter power—P_t

• Increase the transmitter antenna gain, concentrating the focus of the energy—G_t

• Increase the receiver gain so it collects more of the signal—G_r

• Decrease the distance between the transmitter and receiver —R

A few pages back we discussed the concept of signal-to-noise (S/N) ratio in communications. So far in this discussion we've talked about the received signal, S. Earlier, when we discussed communicating across a room, noise came from some rambunctious kids. But where does noise come from for a radio signal? One important source of radio noise is heat. Recall from our discussion of black-body radiation in Chapter 11 that any object having a temperature greater than absolute 0°K puts out EM radiation. While a receiver is running, just like your TV set, it gets hot and produces EM radiation as noise. The noise power is given by

$$N = kTB \tag{15-6}$$

where
  k = Boltzmann's constant = 1.38×10^{-23} Joules/°K
  T = receiver system's temperature in °K
  B = receiving system's bandwidth in Hz

Bandwidth is the range of frequencies the receiver is designed to receive. For example, the range of human eyesight, or the bandwidth of our eyes, is about 3.90×10^{14} Hz to 8.13×10^{14}. This represents the small portion of the EM spectrum we can see. Note that the noise in the receiver increases as the bandwidth increases. This should make sense; the more information you're tuned in to receive, the more likely you'll pick up noise. Ideally, we'd try to reduce the receiver temperature as much as possible and restrict the bandwidth of interest to minimize the noise.

Combining Equation (15-5) and Equation (15-6), we get the signal-to-noise ratio for a radio signal

$$\boxed{\frac{S}{N} = \left(\frac{P_t G_t}{kB}\right)\left(\frac{\lambda}{4\pi R}\right)^2\left(\frac{G_r}{T}\right)} \tag{15-7}$$

where
  S/N = signal-to-noise ratio (unitless)
  P_t = power of transmitter (W)
  G_t = gain of transmitter (unitless)
  k = Boltzmann's constant = 1.38×10^{-23} Joules/°K
  B = receiving system's bandwidth in Hz
  λ = wavelength (m)
  R = range between transmitter and receiver (m)
  G_r = gain of receiver (unitless)
  T = receiver system's temperature in °K

Remember, for effective communication, the signal-to-noise ratio must be greater than or equal to 1. (The voice you hear must be louder than the background noise in the room.) To improve the S/N we can

- Increase the strength of the signal using the methods outlined above
- Reduce the bandwidth of the signal—B
- Reduce the receiver temperature—T

So far we haven't said much about changing the frequency or wavelength of the signal. What effect does this have? Looking at Equation

(15-7), you'd suspect that increasing the wavelength would improve the S/N ratio, but remember the relationship for gain given in Equation (15-2). Both the transmitter and receiver gains are inversely related to wavelength. That is, as wavelength increases (lower frequency), gain decreases. This means the net effect of increasing wavelength is to decrease the antenna gains and thus reduce the S/N ratio. You can see all these relationships in action in Examples 15-1 and 15-2.

Satellite-Control Networks

Now that we've looked at the theoretical aspects of communication networks, let's see how NASA and the United States Air Force use these systems to keep track of the space missions they oversee. NASA has two different networks for tracking and receiving data from space. The spaceflight tracking and data network (STDN) mostly tracks and relays data for the Space Shuttle and other near-Earth missions. STDN includes ground-based antennas at far-flung points on the globe as well as space-based portions which use the Tracking and Data Relay Satellites (TDRS) in geostationary orbits. The deep-space tracking network (DSN) includes very large antennas (more than 70 meters in diameter) which track and receive data from space probes around other planets in the solar system. These antennas are in Spain, Australia and Puerto Rico.

The United States Air Force also has two different networks for tracking space objects and controlling spacecraft. The Space Surveillance Network is a world-wide network of high-power radars that track approximately 7000 objects in Earth orbit. These objects range from the Space Shuttle orbiter to Astronaut Ed White's glove, lost during a Gemini mission. The radar data is transmitted to the Space Surveillance Center in Cheyenne Mountain, Colorado, where orbit analysts maintain the space catalog. This catalog contains the current classical orbital elements of more than 7000 pieces of stuff in orbit large enough to track. One of the U.S. Air Force's tracking sites is shown in Figure 15-10.

The Air Force's Satellite Control Network (AFSCN) consists of stations in such interesting locations as Guam, Thule in Greenland, and Mahe, one of the Seychelles Islands in the Indian Ocean. These stations are connected to control centers at Falcon Air Force Base, Colorado, and Onizuka Air Force Base, Sunnyvale, California, where Air Force engineers and space operations experts command and control almost all Department of Defense satellites. We talked about the missions of these spacecraft in earlier chapters; each type (communications, navigation, etc.) requires a team of specialists. Some missions, such as early warning, require so much ground support that they have their own dedicated control stations. Control of the Defense Support Program (DSP) satellites, which provide early warning of enemy missile launches, requires special ground stations in the United States and in Woomera, Australia.

Figure 15-10. Remote Tracking Site. Remote tracking sites, like this one in Woomera, Australia, form part of the Air Force satellite control network. The "golf ball" shaped structures house large antennas. *(Courtesy of U.S. Air Force Command)*

Section Review

Key Terms

bandwidth
carrier signal
communication
communication architecture
control center
crosslink
data rate
downlink
effective isotropic radiated
  power, EIRP
forward link
gain
ground stations
isotropic
omni-directional
power flux density, F
receiver
relay satellites
return link
signal
signal strength
signal-to-noise ratio
space loss
spacecraft
transmitter
uplink

Key Equations

$$G = \frac{4\pi A \eta}{\lambda^2} = \frac{4\pi A_e}{\lambda^2}$$

$$S = P_t G_t \left(\frac{\lambda}{4\pi R}\right)^2 G_r$$

$$\frac{S}{N} = \left(\frac{P_t G_t}{kB}\right)\left(\frac{\lambda}{4\pi R}\right)^2\left(\frac{G_r}{T}\right)$$

Key Concepts

➤ The command, control, and communication network ties together all the other elements of a space mission

➤ Commands are instructions sent by ground controllers to the spacecraft telling it what to do and when to do it
  • Real-time commands are executed immediately
  • Stored commands are kept in the memory of the spacecraft's data-handling system for later execution

➤ Control is the act of carrying out these real-time or stored commands

➤ Communication architecture is the configuration of satellite and ground stations in a space system and the network that links them together. It has four major elements—spacecraft, ground stations, a control center, and relay satellites.

➤ Information moves between elements of the communication architecture on various links—uplink, downlink, forward link, return link, and cross link

➤ Electromagnetic energy from any source represented by the power flux density decreases in strength with the square of the distance

➤ Electromagnetic energy can be focused in one direction through an antenna. The increase in power flux density achieved using an antenna is the antenna gain.

➤ To increase the received signal strength, we can
  • Increase transmitter power
  • Increase transmitter or receiver antenna gain
  • Decrease distance between transmitter and receiver

➤ Noise in a radio signal can come from the black-body radiation which the receiver temperature emits; it's a function of the signal's bandwidth or range of frequencies

➤ To increase the signal-to-noise ratio we can
  • Increase the signal's strength
  • Reduce the signal's bandwidth
  • Reduce the receiver's temperature

➤ Increasing the communication frequency effectively reduces signal-to-noise ratio

Continued on next page

Section Review

Key Concepts (Continued)

➤ To communicate effectively, whether talking to a friend across a noisy room or to a spacecraft at the edge of the solar system, we must meet four requirements:

• Transmitter and receiver must be close enough to one another

• The language or code the message is in must be common to both transmitter and receiver

• The speed or data rate of the message must be slow enough for the receiver to interpret it

• The volume or strength of the signal must be greater than any noise. That is, the signal-to-noise ratio must be greater than 1.

➤ The basic principle of radio involves accelerating charged particles in the transmitter antenna to generate electromagnetic radiation. The receiver antenna detects this radiation and again accelerates the charged particles, causing the signal to be interpreted.

Example 15-1

Problem Statement

We can think of the Sun as a "perfect light bulb" radiating isotropically. If the Sun's power flux density, F, on the Earth is 1358 W/m², what is the Sun's power output? Distance to the Sun is about 1.496×10^{11} m.

Problem Summary

Given: $F = 1358$ W/m², $R = 1.496 \times 10^{11}$ m
Find: P output

Problem Diagram

$F = 1358$ W/m2

Conceptual Solution

1) Solve Power Flux Density relationship for Power output

$$F = \frac{\text{Power}}{\text{Surface area of a sphere}} = \frac{P}{4\pi R^2}$$

$$P = (F)\, 4\pi R^2$$

Analytical Solution

1) Solve Power Flux Density relationship for Power output

$$P = (F)\, 4\pi R^2$$

$$P = (1358 \text{ W/m}^2)\, (4\pi)\, (1.496 \times 10^{11} \text{ m})^2$$

$$= 3.819 \times 10^{26} \text{ W}$$

Interpreting the Results

The Sun puts out a lot of power—millions of times greater than the outputs of the power plants on Earth. Even at a distance of almost 150 million kilometers (93 million miles), its intensity is still more than 1000 W per square meter.

Example 15-2

Problem Statement

Engineers are trying to design the communication system for FireSat to ensure vital Telemetry, Tracking, and Commanding data gets through to ground controllers. The communication frequency they've chosen is 2 GHz (2×10^9 Hz) with a bandwidth of 2000 Hz. The transmitter antenna will be 2 m in diameter with an efficiency of 0.55. Transmitter power output is 10W. The link will operate over a distance of 10,000 km. If the receiver temperature is 800°K, with an antenna efficiency of 0.55, what receiver-antenna diameter do we need to achieve a signal-to-noise ratio of 10,000?

Problem Summary

Given: $f = 2 \times 10^9$ Hz
$D_t = 2$ m
$B = 2 \times 10^3$ Hz
$\eta = 0.55$
$P_t = 10$ W
$R = 10,000$ km
$T = 800°$ K,
$S/N = 10,000$

Find: Diameter of receiver antenna, D_r

Conceptual Solution

1) Find transmitter-antenna gain, G_t

$$\lambda = \frac{c}{f}$$

$$A = \pi \left(\frac{D}{2}\right)^2$$

$$G = \frac{4\pi A\eta}{\lambda^2}$$

2) Solve S/N for receiver-antenna gain, G_r

$$\frac{S}{N} = \left(\frac{P_t G_t}{kB}\right)\left(\frac{\lambda}{4\pi R}\right)^2 \frac{G_r}{T}$$

$$G_r = \left(\frac{S}{N}\right)\left(\frac{kB}{P_t G_t}\right)\left(\frac{4\pi R}{\lambda}\right)^2 T$$

3) Solve for receiver-antenna area, A

$$G = \frac{4\pi A\eta}{\lambda^2}$$

$$A = \frac{G\lambda^2}{4\pi\eta}$$

4) Solve for receiver-antenna diameter, D_r

$$A = \pi \left(\frac{D_r}{2}\right)^2$$

$$D_r = 2\sqrt{\frac{A}{\pi}}$$

Analytical Solution

1) Find transmitter-antenna gain, G_t

$$\lambda = \frac{c}{f} = \frac{3 \times 10^8 \text{m/s}}{2 \times 10^9 \text{Hz}} = 0.15 \text{ m}$$

$$A = \pi \left(\frac{D}{2}\right)^2 = \pi \left(\frac{2 \text{ m}}{2}\right)^2 = 0.785 \text{ m}^2$$

$$G_t = \frac{4\pi A\eta}{\lambda^2} = \frac{4\pi(3.142 \text{ m}^2)(0.55)}{(0.15 \text{ m})^2} = 965.03$$

2) Solve S/N for receiver-antenna gain, G_r

$$G_r = \left(\frac{S}{N}\right)\left(\frac{kB}{P_t G_t}\right)\left(\frac{4\pi R}{\lambda}\right)^2 T$$

$$G_r = (10,000)\left(\frac{(1.38 \times 10^{-23} \text{J/}°\text{K})(2000 \text{ Hz})}{(10 \text{ W})(965.03)}\right)$$

$$\left(\frac{4\pi(1 \times 10^7 \text{m})}{0.15 \text{ m}}\right)^2 (800° \text{ K})$$

$$G_r = 16.06$$

Example 15-2 (Continued)

3) Solve for receiver-antenna area, A

$$A = \frac{G_r \lambda^2}{4\pi\eta} = \frac{(16.06)(0.15 \text{ m})^2}{4\pi(0.55)} = 0.05 \text{ m}^2$$

4) Solve for receiver-antenna diameter, D_r

$$D_r = 2\sqrt{\frac{A}{\pi}} = 2\sqrt{\frac{0.05 \text{ m}^2}{\pi}} = 0.26 \text{ m}$$

Interpreting the Results

To receive the signal from the remote-sensing spacecraft at a 10,000 km altitude with an acceptable signal-to-noise ratio (S/N = 10,000 isn't unreasonable), we need an antenna diameter of at least 0.52 m at the ground station.

References

Morgan, Walter L., Gary D. Gordon. *Communications Satellite Handbook*. New York, N.Y.: John Wiley & Sons, 1989.

Pratt, Timothy, Charles W. Bostian. *Satellite Communications*. New York, N.Y.: John Wiley & Sons, 1986.

Rockwell International Space Systems Group, Space Shuttle System Summary, 1980.

Mission Problems

15.1   Space Operations

1   What are the three phases of mission operations?

2   List and discuss the operations tasks conducted during Mission Planning and Design.

3   Discuss the operations team members and their tasks.

4   List and discuss the operations tasks during Launch and Orbit Checkout.

5   What are real-time operations?

6   What is telemetry and what is it used for?

7   What are the three ways to collect tracking data?

8   Define spacecraft commands and give three examples of what they can be used for.

9   What is the difference between real-time and stored commands? Give examples of each.

10   Explain why we may not be able to use all physically possible inclinations from a given launch site.

11   List and discuss the operations tasks conducted during Normal Operations.

15.2   Communication Networks

12   Describe the elements of a communication network and the links that tie them together.

13   Explain the four conditions for effective communication.

14   Define signal-to-noise ratio and explain why it's important.

15   A 100 W light bulb is placed at the center of a field. What is the power flux density 100 m away? (Assuming a perfect light source)

16   A parabolic antenna (η=0.55) has a diameter of 3 m. What is its gain at a frequency of 3×10^8 Hz?

17   What is gain?

18   Discuss two ways to increase received signal.

19   Where does noise come from in a communication signal?

20   Satellite A and B both have the same communication equipment (transmitter, antenna, etc.) onboard to communicate with the same ground station. If satellite A is in an orbit 1000 km higher than satellite B, which will have a poorer signal-to-noise ratio? Why?

21   Engineers are planning to use a 2×10^9 Hz link with a 300 Hz bandwidth to communicate with a remote-sensing spacecraft in an orbit at 400 km altitude. The transmitter-antenna gain is 300, and the receiver-antenna diameter is 30 m. Transmitter power will be 13 W. If the receiver temperature is 300° K, will this be an effective link?

For Discussion

22   Why are mission operations so expensive? Suggest some ways to decrease these costs.

23   What type of communication network would be needed to support human exploration of Mars? Describe the number and type of elements needed and the various links.

Projects

24   Lay out the mission control center that would be needed for a manned mission to Mars. What console positions and areas of expertise would be needed by flight controllers?

25   Using the data from Problem 21, develop a computer spreadsheet or program to analyze the change in S/N as you vary frequency and spacecraft altitude.

Mission Profile—Mercury

After the successful launch of Sputnik by the U.S.S.R. in 1957, the U.S. began a big push to be the first country to put a man in space. Programs were proposed by the military and several contractors submitted proposals for manned space vehicles. In the end, it would be the new civilian space agency, NASA, that would be given sole responsibility for manned space flights. NASA decided to pursue a blunt-body capsule to minimize technical challenges of entry and to minimize pilot workload. By April 1959, Project Mercury was in full swing—a contract for 12 capsules had been signed and seven astronauts had been chosen. Six of the seven astronauts would fly in space during the course of the program (the seventh, Deke Slayton, would finally get to fly as commander of the Apollo-Soyuz Test Project in 1975).

Mission Overview

Mercury's mission was to get Americans into space for the first time and if possible, beat the Russians. Another goal was to see how humans would react to the stresses of space flight. The program was plagued by many failures during testing (43% failure rate on 14 test flights), yet all six manned flights were successful.

Mission Data

✓ Mercury 2, January 31, 1961 (Crew: Ham the chimpanzee): First American flight into space with a living occupant.

✓ Mercury 3, May 5, 1961 (Crew: Alan Shepard): First American manned spaceflight. It was a suborbital flight that lasted less than 15 minutes.

✓ Mercury 4, July 21, 1961 (Crew: Virgil "Gus" Grissom): Second and last manned suborbital flight. Capsule sank during recovery attempt— Grissom was recovered safely.

✓ Mercury 5, November 29, 1961 (Crew: Enos the chimpanzee): First orbital flight of a Mercury capsule with a "crew" aboard.

✓ Mercury 6, February 20, 1962 (Crew: John Glenn): First American manned orbital flight. Three orbits completed.

✓ Mercury 7, May 24, 1962 (Crew: Scott Carpenter): Second American manned orbital flight. Four orbits completed.

✓ Mercury 8, October 3, 1962 (Crew: Walter Schirra): Completed six orbits and lasted over nine hours in duration.

✓ Mercury 9, May 15, 1963 (Crew: Gordon Cooper): Last Mercury spaceflight. Completed 22 orbits in over 34 hours.

Mission Impact

Project Mercury provided NASA with valuable flight experience and data on human performance in space. This set the stage for all manned programs to follow.

Mercury 6 blasts off carrying John Glenn, the first American to orbit the Earth. *(Courtesy of NASA)*

For Discussion

• Critics of the early manned program say that our only goal was to beat the Soviets and we should have spent more time developing vehicles that could meet long range goals. Is this true? How might things have been different if we had used a different strategy?

• How did the successes of Mercury help restore America's confidence in space?

References

Baker, David. *The History of Manned Spaceflight.* New York, NY: Crown, 1981.

Yenne, Bill. *The Encyclopedia of US Spacecraft.* New York, NY: Exeter, 1985.

Retrieving and repairing broken satellites for later reuse is one of many potential space businesses of the future. *(Courtesy of NASA)*

The Business of Space

16

Chris Elliott
Smith System Engineering Ltd.

▰ In This Chapter You'll Learn to...

- ☞ Explain some of the basic rules which govern the financial aspects of astronautics
- ☞ Understand the motivation of people who may pay you to work in space and see how their thinking affects design decisions

▰ You Should Already Know...

- ❏ Nothing. You need no specific tools to understand this chapter, but it will make more sense if you're familiar with the rest of the book

There is just one thing I can promise you about the outer space program—your tax dollar will go farther.

Werner VonBraun

Once, an advertisement in the jobs section of a newspaper read "Wanted, one-armed economist." When asked why, the advertiser said he was sick of employing economics advisers who say, "on one hand the answer is yes, but on the other hand, it's no." After 15 chapters based on the rigorous laws of physics, why should this book have a chapter on economics? Despite what you may have inferred from the previous chapters, space is also about money. What does it cost, who pays for it, and why are three questions you have to address if you're going to design, launch, and operate a space mission. This chapter has more to do with commercial enterprise than Starship Enterprise.

We'll begin by reviewing some basic concepts of economics as they apply (or, in some cases, as they *should* apply) to space missions. This leads naturally to describing the big players—the space agencies—and their roles and budgets.

Engineering practice is surprisingly dependent on the laws of economics. As we'll see, the way we spend affects the way we design systems. We'll take another look at system design and how it relates to managing risk and the system's complexity. This will lead us to the procedures for manufacturing and testing space systems. We'll look at launch vehicles, because they're unique to engineering for space missions, and at the costs of space operations.

Next we'll get to the exciting bit—the business opportunities (how to make the big bucks). We'll examine the ways business can play a part in space missions and some of the risks and rewards involved. Three applications of space (communications, Earth observation, and micro-gravity) are often claimed to have business potential. We'll look at each one from conventional and "non-conformist" perspectives. An imaginary business plan will show some of the opportunities and pitfalls. Finally, we'll look at the future and how space missions might be reshaped if the economic constraints were changed.

This chapter is a bit of a ragbag, picking up all of the fuzzy issues of economics and politics that hang around space programs. But if this chapter has one overall message, it's that the laws of physics, engineering, and economics are inextricably linked in space missions. You ignore any of them at your peril.

16.1 Basic Economics

▬ In This Section You'll Learn to...

- ☛ Explain basic concepts of supply and demand, cost, and price, and investigate how they apply in space
- ☛ Explore the role of government agencies in funding

Some General Principles

Classical economics considers the supply of goods in the marketplace. It assumes

- Many customers and many suppliers
- Goods are well defined and understood
- Rational behavior on all sides

With these assumptions, we can understand many simple economic concepts. The most important of these is the relationship between supply and demand and its effect on price.

The classical theory of *supply and demand* says that, as the price rises, more and more suppliers want to enter the market to sell. However, it's also true that as the price increases, customers will demand fewer units of that good. An efficient market will establish an equilibrium price at which the quantity supplied just equals the quantity demanded. Graphically, this is where the supply and demand curves intersect. Figure 16-1 shows this classic curve for supply and demand.

Any change in circumstances can affect the behavior of either the customer or the supplier. Imagine, for example, the impact of a heat wave on the market for ice cream. Suddenly, more customers want to buy, so the demand curve slips outward, and the equilibrium price must also rise as shown in Figure 16-2. Assuming factories running now are at peak capacity, suppliers will bring into service additional ice cream factories to meet the greater demand. Because these factories are probably the least efficient (which is why they were closed in the first place), the price will stabilize at a higher level than it held before the heat wave.

As an aside, a lot of modern analysis of this classical economic model has concentrated on the effects of delay. By the time the extra factories are operational, the weather may have changed, and you don't sell extra ice cream to people in overcoats. The system may literally become unstable, with ever greater fluctuations in the supply, demand, and price of the goods.

The trouble with this classical model is that, although it's very straight-forward, it doesn't apply to space programs. Look back at the first two assumptions in the opening paragraph—many customers and many

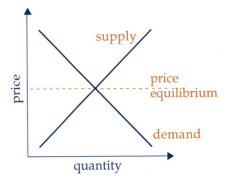

Figure 16-1. Supply and Demand. The laws of supply and demand tell us that if demand increases (like the demand for ice cream on a hot day), price will rise, so the quantity demanded will once again just equal quantity supplied.

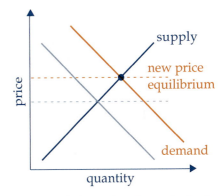

Figure 16-2. Increasing Demand. As the demand curve shifts outward, the equilibrium price increases.

suppliers, and well-defined goods? Neither of these apply to mainstream space programs. In fact, the classical analysis applies only to what economists call "commodities"—goods that are essentially the same wherever they come from and which consumers choose because of price.

To better understand space, let's try a more complicated economic model which looks at developing high-technology products, such as those for the military. This is closer to the world of space. Unlike commodities in the classical supply and demand model, these get cheaper as customers increase the quantity they wish to buy. Through an effect known as the *learning curve* or *economies of scale*, the expertise you get developing the first few prototype products allows you to produce subsequent ones more cheaply. Economies of scale allow cheaper mass production and sharing of development costs over a longer production run. Thus, the cost per item decreases greatly as the order size increases.

We're getting nearer to the world of space programs but we're not quite there. It is unusual to get a repeat order or an order for a batch of space systems. Economies of scale simply don't happen. We are left with one-of-a-kind developments, for which the price paid by the customer is essentially whatever it costs the supplier plus a bit for profits. Later in this chapter we'll come back to these basic principles to see if some of the current trends in space economics might allow a more healthy practice.

Let's take a concrete example. Imagine you run a company that makes television sets and you decide to develop a new model. You start with two people working out the specifications and, by the time you get a prototype of the new set two years later, the team has grown steadily to 50 people. You now call in the production people, and 100 of them work for another year to set up your factory. If you assume that each person costs you $10,000 per month, you have spent nearly $18 million. You can now make TV sets, but the first one off the line appears to have cost you $18,000,100—not a great proposition if you sell it for $300.

However, a year later the 100,000th TV set rolls out. By now they cost you only $280 each. After another year, when the market has grown and you've shipped another 200,000, they will cost you $160 each (but you'll be selling them for $300—that's $140 net profit). Now you're really making money.

The trouble with the space business is that we're always in the position of that *first* television set—the one that costs $18,000,100. It is unusual to make two identical spacecraft; the longest production run might be 10 that are similar (but always different in detail). Space is where the automotive industry stood before Henry Ford came along and said "you can have any color you want as long as it's black."

One more general concept of economics will be useful later, although it hasn't had much influence on the space industry. The *time value of money* says that income in the future is worth less than income today and, similarly, a payment that can be deferred is better than one that must be met today. In our personal affairs we apply this principle by using credit cards. We're so keen to put off paying our bills that we're often happy to pay the credit-card company a good fee to invoice us next month.

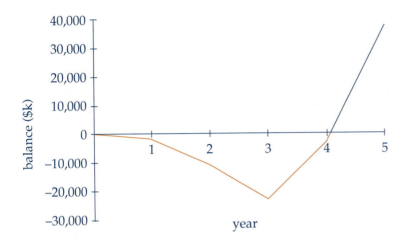

Figure 16-3. Cash Flow. The cash-flow chart for a start-up company making televisions shows they operate "in the red" for the first four years. Later, their investment begins to pay off, and profits soar.

Companies express this in terms of cash flow and discounting. *Cash flow* is the difference between the amount of cash received and paid out by the company during an accounting period. The five-year cash flow for our TV company is shown in Figure 16-3. Notice there was negative cash flow during the first four years, then an increasingly positive cash flow after sales paid off the initial investment. A positive cash flow is as important for the continued success of the company as positive profits (often more important, especially if the spectre of bankruptcy is hanging around).

Discounting is the arithmetic procedure we can use to calculate the present value of future income and payments. The value is discounted by a constant fraction each year into the future. The typical discount rate that an investor might use for a high-tech, high-risk activity like a commercial space project is 30% or 40% per year. The investor is saying that an income two years away is worth only 50%–60% of its apparent cash value today. Similarly, if we can put off expenditure for a couple of years, it's only about half as painful as it would be today.

Governments are yet to take such a strong view, but discount rates of 10%–15% are becoming common in the private sector. The impact of discounting will become clear later when we look at some of the business opportunities in space.

Who Pays? Who Benefits?

Governments fund many space programs. To understand the economics of space, we need to understand *why* as well as *how* they do this. Governments have two roles in space funding—as interveners and as delegated customers. As *interveners*, they act to bring about activities that are good for society as a whole but that wouldn't occur in a free market left to its own devices. A clear example of a space program for the public good was the Apollo moon-landing project. It benefitted everyone in the United States directly through technological spin-offs which spurred the economy and indirectly by inspiring young people to pursue engineering careers.

As *delegated customer*, government acts in a different way. Take defense, for example. If I am threatened directly, I can buy a gun and defend myself. If my neighborhood is threatened, we can all band together to defend ourselves. If my country is threatened by an unknown and distant enemy, this approach is no longer practical. I ask my government to arrange defense on my behalf and to send me the bill through the Internal Revenue Service.

At first glance, these two roles may look pretty similar, but the difference between them is, or at least should be, very important to space programs. As intervener, a government gets involved in space programs to foster the health of the space industry for the public good. It decides which programs to support based on many considerations, including political and foreign pressures. An intervener's decisions aren't based on supply and demand. The intervener is interested in balancing technical gains with project costs to determine the priorities of investments in various projects.

As delegated customer, government takes a different view. It is now able to say to the promoter of a space program, "how does your idea stack up against the alternatives?" Let's look for example at a military program. When the Air Force Space Command asks to buy another communications satellite, it has to persuade the Army to give up 50 tanks or the Navy to forgo a destroyer. Tanks and destroyers aren't given up easily, so the space program has to prove its worth.

This role of government as delegated customer produces some of the most exciting prospects for space programs. Many potential applications of space make *economic* sense, even if they're not necessarily *commercial*. One of the biggest and most exciting of these lies in monitoring the Earth's environment. As the pressure on governments to understand and manage environmental change grows, they will need the kind of objective global data that satellites provide, so the opportunity for space programs will be great. This need justifies the "Mission to Planet Earth" program, but don't think it will be an easy ride. The environmental modelers will use hardnosed economic analysis to guide their purchasing decisions. Nobody owes space workers a living. The nasty rules of economics are applicable to space programs in government as well.

Astro Fun Fact
Spin-Offs

Spending billions of dollars on space programs has caused the public to question what they're getting from this invested money. Numerous spin-offs have come from the space program, with some unusual twists. For example, a California man with a bullet fragment in his brain had little chance for survival. But doctors took him to the centrifuge used for training and experiments at NASA's Ames Research Center. By spinning the patient at six times the normal gravity (6 g's), they shifted the fragment into a less harmful position. In a lighter vein, a company tried to market a paint first developed for satellites. Made to withstand the harsh environment of space, the paint had remarkable qualities. Unfortunately, the public didn't believe the company's claims. Even a 20-year guarantee didn't help market the paint because people thought it was too good to be true. Fortunately, the space program has brought us numerous products—from microwaves to composite materials—that continue to improve the quality of life in the world.

Adams, Carsbie C., Frederick I. Ordway, Mitchell R. Sharpe. Dividends From Space.

Contributed by Steve Crumpton, United States Air Force Academy

▬ Section Review

Key Terms

cash flow
delegated customer
discounting
economies of scale
interveners
learning curve
supply and demand
time value of money

Key Concepts

➤ Classical economics considers the supply of goods in the marketplace and assumes
  • Many customers and many suppliers
  • Rational behavior on all sides
  • Goods are well defined and understood

➤ Supply and demand tells us that, as demand for a product rises, the price will rise until new suppliers step in to stabilize the price

➤ The concept of a learning curve, or economies of scale, tells us that producing more units of a good should drive down the per-unit cost

➤ Elementary ideas of supply and demand or learning curves don't apply to "one-of-a-kind" space projects

➤ Governments fund space projects for two different reasons—as interveners or as delegated customers—and behave differently in each case:
  • As intervener, government is concerned with the public good and the priorities of large investments in space projects
  • As delegated customer, government is concerned with cost-effectiveness

16.2 Space Programs—An Economic Perspective

▬ In This Section You'll Learn to...

- ☞ Describe how much is spent on space, by whom, and why
- ☞ Answer why space is so expensive

Budgets

Congress and the American people decide NASA's budget. Just how much do we spend on space? The graph in Figure 16-4 shows the expenditure by the main space nations from 1981–1991. The numbers look big—$30 billion per year for the United States—and the amount appears to be growing. If we remove inflation, we find there's little real growth, but it still looks like a lot of money.

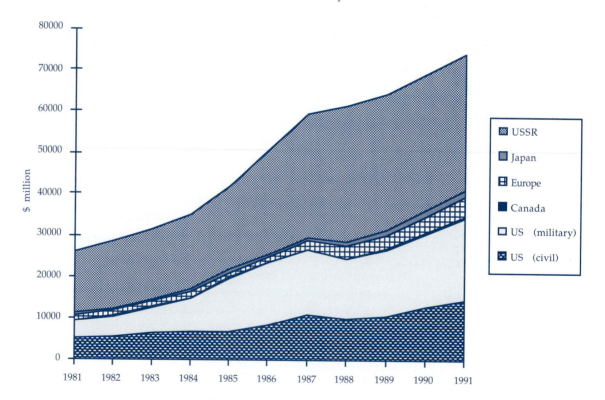

Figure 16-4. Space Budgets. This graph shows the space budgets for major space-faring nations during the 1980s.

But is it? NASA spends about $10 billion per year. Americans spend $6.3 billion per year on potato chips and $1.4 billion on popcorn. Europe spends about $4 billion per year on space, which looks pretty reasonable compared to $1 billion per year to subsidize European farmers who grow tobacco that no one wants to buy.

One of mankind's greatest achievements was to put a person on the Moon. The Apollo program cost the United States the equivalent of $200 billion, less than its citizens spent on cosmetics while the program was underway. Space isn't cheap, but if we want it, we can afford it.

Where Does The Money Go?

If we stick to civilian activities, space agencies spend the space budgets. In the United States that means the National Aeronautics and Space Administration, better known as NASA. Created on October 1, 1958, it remains essentially true to its original purpose. NASA is an investment in America's future. As explorers, pioneers, and innovators, we boldly expand frontiers in air and space to inspire and serve America and to benefit humanity. NASA's purpose is to

- Explore space with an eye toward expanding human presence beyond our planet

- Advance scientific knowledge of the Earth, the solar system, and the universe

- Research, develop, and transfer advanced technologies in aeronautics, space, and related fields

In fulfilling its purpose, NASA contributes to America's goals for economic growth—preserving the environment, educational excellence, peaceful exploration, and discovery.

In Europe the equivalent body is the European Space Agency (ESA), established in 1974 and now representing 15 nations including Canada (don't worry about the geography!). This agency's purpose is to

- Provide for and promote, for exclusively peaceful purposes, cooperation among European States in space research and technology and their space applications, with a view to their being used for scientific purposes and for operational space applications systems

Both of these agencies are charged with administering space programs for peaceful purposes. They define the mission goals and purchase space hardware and services from the industries of the countries paying the bill. In Europe this arrangement is formalized in the principle of *juste retour* or "fair return," which ensures that the amount spent in each country is proportional to that country's contribution. In the United States a process takes place informally; it's sometimes called the "pork barrel," where congressmen try to balance the dividends from space activists among their states and districts.

The Vicious Circle of Space Planning

Why do space projects cost so much? NASA and Department of Defense officials have concluded that they couldn't repeat the successes of the 60s and 70s within the same time or budget. Perhaps, the problem starts with the commonly held belief that space projects are expensive, which limits their number. Few projects mean

- They must be planned carefully to get the most out of them
- They must be reliable
- They need to be large to achieve a lot from each project
- There will be little competition (of ideas or between suppliers) if only one project occurs every few years

Each truth brings consequences for conducting the projects.

Extensive planning can lead to delays. High reliability when not building many systems means that integrity must be achieved by design. This precludes using the latest, unproven technology and thus, taken with the planning delays, means that spacecraft are built with obsolete parts. Large projects need large launch vehicles which, when combined with the lack of competition and the need for high reliability, means that launches become expensive.

Large projects also require large payloads and justify a large (and expensive) ground infrastructure. Large payloads bring a twist of their own. The space systems become too big to be built by a single contractor, so management structures become complex and expensive. As a result, trade-offs to optimize parts of the system are difficult because of the rigid contractual and organizational boundaries. Thus, we don't get the best designs and often have to live with very poor performance per kilogram. Now we enter the first *vicious circle*, as shown in Figure 16-5, because poor performance per kg requires even bigger payloads.

When all of these arguments come together, the resulting vicious circle reinforces the opening premise—"space projects are expensive."

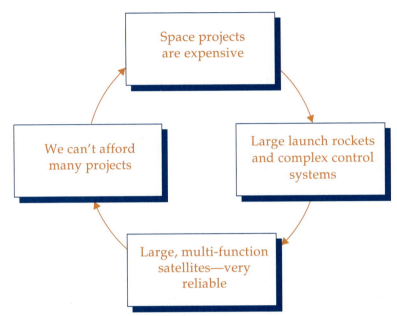

Figure 16-5. The Vicious Circle. Once we assume space is expensive, we make several key decisions which guarantee it *will be* expensive.

Section Review

Key Terms

juste retour
vicious circle

Key Concepts

➤ Governments fund civilian space programs largely through space agencies

➤ Space agencies ensure that spending reflects political interests

➤ If you start believing that space is expensive, a vicious circle of decisions makes sure it is

16.3 Consequences for Space Mission Engineering

▬ In This Section You'll Learn to...

☛ Examine the effect of economics on mission engineering

☛ Review the need for reliability and the pressure to "over-engineer"

☛ Summarize the manufacturing issues

☛ Describe the problems of launch and operations

The economic issues directly affect all aspects of space mission engineering, from design through manufacture to ground operations. Let's see how.

Design

Let's start with the engineering of spacecraft and see how economic issues affect the design. First, realize that spacecraft are very complex beasts. A TV set might have 250 parts, and a typical automobile can have a few thousand. But a communications satellite might have 50,000. What happens if something goes wrong? It's not easy to send a service engineer to a spacecraft in orbit. But NASA did just that! In 1992, the Shuttle was used to rescue a faulty communications satellite belonging to Intelsat. Cost of the Shuttle flight—approximately $500 million. Cost of the satellite saved—approximately $50 million. (Although this may appear to be a poor tradeoff, the *value* of the satellite was much greater if you consider potential revenues for the corporation and the insurance cost.)

To ensure the overall system is reliable, the first step is to make sure individual components are as reliable as possible. Space projects usually employ only components that have been through *space qualification*—an exhaustive program that checks every step in making the component, from the raw materials to packing and storage. Sounds expensive? It is. A space component typically must meet a military standard (MIL STD), which makes it cost up to ten times more than the commercial version.

The next step is redundancy. Spacecraft use duplicate or triplicate parts so that, wherever possible, a single failure can't cause the whole system to fail. If the spacecraft carries a crew, more redundancy may be required, just like in an airplane.

Let's illustrate this with a simple example. Suppose we want to unfold a solar panel from the side of the spacecraft when it's in orbit. The panel will be hinged with at least two springs to make it fold out. A strap with an exploding bolt at each end will hold it in place. To deploy the panel, we send a signal to both bolts to make them explode. If either works, the strap is free, and the panel unfolds.

Exploding bolts are an example of an interesting component—the sort we can't test after it's installed on the flight vehicle. Instead, through space qualification, we can tightly control the manufacture of the bolts so we can safely assume that, if a few bolts drawn from a batch go "bang" on command, they all will. Exploding bolts are very reliable, but we still try to duplicate them in case one should fail.

We do all this in a quest for *reliability*, which is the probability that a given component will do a certain job under certain conditions for a certain amount of time. This enthusiasm for reliability can lead to over-specification and hence higher costs. One example is specifying the ability of electronic devices to resist the effects of the space environment. As we saw in Chapter 3, radiation and charged particles can harm spacecraft components. Special computers have been developed to survive the space environment, but because so few are needed, they are *very* expensive and can lag about ten years behind the state-of-the-art. Result—spacecraft costing hundreds of millions of dollars that can compute no better than a washing machine. This makes it very desirable to be able to replace computers on orbit!

In many cases, these specifications are unnecessarily demanding. They represent a combination of "worst cases" that never occurs for most operational spacecraft in a low-Earth orbit. Many space missions could use off-the-shelf microchips, with a great improvement in performance and cost, but so far few have been willing to take the risk.

The main reason for these stringent requirements is the quest for "zero defects" or nearly perfect performance to protect people and ensure mission success. For instance, the public furor over the Challenger accident underscores the widely held belief that any failure is unacceptable. This demand for "perfect" systems inevitably drives up cost. In general, cost and reliability have an exponential relationship as shown in Figure 16-6. For example, going from 80%–90% reliability costs significantly more than going from 50%–60%. As reliabilities move to 95% or 98%, the design costs can get excessively high.

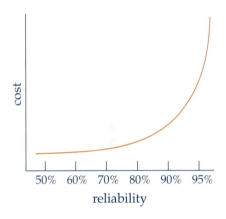

Figure 16-6. Cost Versus Reliability. The cost of reliability increases exponentially. A 100% reliable system isn't possible.

Manufacture

Although we may need to build a one-of-a-kind spacecraft, getting one that flies could require us to manufacture at least four spacecraft:

- An engineering model
- A test model
- A flight model
- A flight spare

Usually cost constraints don't allow us to build this many models. We can afford to build one, maybe two. We use *engineering models* to make sure all the subsystems work together and for non-destructive tests. The *test models* if built, will be subject to the sort of testing that might cause damage, such as vibration and shock or exposure to radiation vacuum and heat. In Chapter 13 we learned that a spacecraft can eject heat only through radiation. Thus, we can use a *thermal vacuum test* to sample the temperature of the spacecraft's components in a vacuum and determine whether they'll be able to rid themselves of heat in space. After the components test out, the entire spacecraft often undergoes a complete vehicle test; when money is tight, analysis may replace this test.

The *flight model* is the one that actually gets launched, and the *flight spare* may be built in case something goes wrong just before launch, to have a duplicate on the ground for testing if the on-orbit vehicle experiences problems. In practice it's sometimes possible to refurbish the engineering or test models as a flight spare.

The manufacturing procedures are as important as the design itself. Spacecraft are usually assembled in a clean room. A *clean room* is a dust-free environment, fed with filtered air and maintained scrupulously clean. The assembly staff in the clean room look more like the staff of an operating room. They wear gowns, hats, and masks to prevent them from introducing dust or dirt. The clean-room procedures do more than keep a dust particle from ruining the pictures from $500 million worth of Earth-observing satellite. They also help to instill in the assembly staff the sense of discipline needed to maintain work of the highest quality.

Before you ask, yes, the Hubble space telescope *was* assembled in a clean room and yes, it did end up with an optical fault. It illustrates the need to carry out an end-to-end test of a spacecraft system, not just to rely on separately testing the subsystems. Of course, this complete test is easier said than done, especially if the system's behavior depends on some feature of the space environment we can't reproduce on Earth, such as free-fall. Actually, NASA had planned a test that could have picked up Hubble's flaw, but it was removed to cut cost.

Launch and Operations

What happens to the result of all this tender loving care? We bolt it on the top of a massive rocket, containing hundreds of tons of explosive fuel, and fire it into space. In fact, we do a bit more than that. Launch, and the spacecraft's operation and control on orbit, take the same care and attention to detail and therefore are also expensive and complicated.

The cost of launching a spacecraft varies greatly, from around $5000 per kg for a proposed advanced launch system up to $30,000 per kg for a small commercial rocket. As with other economic activities, the price per kg appears to be less as more kilograms are launched. But this may be a false bargain. The cost of assembling several tons of payload to take advantage of the cheap launch may exceed the apparent savings.

In addition to the cost of the launch vehicle itself, we must consider all the incidental costs of running the launch site, including those for operations, monitoring, and telemetry before, during, and immediately after launch. Someone must also pay the capital cost of the launch site. A site like the European Space Agency's launch center in Kourou, South America, costs on the order of $1 billion to set up. If it's used for ten launches per year, the interest on that money alone is $10 million per launch!

A further expense is insurance. Typically one in ten launches fails in some way; that is, the payload doesn't end up in the right orbit or the booster explodes. Insurance companies cover failure of the rocket or the satellite, typically offering a free relaunch and a new satellite. They charge a premium of around 15% of the total costs for the satellite and launch. Coverage for lost revenue if the satellite doesn't enter service may be available but is often prohibitively expensive. The case study at the end of this chapter illustrates the dubious benefits of insurance.

After successfully launching the satellite, someone must manage it. The famous pictures of "Mission Control, Houston" show a team of nearly 100 people managing one spacecraft. A center like that would cost about $100 million per year to run, so we have to seek more efficient approaches. As discussed in Chapter 13, the more the spacecraft can do for itself, the less expensive an operations infrastructure can be to support it.

Another reason for automation is to prevent human errors. For example, when the Olympus communications satellite went into its "safe mode" just before a politically important demonstration, the operators tried to get it up and running quickly. The procedure went wrong, and it took months of work to get the satellite back under control, with most of its fuel reserves being consumed in the process. Rumors suggest that a Russian operator sent a command which the planetary-exploration satellite, halfway to Venus, interpreted as "ignore all future commands." The satellite did exactly that, so the controller's career was presumably short-lived. Automation has a lot to offer, but it's very expensive to set-up initially. You'll see in the case study in Section 16.5 that we need only a small control team if most ground operations can be fully automated.

Section Review

Key Terms	Key Concepts
clean room engineering models flight model flight spare space qualification test models thermal vacuum test value	➤ Spacecraft are complex and designed to use reliable components and reliable systems, but obsessive pursuit of reliability can be very expensive ➤ Manufacturing must be set at very high standards, but even then success is not guaranteed ➤ Launch costs are high. Large launchers reduce cost per kg, but the cost of building payloads to exploit them may more than erase this savings. ➤ Costs for operations can be high unless the spacecraft is mostly autonomous

16.4 The Space Business

In This Section You'll Learn to...

- Identify the areas in which it's possible to make money from space
- Describe conventional and unconventional attempts to do so

In the first section of this chapter, we saw the concept of "private goods," whose benefits go to a particular purchaser, and "public goods," whose benefits go to society as a whole. Both offer business opportunities. In the first area, there are opportunities in communications, Earth observation, and free-fall. In the second area, contractors may carry out tasks traditionally reserved for government-owned agencies.

Communications

Communication by satellites has been *the* great commercial success of space. We now take for granted world-wide telephone services, live television by CNN from the middle of a war, and direct broadcast of satellite television to our homes. The "Live Aid" charity concert in 1985 took place simultaneously in five countries and was watched live in 88 countries through relays by eight satellites. The 1992 Olympic Games were broadcast live to 150 countries using nine satellites.

The workhorse of satellite communications has been the geostationary satellite. Conventional economics has driven these satellites to be larger and more powerful (and more expensive)—to the point where, for example, an Intelsat V satellite can transmit 25,000 simultaneous telephone calls. The Direct Broadcast Satellites are so powerful that you can receive color television signals with a dish hardly bigger than a soup bowl. Most communications satellites are owned and operated by international agencies (Intelsat world-wide, Eutelsat in Europe, Inmarsat for mobile communications), which manage to be a bit more commercial than the space agencies.

Geostationary satellites have one big disadvantage besides their high cost. The geostationary orbit is a long way above the Earth (36,000 km) and it needs a lot of transmitter power to communicate over that distance, so we can't use hand-held terminals for telephone calls to and from these satellites. Also, the time it takes for the radio signal to get to the satellite and back again (about one third of a second) makes conversation stilted. A further problem is that geostationary satellites provide poorer coverage above 70° latitude.

These disadvantages have led to using many satellites in low-Earth orbits. With enough satellites we can ensure one is always in range, at an altitude of only a few hundred kilometers. Therefore, Motorola has proposed an ambitious commercial system known as Iridium. It originally was to have 77 satellites (the atomic number for Iridium), but now is designed for 66. By passing calls among these satellites, Iridium will be able to provide world-wide telephone service. The system is still in planning as we go to press, but Motorola predicts a $3.4 billion cost for the system, which will provide a three-minute call for $3.00. Less ambitious proposals are for systems that can provide data-relay services or telephone coverage to certain areas of the world.

Although large, international, geostationary communication satellites have enjoyed wide success, their economics raise questions. They emerged from the vicious circle in which presuming that space projects are expensive leads to decisions that make them more expensive. Communications satellites may be uneconomical, particularly when we consider the cost of borrowing the money to build and launch them before receiving any revenue. The case study at the end of this chapter illustrates the thinking that leads to smaller, cheaper, and, above all, quicker-to-build satellites. In the next section, where we look at Earth observation, we can see this thinking in action.

Earth Observation

In 1986, the French space agency CNES launched its satellite SPOT-1 (Satellite Pour l'Observation de la Terre). This satellite is funded and owned by the government but operated to provide images having moderately high resolution for economic uses. By steering the satellite's telescope, it's possible to obtain frequent images of specific points of interest. A commercial company, SPOT Image, was set up to exploit the data effectively. SPOT Image markets, distributes, and archives the calibrated raw data it receives from the CNES receiving station.

Although space agencies own most of SPOT Image, its behavior is very commercial. Its gross income was only $500,000 in 1986, but it has risen steadily to $30 million in 1991 and is still rising. A second satellite (SPOT-2) was launched in 1990, so SPOT-1 was retired. SPOT Image has been so successful that it was unable to meet the demand and, in 1992, SPOT-1 was recommissioned. SPOT-3 is built and ready to launch when needed, and SPOT-4 is planned. Planners hope to use revenue from selling data to fund later satellites.

SeaStar, which will be the first privately owned Earth-observation satellite, shows us a different approach. Orbital Sciences Corporation, a private company, will build, launch, and operate the satellite. Hughes' Santa Barbara Research Center will supply the sensing instrument. NASA will pay the corporation $43 million to buy, for research uses only, the data that SeaStar will generate during its five-year life. The corporation retains the rights to sell data to operational users in the public sector and to commercial customers.

SeaStar represents a very significant step towards making the business of space more like other types of business. Orbital Sciences Corporation won a competition for the contract and is taking the technical risks of the mission. In exchange, it can profit from selling the data.

Free-Fall

As we learned in Chapter 3, satellites in orbit are in free-fall. We're used to the sight of astronauts drifting around in space, and we've discussed the difficulty of going to the restroom up there, but free-fall causes many more important and subtle effects. The Cuban cosmonaut, Arnaldo Tamayo Méndez, summed this up very well

> At home you use a spoon [to stir in sugar] and there you have it— sweet tea. In space you shake the flask until you're blue in the face.

The point is that convection, driven by slight differences in temperature and hence, density and buoyancy, causes much of the mixing in a cup of tea. In free-fall, there is no natural mixing.

This attribute allows us to explore and exploit many new phenomena. The first major commercial investigation was the Electrophoresis Operations in Space project by McDonnell Douglas in the mid 1980s. *Electrophoresis* is a technique for separating proteins using an electric field. Scientists found that it could be up to 700 times more efficient in space than on the ground. Other proposed uses of free-fall include growing viruses and proteins for medical research and producing large single crystals of semiconductors to manufacture integrated circuits.

Unfortunately, free-fall has yet to prove itself commercially attractive, although we hope for breakthroughs in development of power metals, vaccines, and electronic chips. The high costs of launch and operations have prevented many of the scientifically attractive ideas from coming to fruition. (A realistic cost for astronaut time is $10.00 per person per second, which is even more expensive than a New York lawyer or a plumber.)

▀▀▀ Section Review

Key Terms	Key Concepts
electrophoresis	➤ Many business opportunities exist in communications, Earth observation, and free-fall, as well as in projects for which government agencies contract out their activities
	➤ Conventional communication uses geostationary satellites; unconventional approaches use constellations of low-orbiting satellites
	➤ Conventional Earth observation involves selling the data from satellites funded by space agencies; the unconventional approach sells data from private satellites to space agencies (and others)
	➤ Research into free-fall manufacturing continues but has yet to prove its commercial value

16.5 A Study in Space Business

▬ In This Section You'll Learn to...

☞ Examine the elements of a commercial space venture

☞ Show the effect of a realistic rate of return

☞ Show how vulnerable such ventures are to something going wrong

Let's bring all of the issues of space business together and look at an imaginary commercial venture. By putting in realistic estimates of costs, risks, and rewards, we can highlight how space is different from other kinds of business, even though it's based on the same laws of economics.

We have contrived the example so the cost to fly the manufacturing equipment is about half of the value of the product brought back. This value represents a typical level for a commercial project.

The Project

We'll use on-orbit manufacture as our example commercial project. To have any hope of making a profit, the product must have a high ratio of value to mass. Contact lenses are such a product. With a current wholesale price of about $2.5 million per kg, manufacturing lenses on orbit might make money.

Spin casting is one of the standard ways to manufacture contact lenses on Earth. In this technique, a mold formed like the front face of the finished lens is loaded with a drop of liquid monomer and spun about its axis of symmetry. The liquid forms a parabolic surface when inertia forces from spinning balance gravitational forces, as shown in Figure 16-7. In this state the monomer polymerizes (hardens), and the lens forms. The thickness of the lens and the radius of the back face depend on the amount of monomer and the rate of rotation.

Suppose a spherical back face would have advantages over a parabolic one. Let's assume we've discovered a new material which, in free-fall, forms perfect contact lenses. If we place a drop of monomer in a mold in free-fall, it will form a spherical back face due to the pull of surface tension alone, as shown in Figure 16-8. The amount of monomer and surface tension determine the thickness and radius. If the latter strongly depend on the temperature of the mold, we can vary the temperature to adjust both parameters.

Some overall assumptions in planning the project are:

• Manned operation won't work (astronaut time costs $10 per second)

• We can recover the manufacturing equipment with the lenses and then refurbish and reuse it

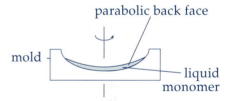

Figure 16-7. Making Contact Lenses on Earth. On Earth, a liquid monomer material is placed in a mold and then spun. As the material accelerates away from the center, it bonds to the sides of the mold and hardens to form a contact lens.

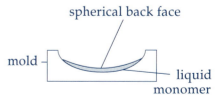

Figure 16-8. Making Contact Lenses in Space. In space, a liquid monomer material is placed into a mold. In free-fall, surface tension forces dominate, so the material bonds to the mold and forms a perfect contact lens.

- Staff are available as required and are not charged to the project when not needed. (Thus, the project may be more appropriate to a large company which can redeploy the staff onto other projects.)

We will also assume the manufacturing equipment has a mass of 100 kg, carries 6 kg of lens material, and consists of ten manufacturing cells, each of which can make a lens worth $100 every 100 seconds. This corresponds to manufacturing $15 million worth of lenses in six months.

To work out the costs and income, we'll have to break the project down into phases of work.

- Phase 1—to demonstrate the behavior of the new material in microgravity by means of simple "lash-up" experiments onboard an aircraft in parabolic flight. Costs include
  - Six months with three staff members, five aircraft flights, and $100,000 for materials and expenses

- Phase 2—to develop equipment for a ground test of on-orbit manufacturing and qualify it for space use. Costs include
  - 24 months with 12 staff members to develop and ground test, three months with six additional staff to qualify, and $500,000 for materials and expenses

- Phase 3—to test the manufacturing equipment in microgravity, using aircraft in parabolic flight and sounding rockets. Costs include
  - Three months with six staff members for aircraft tests (ten flights), plus three months with six staff for rocket tests (two flights)

- Phase 4—to integrate the manufacturing equipment with the spacecraft, insure, launch, and commission. Costs include
  - Five months with six staff members for integration, plus one month with two staff for commissioning

- Phase 5—to operate the equipment (by means of telecontrol) and recover the spacecraft. Costs include
  - Six months with three staff members for operation and a delay of three months before selling the lenses

- Phase 6—to refurbish the equipment and repeat. Costs include
  - Three months with six staff members to refurbish, two months with six staff members for integration, one month with three staff members for commissioning, six months with three staff for operation and recovery, and again, a delay of three months before selling the lenses

To replace the manufacturing equipment, we'd expect to need eight staff members for 12 months (half of the time to build the original equipment) and spend another $500,000 for materials and expenses. The total cost would be $1,940,000; insurance would cost about $388,000. We'll assume other costs are as listed in Table 16-1.

Table 16-1. Cost for Contact-Lens Project.

Expense	Cost
Staff	$10,000 per person per month
Parabolic flights by aircraft	$10,000 each
Sounding flights by rockets	$50,000
Launch	$7 million
Launch insurance	20% of the cost of replacing the manu-facturing equipment ~$388,000

Project Budget

Now we can plot the project balance sheet, as shown in Figure 16-9, as the project unfolds. Like any development project, it starts off going negative as we invest in research and development, and then goes positive as we start selling lenses. After just over four years, the project gets its first income and has $2 million in profit. Each successive flight generates just over $7 million additional profit so that, after fewer than eight years, the investor has nearly $25 million in his bank account with all of the investment repaid. Not a bad proposition?

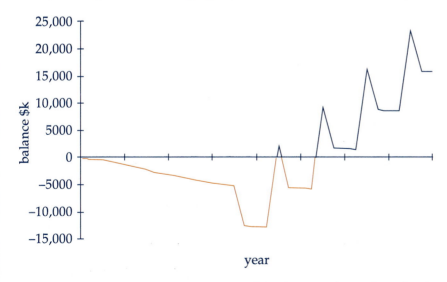

Figure 16-9. Cash Flow for Contact-Lens Project. Assuming money has no time value, the project becomes profitable during the fifth year.

Time Value of Money

The trouble with the cash flow in Figure 16-9 is the issue that we saw right at the beginning of this chapter—the time value of money. Any commercial investor knows that money costs money and wants to see a realistic return on his investment. It's normal for speculative venture capitalists to seek a return on investment (ROI) of 30% per year for high-risk projects such as this one. We can see the effect on the project by charging interest at the ROI on the outstanding debt. The investor sees a worthwhile return only when the balance after interest is positive.

The project's revised cash flow appears in Figure 16-10 for ROI rates of 0% (solid line) and 30% (broken line). A number of interesting results emerge. When we ignored the need to show an ROI, the project was in profit after the first launch. At 30% ROI, it's showing a return only after three launches—more than six years after the start rather than the 4½ years we obtained from a simple cash model. The proposition is starting to look less attractive.

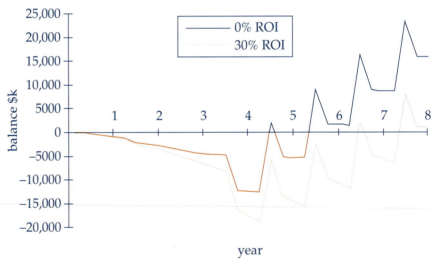

Figure 16-10. Cash Flow for the Contact-Lens Project with Time Value of Money. This graph shows the effect of considering the time value of money. The dashed line assumes a 30% discount rate and shows no profitability until the seventh year.

If Disaster Strikes?

What happens if disaster strikes? Figure 16-11 shows the project's cash flow if the first launch is a failure and it's necessary to claim on the insurance policy to rebuild the equipment and take a free reflight from the launch contractor. The 0% ROI curve returns to the same point, one year later, but the 30% ROI curve is still negative after eight years. To a venture capitalist, the result is disastrous, with no return for almost 12 years rather than six years after starting to invest. This delay may cause an investor to drop the project.

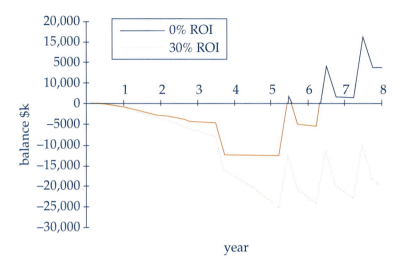

Figure 16-11. Contact-Lens Project with a Failure. The effect of a single launch failure on profit. The solid line assumes 0% discount rate, and the dashed line assumes 30% discount.

Buy Insurance?

This leads to an interesting conclusion about insurance for a high-risk venture. Insurance underwriters often complain that new, entrepreneurial companies object strongly to insurance costs and are prepared to "bet the company" on an uninsured flight. From our analysis, however, the venture is knocked out by a launch failure, insured or not.

In that case, why bother to pay the premium? Let's look at the original project plan but exclude any payments for insurance in Figure 16-12. After three launches, the project is in credit by $4 million instead of $2 million, even after paying 30% ROI. Thus, insurance may not be as good a deal as it looks.

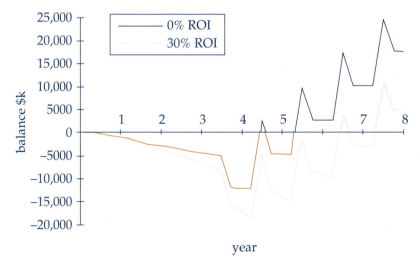

Figure 16-12. Contact-Lens Project without Insurance. By *not* paying for insurance, the net profit is higher.

Project Delays?

Even a problem less dramatic than launch failure may bring the project down. For example, a delay of six months before the first launch can have a big effect. As you would expect, the 0% ROI curve returns to the same value six months later. However, the project now takes eight years to show a return at a 30% ROI, a position that it would achieve after only six years without that six-month hold.

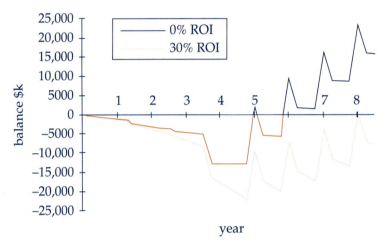

Figure 16-13. Contact-Lens Project with Delay. With a six-month delay, early in the project, it takes eight years to show a return at 30% ROI.

Increased Costs?

How robust is the project against relatively small changes in the costs? Consider having to spend another $3 million on a recovery. If we ignore the time value of money, it's still profitable after two launches, but at 30% ROI, the project is diverging and accumulating ever greater losses.

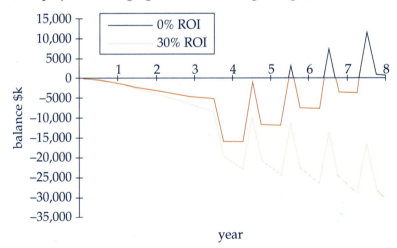

Figure 16-14. Contact-Lens Project with Increased Costs. An increase in costs of only $3 million means the project *never* becomes profitable (assuming 30% ROI).

Conclusions

What can we conclude from this simple analysis? The project looked good at first ($15 million worth of lenses from a $7 million flight), but as soon as we account for development costs and, above all, the return on investment a commercial investor will seek, it becomes a lot less appealing. The effects of any problem (launch failure or delay, marginal increase in costs) become much greater when we must support a "marching army" of debt.

This highlights one of the essential differences between a government-funded project and a private project. Government agencies, such as the military or the meteorological service, may truly trade off costs and benefits to establish whether a project is worth doing. But they are almost always able to work in terms of current costs and don't have to allow for debt costs. Indeed, they often prefer a project with a flat spending profile so they can budget so much per year for the next five years. The 0% ROI curves are the ones they consider. In comparison, a private investor looking for a profit has no difficulty finding other high-risk ventures offering an ROI of 30% or more. The investor wants a project with large expenses pushed into the future rather than a flat spending profile.

This has very great implications for design and development procedures. If a six-month delay at a late stage in development can lead to a two-year delay in returning profit to the investor, we don't have much time to space-qualify expensive new components or sub-systems. If a change in the gross margin per flight from 114% to 50% can keep the project from ever showing a return, each step must be as inexpensive as possible. Conventional thinking, which says we should seek perfection in space projects, has no place in the grubby world of projects funded by venture capital.

▮ Section Review

Key Concepts

➤ Manufacturing high-value products in space might make money

➤ Even so, this type of project is difficult to finance with venture capital

➤ Even if we can get venture capital, there is a difficult balance between under-engineering (risking something failing) and over-engineering (delaying return on investment)

➤ Quite small increases in costs can sink the venture

➤ Communications and Earth-observation projects offer similar lessons

16.6 The Future

▤ In This Section You'll Learn to...

- ☛ Look at the changes that economic pressures might bring to mission design
- ☛ Describe the virtuous circle that could result from these changes

The business case study we've just seen illustrates very clearly that space is a difficult area for private enterprise. It's equally difficult for the public sector, with pressure to reduce spending and to seek more cost-effective ways of achieving the goals. How might these *economic* pressures affect *engineering* decisions? We can expect five key changes:

- Space agencies will be forced to establish priorities based on user needs
- Failures will be tolerated more in smaller, less expensive missions
- Appropriate technology will be used
- Missions will be dedicated to fewer payloads
- Projects will have a shorter duration and a smaller, integrated team

Let's examine each of these to see the implications.

Users determine needs—the essential difference between a project driven by a space agency and one driven by the user is the measure of success. Space agencies tend to measure success by inputs, such as advancing technology or completing the project, not by how well it satisfies user needs. Users measure success by outputs, such as the data or service they receive.

Space agencies are able to take a more constructive view and behave like users. The most quoted example is the Apollo programs in the 1960s. A similar success, although on a smaller scale, was the European Giotto mission to Comet Halley. In both cases, the space agency received a firm target date for the mission and a clear objective and, in both cases, the agency responded beautifully. This was possible because they had a stable budget that was appropriate for the task.

Failures tolerated—top tennis players try to win matches, not just points, and typically hit one in ten first services out or into the net. They know that, to be competitive, they must take calculated risks and accept some failures. Insisting on perfection in space systems is fundamentally wrong. The perfect system, whether it's a spacecraft or a toaster, will never be built.

The description of the vicious circle in Section 16.2 concluded that everyone behaves reasonably when making the individual decisions that collectively lead to such an unreasonable consequence. To introduce the acceptance of failure, we must change the motivation that drives each decision. As David McLelland, a business researcher, said in the 1950s,

> If the penalty for failure exceeds the reward for success, then people will plan not to fail, not plan to succeed.

Appropriate technology used—it's normal to demand full space qualification for all components and subsystems, but this decision should depend on how critical they are to the mission and the stress they must undergo.

Generally, planners should use existing military or civilian devices whenever possible. By choosing specifications properly, they can use reliable, well engineered products while avoiding the expense of special-purpose subsystems.

Missions dedicated to fewer payloads—large, multi-payload satellites are like a convoy, only able to move at the speed of the slowest element. If any part is delayed, the overall project is delayed. They often don't meet users' needs and drive up costs. This leads to the last of five changes needed to get onto the virtuous circle.

Projects with short duration and small team—space technology often lags by ten or more years because a long-lasting project requires an early freeze on the design. In Chapter 11, we learned that design is iterative, with changes in one subsystem affecting the design of another. To keep from endlessly redesigning every subsystem, designs must be "frozen" at some point. For a complex project with a large team, an early freeze is necessary to define the interfaces between elements of the spacecraft, so the distributed team can start their detailed design work.

Project costs usually go up proportionally with duration and team size. Any delay in the project means that the team keeps working, and spending, and any added complexity or management interfaces mean more people generating more documentation.

Project management and changes driven by management represent the largest cost element for many projects, especially if international collaboration is involved. Small teams, preferably collocated, working intensively on a short project, bring this cost down.

The Virtuous Circle

What could happen if all of these changes in approach were to occur? If we assume that space projects are cheap, a different positive-feedback picture emerges—a *virtuous circle*, as shown in Figure 16-15.

Could it happen? An earlier writer identified the difficulty of bringing about revolutionary changes and effectively anticipated the problems of the space industry by four hundred years. In "The Prince," Machiavelli wrote

> Nothing is more difficult to undertake, more perilous to conduct or more uncertain in its outcome than to take the lead in introducing a new order of things. For the innovator has for enemies all those who

have done well under the old and lukewarm defenders who may do well under the new.

Only time will tell if space remains the playground of governments and large corporations or becomes the new frontier of the entrepreneur.

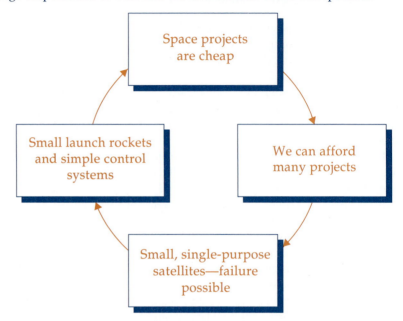

Figure 16-15. The Virtuous Circle. By assuming that space projects are cheap, we can take actions and make decisions that will ensure it will be cheap.

The challenges for you are in

- Science and technology—develop technology in all disciplines that helps us do space missions better and cheaper
- Business—create an environment and opportunities that facilitate quicker, cheaper, and more profitable systems
- Politics—nurture cooperation in space ventures within space missions and among nations

▓ Section Review

Key Terms	Key Concepts

Key Terms

virtuous circle

Key Concepts

➤ A very different approach to space missions can make them more like other engineering projects

➤ Space missions need five key changes:
  - Space agencies must establish priorities based on user needs
  - Smaller, less expensive missions must tolerate more failures
  - Missions must use appropriate technology
  - Missions must carry fewer payloads
  - Projects must be shorter and use smaller, integrated teams

➤ The consequence would be many more projects, carried out quickly and at much lower cost

➤ It's hard to bring this about against the resistance of the established space community

Mission Problems

16.1   Basic Economics

1   What are the basic assumptions of classical economics?

2   Describe the law of supply of demand and give a brief example of it in your every day life.

3   Describe economies of scale. Why doesn't this principle necessarily apply to space projects?

4   What is the difference between positive and negative cash flow?

5   What are the two roles of government in space programs?

16.2   Space Programs—An Economic Perspective

6   Describe the vicious circle of space-mission economics.

16.3   Consequences for Space Mission Engineering

7   Explain why space-qualified hardware is more expensive than hardware tested on Earth.

8   Define reliability. What is the relationship between it and cost?

16.4   The Space Business

9   What are some of the areas of opportunity for business in space?

16.5   A Study in Space Business

10   Describe the effects of ROI, launch failure, and delay on a business venture in space.

16.6   The Future

11   Describe the virtuous circle and how it could reduce the cost of doing business in space.

For Discussion

12   Write to NASA and ask for information on the budget and how it is allocated, also for details of some spacecraft and missions. Analyze what you get back to check some of the statements in this chapter. For example

- What fraction goes to missions and what to administration?

- What fraction of the mission money goes to the space segment and what to the ground segment?

- How is modern technology used in the spacecraft? (e.g., How fast is the computer compared with your PC? Does it use carbon-fiber materials?)

- How long does a typical mission take, from concept to operations? How does the spending vary with time?

13   Study some of the everyday machines you encounter (automobile, photocopier, toaster, washing machine, etc.) How would the design and manufacture be different if no-one could service the machine after it was built?

14   Can you think of any other products, like contact lenses, which are so valuable per kg that it would be economical to manufacture them in space?

15   How would the planning and execution of a space project change as you vary the discount rate applied to the funding?

16   Can you think of "users" who could determine the needs of programs space agencies are planning?

17   Air freight costs around a few dollars per kg. Space launch costs several thousand dollars per kg. How might we use space if the costs of launch fell to approach those of air freight?

Mission Profile—UoSat

Since the very first satellite, Sputnik, weighing in at 84 kg (185 lb.), satellites have grown in size, weight, and corresponding cost. Billion-dollar satellites taking ten years to complete are common. Beginning in 1979, researchers at the University of Surrey, U.K.,—inspired by advances in computer technology—decided to buck this trend. They've built and launched a series of "microsatellites" (all under 100 kg [220 lb.]), proving that spacecraft don't have to be big and expensive.

Mission Overview

The objectives of the University of Surrey Satellite (UoSat) program include investigating cost-effective engineering techniques, providing an on-orbit test bed for new space technology, demonstrating small-scale space science, exploring commercial and industrial applications for microsatellites, and stimulating space education in schools around the world.

Mission Data

The primary mission of most of the satellites has been to provide "store-and-forward" communications to Earth-based users such as amateur radio operators. This means a user at one location sends an electronic message or "mail" along with an address to the satellite, which stores the mail and then delivers it to intended recipient.

✓ All launches are "piggy-back" aboard boosters carrying large, expensive primary spacecraft such as SPOT or Earth-resource satellite (ERS).

✓ UoSat 1: Launched 1981 aboard a NASA Delta booster into a sun-synchronous orbit. Payload included auroral radiation and Earth-imaging experiments. Reentered October 13, 1981; still operational after eight years.

✓ UoSat 2: Launched 1984 aboard a NASA Delta booster into a sun-synchronous orbit. Payload included digital communications and gravity-gradient stabilization experiments. Still operational as of 1993.

✓ UoSat 3 & 4: Launched together in 1990 on an ESA Ariane booster into a sun-synchronous orbit. Mass of 50 kg (110 lb.) each. These flights qualified the University of Surrey's multi-mission microsatellite platform for future flights. UoSat 4 failed, but UoSat 3 was still operational as of 1993.

✓ UoSat 5: Launched 1991 aboard ESA Ariane booster into a sun-synchronous orbit. Mass of 48.5 kg (107 lb.) and peak production of 49 W of electrical power. Payload included a store-and-forward communications transponder, and an attitude-control experiment using magnetic torquers.

A spacecraft engineer prepares the UoSat 5 spacecraft for launch. *(Courtesy of Surrey Satellite Technology, Ltd.)*

Mission Impact

The UoSat series has proven the worth of microsatellites for space experimentation and as communication platforms. Microsatellites such as UoSat can go from mission definition to orbital commissioning in about one year, at a cost of under $3 million, and have an on-orbit life of more than eight years. Building on their success, the University of Surrey has embarked on several commercial missions using UoSat technology, including S80/T, KITSat-1, PoSat-1, and HealthSat-2.

For Discussion

• When does it make sense to do space missions with lots of small spacecraft rather than one big one?

• How could microsatellites be used to explore the solar system?

References

Adapted from "University of Surrey (UoSat) Satellite Programme." A briefing prepared by Jacky Radbone. University of Surrey, U.K.: Surrey Satellite Technology, Ltd., 1992.

Math Review

A

A.1 Trigonometry

Trigonometric Functions

Trigonometric functions allow us to compute the sizes of angles and the lengths of sides in geometric figures. If we have a right triangle ABC in Figure A-1 with hypotenuse (longest side) of length C and angle θ, we can find the lengths of the other sides by using the *sine* and *cosine* functions. Basic trigonometric functions start by defining the sides and angles in a right triangle

- cosine θ = cos θ = B/C
- sine θ = sin θ = A/C
- tangent θ = tan θ = A/B

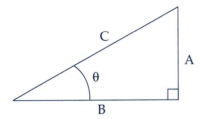

Figure A-1. Right Triangle. We can determine the angles and lengths of the sides of a right triangle using trigonometric functions.

Thus,

- B = C cos θ
- A = C sin θ

One way to remember this relationship is SOH-CAH-TOA. That is, sine is opposite side over hypotenuse, cosine is adjacent side over hypotenuse, and tangent is opposite side over adjacent side.

Example

Find the length of side A if C = 2 and θ = 15°

$$A = C \sin \theta = 2 \sin (15°) = (2)(0.2588) = 0.5176$$

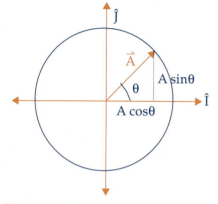

Now if we take a different perspective and look at a vector in an $\hat{I}\hat{J}$ coordinate system, as shown in Figure A-2, we can see that $A_I = A\cos\theta$ and $A_J = A\sin\theta$.

We can also see that as θ goes to 0°, sin θ goes to 0 and cos θ goes to 1. Also, as θ goes to 90°, sin θ goes to 1 and cos θ goes to 0. Summarizing,

$$\cos 0° = \sin 90° = 1$$
$$\cos 90° = \sin 0° = 0$$
$$\cos 180° = \sin 270° = -1$$
$$\cos 45° = \sin 45°$$

A graph of the cosine and sine functions is in Figure A-3.

Figure A-2. Vector Components as a Function of Angle. Here the vector A is shown represented by its I and J components on a circle. As t changes, A marches around the circle like the second hand of a clock in reverse.

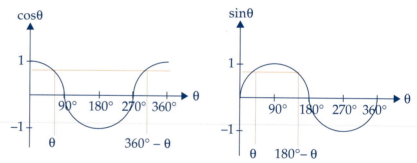

Figure A-3. Trigonometric Functions. The cosine and sine function both repeat periodically as shown.

Also notice the following important relationships

$$\cos \theta = \cos (360° - \theta) \tag{A-1}$$

$$\sin \theta = \sin (180° - \theta) \tag{A-2}$$

This means when we take the inverse of a trigonometric function we get two possible answers! That is

$$\arccos x = \cos^{-1} x = \theta \; and \; (360° - \theta)$$

$$\arcsin y = \sin^{-1} y = \theta \; and \; (180° - \theta)$$

Example

Find the angle(s) whose sine is 0.65.

$$\sin^{-1}(0.65) = 40.54° \text{ and } (180° - 40.54° = 139.46°)$$

Find the angle(s) whose cosine is 0.65.

$$\cos^{-1}(0.65) = 49.46° \text{ and } (360° - 49.46° = 310.54°)$$

Angle Measurements

Angles are normally measured in one of two ways: degrees or radians. Of course there are 360° in a circle. The measure of radians came about by looking at the relationship between the diameter and the circumference of a circle. The constant number Pi (π) is exactly equal to the ratio of a circle's circumference (C) to its diameter (D).

$$\pi = C/D = 3.14159...$$

We then say a circle contains 2π radians.

$$2\pi \text{ rads} = 360°$$

Spherical Trigonometry

The preceding discussion develops from angles and sides measured on a plane. However, when measuring angles and sides on the surface of a sphere such as the Earth, things are different. For example, the sum of the angles in a spherical triangle can be greater than 180° as shown in Figure A-4.

The relationship for right spherical triangles is shown in Table A-1

Table A-1. Right Spherical Triangles. Adapted from Space Mission Analysis and Design [1992].

The line below each formula indicates the quadrant of the answer. Q (A) = Q (a) means that the quadrant of angle A is the same as that of side a. "Two possible solutions" means that either quadrant provides a correct solution to the triangle defined.

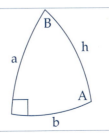

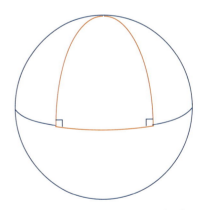

Figure A-4. Spherical Triangles. A triangle drawn on the surface of a sphere requires the use of spherical trigonometry.

<u>Given</u>   <u>Find</u>

a, b $\cos h = \cos a \cos b$ $\tan A = \tan a / \sin b$ $\tan B = \tan b / \sin a$
           $Q(h) = \{Q(a) Q(b)\}*$ $Q(A) = Q(a)$ $Q(B) = Q(b)$

a, h $\cos b = \cos h / \cos a$ $\sin A = \sin a / \sin h$ $\cos B = \tan a / \tan h$
           $Q(b) = \{Q(a) / Q(h)\}**$ $Q(A) = Q(a)$ $Q(B) = \{Q(a) / Q(h)\}**$

b, h $\cos a = \cos h / \cos b$ $\cos A = \tan b / \tan h$ $\sin B = \sin b / \sin h$
           $Q(a) = \{Q(b) / Q(h)\}**$ $Q(A) = \{Q(b) / Q(h)\}**$ $Q(B) = Q(b)$

a, A $\sin b = \tan a / \tan A$ $\sin h = \sin a / \sin A$ $\sin B = \cos A / \cos a$
           Two possible solutions   Two possible solutions   Two possible solutions

a, B $\tan b = \sin a \tan B$ $\tan h = \tan a / \cos B$ $\cos A = \cos a \sin B$
           $Q(b) = Q(B)$ $Q(h) = \{Q(a) Q(B)\}*$ $Q(A) = Q(a)$

b, A $\tan a = \sin b \tan A$ $\tan h = \tan b / \cos A$ $\cos B = \cos b \sin A$
           $Q(a) = Q(A)$ $Q(h) = \{Q(b) Q(A)\}*$ $Q(B) = Q(b)$

b, B $\sin a = \tan b / \tan B$ $\sin h = \sin b / \sin B$ $\sin A = \cos B / \cos b$
           Two possible solutions   Two possible solutions   Two possible solutions

h, A $\sin a = \sin h \sin A$ $\tan b = \tan h \cos A$ $\tan B = 1 / \cos h \tan A$
           $Q(a) = Q(A)$ $Q(b) = \{Q(A) / Q(h)\}**$ $Q(B) = \{Q(A) / Q(h)\}**$

h, B $\sin b = \sin h \sin B$ $\tan a = \tan h \cos B$ $\tan A = 1 / \cos h \tan B$
           $Q(b) = Q(B)$ $Q(a) = \{Q(B) / Q(h)\}**$ $Q(A) = \{Q(B) / Q(h)\}**$

A, B $\cos a = \cos A / \sin B$ $\cos b = \cos B / \sin A$ $\cos h = 1 / \tan A \tan B$
           $Q(a) = Q(A)$ $Q(b) = Q(B)$ $Q(h) = \{Q(A) Q(B)\}*$

* $\{Q(x) Q(y)\} \equiv$ 1st quadrant if $Q(x) = Q(y)$, 2nd quadrant if $Q(x) \neq Q(y)$
** $\{Q(x) / Q(h)\} \equiv$ quadrant of x if $h \leq 90$ deg, opposite quadrant of x if $h > 90$ deg.

A.2 Vector Math

Definitions

Scalar. A *scalar* is a quantity that has magnitude only. Speed, energy, and temperature are examples of scalars. None of these quantities has a unique meaning in any certain direction. A single letter, such as E for Total Mechanical Energy, denotes a scalar quantity.

Vector. A *vector* is a quantity that has both magnitude and direction. For example, if I ask you where you went on your hike, you could say, "I went north." But this wouldn't tell me much. "How far?" If instead you answered, "I went five miles," I still wouldn't know much. "Five miles in what direction?" If, however, you answered "I went five miles north," I would know both the distance you hiked (magnitude) and the direction (north). Position, velocity, and angular momentum are examples of vector quantities. A letter with an arrow over it, such as $\vec{R}$ for the position vector, denotes a vector quantity.

Unit Vector. A *unit vector* is a vector having a magnitude of one and is used to determine direction only. For example, when we define a three-dimensional coordinate system, we do so using three orthogonal (mutually perpendicular) unit vectors such as $\hat{X}\hat{Y}\hat{Z}$ or $\hat{I}\hat{J}\hat{K}$. A letter with a caret or hat over it, such as $\hat{I}$ for the I unit vector, denotes a unit vector.

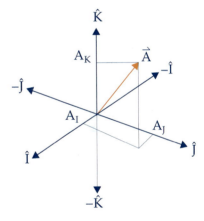

Figure A-5. Components of a Vector. Here you see the three components of a vector.

Vector Components

A vector can be resolved into components along each of three directions in an orthogonal coordinate system. If we have the vector $\vec{A}$ in the $\hat{I}\hat{J}\hat{K}$ coordinate system, as shown in Figure A-5, we can resolve it into its three components as

$$\vec{A} = A_I\hat{I} + A_J\hat{J} + A_K\hat{K}$$

Vector Operations

Magnitude of a Vector. Magnitude is the scalar part of a vector. We find it by taking the square root of the sum of the squares of each of its components. That is,

$$|\vec{A}| = A = \sqrt{A_I^2 + A_J^2 + A_K^2} \qquad \text{(A-3)}$$

Note: We use both $|\vec{A}|$ and A to denote the magnitude of the vector $\vec{A}$.

Example

Find the magnitude of the vector $\vec{B} = 3\hat{I} + 1\hat{J} - 2\hat{K}$

$$B = \sqrt{3^2 + 1^2 + (-2)^2} = \sqrt{9+1+4} = \sqrt{14} = 3.74$$

To create a unit vector (of magnitude one), divide each component by the magnitude of the original vector.

Example

Find a unit vector in the direction of $\hat{B}$

$$\hat{B} = \frac{3}{3.74}\hat{I} + \frac{1}{3.74}\hat{J} - \frac{2}{3.74}\hat{K}$$

$$\hat{B} = 0.8\hat{I} + 0.27\hat{J} - 0.53\hat{K}$$

$$|\hat{B}| = \sqrt{(0.8)^2 + (0.27)^2 + (-0.53)^2} = 1.0$$

Vector Addition. The result of vector addition is a vector. To add or subtract vectors you must add or subtract individual like components. That is, for two vectors, $\vec{A}$ and $\vec{B}$,

$$\vec{A} + \vec{B} = (A_I + B_I)\hat{I} + (A_J + B_J)\hat{J} + (A_K + B_K)\hat{K} \qquad \text{(A-4)}$$

Example

Find the sum of two vectors $\vec{A}$ and $\vec{B}$ where

$$\vec{A} = 4\hat{I} - 3\hat{J} + 1\hat{K} \text{ and } \vec{B} = 3\hat{I} + 1\hat{J} - 2\hat{K}$$

$$\vec{A} + \vec{B} = (4+3)\hat{I} + (-3+1)\hat{J} + (1-2)\hat{K} = 7\hat{I} - 2\hat{J} - 1\hat{K}$$

Vector Multiplication. There are two ways to multiply vectors. The first way results in a scalar and is called the scalar or *dot product*. The second way results in a vector and is called the vector or *cross product*.

Scalar or Dot Product. *The result of a scalar or dot product is a scalar quantity.* The dot product of two vectors $\vec{A}$ and $\vec{B}$ lets us multiply the amount of $\vec{A}$ which is in the direction of $\vec{B}$ by the magnitude of $\vec{B}$. Thus, if we have two vectors $\vec{A}$ and $\vec{B}$, we can use trigonometry to find the amount of $\vec{A}$ which is in the direction of $\vec{B}$, is found from trigonometry as shown in Figure A-6.

$$\vec{A} \text{ which is in the direction of } \vec{B} = A\cos\theta$$

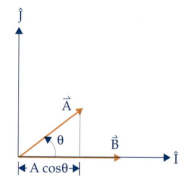

Figure A-6. The Dot Product. The dot product of two vectors, $\vec{A}$ and $\vec{B}$, represents the amount of A in the direction of B. This is also the "projection" of $\vec{A}$ onto $\vec{B}$.

When we multiply this value by the amount of $\vec{B}$ in the direction of $\vec{B}$ (which is simply the magnitude of $\vec{B}$), we get the value for the dot product:

$$\boxed{\vec{A} \cdot \vec{B} = AB\cos\theta} \qquad (A\text{-}5)$$

We can also find this value by multiplying the individual like components of the two vectors and adding the result, as follows:

$$\boxed{\vec{A} \cdot \vec{B} = (A_I B_I) + (A_J B_J) + (A_K B_K)} \qquad (A\text{-}6)$$

By combining these two approaches to finding the dot product, we have a way of determining the angle between two vectors. If we know the components of the two vectors, we can use Equation (A-6) to find the dot product. Then, by rearranging Equation (A-5), we can get a relationship which will tell us the angle between the two vectors.

$$\boxed{\theta = \cos^{-1}\frac{\vec{A} \cdot \vec{B}}{AB}} \qquad (A\text{-}7)$$

Example

Find the angle between the two vectors $\vec{A}$ and $\vec{B}$ where:

$$\vec{A} = 4\hat{I} - 3\hat{J} + 1\hat{K} \text{ and } \vec{B} = 3\hat{I} + 1\hat{J} - 2\hat{K}$$

First we find the dot product of the two vectors using Equation (A-4):

$$\vec{A} \cdot \vec{B} = (4)(3) + (-3)(1) + (1)(-2) = 12 - 3 - 2 = 7$$

Next we need the magnitudes of $\vec{A}$ and $\vec{B}$ which we find using Equation (A-1)

$$A = \sqrt{A_I^2 + A_J^2 + A_K^2} = \sqrt{4^2 + (-3)^2 + 1^2} = \sqrt{16 + 9 + 1} = \sqrt{26} = 5.1$$

$$B = \sqrt{3^2 + 1^2 + (-2)^2} = \sqrt{9 + 1 + 4} = \sqrt{14} = 3.74$$

Now we can use Equation (A-5) to solve for the angle between the two vectors.

$$\theta = \cos^{-1}\frac{\vec{A} \cdot \vec{B}}{AB} = \cos^{-1}\frac{7}{(5.1)(3.74)} = \cos^{-1}(0.37) = 68.5° \text{ (or } 291.5°)$$

(Note the inverse cosine produces two possible results.)

Properties of the Dot Product:

• The dot product is commutative: $\vec{A} \cdot \vec{B} = \vec{B} \cdot \vec{A}$

• The dot product of like vectors is equal to the square of the magnitude of that vector: $\vec{A} \cdot \vec{A} = A^2$. This implies

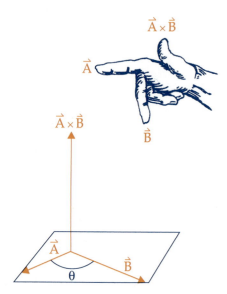

$\vec{A} \times \vec{B}$

$\vec{A}$

$\vec{A} \times \vec{B}$

$\vec{B}$

$\vec{A} \times \vec{B}$

$\vec{A}$

$\vec{B}$

θ

Figure A-7. The Cross Product. The cross product of two vectors $\vec{A}$ and $\vec{B}$ (both in the plane of the page) results in a new vector which is perpendicular to both A and B and thus comes out of the page.

$$\hat{I} \cdot \hat{I} = \hat{J} \cdot \hat{J} = \hat{K} \cdot \vec{K} = 1 \,(\text{recall } \cos 0° = 1)$$

- The dot product of perpendicular vectors is zero. This means

$$\hat{I} \cdot \hat{J} = \hat{J} \cdot \hat{K} = \hat{I} \cdot \vec{K} = 0 \,(\text{recall } \cos 90° = 0)$$

Vector or Cross Product. The vector or cross product of two vectors results in a vector quantity. Assume two vectors lie in the same plane. The cross product of these two vectors forms a third vector which is perpendicular to this plane. We use the "right-hand rule" to find the direction of this third vector. If you point your index finger along the first vector and your middle finger along the second vector, your thumb points in the direction of the resultant vector, as shown in Figure A-7.

We find the cross product by solving the determinant

$$\vec{A} \times \vec{B} = \begin{vmatrix} \hat{I} & \hat{J} & \hat{K} \\ A_I & A_J & A_K \\ B_I & B_J & B_K \end{vmatrix} \quad \text{(A-8)}$$

Then, we can evaluate the solution component by component as follows:

- The $\hat{I}$ component is

$$\begin{vmatrix} \hat{I} & & \\ & A_J & A_K \\ & B_J & B_K \end{vmatrix} = [(A_J)(B_K) - (B_J)(A_K)]\hat{I}$$

- The $\hat{J}$ component is

$$\begin{vmatrix} & \hat{J} & \\ A_I & & A_K \\ B_I & & B_K \end{vmatrix} = -[(A_I)(B_K) - (B_I)(A_K)]\hat{J}$$

[Note the minus sign!]

- The $\hat{K}$ component is

$$\begin{vmatrix} & & \hat{K} \\ A_I & A_J & \\ B_I & B_J & \end{vmatrix} = [(A_I)(B_J) - (B_I)(A_J)]\hat{K}$$

The result is

$$\vec{A} \times \vec{B} = [(A_J)(B_K) - (B_J)(A_K)]\hat{I}$$
$$- [(A_I)(B_K) - (B_I)(A_K)]\hat{J} + [(A_I)(B_J) - (B_I)(A_J)]\hat{K} \quad \text{(A-9)}$$

Example

Find the cross product of two vectors $\vec{A}$ and $\vec{B}$ where

$$\vec{A} = 4\hat{I} - 3\hat{J} + 1\hat{K} \text{ and } \vec{B} = 3\hat{I} + 1\hat{J} - 2\hat{K}$$

First we set up the determinant per Equation (A-8).

$$\vec{A} \times \vec{B} = \begin{vmatrix} \hat{I} & \hat{J} & \hat{K} \\ 4 & -3 & 1 \\ 3 & 1 & -2 \end{vmatrix}$$

Evaluating as in Equation (A-9) we get

$$\vec{A} \times \vec{B} = [\,(-3)\,(-2) - (1)\,(1)\,]\hat{I} - [\,(4)\,(-2) - (3)\,(1)\,]\hat{J}$$
$$+ [\,(4)\,(1) - (3)\,(-3)\,]\hat{K}$$

$$\vec{A} \times \vec{B} = [\,(6) - (1)\,]\hat{I} - [\,(-8) - (3)\,]\hat{J} + [\,(4) - (-9)\,]\hat{K}$$

$$\vec{A} \times \vec{B} = 5\,\hat{I} + 11\,\hat{J} + 13\,\hat{K}$$

If we know the angle between the two vectors, we can also find the magnitude of the cross-product vector. That is

$$\left|\vec{A} \times \vec{B}\right| = AB\sin\theta \qquad (A\text{-}10)$$

This allows us to rewrite the cross product as

$$\vec{A} \times \vec{B} = AB\sin\theta\ \hat{n}$$

where $\hat{n}$ is a unit vector formed by the right-hand rule in the direction of $\vec{A} \times \vec{B}$.

- Properties of the cross product:

  - Cross product is *not* commutative: $\vec{A} \times \vec{B} \neq \vec{B} \times \vec{A}$.
     Note: $(\vec{A} \times \vec{B}) = -(\vec{B} \times \vec{A})$

  - Distributive Law: $\vec{A} \times (\vec{B} + \vec{C}) = \vec{A} \times \vec{B} + \vec{A} \times \vec{C}$

  - Associative Law: $c\,(\vec{A} \times \vec{B}) = (c\vec{A}) \times \vec{B} = \vec{A} \times (c\vec{B})$ (where c is a scalar)

  - Because the angle between parallel vectors is zero,

$$\hat{I} \times \hat{I} = \hat{J} \times \hat{J} = \hat{K} \times \hat{K} = 0 \quad \sin(0°) = 0$$

  - Because the angle between perpendicular vectors is 90°,

$$\hat{I} \times \hat{J} = \hat{K} \qquad \hat{J} \times \hat{I} = -\hat{K} \qquad \text{note that } \sin 90° = 1$$

$$\hat{J} \times \hat{K} = \hat{I} \qquad \hat{K} \times \hat{J} = -\hat{I}$$

$$\hat{K} \times \hat{I} = \hat{J} \qquad \hat{I} \times \hat{K} = -\hat{J}$$

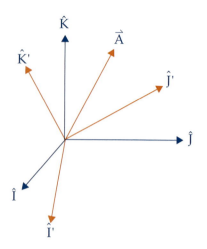

Figure A-8. The vector $\vec{A}$ can be expressed in either the $\hat{I}\,\hat{J}\,\hat{K}$ system or the $\hat{I}'\hat{J}'\hat{K}'$ system. In either case, the vector $\vec{A}$ is the same.

Transforming Vector Coordinates

A vector can be written in different coordinate frames to make writing equations of motion easier. For example, as shown in Figure A-8, $\vec{A}$ can be written as

$$\vec{A} = x\hat{I} + y\hat{J} + z\hat{K} = a\hat{I}' + b\hat{J}' + c\hat{K}'$$

These two descriptions both represent the same vector. Therefore, we need a method of rotating or transforming the descriptions from one frame to another. To do so, we can use

- Positive rotation about the $\hat{I}$ axis through an angle α (ROT1):

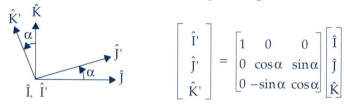

$$\begin{bmatrix} \hat{I}' \\ \hat{J}' \\ \hat{K}' \end{bmatrix} = \begin{bmatrix} 1 & 0 & 0 \\ 0 & \cos\alpha & \sin\alpha \\ 0 & -\sin\alpha & \cos\alpha \end{bmatrix} \begin{bmatrix} \hat{I} \\ \hat{J} \\ \hat{K} \end{bmatrix}$$

- Positive rotation about the $\hat{J}$ axis through an angle β (ROT2):

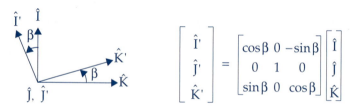

$$\begin{bmatrix} \hat{I}' \\ \hat{J}' \\ \hat{K}' \end{bmatrix} = \begin{bmatrix} \cos\beta & 0 & -\sin\beta \\ 0 & 1 & 0 \\ \sin\beta & 0 & \cos\beta \end{bmatrix} \begin{bmatrix} \hat{I} \\ \hat{J} \\ \hat{K} \end{bmatrix}$$

- Positive rotation about the $\hat{K}$ axis through an angle γ (ROT3):

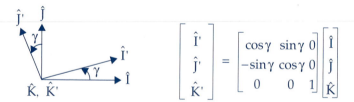

$$\begin{bmatrix} \hat{I}' \\ \hat{J}' \\ \hat{K}' \end{bmatrix} = \begin{bmatrix} \cos\gamma & \sin\gamma & 0 \\ -\sin\gamma & \cos\gamma & 0 \\ 0 & 0 & 1 \end{bmatrix} \begin{bmatrix} \hat{I} \\ \hat{J} \\ \hat{K} \end{bmatrix}$$

- Multiple rotations

    For multiple rotations about two or more axes, we find the total transformation matrix by multiplying the transformation matrices for each axis. For example, for a transformation involving a ROT1 followed by a ROT2, the transformation from the vector $\hat{I}\,\hat{J}\,\hat{K}$ to $\hat{I}''\hat{J}''\hat{K}''$ would be

$$[ROT\,2]\,[ROT\,1] \begin{bmatrix} \hat{I} \\ \hat{J} \\ \hat{K} \end{bmatrix} = \begin{bmatrix} \hat{I}'' \\ \hat{J}'' \\ \hat{K}'' \end{bmatrix}$$

A.3 Calculus

Definitions

Derivative. The *derivative* represents the rate of change of one parameter with respect to another. Calculus was developed to analyze changing parameters. For example, if you're travelling north in your car, your position vector is changing over time. The rate at which your position changes over time is your velocity. Thus, if you go 25 miles north in 30 minutes, your velocity is simply

$$\text{velocity} = \vec{V} = \frac{\text{change in position}}{\text{change in time}} = \frac{\Delta\vec{R}}{\Delta t} = \frac{25 \text{ miles}}{30 \text{ minutes}}$$

$$= 0.83 \left(\frac{\text{mi}}{\text{min}}\right) = 50 \text{ mph north}$$

Note that in mathematics and engineering, we use the Greek letter Δ (delta) to represent a change in something. Here we say that velocity is the derivative of position with respect to time. Likewise, the rate of change of velocity with respect to time (like when you step on the gas) is the derivative of velocity, which is acceleration.

The derivative is really an *instantaneous* rate of change rather than a change over time. For this reason, we use the letter d to represent a change over a tiny amount of time.

$$\text{acceleration} = \vec{a} = \frac{\text{change in velocity}}{\text{change in time}}$$

$$= \lim_{\Delta t \to 0} \frac{\Delta\vec{V}}{\Delta t}$$

$$= \frac{d\vec{V}}{dt} = \dot{\vec{V}} \text{ or } "\vec{V} \text{ dot}"$$

Note: In this text we'll use the "dot" notation to denote a derivative with respect to time. That is

$$dy/dt = \dot{y} \text{ for a first derivative with respect to time and}$$

$$d^2y/dt^2 = \ddot{y} \text{ for a second derivative with respect to time, etc.}$$

Because acceleration is the derivative of velocity and velocity is the derivative of position, we can say that acceleration is the *second* derivative of position.

$$\vec{a} = \frac{d\vec{V}}{dt} = \frac{d}{dt}\left(\frac{d\vec{R}}{dt}\right) = \frac{d^2\vec{R}}{dt^2} = \ddot{\vec{R}}$$

To better illustrate the meaning of the derivative, consider the function which describes the distance an object travels at constant acceleration:

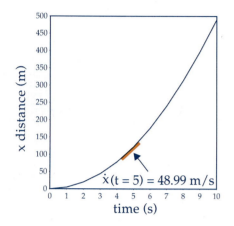

Figure A-9. Distance Over Time. Here we see a plot of the function x = 1/2 (9.798) t^2. The derivative of the function at t = 5 represents the velocity at this time and can be seen as the slope of the curve at that point.

$$x = 1/2\ a\ t^2 \qquad \text{(A-11)}$$

where

x = position at time t

a = acceleration

t = time

If we let a = 9.798 m/s^2, we can graph the position with respect to time as shown in Figure A-9. If we wanted to know the velocity after five seconds, we would take the derivative of the function, which would be

$$\dot{x}\ =\ a\,t\ =\ (9.798\ \text{m/s}^2)\,t \qquad \text{(A-12)}$$

and substitute t = 5 to get

$$\dot{x}\ (t = 5) = 48.99\ \text{m/s} \qquad \text{(A-13)}$$

But the derivative is also the slope of the curve at this point, as you can see in Figure A-9.

Integral. The *integral* represents the cumulative effect of one parameter changing with respect to another. On a graph of one parameter vs. another, the integral is the area under the curve. For example, if you're traveling in a car at some velocity for some amount of time, your change in position would be the integral of velocity over the period. That is, you're adding up all the changes in position over time to get the total change in position. The integral is essentially the reverse of the derivative. Because acceleration is the derivative of velocity over time, we can say that velocity is the integral of acceleration over time.

Velocity = sum of accelerations over time

$$\vec{V}\ =\ \int \vec{a}\,dt\ =\ \int \frac{d\vec{V}}{dt}dt\ =\ \int d\vec{V}$$

If we want to know our velocity after constantly accelerating over some time, we simply integrate the acceleration from the initial time (0) to the final time (t) to get the familiar relationship

$$\vec{V}\ =\ \int_{0}^{t} \vec{a}\,dt\ =\ \vec{a}\,(t - 0) + V_0\ =\ \vec{a}t + V_0$$

Note that the V_0 representing the initial velocity shows up because a constant of integration must always be added to the result of the integration. To get our position after this acceleration we simply integrate our result from above one more time (position is the second integral of acceleration because acceleration is the second derivative of position).

$$\vec{R}\ =\ \int_{0}^{t} (\vec{a}t + V_0)\,dt\ =\ \frac{1}{2}\vec{a}t^2 + V_0 t + R_0$$

One way to look at the integral is as if we were adding together tiny little slices of areas under a curve. If we can approximate each little slice as a rectangle, we can easily find the area of each slice and then add them together to get the entire area under the curve.

Returning to our simple example of a falling object's velocity from Equation (A-12), we can plot this function as shown in Figure A-10. We can then approximate the distance traveled after five seconds using one-second-wide rectangles, as shown under the curve in Figure A-10, to get

$$x \cong (1)(0) + (1)(9.798) + (1)(19.596) + (1)(29.394) + (1)(39.192)$$

$$\cong 97.980 \text{ m}$$

To get the exact distance traveled, we take the integral of Equation (A-12) and solve at $t = 5$ to get

$$x = \int_0^5 \dot{x} \, dt = \int_0^5 (9.798 \text{ m/s}^2) t \, dt = 1/2 \, (9.798 \text{ m/s}^2) \, t^2 \Big|_0^5$$

$$= 4.899 \, (5)^2 - 4.899 \, (0)$$

$$= 122.475 \text{ m}$$

Integration uses narrower and narrower rectangles, until they essentially equal the real curve.

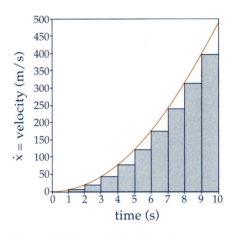

Figure A-10. Integral. The integral represents the cumulative effect of something over time. Here the integral represents the area under the curve found by adding together each little rectangle.

Units and Constants

B

B.1 Canonical Units

When analyzing spacecraft motion across literally astronomical distances, using traditional measures (meters and kilometers or feet and miles) becomes cumbersome. The gravitational parameter of the central body, μ, is also an unwieldy number when expressed in conventional units (3.986005×10^5 km^3/s^2 for example). Therefore, when solving astronautics problems it's sometimes convenient to use a system of normalized units called *canonical units*.

We'll define canonical units for the Earth and the Sun, but we could easily do so for any other central body.

Canonical Units for the Earth

1 Distance Unit (DU) = Radius of the Earth

$$1 \text{ DU} = 6378.137 \text{ km} = 3963.190592 \text{ miles}$$

1 Earth Time Unit (TU$_{Earth}$ or TU$_\oplus$) = time for a satellite in a circular orbit of 1 DU radius (just skimming the Earth's surface) to travel 1 DU (1 radian of arc)

$$1 \text{ TU}_\oplus = 13.44685116 \text{ minutes} = 806.8110953 \text{ s}$$

1 DU/TU$_\oplus$ (Earth speed unit) = speed of a satellite in a circular orbit

$$1 \text{ DU/TU}_\oplus = 7.90536599 \text{ km/s} = 25936.24015 \text{ ft/s}$$

Gravitational Parameter for the Earth

$$\mu_\oplus \equiv 1 \frac{\text{DU}^3}{\text{TU}_\oplus^2} = 3.986005 \times 10^5 \frac{\text{km}^3}{\text{s}^2} = 1.407644381 \times 10^{16} \frac{\text{ft}^3}{\text{s}^2}$$

All data adapted from
Space Mission Analysis and Design
and *Fundamentals of Astrodynamics*

Solar Canonical Units

1 Astronomical Unit (AU) = Mean distance from the Earth to the Sun (radius of Earth's orbit)

$$1 \text{ AU} = 149.59787 \times 10^6 \text{ km} = 4.908125 \times 10^{11} \text{ ft}$$

1 Solar Time Unit (TU_{Sun}) = time for the Earth in its (nearly) circular orbit around the Sun at 1 AU radius to travel 1 AU (1 radian of arc)

$$1 \text{ } TU_{Sun} = 58.132441 \text{ days} = 5.0226429 \times 10^6 \text{ seconds}$$

1 AU/TU_{Sun} (Solar speed unit) = speed of the Earth in its 1 AU radius circular orbit

$$1 \text{ AU}/TU_{Sun} = 29.784692 \text{ km/s} = 9.7719329 \times 10^4 \text{ ft/s}$$

Gravitational Parameter of the Sun

$$\mu_{Sun} \equiv 1 \frac{\text{AU}^3}{TU_{Sun}^2} = 1.327124 \times 10^{11} \frac{\text{km}^3}{\text{s}^2} = 4.6868016 \times 10^{21} \frac{\text{ft}^3}{\text{s}^2}$$

B.2 Unit Conversions

In this book, we use the metric system of units, officially known as the *International System of Units*, or *SI*, except that we sometimes express angular measurements in degrees rather than the SI unit of radians. By international agreement, the fundamental SI units of length, mass, and time are as follows (see National Bureau of Standards Special Publication 330, 1986):

- The *meter* is the length of the path traveled by light in a vacuum over $1/299,792,458$ of a second

- The *kilogram* is the mass of the international prototype of the kilogram

- The *second* is the duration of $9,192,631,770$ periods of the radiation corresponding to the transition between two hyperfine levels of the ground state of the cesium-133 atom

Additional base units in the SI system are the *ampere* for electric current, the *kelvin* for thermodynamic temperature, the *mole* for amount of substance, and the *candela* for intensity of light. Mechtly [1977] neatly summarizes SI units for science and technology.

The names of multiples and submultiples of SI units take the following prefixes:

Factor by which unit is multiplied	Prefix	Symbol
10^{18}	exa	E
10^{15}	peta	P
10^{12}	tera	T
10^{9}	giga	G
10^{6}	mega	M
10^{3}	kilo	k
10^{2}	hecto	h
10	deka	da
10^{-1}	deci	d
10^{-2}	centi	c
10^{-3}	milli	m
10^{-6}	micro	μ
10^{-9}	nano	n
10^{-12}	pico	p
10^{-15}	femto	f
10^{-18}	atto	a

For each quantity listed below, the SI unit and its abbreviation are in brackets. For convenient use in computers, we've listed conversion factors with the greatest available accuracy. Note that some conversions are exact definitions and some (speed of light, astronomical unit) depend on the value of physical constants. All notes are on the last page of the list.

588

To convert from	To	Multiply by	Notes
Acceleration [meters/second², m/s²]			
Foot/second²	m/s²	3.048×10^{-1}	E
Free fall (standard), g	m/s²	9.80665	E
Angular Acceleration [radians/second², rad/s²]			
Degrees/second²	rad/s²	$\pi/180$ $\approx 1.745329251994329577 \times 10^{-2}$	E
Revolutions/second², rev/s²	rad/s²	2π $\approx 6.283185307179586477$	E
Revolutions/minute²	rad/s²	$\pi/1800$ $\approx 1.745329251994329577 \times 10^{-3}$	E
Revolutions/minute²	deg/s²	0.1	E
Radians/second², rad/s²	deg/s²	$180/\pi$ $\approx 5.729577951308232088 \times 10^{1}$	E
Revolutions/second², rev/s²	deg/s²	3.6×10^{2}	E
Angular Measure [radian, rad]. This book uses degree (abbreviated deg) as the basic unit.			
Degree	rad	$\pi/180$ $\approx 1.745329251994329577 \times 10^{-2}$	E
Minute (of arc)	rad	$\pi/10,800$ $\approx 2.908882086657216 \times 10^{-4}$	E
Radian	deg	$= 180/\pi$ $\approx 5.729577951308232088 \times 10^{1}$	E
Second (of arc)	rad	$\pi/648,000$ $\approx 4.8481368110954 \times 10^{-6}$	E
Angular Momentum [kilogram · meter²/second, kg · m²/s]			
Gram · cm²/second	kg · m²/s	1.0×10^{-7}	E
lbm · inch²/second	kg · m²/s	2.926397×10^{-4}	
Slug · inch²/second	kg · m²/s	9.415402×10^{-3}	
lbm · foot²/second	kg · m²/s	4.214011×10^{-2}	
Inch · lbf . second	kg · m²/s	1.129848×10^{-1}	
Slug · foot²/second = ft · lbf · second	kg · m²/s	1.355818	
Angular Velocity [radian/second, rad/s].			
Degrees/second	rad/s	$\pi/180$ $\approx 1.745329251994329577 \times 10^{-2}$	E
Radians/second	deg/s	$180/\pi$ $\approx 5.729577951308232088 \times 10^{1}$	E
Revolutions/minute, rpm	rad/s	$\pi/30$ $\approx 1.047197551196597746 \times 10^{-1}$	E
Revolutions/second	rad/s	2π $\approx 6.283185307179586477$	E
Revolutions/minute, rpm	deg/s	6.0	E
Revolutions/second	deg/s	3.6×10^{2}	E

To convert from	To	Multiply by	Notes
Area [meter^2, m^2]			
Foot^2, ft^2	m^2	9.290304×10^{-2}	E
Inch^2, in^2	m^2	6.4516×10^{-4}	E
Mile^2 (U.S. statute)	m^2	2.589998×10^6	M
Yard^2, yd^2	m^2	8.3612736×10^{-1}	E
(Nautical mile)^2	m^2	3.429904×10^6	E
Density [kilogram/meter^3, kg/m^3]			
g/cm^3	kg/m^3	1.00×10^3	E
lbm/inch^3	kg/m^3	2.7679905×10^4	M
lbm/ft^3	kg/m^3	1.6018463×10^1	M
Slug/ft^3	kg/m^3	5.15379×10^2	M
Energy or Torque [joule $\equiv$ newton $\cdot$ meter $\equiv$ kilogram $\cdot$ meter^2/s^2,			
J $\equiv$ N $\cdot$ m $\equiv$ kg $\cdot$ m^2/s^2]			
British thermal unit, Btu (mean)	J	1.05587×10^3	M
Calorie (mean), cal	J	4.19002	M
Kilocalorie (mean), kcal	J	4.19002×10^3	M
Electron volt, eV	J	$1.60217733 \times 10^{-19}$	C
Erg $\equiv$ gram $\cdot$ cm^2/s^2 = pole $\cdot$ cm $\cdot$ oersted	J	1.0×10^{-7}	E
Foot poundal	J	4.2140110×10^{-2}	M
Foot lbf = slug $\cdot$ foot^2/s^2	J	1.3558179	M
Kilowatt hour, kW $\cdot$ hr	J	3.60×10^6	E
Ton equivalent of TNT	J	4.184×10^9	E
Length [meter, m]			
Angstrom	m	1.00×10^{-10}	E
Astronomical unit, (IAU)	m	$1.49597870 \times 10^{11}$	M
Astronomical unit, (radio)	m	1.4959789×10^{11}	M
Earth equatorial radius	m	6.378140×10^6	AA
Foot	m	3.048×10^{-1}	E
Inch	m	2.54×10^{-2}	E
Light year	m	9.46055×10^{15}	M
Micron	m	1.0×10^{-6}	E
Mil (10^{-3} inch)	m	2.54×10^{-5}	E
Mile (U.S. statute)	m	1.6093×10^3	M
Nautical mile (U.S.)	m	1.852×10^3	E
Parsec (IAU)	m	3.0857×10^{16}	M
Yard	m	9.144×10^{-1}	E
Mass [kilogram, kg]			
Metric ton	kg	1.0×10^3	E
Pound mass, lbm (avoirdupois)	kg	4.5359237×10^{-1}	E
Slug	kg	1.45939029×10^1	M

To convert from	To	Multiply by	Notes
Short ton (2000 lbm)	kg	9.0718474×10^2	E
Solar mass	kg	1.991×10^{30}	H

Force [newton $\equiv$ kilogram $\cdot$ meter/s^2, N $\equiv$ kg $\cdot$ m /s^2]

Dyne	N	1.0×10^{-5}	E
Pound force (avoirdupois),			
lbf $\equiv$ slug $\cdot$ foot/s^2)	N	4.4482216152605	E

Moment of Inertia [kilogram $\cdot$ meter^2, kg $\cdot$ m^2]

Gram $\cdot$ centimeter^2	kg $\cdot$ m^2	1.0×10^{-7}	E
lbm $\cdot$ inch^2	kg $\cdot$ m^2	2.926397×10^{-4}	
lbm $\cdot$ foot^2	kg $\cdot$ m^2	4.214011×10^{-2}	
Slug $\cdot$ inch^2	kg $\cdot$ m^2	9.415402×10^{-3}	
Inch $\cdot$ lbf $\cdot$ s^2	kg $\cdot$ m^2	1.129848×10^{-1}	
Slug $\cdot$ foot^2 = ft $\cdot$ lbf $\cdot$ s^2	kg $\cdot$ m^2	1.355818	

Power [watt $\equiv$ joule/second $\equiv$ kilogram $\cdot$ meter^2/s^3, W $\equiv$ J/s $\equiv$ kg $\cdot$ m^2/s^3]

Foot lbf/second	W	1.355817	M
Horsepower (550 ft lbf/s)	W	7.456998×10^2	M
Horsepower (electrical)	W	7.46×10^2	E
Solar luminosity	W	3.826×10^{26}	W

Pressure or Stress [pascal $\equiv$ newton/meter^2 $\equiv$ kilogram/(meter $\cdot$ second^2), Pa $\equiv$ N/m^2 $\equiv$ kg $\cdot$ m^{-1} $\cdot$ s^{-2}]

Atmosphere	Pa	1.01325×10^5	E
Bar	Pa	1.0×10^5	E
Centimeter of mercury (0°C)	Pa	1.33322×10^3	M
Dyne/centimeter^2	Pa	1.0×10^{-1}	E
Inch of mercury (32°F)	Pa	3.386389×10^3	M
lbf/foot^2	Pa	4.7880258×10^1	M
lbf/inch^2, psi	Pa	6.8947572×10^3	M
Torr (0°C)	Pa	1.33322×10^2	M

Solid Angle [steradian, sr]

Degree^2, deg^2	sr	$(\pi/180)^2$	E
		$\approx 3.046174197867085993 \times 10^{-4}$	
Steradian, Sr	deg^2	$(180/\pi)^2$	E
		$\approx 3.282806350011743794 \times 10^3$	

Stress (see Pressure)

Temperature [kelvin, K]

Celsius, °C	K	$t_K = t_C + 273.15$	E
Fahrenheit, °F	K	$t_K = (5/9)(t_F + 459.67)$	E
Fahrenheit, °F	C	$t_C = (5/9)(t_F - 32.0)$	E

Torque (see Energy)

To convert from	To	Multiply by	Notes
Velocity [meter/second, m/s]			
Foot/minute, ft/min	m/s	5.08×10^{-3}	E
Kilometer/hour, km/hr	m/s	$(3.6)^{-1} = 0.277777...$	E
Foot/second, fps or ft/s	m/s	3.048×10^{-1}	E
Miles/hour, mph	m/s	4.4704×10^{-1}	E
Knot (international)	m/s	$5.144444444 \times 10^{-1}$	M
Miles/minute	m/s	2.68224×10^{1}	E
Miles/second	m/s	1.609344×10^{3}	E
Astronomical unit/ sidereal year	m/s	$4.740388554 \times 10^{3}$	W
Velocity of light, c	m/s	2.99792458×10^{8}	E
Volume [meter³, m³]			
Foot³	m³	$2.8316846592 \times 10^{-2}$	E
Gallon (U.S. liquid)	m³	$3.785411784 \times 10^{-3}$	E
Liter	m³	1.0×10^{-3}	E

NOTES:

E   (*Exact*) indicates that the conversion is exact by definition of the non-SI unit or that we've obtained it from other exact conversions

M   Values from Mechtly

W   Values from Wertz

AA   Values from *Astronomical Almanac*

H   Values from Weast

B.3 Constants

Table B-1. Fundamental Physical Constants.

Quantity	Symbol	Value	Units	Relative Uncertainty (1 σ, ppm)
Speed of light in a vacuum	c	299,792,458	m/s	(exact)
Universal Gravitational Constant	G	6.67259×10^{-11}	$N \cdot m^2/kg^2$	128
Planck constant	h	$6.6260755 \times 10^{-34}$	J·s	0.60
Elementary charge	e	$1.60217733 \times 10^{-19}$	C	0.30
Electron mass	m_e	$9.11093897 \times 10^{-31}$	kg	0.59
Proton mass	m_p	$1.6726231 \times 10^{-27}$	kg	0.59
Proton-electron mass ratio	m_p/m_e	1836.152701	--	0.020
Neutron mass	m_n	$1.6749286 \times 10^{-27}$	kg	0.59
Boltzmann constant, R/NA	k	1.380658×10^{-23}	J/K	8.5
Stefan-Boltzmann Constant	σ	5.67051×10^{-8}	$W \cdot m^{-2} \cdot K^{-4}$	34
Electron volt	eV	$1.60217733 \times 10^{-19}$	J	0.30
Atomic mass unit	u	$1.6605402 \times 10^{-27}$	kg	0.59

Table B-2. Spaceflight Constants.

Quantity	Value	Units
$\mu_{Earth} = G\, m_{Earth}$	3.986005×10^5	km^3/s^2
$\mu_{Sun} = G\, m_{Sun}$	$1.32712438 \times 10^{11}$	km^3/s^2
$\mu_{Moon} = G\, m_{Moon}$	4.902794×10^3	km^3/s^2
$\mu_{Earth\ and\ Moon}$ $= G\, m_{Earth\ and\ Moon}$	4.035033×10^5	km^3/s^2
Obliquity of the ecliptic at Epoch 2000	23.4392911	deg
Precession of the equinox	1.39697128	deg/century
Flattening factor for Earth	1/298.257	
Earth's equatorial radius	6378.137	km
1 AU	1.4959787×10^{11}	m
Mean lunar distance	3.84401×10^8	m
Solar constant	1358	W/m^2 at 1 AU
Acceleration due to gravity at Earth's equatorial radius	9.798	m/s^2
Acceleration due to gravity at standard sea level	9.81	m/s^2
1 solar day	1.00273790935	sidereal days
Earth's rotation rate	15 0.25 0.25068447733746215	deg/sidereal hr deg/sidereal min deg/solar min

B.4 Greek Alphabet

Table B-1. Greek Alphabet. [Adapted from The American Heritage Dictionary]

Symbol	Name	Symbol	Name
A α	alpha	N ν	nu
B β	beta	Ξ ξ	xi
Γ γ	gamma	O o	omicron
Δ δ	delta	Π π	pi
E ε	epsilon	P ρ	rho
Z ζ	zeta	Σ σ	sigma
H η	eta	T τ	tau
Θ θ	theta	Y υ	upsilon
I ι	iota	Φ φ	phi
K κ	kappa	X χ	chi or khi
Λ λ	lambda	Ψ ψ	psi
M μ	mu	Ω ω	omega

▬ References

American Heritage Dictionary. Boston, MA: Houghton Mifflin Company, 1985.

Cohen, E. Richard and Taylor, B.N. *CODATA Bulletin No. 63*, Pergamon Press, Nov. 1986.

Hagen, James B. and Boksenberg, A., eds. *Astronomical Almanac.* Washington, D.C.: U.S. Government Printing Office. 1991.

Mechtly, E. A. *The International System of Units.* Champaign, IL: Stipes Publishing Company, 1977.

Weast, R. C., ed. *CRC Handbook of Chemistry and Physics.* Boca Raton, FL: CRC Press, 1985.

Wertz, James R., ed. *Spacecraft Attitude Determination and Control.* Holland: D. Reidel Publishing Company, 1978.

Wertz, James R., Wiley J. Larson (ed.) *Space Mission Analysis and Design.* Netherlands: Kluwer Academic Publishers, 1991.

Derivations

C.1 Restricted Two-Body Equation of Motion

In Chapter 4 we developed the restricted two-body equation of motion

$$\ddot{\vec{R}} + \frac{\mu}{R^2}\hat{R} = 0 \tag{C-1}$$

knowing that $\hat{R} = \dfrac{\vec{R}}{R}$, we can write it as

$$\ddot{\vec{R}} + \frac{\mu}{R^3}\vec{R} = 0 \tag{C-2}$$

Several of the following derivations will use this fundamental relationship.

C.2 Constants of Motion

Proving Specific Mechanical Energy is Constant

We can prove that the specific mechanical energy of an orbit is constant by beginning with the restricted two-body equation of motion in Equation (C-2). We then take the dot product of both sides with $\dot{\vec{R}}$.

$$\ddot{\vec{R}} + \frac{\mu}{R^3}\vec{R} = 0$$

dot with $\dot{\vec{R}}$

$$\dot{\vec{R}} \cdot \left(\ddot{\vec{R}} + \frac{\mu}{R^3}\vec{R}\right) = \dot{\vec{R}} \cdot 0$$

or

$$\dot{\vec{R}} \cdot \ddot{\vec{R}} + \frac{\mu}{R^3}\vec{R} \cdot \dot{\vec{R}} = 0 \cdot \vec{R} = 0$$

Note: $\dot{\vec{R}} = \vec{V}$ and $\ddot{\vec{R}} = \dot{\vec{V}}$ so

$$\vec{V} \cdot \dot{\vec{V}} + \frac{\mu}{R^3}\vec{R} \cdot \dot{\vec{R}} = 0 \qquad\qquad \text{(C-3)}$$

From the definition of the dot product, we know for any two vectors $\vec{a}$ and $\vec{b}$

$$\vec{a} \cdot \vec{b} = a\, b\, \cos\theta$$

where
  θ = angle between the two vectors

thus

$$\vec{a} \cdot \vec{a} = a\, a\, \cos\theta$$

but $\theta = 0$ because $\vec{a}$ is parallel to $\vec{a}$, so $\cos\theta = 1$. Thus

$$\vec{a} \cdot \vec{a} = a^2 \qquad\qquad \text{(C-4)}$$

Taking the derivative of both sides of Equation (C-4), we get

$$\frac{d}{dt}(\vec{a} \cdot \vec{a}) = \frac{d}{dt}(a^2)$$

Applying the chain rule from calculus

$$\vec{a} \cdot \dot{\vec{a}} + \dot{\vec{a}} \cdot \vec{a} = 2a\,\dot{a}$$

$$2\,(\vec{a} \cdot \dot{\vec{a}}) = 2a\,\dot{a}$$

thus

$$\vec{a} \cdot \dot{\vec{a}} = a\,\dot{a} \tag{C-5}$$

[Note: $|\dot{\vec{a}}| \neq \dot{a}$]

Therefore, we can rewrite Equation (C-3) as

$$V\dot{V} + \frac{\mu}{R^3}R\dot{R} = 0$$

or

$$V\dot{V} + \frac{\mu}{R^2}\dot{R} = 0 \tag{C-6}$$

Now imagine there are two variables x and y such that

$$x = \frac{V^2}{2} \tag{C-7}$$

and

$$y = \frac{-\mu}{R} \tag{C-8}$$

Taking their derivatives, we see

$$\frac{dx}{dt} = V\dot{V}$$

and

$$\frac{dy}{dt} = \frac{\mu}{R^2}\dot{R}$$

Therefore, Equation (C-6) could be written as

$$\frac{dx}{dt} + \frac{dy}{dt} = \frac{d}{dt}(x+y) = 0 \tag{C-9}$$

or

$$\frac{d}{dt}\left(\frac{V^2}{2} - \frac{\mu}{R}\right) = 0 \tag{C-10}$$

From Chapter 4, we know the term in parenthesis is defined to be the specific mechanical energy, ε

$$\varepsilon = \frac{V^2}{2} - \frac{\mu}{R} \tag{C-11}$$

And if $\dfrac{d}{dt}(\varepsilon) = 0$, then

$$\varepsilon = \text{constant}$$

Proving Specific Angular Momentum is Constant

We can prove the specific angular momentum of an orbit is constant by taking the cross product of the two-body equation of motion, Equation (C-2) with the position vector, $\vec{R}$

$$\vec{R} \times \left(\ddot{\vec{R}} + \frac{\mu}{R^3} \vec{R} \right) = \vec{R} \times 0$$

$$\vec{R} \times \ddot{\vec{R}} + \frac{\mu}{R^3} (\vec{R} \times \vec{R}) = 0$$

Because the cross product of parallel vectors is zero, the second term goes to zero, and we're left with

$$\vec{R} \times \ddot{\vec{R}} = 0 \qquad\qquad \text{(C-12)}$$

Now realize that

$$\frac{d}{dt} (\vec{R} \times \dot{\vec{R}}) = (\dot{\vec{R}} \times \dot{\vec{R}}) + (\vec{R} \times \ddot{\vec{R}})$$

where $\dot{\vec{R}} \times \dot{\vec{R}} = 0$, so

$$\frac{d}{dt} (\vec{R} \times \dot{\vec{R}}) = \vec{R} \times \ddot{\vec{R}}$$

Substituting this identity into Equation (C-12), we get

$$\frac{d}{dt} (\vec{R} \times \dot{\vec{R}}) = 0$$

but $\dot{\vec{R}} = \vec{V}$, so we get

$$\frac{d}{dt} (\vec{R} \times \vec{V}) = 0$$

Recall from Chapter 4 that the specific angular momentum is

$$\vec{h} = \vec{R} \times \vec{V}$$

Thus,

$$\frac{d}{dt} (\vec{h}) = 0$$

This implies

$$\vec{h} = \text{constant}$$

C.3 Solving the Two-Body Equation of Motion

Begin with the two-body equation of motion from Equation (C-2)

$$\ddot{\vec{R}} + \frac{\mu}{R^3} \vec{R} = 0 \qquad \text{(C-2)}$$

We can't solve for $\vec{R}$ as a function of time in closed form, but we can find an exact solution using variable substitution. Cross both sides of the equation with the specific angular momentum vector, $\vec{h}$

$$\ddot{\vec{R}} \times \vec{h} + \frac{\mu}{R^3}(\vec{R} \times \vec{h}) = 0 \times \vec{h} = 0$$

using the cross-product identity $\vec{a} \times \vec{b} = -(\vec{b} \times \vec{a})$, we get

$$\ddot{\vec{R}} \times \vec{h} = \frac{\mu}{R^3}(\vec{h} \times \vec{R}) \qquad \text{(C-13)}$$

Beginning with the left-hand side of Equation (C-13), we can show

$$\frac{d}{dt}(\dot{\vec{R}} \times \vec{h}) = \ddot{\vec{R}} \times \vec{h} + \dot{\vec{R}} \times \dot{\vec{h}}$$

But $\vec{h} =$ constant so $\dot{\vec{h}} = 0$, thus

$$\frac{d}{dt}(\dot{\vec{R}} \times \vec{h}) = \ddot{\vec{R}} \times \vec{h} \qquad \text{(C-14)}$$

Now we turn to the right-hand side of Equation (C-13). From the vector identity

$$(\vec{a} \times \vec{b}) \times \vec{c} = \vec{b}(\vec{a} \cdot \vec{c}) - \vec{a}(\vec{b} \cdot \vec{c})$$

we can say

$$\vec{h} \times \vec{R} = (\vec{R} \times \vec{V}) \times \vec{R} = \vec{V}(\vec{R} \cdot \vec{R}) - \vec{R}(\vec{V} \cdot \vec{R})$$

From Equation (C-4), we know $\vec{R} \cdot \vec{R} = R^2$, so

$$\vec{h} \times \vec{R} = \vec{V}R^2 - \vec{R}(\dot{\vec{R}} \cdot \vec{R})$$

From Equation (C-5), we know $\dot{\vec{R}} \cdot \vec{R} = R\dot{R}$, so we get

$$\vec{h} \times \vec{R} = \vec{V}R^2 - \vec{R}R\dot{R}$$

Multiplying by μ/R^3, we get

$$\frac{\mu}{R^3}(\vec{h} \times \vec{R}) = \frac{\mu}{R^3}(\vec{V}R^2 - \vec{R}R\dot{R}) \qquad \text{(C-15)}$$

Now, realize

$$\mu \frac{d}{dt}\left(\frac{\vec{R}}{R}\right) = \mu \frac{d}{dt}(\vec{R}\, R^{-1})$$

$$= \mu\, (\dot{\vec{R}}\, R^{-1} - \vec{R}\, R^{-2}\dot{R}\,)$$

Factor out $1/R^3$

$$\mu \frac{d}{dt}\left(\frac{\vec{R}}{R}\right) = \frac{\mu}{R^3}(\vec{V}R^2 - \vec{R}\, R\, \dot{R}\,) \qquad \text{(C-16)}$$

Equating Equation (C-15) and Equation (C-16), we get

$$\frac{\mu}{R^3}(\vec{h}\times\vec{R}) = \mu \frac{d}{dt}\left(\frac{\vec{R}}{R}\right) \qquad \text{(C-17)}$$

Substituting Equation (C-14) and Equation (C-17) into Equation (C-13), we end up with

$$\frac{d}{dt}(\dot{\vec{R}}\times\vec{h}) = \mu \frac{d}{dt}\left(\frac{\vec{R}}{R}\right)$$

Integrating both sides, we get

$$\dot{\vec{R}}\times\vec{h} = \mu\frac{\vec{R}}{R}+\vec{B} \qquad \text{(C-18)}$$

where
$\vec{B}$ = constant vector of integration

Now dot both sides of Equation (C-18) with $\vec{R}$

$$\vec{R}\cdot(\dot{\vec{R}}\times\vec{h}) = \vec{R}\cdot\left(\mu\frac{\vec{R}}{R}+\vec{B}\right) \qquad \text{(C-19)}$$

From the vector identity

$$\vec{a}\cdot(\vec{b}\times\vec{c}) = (\vec{a}\times\vec{b})\cdot\vec{c}$$

we have

$$\vec{R}\cdot(\dot{\vec{R}}\times\vec{h}) = (\vec{R}\times\dot{\vec{R}}\,)\cdot\vec{h}$$

$$= (\vec{R}\times\vec{V})\cdot\vec{h}$$

$$= \vec{h}\cdot\vec{h}$$

$$= h^2$$

so, Equation (C-19) becomes

$$h^2 = \vec{R}\cdot\left(\mu\frac{\vec{R}}{R}+\vec{B}\right) \qquad \text{(C-20)}$$

Looking at the right-hand side of Equation (C-20)

$$\vec{R} \cdot \left(\mu \frac{\vec{R}}{R} + \vec{B} \right) = \frac{\mu}{R} (\vec{R} \cdot \vec{R}) + (\vec{R} \cdot \vec{B})$$

$$= \frac{\mu}{R} (R^2) + \vec{R} \cdot \vec{B}$$

$$= \mu R + R B \cos v$$

where
v = angle between $\vec{R}$ and $\vec{B}$

Thus, we end up with

$$h^2 = \mu R + R B \cos v$$

Solving for R, we get

$$R = \frac{h^2/\mu}{1 + (B/\mu) \cos v}$$

Now, let $h^2/\mu = k_1$ and $B/\mu = k_2$. So, we have

$$R = \frac{k_1}{1 + k_2 \cos v} \qquad (C\text{-}21)$$

which is the solution of the restricted two-body equation of motion in terms of two constants k_1, k_2, and angle v. This solution derives purely from the dynamics of the problem. From geometry, we know the polar form of the equation for a conic section is also

$$R = \frac{k_1}{1 + k_2 \cos v}$$

where
$k_1 = p$ = semi-latus rectum shown in Figure C-1
$k_2 = e$ = eccentricity

so,

$$R = \frac{p}{1 + e \cos v}$$

Because

$$p = a (1 - e^2)$$

where
a = semi-major axis

we get the familiar solution to the two-body equation of motion which relates dynamics to geometry and shows that all objects moving under the influence of gravity travel along conic sections.

$$\boxed{R = \frac{a (1 - e^2)}{1 + e \cos v}} \qquad (C\text{-}22)$$

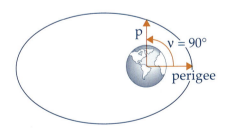

Figure C-1. Semi-Latus Rectum. Semi-latus rectum, p, is defined as the distance from the center of the Earth to the orbit where true anomaly, v, is 90°.

C.4 Relating the Energy Equation to Semi-Major Axis

Given the equation for specific angular momentum

$$\vec{h} = \vec{R} \times \vec{V} = RV\cos\phi$$

apply it at perigee, where flight-path angle $\phi = 0$:

$$h = R_{perigee} V_{perigee}$$

Now recall the relationship for specific mechanical energy

$$\varepsilon = \frac{V^2}{2} - \frac{\mu}{R}$$

Because ε is constant everywhere along an orbit, we can examine ε at perigee. Realize

$$V^2_{perigee} = \frac{h^2}{R^2_{perigee}}$$

Substituting

$$\varepsilon = \frac{h^2}{2R^2_{perigee}} - \frac{\mu}{R_{perigee}} \tag{C-23}$$

From Equation (C-22)

$$R = \frac{a\,(1 - e^2)}{1 + e\cos\nu}$$

at perigee $\nu = 0$, so

$$R_{perigee} = \frac{a\,(1 - e^2)}{1 + e} = \frac{a\,(1 - e)\,(1 + e)}{1 + e}$$

$$R_{perigee} = a\,(1 - e) \tag{C-24}$$

Note that $p = a\,(1 - e^2) = h^2/\mu$. Thus,

$$h^2 = \mu\,a\,(1 - e^2)$$

Substituting into Equation (C-23)

$$\varepsilon = \frac{\mu a\,(1 - e^2)}{2R^2_{perigee}} - \frac{\mu}{R_{perigee}}$$

and from Equation (C-24), we get

$$\varepsilon = \frac{\mu a\,(1 - e^2)}{2a^2\,(1 - e)^2} - \frac{\mu}{a\,(1 - e)}$$

$$= \frac{\mu(1-e^2)}{2a(1-e)^2} - \frac{2\mu(1-e)}{2a(1-e)(1-e)}$$

$$= \frac{\mu(1-e^2) - 2\mu(1-e)}{2a(1-e)^2}$$

$$= \frac{-\mu[2(1-e) - (1-e^2)]}{2a(1-e)^2}$$

$$= \frac{-\mu}{2a}\left[\frac{2-2e-1+e^2}{1-2e+e^2}\right]$$

$$= \frac{-\mu}{2a}\left[\frac{1-2e+e^2}{1-2e+e^2}\right]$$

$$\boxed{\therefore \varepsilon = \frac{-\mu}{2a}}$$

This relationship says that specific mechanical energy is inversely proportional to the orbit's semi-major axis (its size) and vice-versa. It's valid for all conic sections.

C.5 The Eccentricity Vector

In derivating the solution to the two-body equation of motion, we developed a constant vector, $\vec{B}$. We know the magnitude of this vector is related to eccentricity by $B = \mu e$. We can also define an eccentricity vector, $\vec{e}$, in the same direction as $\vec{B}$:

$$\vec{e} = \vec{B}/\mu \tag{C-25}$$

To develop a more useful relationship for $\vec{e}$, we begin with the relationship we developed in Equation (C-18)

$$\dot{\vec{R}} \times \vec{h} = \mu\frac{\vec{R}}{R} + \vec{B} \tag{C-18}$$

solving for $\vec{B}$

$$\vec{B} = (\dot{\vec{R}} \times \vec{h}) - \mu\frac{\vec{R}}{R}$$

Substituting into Equation (C-25)

$$\vec{e} = \frac{\left[(\dot{\vec{R}} \times \vec{h}) - \mu\dfrac{\vec{R}}{R}\right]}{\mu}$$

$$\vec{e} = \frac{\dot{\vec{R}} \times \vec{h}}{\mu} - \frac{\vec{R}}{R} \tag{C-26}$$

Substituting for $\vec{h} = \vec{R} \times \vec{V}$ and $\dot{\vec{R}} = \vec{V}$

$$\vec{e} = \frac{\vec{V} \times (\vec{R} \times \vec{V})}{\mu} - \frac{\vec{R}}{R}$$

Knowing that $\vec{a} \times (\vec{b} \times \vec{c}) = \vec{b}\,(\vec{a} \cdot \vec{c}) - \vec{c}\,(\vec{a} \cdot \vec{b})$, we get

$$\vec{e} = \frac{\vec{R}\,(\vec{V} \cdot \vec{V}) - \vec{V}\,(\vec{V} \cdot \vec{R})}{\mu} - \frac{\vec{R}}{R}$$

From Equation (C-4) $\vec{V} \cdot \vec{V} = V^2$, so

$$\vec{e} = \frac{\vec{R}V^2 - \vec{V}\,(\vec{V} \cdot \vec{R})}{\mu} - \frac{\vec{R}}{R}$$

Multiply by μ

$$\vec{e}\mu = \vec{R}V^2 - \vec{V}\,(\vec{V} \cdot \vec{R}) - \mu\frac{\vec{R}}{R}$$

Arrange terms

$$\grave{e}\mu = \vec{R}\left(V^2 - \frac{\mu}{R}\right) - \vec{V}(\vec{V} \cdot \vec{R})$$

Divide by μ

$$\grave{e} = \frac{1}{\mu}\left[\left(V^2 - \frac{\mu}{R}\right)\vec{R} - (\vec{R} \cdot \vec{V})\vec{V}\right]$$

How do we know what direction $\grave{e}$ points? Beginning with the relationship for $\grave{e}$ in Equation (C-26)

$$\grave{e} = \frac{\dot{\vec{R}} \times \vec{h}}{\mu} - \frac{\vec{R}}{R} \qquad \text{(C-27)}$$

we can express $\vec{R}$ and $\vec{V}$ in the perifocal coordinate system, $\hat{P}\hat{Q}\hat{W}$ shown in Figure C-2, where

- Origin is the Earth's center
- Fundamental plane is the orbit plane
- Principle direction is perigee

$\vec{R}$ and $\vec{V}$ then become

$$\vec{R} = R\cos v\hat{P} + R\sin v\hat{Q}$$

$$\vec{V} = \sqrt{\frac{\mu}{p}}[-\sin v\hat{P} + (e + \cos v)\hat{Q}]$$

In terms of $\vec{R} \times \vec{V}$, we start by solving for $\vec{h}$

$$\vec{h} = \vec{R} \times \vec{V} = \begin{vmatrix} \hat{P} & \hat{Q} & \hat{W} \\ R\cos v & R\sin v & 0 \\ -\sqrt{\frac{\mu}{p}}\sin v & \sqrt{\frac{\mu}{p}}(e + \cos v) & 0 \end{vmatrix}$$

$$= \sqrt{\frac{\mu}{p}}[R\cos v((e + \cos v) + R\sin^2 v)]\hat{W}$$

$$= \sqrt{\frac{\mu}{p}}[eR\cos v + R\cos^2 v + R\sin^2 v]\hat{W}$$

Applying the trigonometric identity $\cos^2\theta + \sin^2\theta = 1$

$$\vec{h} = \sqrt{\frac{\mu}{p}}[eR\cos v + R]\hat{W}$$

$$\vec{h} = \sqrt{\frac{\mu}{p}}[R(1 + e\cos v)]\hat{W}$$

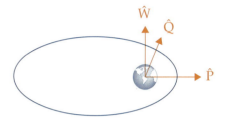

Figure   C-2. $\hat{P} - \hat{Q} - \hat{W}$ System.   The $\hat{P} - \hat{Q} - \hat{W}$ system has its origin at the center of the Earth. $\hat{P}$ points in the direction of the eccentricity vector, $\hat{W}$ is perpendicular to the orbit plane in the direction of the angular momentum vector, and $\hat{Q}$ completes the right-hand rule.

Now look at the first cross product in Equation (C-26)

$$\vec{R} \times \vec{h} = \vec{V} \times \vec{h} = \begin{vmatrix} \hat{P} & \hat{Q} & \hat{W} \\ \sqrt{\dfrac{\mu}{p}}(-\sin v) & \sqrt{\dfrac{\mu}{p}}(e+\cos v) & 0 \\ 0 & 0 & \sqrt{\dfrac{\mu}{p}}[R(1+e\cos v)] \end{vmatrix}$$

$$= \left(\sqrt{\dfrac{\mu}{p}}\right)^2 \{ [R(e+\cos v)(1+e\cos v)]\hat{P} - [(-\sin v)(R)(1+e\cos v)]\hat{Q}\}$$

From the solution to the two-body equation of motion in Equation (C-22) and the definition of semi-latus rectum, p

$$p = R(1+e\cos v)$$

Substituting

$$\vec{V} \times \vec{h} = \dfrac{\mu}{p}\{[(e+\cos v)p]\hat{P} + [p\sin v]\hat{Q}\}$$

$$\vec{V} \times \vec{h} = \mu(e+\cos v)\hat{P} + \mu\sin v\hat{Q}$$

Substituting into Equation (C-26)

$$\grave{e} = \dfrac{\mu(e+\cos v)\hat{P} + \mu\sin v\hat{Q}}{\mu} - \dfrac{R\cos v\hat{P} + R\sin v\hat{Q}}{R}$$

$$= (e+\cos v)\hat{P} + \sin v\hat{Q} - \cos v\hat{P} - \sin v\hat{Q}$$

$$\grave{e} = e\hat{P} + \cos v\hat{P} + \sin v\hat{Q} - \cos v\hat{P} - \sin v\hat{Q}$$

$$\grave{e} = e\hat{P}$$

Therefore, $\grave{e}$ points at perigee.

C.6 Deriving the Period Equation for an Elliptical Orbit

The orbit geometry in Figure C-3 shows the horizontal component of velocity is

$$V \cos \phi \text{ or simply } R\dot{v}$$

Knowing $|\vec{h}| = |\vec{R} \times \vec{V}| = RV\cos\phi$, we can express h as

$$h = \frac{R^2 dv}{dt} \Rightarrow dt = \frac{R^2}{h} dv$$

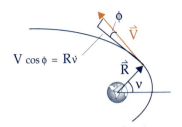

Figure C-3. Components of $\vec{V}$. We can break out the components of the velocity vector, $\vec{V}$, as shown.

From elementary calculus we know that the differential element of area, dA, as shown in Figure C-4, swept out by the radius vector as it moves through an angle, dv, is given by the expression

$$dA = \frac{1}{2} R(R dv) = \frac{1}{2} R^2 dv$$

So, we can rewrite the above equation as

$$dt = \frac{2}{h} dA$$

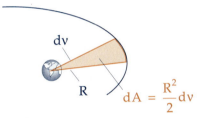

Figure C-4. A satellite sweeps out a small area, dA, per unit time.

which proves Kepler's Second Law that "equal areas are swept out by the radius vector in equal time intervals" because h is constant for an orbit. Integrating this equation through one period yields the following:

$$P = \frac{2\pi ab}{h}$$

where
P   = period
πab = total area of an ellipse

Using $b = \sqrt{a^2 - c^2} = \sqrt{a^2(1-e^2)} = \sqrt{ap}$ and because $h = \sqrt{\mu p}$

$$P = \frac{2\pi a \sqrt{ap}}{\sqrt{\mu p}} = \frac{2\pi}{\sqrt{\mu}} a^{3/2} = 2\pi \sqrt{\frac{a^3}{\mu}}$$

C.7 Finding Position and Velocity Vectors from COEs

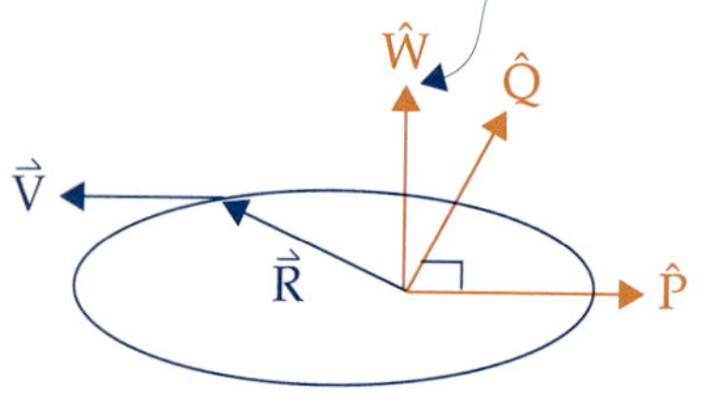

We start by defining a new coordinate system, the *Perifocal System (PQW)*, as shown in Figure C-5

- Origin is the Earth's center
- Fundamental plane is the orbit plane
- Principal direction is perigee

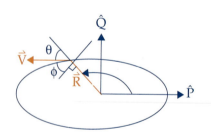

Figure C-5. The Perifocal Coordinate System.

Writing $\vec{R}$ and $\vec{V}$ in terms of $\hat{P}\ \hat{Q}\ \hat{W}$

$$\vec{R} = R\cos v \hat{P} + R\sin v \hat{Q}$$

$$\vec{V} = \frac{d\vec{R}}{dt} = (\dot{R}\cos v - R\dot{v}\sin v)\hat{P} + (\dot{R}\sin v + R\dot{v}\cos v)\hat{Q}$$

What about $\dot{\hat{P}}$ and $\dot{\hat{Q}}$? Answer: PQW is an inertial reference frame, so they equal zero. Although we know R and v, we don't know $\dot{R}$ and $\dot{v}$. From the solution to the two-body equation of motion

$$R = \frac{p}{1 + e\cos v} = p(1 + e\cos v)^{-1}$$

$$\dot{R} = -p(-e\dot{v}\sin v)(1 + e\cos v)^{-2} = \frac{pe\dot{v}\sin v}{(1 + e\cos v)^2}$$

To find $\dot{v}$, we must look at orbital geometry:

$$|\vec{h}| = |\vec{R} \times \vec{V}| = RV\sin\theta = RV\cos\phi$$

The component of $\vec{V}$ normal to $\vec{R}$ is R R$\dot{v}$ (tangential velocity)

$$R\dot{v} = V\cos\phi$$

so,

$$h = R^2\dot{v}$$

Therefore,

$$\dot{v} = \frac{h}{R^2}$$

Figure C-6. $\vec{R}$ and $\vec{V}$ in the Perifocal System.

Substituting into the equation for $\dot{R}$

$$\dot{R} = \frac{p\ e\ h\sin v}{R^2(1 + e\cos v)^2} \text{ but } R = \frac{p}{1 + e\cos v}$$

so

$$\dot{R} = \frac{pe \, \sin v \, (1 + e\cos v)^2}{p^2 (1 + e\cos v)^2} = \frac{e\sqrt{\mu\rho} \, \sin v}{p}$$

$$\dot{R} = \sqrt{\frac{\mu}{p}} e \sin v$$

Now

$$R\dot{v} = \frac{p}{1 + e\cos v} \frac{h}{R^2} = \frac{ph \, (1 + e\cos v)^2}{p^2 (1 + e\cos v)} = \frac{\sqrt{\mu p} \, (1 + e\cos v)}{p}$$

$$R\dot{v} = \sqrt{\frac{\mu}{p}} (1 + e\cos v)$$

Going back to $\vec{V}$

$$\vec{V} = \left[\sqrt{\frac{\mu}{p}} e \sin v \cos v - \sqrt{\frac{\mu}{p}} (1 + e\cos v) \sin v \right] \hat{P}$$

$$+ \left[\sqrt{\frac{\mu}{p}} e \sin v \sin v + \sqrt{\frac{\mu}{p}} (1 + e\cos v) \cos v \right] \hat{Q}$$

$$\vec{V} = \sqrt{\frac{\mu}{p}} (e \sin v \cos v - \sin v - e \cos v \sin v) \hat{P}$$

$$+ \sqrt{\frac{\mu}{p}} (e \sin^2 v + \cos v + e \cos^2 v) \hat{Q}$$

Finally, we can write $\vec{R}$, $\vec{V}$ in the perifocal coordinate frame as

$$\vec{R}_{PQW} = \frac{a \, (1 - e^2)}{1 + e\cos v} [\cos v \hat{P} + \sin v \hat{Q}]$$

$$\vec{V}_{PQW} = \sqrt{\frac{\mu}{p}} [-\sin v \hat{P} + (e + \cos v) \hat{Q}]$$

which are the position and velocity vectors entirely in terms of the Classical Orbital Elements (COEs).

The next step in this problem is to transform the coordinates of the $\vec{R}$ and $\vec{V}$ vectors from the $\hat{P} - \hat{Q} - \hat{W}$ system to the $\hat{I} - \hat{J} - \hat{K}$ system. This will require three separate transformation matrices using the remaining COEs, i, ω, and Ω. To get the $\hat{P} - \hat{Q} - \hat{W}$ system into the $\hat{I} - \hat{J} - \hat{K}$, we begin with a rotation about the $\hat{W}$ axis (an ROT3) through a negative argument of perigee angle, $-\omega$, to bring $\hat{P}$ into the equatorial plane as shown in Figure C-7. The matrix for this operation is

$$\text{ROT3} \, (-\omega) = \begin{bmatrix} \cos\omega & -\sin\omega & 0 \\ \sin\omega & \cos\omega & 0 \\ 0 & 0 & 1 \end{bmatrix}$$

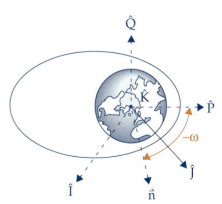

Figure C-7. Rotation 3 of $-\omega$ about $\hat{W}$. (Note: $\hat{W}$ is out of the page.)

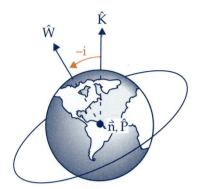

Figure C-8. Rotation 1 of –i about new $\hat{P}$ or $\mathring{n}$.

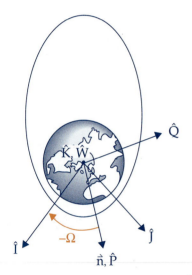

Figure C-9. Rotation 3 of – Ω about new $\hat{W}$ or $\hat{K}$.

This rotation aligns $\hat{P}$ to the ascending node vector, $\mathring{n}$ Next, we rotate about this new $\hat{P}/\mathring{n}$ axis through an angle of minus inclination, –i, to bring $\hat{Q}$ to the equatorial plane, as shown in Figure C-8, to get an ROT1 matrix

$$ROT1\ (-i)\ =\ \begin{bmatrix} 1 & 0 & 0 \\ 0 & \cos i & -\sin i \\ 0 & \sin i & \cos i \end{bmatrix}$$

This rotation aligns $\hat{W}$ to $\hat{K}$. Finally, we rotate about the $\hat{W}/\hat{K}$ axis through a negative longitude of ascending node angle, – Ω, to align $\hat{P}$ with $\hat{I}$ and $\hat{Q}$ with $\hat{J}$, as shown in Figure C-9 to get

$$ROT3\ (-\Omega)\ =\ \begin{bmatrix} \cos\Omega & -\sin\Omega & 0 \\ \sin\Omega & \cos\Omega & 0 \\ 0 & 0 & 1 \end{bmatrix}$$

Putting it all together we have

$$\vec{R}_{IJK}\ =\ [A]\,\vec{R}_{PQW}$$

$$\vec{V}_{IJK}\ =\ [A]\,\vec{V}_{PQW}$$

where

$\vec{R}_{IJK},\ \vec{R}_{PQW}$ = position vectors in $\hat{I}-\hat{J}-\hat{K}$ and $\hat{P}-\hat{Q}-\hat{W}$ systems

$\vec{V}_{IJK},\ \vec{V}_{PQW}$ = velocity vectors in $\hat{I}-\hat{J}-\hat{K}$ and $\hat{P}-\hat{Q}-\hat{W}$ systems

$[A]$ = transformation matrix from $\hat{P}-\hat{Q}-\hat{W}$ to $\hat{I}-\hat{J}-\hat{K}$

$$[ROT3\ (-\Omega)\,]\,[ROT1\ (-i)\,]\,[ROT3\ (-\omega)\,]\ =\ \begin{bmatrix} A_{11} & A_{12} & A_{13} \\ A_{21} & A_{22} & A_{23} \\ A_{31} & A_{32} & A_{33} \end{bmatrix}$$

where
$A_{11} = \cos\Omega\,\cos\omega - \sin\Omega\,\sin\omega\,\cos i$
$A_{12} = -\cos\Omega\,\sin\omega - \sin\Omega\,\cos\omega\,\cos i$
$A_{13} = \sin\Omega\,\sin i$
$A_{21} = \sin\Omega\,\cos\omega + \cos\Omega\,\sin\omega\,\cos i$
$A_{22} = -\sin\Omega\,\sin\omega + \cos\Omega\,\cos\omega\,\cos i$
$A_{23} = -\cos\Omega\,\sin i$
$A_{31} = \sin\omega\,\sin i$
$A_{32} = \cos\omega\,\sin i$
$A_{33} = \cos i$

C.8 V_{needed} in SEZ Coordinates

Using the flight-path angle, ϕ, and the launch azimuth angle, β, we can derive the components for the burnout velocity, $\vec{V}_{burnout}$, as shown in Figure C-10. We measure ϕ from the horizon to the velocity vector and β clockwise from due North to the projection of the velocity on the horizon plane. For the zenith component

$$V_{burnout_{zenith}} = V_{burnout} \sin\phi$$

To get the south and east components, we must project the magnitude of the burnout velocity into the horizon plane again using the flight-path angle

$$V_{burnout_{South and East}} = V_{burnout} \cos\phi$$

Then, using the azimuth angle, we get the East component

$$V_{burnout_{East}} = V_{burnout} \cos\phi \sin(180° - \beta)$$

We can simplify using the trigonometric identity

$$\sin(180° - \beta) = \sin\beta$$

$$V_{burnout_{East}} = V_{burnout} \cos\phi \sin\beta$$

And for the south component

$$V_{burnout_{South}} = V_{burnout} \cos\phi \cos(180° - \beta)$$

Then simplifying with the trigonometric identity

$$\cos(180° - \beta) = -\cos\beta$$

$$V_{burnout_{South}} = -V_{burnout} \cos\phi \cos\beta$$

Substituting in these component values for the burnout velocity, we can get the pieces of the needed velocity vector:

$$V_{needed_{South}} = V_{burnout_{South}} = -V_{burnout} \cos\phi \cos\beta$$

$$V_{needed_{East}} = V_{burnout}^{East} - V_{launch\,site} = V_{burnout} \cos\phi \sin\beta - \left(465.1 \frac{m}{s}\right) \cos L_o$$

$$V_{needed_{zenith}} = V_{burnout_{zenith}} = V_{burnout} \sin\phi$$

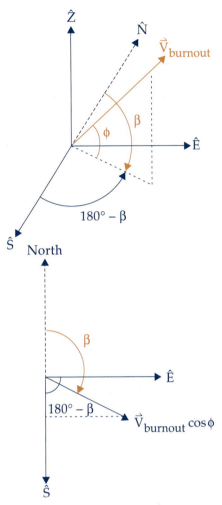

Figure C-10. We can analyze the geometry of the launch site with respect to known burnout conditions such as flight-path angle, ϕ, and launch azimuth, β, to determine what our burnout velocity vector should be.

C.9 Deriving the Rocket Equation

Newton's Second Law states

$$\sum F_{ext} = d\frac{(mV)}{dt} = d\frac{(p)}{dt}$$

where
$$p = mV$$

Consider a rocket expelling mass at a constant velocity, V_{exit}, relative to the vehicle. If we ignore gravity, drag, and air pressure, no external forces acting on the body.

$$\therefore \sum F_{ext} = 0 = d\frac{(p)}{dt}$$

But the real question is, "what is the momentum of the rocket?"

$$p(t) = \underbrace{m(t)V(t)}_{(1)} + \underbrace{\int_o^t dm(t)(V(t) - V_{exit})}_{(2)}$$

(1)   = current momentum of the rocket

(2)   · = sum of the momentum associated with all of the mass ejected as fuel from the rocket from time zero to the present

If we assume the rocket shown in Figure C-11 is burning fuel at some rate, $\dot{m}$, the time rate of change of the vehicle mass is $\dot{m} < 0$ (negative because the vehicle is losing mass). But $dm(t) = -\dot{m}\,dt > 0$ (the mass of the fuel expelled by the rocket is increasing with time).

$$\therefore p(t) = m(t)V(t) - \int_0^t \dot{m}(V(t) - V_{exit})\,dt$$

Differentiating

$$\frac{dp(t)}{dt} = \cancel{\dot{m}(t)V(t)} + m(t)\dot{V}(t) - \cancel{\dot{m}(t)V(t)} + \dot{m}(t)V_{exit} = 0$$

Notice $\dot{m}(t)V(t)$ terms cancel

$$\therefore m(t)\dot{V}(t) = -\dot{m}(t)V_{exit}$$

$$\dot{V}(t) = -\frac{\dot{m}(t)}{m(t)}V_{exit} = -\frac{\dot{m}}{m}V_{exit}$$

Integrating both sides

$$\int dV = \int \frac{-dm}{m}V_{exit}$$

$$\therefore \Delta V = -V_{exit}\,\ln m\,]_{m_o}^{mf}$$

$$\Delta V = V_{exit}\,\ln\frac{m_o}{m_f}$$

From Chapter 14, $V_{exit} = c$, so

$$\boxed{\Delta V = c\ln\left(\frac{m_o}{m_f}\right)}$$

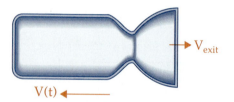

Figure C-11. Absolute velocity. The absolute velocity of the ejected mass, dm (t) is V(t) − V_{exit}.

Solar and Planetary Data

D

D.1 Physical Properties of the Sun

Table D-1. Physical Properties of the Sun. [Larson & Wertz, *Space Mission Analysis and Design*, 1992.]

Quantity	Value
Radius of the photosphere	6.9599×10^5 km
Angular diameter of the photosphere at 1 AU	0.53313 deg
Mass	1.9891×10^{30} kg
Mean density	1.410 g/cm^3
Total radiation emitted	3.826×10^{26} J/s
Total radiation per unit area at 1 AU	1358 W/m^2
Apparent visual magnitude at 1 AU	−26.74
Absolute visual magnitude (magnitude at distance of 10 parsecs)	+4.83
Color index, B-V	+0.65
Spectral type	G2 V
Effective temperature	5770 K
Inclination of the equator to the ecliptic	7.25 deg
Adopted period of sidereal rotation (L = 17°)	25.38 days
Corresponding synodic rotation period (relative to the Earth)	27.275 days
Mean sunspot period	11.04 years
Dates of former maxima	1968.9, 1980.0
Mean time from maximum to subsequent minimum	6.2 years

D.2 Physical Properties of the Earth

Table D-2. Physical Properties of the Earth. [Larson & Wertz, *Space Mission Analysis and Design*, 1992.]

Quantity	Value
Equatorial radius, a	6378.137 km
Flattening factor (ellipticity), $f \equiv (a - c) / a$	$1/298.257 \approx 0.00335281$
Polar radius,* c	6356.755 km
Mean radius,* $(a^2 c)^{1/3}$	6371.00 km
Eccentricity,* $(a^2 - c^2)^{1/2}/a$	0.0818192
Surface area	5.100645×10^8 km^2
Volume	1.08321×10^{12} km^3
Ellipticity of the equator $(a_{max} - a_{min})/a_{mean}$	$\sim 1.6 \times 10^{-5}$
Longitude of the maxima	20° W, 160°E
Ratio of the mass of the Sun to the mass of the Earth	332946.0
Gravitational parameter, $Gm_{Earth} \equiv \mu_{Earth}$	3.986005×10^{14} m^3/s^2
Mass of the Earth	5.9742×10^{24} kg
Mean density	5.15 g/cm^3
Gravitational field constants	$J2 = +1.08263 \times 10^{-3}$ $J3 = -2.54 \times 10^{-6}$ $J4 = -1.61 \times 10^{-6}$
Mean distance of Earth center from Earth-Moon barycenter	4671 km
Average lengthening of the day	0.0015 s/century

* Based on adopted values of f and a.

D.3 Physical Properties of the Moon

Table D-3. Physical Properties of the Moon. [Adapted from *The Astronomical Almanac*, Nautical Almanac Office, U.S. Naval Observatory, Government printing office, 1990, except where noted.]

Quantity	Value
Equatorial radius	1738 km
Surface area	37.9×10^6 km^2 [*]
Ratio of the mass of the Moon to the mass of the Earth	0.00123
Mass of the Moon	7.3483×10^{22} kg
Mean density	3.34 s/cm^3 [*]
Gravitational field parameters	J2 = $+0.2027 \times 10^{-3}$
Semi-major axis of lunar orbit	384,400 km
Gravitational parameter, $Gm_{Moon} \equiv \mu_{Moon}$	4.902794×10^3 km^3/s^2 [**]
Sidereal orbit period	27.321661 solar days
Sidereal rotation period	27.321661 solar days
Orbit eccentricity	0.054900489
Orbit inclination with respect to Earth's equator	$18.28° - 28.58°$

[*] Heiken, Grant H., David T. Vaniman, Bevan M. French. *Lunar Sourcebook.* Cambridge, U.K.: Cambridge University Press, 1991.

[**] Wertz, James R. and Wiley J. Larson (ed.). *Space Mission Analysis and Design.* Dordrecht, Netherlands: Kluwer Academic Publishers, 1991.

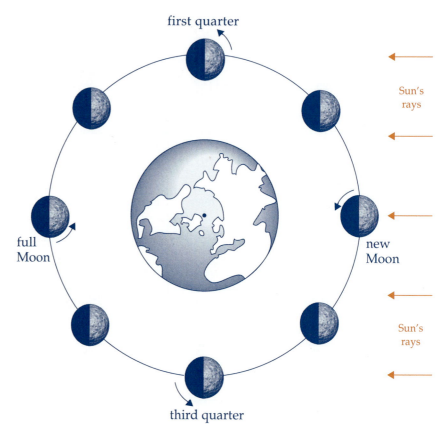

Figure D-1. Revolution of the Moon Around the Earth.

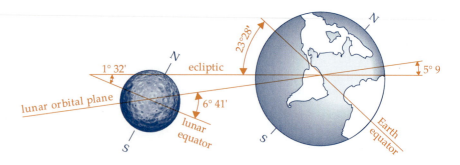

Figure D-2. Relationship Between Inclinations of the Earth and Moon.

D.4 Planetary Data

Table D-4. Planetary Data. [Adapted from *McGraw-Hill Encyclopedia of Science and Technology*. 7th ed., Vol. 13. New York, NY: McGraw-Hill, Inc., 1992. Note: 1 AU = 149.6×10^6 km.

Planet	Diameter (km)	Mass (kg)	Distance from Sun (AU)	Orbit Period Years	Orbit Eccen-tricity	Orbit Inclin-ation	Atmos-phere	Surface Gravity Earth=1 g	Gravita-tional Parameter μ (km^3/s^2)
Mercury	4878	3.169×10^{23}	0.387	0.241	0.206	7.0	None	0.352	2.094×10^4
Venus	12,102	4.87×10^{24}	0.723	0.615	0.007	3.39	CO_2	0.8874	3.249×10^5
Earth	12,756	5.98×10^{24}	1.0	1.0	0.017	0	$N_2 + O_2$	1.0	3.986×10^5
Mars	6790	6.399×10^{23}	1.524	1.881	0.093	1.85	CO_2	0.37	4.269×10^4
Jupiter	142,984	1.90×10^{27}	5.203	11.862	0.048	1.30	$H_2 + H_e$	N/A	1.267×10^8
Saturn	120,536	5.69×10^{26}	9.539	29.458	0.056	2.49	$CH_4 + NH_3$	N/A	3.7967×10^7
Uranus	51,118	8.68×10^{25}	19.19	84.014	0.046	0.77	$H_2 + H_e$	N/A	5.7918×10^6
Neptune	49,600	1.021×10^{26}	30.06	164.79	0.009	1.77	$H_2 + H_e$	N/A	6.806×10^6
Pluto	~2300	$\sim 1.196 \times 10^{22}$	39.53	247.7	0.25	17.2	thin CH_4	0.0603	798.04

D.5 Spheres of Influence for the Planets

Table D-5. Sphere of Influence Radii for the Planets. [Computed using method from Chapter 7.]

Planet	Distance from Sun ($\times 10^6$ km)	Mass of Planet ($\times 10^{24}$ kg)	Radius of SOI ($\times 10^3$ km)
Mercury	57.9	0.3169	111
Venus	108.1	4.87	616
Earth	149.6	5.98	925
Mars	227.9	0.6399	576
Jupiter	778	1900	48,200
Saturn	1427	569	54,600
Uranus	2870	86.8	51,700
Neptune	4497	102.1	86,500
Pluto	5913	0.0196	3710

Table D-6. Synodic Periods for Missions from Earth to the Planets.

Planet	Synodic Period (Years)
Mercury	0.32
Venus	1.60
Mars	2.13
Jupiter	1.09
Saturn	1.04
Uranus	1.01
Neptune	1.01
Pluto	1.00

References

Allen, C. W. *Astrophysical Quantities*. Third Edition. London, England: The Athlene Press, 1973.

American Ephemeris and Nautical Almanac. London, England: Her Majesty's Stationery Office, 1961.

H. M. Nautical Almanac Office. *Explanatory Supplement to the Astronomical Ephemeris*. London, England: Her Majesty's Stationery Office, 1961.

Hagen, James B. and Boksenberg, A., eds. *The Astronomical Almanac, 1992*. Nautical Almanac Office, U.S. Naval Observatory, and H. M. Nautical Almanac Office. Washington, DC: U.S. Government Printing Office, 1991.

Hartman, William K. *Moon and Planets*. Belmont, CA: Wadsworth, Inc., 1983.

Hedgley, David R., Jr. *An Exact Transformation from Geocentric to Geodetic Coordinates for Nonzero Altitudes*. NASA TRR-458, Flight Research Center, 1976.

Hedman, Edward L., Jr. A High Accuracy Relationship Between Geocentric Cartesian coordinates and Geodetic Latitude and Altitude. *J. Spacecraft*. 7: 993-995, 1970.

Heiken, Grant, David Vaniman, and Bevan M. French. *Lunar Sourcebook*. Cambridge, U.K.: Cambridge University Press, 1991.

Muller, Edith A. and Jappel Arndst, eds. *International Astronomical Union Proceedings of the Sixteenth General Assembly, Grenoble, 1976*. Dordrecht, Holland: D. Reidel Publishing Co., 1977.

Wertz, James R. and Wiley J. Larson (ed.) *Space Mission Analysis and Design*. Netherlands: Kluwer Academic Publishers Group, 1992.

Booster Systems

Table E-1. Booster Systems. (Isakowitz, Steven J., *International Reference Guide to Space Launch Systems.* American Institute of Aeronautics and Astronautics. 1991.)

System	Length/Mass at Lift-Off	Stages	Engines	Payloads
Atlas II USA	47.5 m (156 ft) 1870,000 kg (413,000 lb)	1st: 2 boosters + 1 sustainer 2nd:1 sustainer 3rd:2 Centaur RL-10A-3-3A	• **Booster** - liquid: LOX/RP1 - $F_{thrust} = 1.84 \times 10^6$ N (sl) - $I_{sp} = 261$ s (sl) • **Sustainer** - liquid: LOX/RP1 - $F_{thrust} = 269,000$ N (60,500 lb) (sl) - $I_{sp} = 220$ s (sl) • **Centaur** - liquid: LOX/LH2 - $F_{thrust} = 147,000$ N(33,000 lb) (vac) - $I_{sp} = 442$ s	LEO: 6395 kg (14,100 lb) GTO: 2680 kg (5900 lb)
Pegasus USA	15.5 m (50.9 ft) 19,000 kg (42,000 lb)	0: air launch 1st: 1 solid rocket motor (SRM) 2nd: 1 SRM 3rd: 1 SRM	Air launch from B-52 or other aircraft at an altitude of 12.2 km (40,000 ft) • **SRM Stage 1** - Solid: HTPB - $F_{thrust} = 486,700$ N (109,420 lb)(vac) - $I_{sp} = 295$ s (vac) • **SRM Stage 2** - Solid: HTPB - $F_{thrust} = 122,800$ N (27,600 lb) (vac) - $I_{sp} = 295$ s (vac) • **SRM Stage 3** - Solid: HTPB - $F_{thrust} = 34,560$ N (7770 lb) (vac) - $I_{sp} = 291$ s (vac)	LEO: 455 kg (1000 lb) GTO: 125 kg (275 lb) + PKM

sl = sea level
HTPB = hydroxyl-terminated polybutadiene
vac = vacuum

LEO = low-Earth orbit
GTO = geosynchronous transfer orbit
PKM = perigee kick motor

Table E-1. Booster Systems.(Continued) [Isakowitz, Steven J., *International Reference Guide to Space Launch Systems*. American Institute of Aeronautics and Astronautics. 1991.]

System	Length/Mass at Lift-Off	Stages	Engines	Payloads

Ariane 44L ESA	58.4 m (192 ft) 470,000 kg (1,040,000 lb)	1st: 4 Viking 6 + 4 Viking SC
		2nd: 1 Viking 4B
		3rd: 1 HM 7B
			• **Viking 6**
			 - Liquid: N_2O_4/UH25
			 - F_{thrust} = 667,000 N (150,000 lb) (vac)
			   = 737,000 N (167,000 lb) (sl)
			 - I_{sp} = 278 s (vac) = 248 s (sl)
			• **Viking 5C**
			 - Liquid: N_2O_4/UH25
			 - F_{thrust} = 758,500 N (171,000 lb) (vac)
			   = 676,900 N (152,000 lb) (sl)
			 - I_{sp} = 278 s (vac) = 248 s (sl)
			• **Viking 4B**
			 - Liquid N_2O_4/UH25
			 - F_{thrust} = 785,000 N (177,000 lb) (vac)
			 - I_{sp} = 293 s (vac)
			• **HM7B**
			 - Liquid: LOX/LH2
			 - F_{thrust} = 62,700 N (14,100 lb) (vac)
			 - I_{sp} = 444 s (vac)
				LEO: 9600 kg (21,000 lb)
				GTO: 4200 kg (9260 lb)

STS USA	56.14 m (184.2 ft) 2,040,000 kg (4,500,000 lb)	1st: 2 SRB + 3 SSME 2nd: 3 SSME 3rd: 2 OMS	• **SRB** - Solid: ammonium perclorate - F_{thrust} = 11.79×10^6 N (2.65×10^6 lb) - I_{sp} = 267 s (vac) • **SSME** - Liquid: LOX/LH_2 - F_{thrust} = 1.67×10^6 N (375,000 lb)(sl) = 2.10×10^6 N (470,000 lb)(vac) - I_{sp} = 363 s (sl) = 455 s (vac) • **OMS** - Liquid: MMH/N_2O_4 - F_{thrust} = 26,700 N (6000 lb) (vac) - I_{sp} = 313 s (vac)	LEO: 24,400 kg (53,700 lb) GTO: 5900 kg (13,000 lb) + PKM

SRB = solid-rocket boosters
SSME = Space Shuttle main engines
OMS = orbital maneuvering system engines

LEO = low-Earth orbit
GTO = geosynchronous transfer orbit
PKM = perigee kick motor

Table E-1. Booster Systems. (Continued) [Isakowitz, Steven J., *International Reference Guide to Space Launch Systems.* American Institute of Aeronautics and Astronautics. 1991.]

System	Length/Mass at Lift-Off	Stages	Engines	Payloads
Titan IV **USA**	62.2 m (204 ft) 860,000 kg (1,900,000 lb)	0: SRM 1st: 2 LR87 engines 2nd: 1 LR91 engine	• **SRM** - Solid: ammonium perchlorate - $F_{thrust} = 7.0 \times 10^6\,N$ (1.6×10^6 lb) (vac) - $I_{sp} = 271$ s (sl) • **LR87** - Liquid: N_2O_4/aerozine 50 - $F_{thrust} = 2.41 \times 10^6\,N$ (548,000 lb) (vac) - $I_{sp} = 302$ s (vac) • **LR91** - Liquid: N_2O_4/aerozine - $F_{thrust} = 462,000\,N$ (105,000 lb) (vac) - $I_{sp} = 316$ s (vac)	LEO: 17,700 kg (39,000 lb) GEO: 4450 kg (10,000 lb)
Delta II **7925** **USA**	38.1 m (125 ft) 230,000 kg (506,000 lb)	1st: 9 SRM + 1 RS-27A engine 2nd: 1 AJ-10-118K engine 3rd: 1 PAM-D motor	• **SRM** - Solid: HTPB - $F_{thrust} = 435,000\,N$ (98,900 lb) (sl) = 487,600\,N (110,800 lb) (vac) - $I_{sp} = 245$ s (sl) = 273 s (vac) • **RS-27A** - Liquid: LOX/RP-1 - $F_{thrust} = 1.043 \times 10^6\,N$ (237,000 lb) (sl) = 884,000\,N (201,000 lb) (vac) - $I_{sp} = 255$ s (sl) = 301 s (vac) • **AJ-10-118K** - Liquid: N_2O_4/A50 - $F_{thrust} = 42,430\,N$ (9645 lb) (vac) - $I_{sp} = 319$ s (vac) • **PAM-D** - Solid: HTPB - $F_{thrust} = 66.440\,N$ (15,100 lb) (vac) - $I_{sp} = 292$ s (vac)	GTO: 1820 kg (4010 lb) LEO: 5045 kg (11,100 lb)

* SRM = solid-rocket motors
LEO = low-Earth orbit
GEO = geosynchronous orbit
GTO = geosynchronous transfer orbit

Table E-1. Booster Systems. (Continued) [Isakowitz, Steven J., *International Reference Guide to Space Launch Systems.* American Institute of Aeronautics and Astronautics. 1991.]

System	Length/Mass at Lift-Off	Stages	Engines	Payloads

Soyuz SL-4 Russia/ USSR	49.3 m (162 ft)
	290,000 kg (639,000 lb)
		1st: 4 strap-on RD107 + 1 sustainer core RD108
		
		2nd: 4 RD-461
			• **RD107**
			 - Liquid: LOX/kerosene
			 - $F_{thrust} = 821{,}000\,N\,(185{,}000\,lb)\,(sl)$
			   $= 1 \times 10^6\,N$ (225,000 lb) (vac)
			 - $I_{sp} = 257$ s (sl) = 314 s (vac)
			• **RD108**
			 - Liquid: LOX/kerosene
			 - $F_{thrust} = 745{,}000\,N$ (167,000 lb) (sl)
			   $= 941{,}000\,N$ (211,000 lb) (vac)
			 - $I_{sp} = 248$ s (sl) = 315 s (vac)
			• **RD-461**
			 - Liquid: LOX/kerosene
			 - $F_{thrust} = 298{,}000$ N (67,000 lb) (vac)
			 - $I_{sp} = 330$ s (vac)
				LEO: 7000 kg (15,400 lb)

Proton SL-13 Russia/ USSR	60 m (197 ft) 705,000 kg (1,550,000 lb)	1st: 6 RD-253 engines 2nd: 4 RD-*engines 3rd: 1 DR-* engine	• **RD-253** - Liquid: N_2O_4/UDMH - $F_{thrust} = 1.47 \times 10^6$ N (331,000 lb) $(sl) = 1.635 \times 10^6$ N (368,000 lb) (vac) - $I_{sp} = 285$ s (sl) = 316 s (vac) • **RD-* 2nd stage** - Liquid: N_2O_4/UDMH - $F_{thrust} = 600{,}000\,N$ (135,000 lb) (vac) - $I_{sp} = 316$ s (vac) • **RD-* 3rd stage** - Liquid: N_2O_4/UDMH - $F_{thrust} = 600{,}000\,N$ (135,000 lb) (vac) - $I_{sp} = 316$ s (vac)	LEO: 20,000 kg (44,100 lb)

* = designation unknown
LEO = low-Earth orbit

Table E-1. Booster Systems.(Continued) [Isakowitz, Steven J., *International Reference Guide to Space Launch Systems.* American Institute of Aeronautics and Astronautics. 1991.]

System	Length/Mass at Lift-Off	Stages	Engines	Payloads

Long March CZ-2E China	51.2 m (168 ft)
	464,000 kg
	$(1.023 \times 10^6$ lb)
		1st: 4 liquid strap-on YF-20 + 4 YF-20 core
		2nd: 1 YF-22 engine
			• **YF-20 strap-on/core**
			  - Liquid: UDMH/N_2O_4
			  - F_{thrust} = 741,000 N (166,000 lb) (sl)
			  - I_{sp} = 289 s (vac) = 259 s (sl)
			• **YF-22**
			  - Liquid: UDMH/N_2O_4
			  - F_{thrust} = 720,000 N (162,000 lb) (vac)
			  - I_{sp} = 296 s (vac)
				LEO: 9265 kg
				(20,430 lb)
				GTO: 3370 kg
				(7430 lb) +
				PKM

H-1 Japan	40.3 m (132 ft)
	140,000 kg
	(308,000 lb)
		1st: 9 SRM + 1 MB-3 block III engine
		2nd: 1 LE-5 engine
			• **SRM**
			  - Solid: CTPB
			  - F_{thrust} = 220,000 N (49,600 lb) (sl)
			  - I_{sp} = 235 s (sl)
			• **MB-3 Block III**
			  - Liquid: LOX/RJ-1
			  - F_{thrust} = 756,000 N (170,000 lb) (sl)
			  - I_{sp} = 253 s (sl)
			• **LE-5**
			  - Liquid: LOX/LH2
			  - F_{thrust} = 103,000 N (23,150 lb) (vac)
			  - I_{sp} = 447.8 s (vac)
				LEO: 3200 kg
				(7000 lb)
				GTO: 1100 kg
				(2400 lb)
				GEO: 550 kg
				(1200 lb)

LEO = low-Earth orbit
GTO = geosynchronous transfer orbit
GEO = geosynchronous orbit
PKM = perigee kick motor

Motion of Ballistic Vehicles

F.1 Equation of Motion

Ballistic trajectories are the paths followed by nonthrusting objects, such as baseballs or intercontinental ballistic missiles (ICBMs), moving under the influence of gravity. We can use the geocentric-equatorial coordinate system to describe this motion. The equation of motion is

$$\ddot{\vec{R}} + \frac{\mu}{R^2}\hat{R} = 0 \tag{F-1}$$

using three assumptions:

- Most of the trajectory is outside the Earth's atmosphere: $\hat{F}_{Drag} = 0$

- Start time at burnout: $\hat{F}_{Thrust} = 0$

- Other forces are negligible compared to gravity: $\hat{F}_{Other} = 0$

A ballistic trajectory, like an orbit, is defined by six initial conditions (IC):

- Radius at burnout, $R_{burnout}$
- Velocity at burnout, $V_{burnout}$
- Flight-path angle at burnout, $\phi_{burnout}$
- Azimuth angle at burnout, $\beta_{burnout}$
- Latitude at burnout, $L_{burnout}$
- Longitude at burnout, $l_{burnout}$

The shape of a ballistic trajectory is an ellipse which intersects the Earth's surface at two points—launch and impact, as shown in Figure F-1.

Figure F-1. Basic Trajectory for an ICBM.

Ground-Track Geometry

The ground track of a ballistic trajectory is the arc of a great circle. To determine this range angle, Λ, from the launcher to the target, we start with L_o and l_o, the latitude and longitude of the launcher, and L_T and l_T, the latitude and longitude of the target. Then, using spherical trigonometry as shown in Figure F-2, we get

$$\cos \Lambda = \sin L_o \sin L_T + \cos L_o \cos L_T \cos \Delta l \qquad \text{(F-2)}$$

where
$$\Delta l = l_T - l_o$$

Note that this assumes the Earth isn't rotating. This equation gives two values for the range angle, Λ. The smaller one, Λ, is the short way around the Earth; the larger value, $360° - \Lambda$, is for the long way. To convert from a range angle in degrees to range in kilometers, multiply by $10{,}000 \text{ km}/90°$. The range-angle geometry is shown in Figure F-2.

Figure F-2. Ballistic Trajectories. To visualize ballistic trajectories, it's helpful to slice open the Earth like an apple revealing a launch site (launch site latitude, L_o, target latitude, L_T, and the range angle, Λ. The range, R, is traced over the Earth's surface.

One of the initial conditions to locate the trajectory is the burnout azimuth angle. This angle is measured clockwise from true north to the trajectory as shown in Figure F-3.

$$\cos \beta = \frac{\sin L_T - \sin L_o \cos \Lambda}{\cos L_o \sin \Lambda} \qquad \text{(F-3)}$$

A polar plot of the trajectory is useful to determine the correct quadrant for the azimuth angle as shown in Figure F-4.

Figure F-3. Launch Azimuth. Launch azimuth, β, is the angle measured from true north at the launch site clockwise to the launch direction.

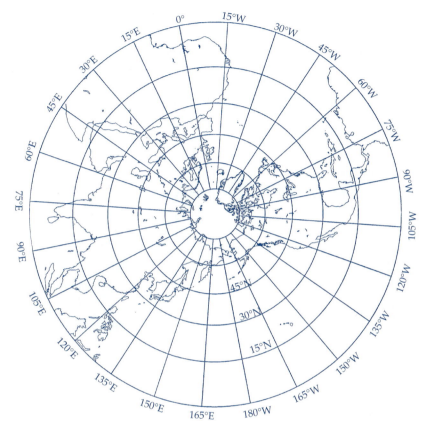

Figure F-4. Polar Plot. Polar plots help us visualize the ground tracks of ballistic objects.

Trajectory Geometry

Define a trajectory parameter

$$Q_{burnout} = \frac{V_{burnout}^2}{V_{circular}^2} = \frac{V_{burnout}^2 R_{burnout}}{\mu} \qquad (F\text{-}4)$$

- $Q_{burnout} < 1.0$: This restricts the booster to go only the short way to a target. Because most ballistic rockets use the short way to a target, they need a $Q_{burnout}$ less than one.

- $Q_{burnout} \geq 1.0$: This implies $V_{burnout} \geq V_{circular}$, which means the rocket can place a payload into orbit at a radius, $R_{burnout}$. This also means the booster can reach any point on the Earth using either the short or long way.

- $Q_{burnout} = 2.0$: This implies $V_{burnout} = \sqrt{\dfrac{2\mu}{R_{burnout}}}$ This is the escape velocity for the Earth. A booster with $Q_{burnout}$ 2.0 would leave on an

useless for getting from one point on the Earth to another.

Another of the angles describing the trajectory is the flight-path angle, ϕ, defined as the angle between the local horizon and the velocity vector. Based on the value of the trajectory parameter, $Q_{burnout}$, the ballistic vehicle can follow either a high or low path, as Figure F-5 shows. The following figures show the various paths available

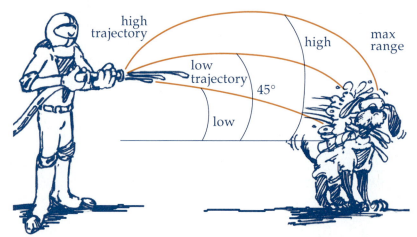

Figure F-5. Flight-Path Angle and Trajectory. Whether you're squirting a hose or launching a missile, the effect of flight-path angle is the same. Maximum range is achieved with a flight-path angle of 45° (for very short trajectories). Two other angles will get you to the same spot— a low, direct trajectory or a high, arcing trajectory.

To solve for the flight-path angle

$$\phi_{burnout_{low}} = \frac{1}{2} \{ \sin^{-1} \left[\left(\frac{2 - Q_{burnout}}{Q_{burnout}} \right) \sin \frac{\Lambda}{2} \right] - \frac{\Lambda}{2} \} \tag{F-5}$$

This angle results in a low, direct trajectory.

$$\phi_{burnout_{high}} = \frac{1}{2} \{ 180° - \sin^{-1} \left[\left(\frac{2 - Q_{burnout}}{Q_{burnout}} \right) \sin \frac{\Lambda}{2} \right] - \frac{\Lambda}{2} \} \tag{F-6}$$

This angle gives a high, arcing trajectory.

Maximum Range

For a specified value of the trajectory parameter, we can determine the maximum range achievable for that ballistic vehicle. Given the trajectory parameter, $Q_{burnout}$

$$\Lambda_{max} = 2 \sin^{-1} \left(\frac{Q_{burnout}}{2 - Q_{burnout}} \right) \tag{F-7}$$

To find the flight-path angle for launch to achieve the maximum range angle, Λ_{max}

$$\left(\phi_{\text{burnout}}\right)_{\text{max range}} = 45° - \frac{\Lambda_{\text{max}}}{4} \qquad \text{(F-8)}$$

Similarly, to avoid over-designing a missile, we can solve for the minimum value of Q_{burnout} needed to reach some range angle, Λ

$$Q_{\text{burnout}_{\text{min}}} = \frac{2\sin\dfrac{\Lambda}{2}}{1 + \sin\dfrac{\Lambda}{2}} \qquad \text{(F-9)}$$

Time of Flight

Time of flight can be determined in two ways. The first way was previously discussed in Chapter 8. This involves using the definition of the ballistic trajectory as an elliptical path and solving Kepler's equation. The second method uses two charts based on these equations.

Figure F-6 shows a chart that relates the ratio of time of flight (TOF) to the period of a circular orbit and to the total range angle, Λ. The graph also contains lines for the trajectory parameter, Q_{burnout}, and for the flight-path angle, ϕ.

To find the TOF for the trajectory you must first compute the range angle, Λ, and have the value for the radius at burnout, R_{burnout}. Looking at Figure F-6, you can see that the vertical axis is a ratio, TOF/P_{circular}. We

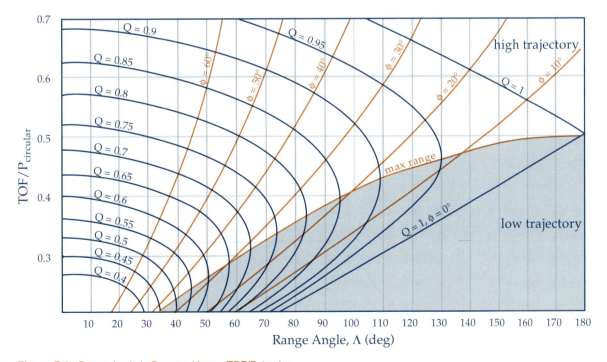

Figure F-6. Range Angle in Degrees Versus TOF/Pcircular.

earlier defined $Q_{burnout}$ as the ratio of the square of $V_{burnout}$ to the square of $V_{circular}$. Similarly, we set up a ratio of the TOF of a trajectory to the period of a circular orbit at that radius of burnout. Let's step through how we use this chart to find TOF:

- Find the value of Λ on the horizontal axis. Move vertically until you intersect the given value of $Q_{burnout}$ for the problem. The values for $Q_{burnout}$ are in increments of 0.05, so if your value is between curves, you must estimate.

- Find the intersection with the appropriate $Q_{burnout}$ curve to get two possibilities—a high and low trajectory. The one above the max range line is the high path, and the one below is the low path.

- Estimate the value for flight-path angle from the lines for ϕ. These lines are in 10° increments, so you may need to interpolate.

- Move to the left of the intersection of the range-angle value and the $Q_{burnout}$ curve to find the value of the ratio $TOF/P_{circular}$.

- Find the appropriate value of the circular orbit period at $R_{burnout}$ by using the equation for period

$$P_{circular} = 2\pi \sqrt{\frac{R_{burnout}^3}{\mu}}$$

Be careful of units. The value of $P_{circular}$ needs to be in *minutes*.

- Multiply the ratio by the circular orbit period to find

$$TOF = \left(\frac{TOF}{P_{circular}}\right) P_{circular}$$

Recall that all trajectories with $Q_{burnout} < 1.0$ have two options: a high trajectory and a low trajectory. We must solve each case separately for the time of flight.

Rotating-Earth Correction

The rotation of the Earth at 15°/hr has the following effect on trajectories

- Eastward launches
  - Target moves away from the launcher
  - Range angle, Λ, *increases* (from nonrotating solution)
  - Flight-path angle, ϕ, must *increase* on the *low* trajectory and *decrease* on the *high* trajectory
- Westward launches
  - Target moves toward the launcher
  - Range angle, Λ, *decreases* (from nonrotating solution)
  - Flight-path angle, ϕ, must *decrease* on the *low* trajectory and *increase* on the *high* trajectory

To account for the rotation we adjust the range-angle equation to

$$\cos \Lambda = \sin L_o \sin L_T + \cos L_o \cos L_T \cos (\Delta l + \omega_{Earth} TOF) \qquad \text{(F-10)}$$

We can't solve directly for the range angle, Λ, so we must start with a "guess" and then iterate until we reach a solution. We're given L_o and l_o, as well as L_T and l_T. To find Λ, β, ϕ, we must go through the following algorithm:

- Solve for the nonrotating range angle

$$\cos \Lambda = \sin L_o \sin L_T + \cos L_o \cos L_T \cos\Delta l$$

- Find the time of flight, TOF, with $\Lambda_{nonrotating}$ by using the chart in Figure F-6 for *high* trajectory

- Plug this time into the rotating range-angle equation and get a value for $\Lambda_{rotating}$

$$\cos \Lambda = \sin L_o \sin L_T + \cos L_o \cos L_T \cos(\Delta l + \omega_{Earth} TOF)$$

- Compute a new TOF using the new value of $\Lambda_{rotating}$

- Repeat the last two steps until the difference between successive values of range angle is small enough (usually 0.5° to 1°)

- Solve for β and ϕ for the chosen trajectory

- Repeat all of the above for low trajectory

Error Analysis

Errors in any of the six initial conditions for a ballistic trajectory will cause it to miss the target. We categorize how we miss the target in terms of downrange errors (in the direction of motion) and crossrange errors (perpendicular to the direction of motion). We use the following sign convention, as shown in Figure F-7.

- Landing long or to the right of the target is a positive error

- Landing short or to the left of the target is a negative error

Downrange error ($\Delta\Lambda$) has three causes

- Burning out at higher or lower altitude

$$\Delta\Lambda = \left(\frac{\partial \Lambda}{\partial R_{burnout}}\right)\Delta R_{burnout}$$

$$\frac{\partial \Lambda}{\partial R_{burnout}} = \frac{4\mu}{V_{burnout}^2 R_{burnout}^2} \frac{\sin^2 \frac{\Lambda}{2}}{\sin 2\phi} \frac{180°}{\pi} \frac{\deg}{km} \qquad \text{(F-11)}$$

- Burning out at higher or lower velocity

$$\Delta\Lambda = \left(\frac{\partial \Lambda}{\partial V_{burnout}}\right)\Delta V_{burnout}$$

Figure F-7. Conventions for Error Analysis.

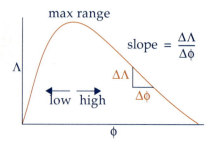

Figure F-8. Plot of Flight-Path Angle Versus Range Angle.

$$\frac{\partial \Lambda}{\partial V_{burnout}} = \frac{8\mu}{(V_{burnout})^3 (R_{burnout})} \frac{\sin^2 \frac{\Lambda}{2}}{\sin 2\phi} \frac{180°}{\pi} \frac{deg}{m/s} \qquad (F-12)$$

- Burning out at a different flight-path angle. This is also shown in

$$\Delta \Lambda = \left(\frac{\partial \Lambda}{\partial \phi_{burnout}} \right) \Delta \phi_{burnout}$$

$$\frac{\partial \Lambda}{\partial \phi_{burnout}} = \frac{2 \sin (2\phi_{burnout} + \Lambda)}{\sin (2\phi_{burnout})} - 2 \frac{deg}{deg} \qquad (F-13)$$

These three ratios showing the change in range angle due to a change in some initial condition are called influence coefficients. Because they are rather complicated to compute, we can use estimates of the influence coefficients called Rule of Thumb values.

Crossrange errors (ΔC) have two causes

- Displacing the launch site left or right of the trajectory

$$\Delta C = (\Delta y_{burnout}) \cos \Lambda \qquad (F-14)$$

- Burning out with a larger or a smaller azimuth angle

$$\Delta C = \Delta \beta \left(111.1 \frac{km}{deg} \right) \sin \Lambda \qquad (F-15)$$

References

Bate, Roger R., Donald D. Mueller, and Jerry E. White. *Fundamentals of Astrodynamics.* New York, N.Y.: Dover Publications, Inc., 1971

Answers to Numerical Mission Problems

Chapter 1
Why Space

N/A

Chapter 2
Exploring Space

N/A

Chapter 3
The Space Environment

N/A

Chapter 4
Understanding Orbits

8) $H = 0.25 \text{ kg} \cdot \text{m}^2/\text{s}$

9) $H = 0.006283 \text{ kg} \cdot \text{m}^2/\text{s}$

10) $F_g = 0.05336 \text{ N}$

11) $g = 9.722 \text{ m/s}^2$

12) $V = 76.669 \text{ m/s}$

   $t = 7.825 \text{ s}$

24)

   a) $a = 7565.5 \text{ km}$

   b) $e = 0.1074$

   c) $R = 8374.926 \text{ km}$

      $\text{Alt} = 1996.926 \text{ km}$

   d) Negative

26)

   b) $\vec{h} = -49,441 \ \hat{K} \ \text{km}^2/\text{s}$

   d) $\varepsilon = -31.999 \text{ km}^2/\text{s}^2$

27) $KE_{\text{truck}} = 0.844 \text{ kg} \cdot \text{km}^2/\text{s}^2$

   $V_{\text{space}} = 7.473 \text{ km/s}$

   $KE_{\text{space}} = 279,249 \text{ kg} \cdot \text{km}^2/\text{s}^2$

28) $\text{Alt} = 35,863.08 \text{ km}$

30)

   a) $\Delta V = 1.66 \text{ km/s}$

   b) $\Delta KE = 8 \text{ km}^2/\text{s}^2$

   c) $\varepsilon_{\text{circle}} = -7.97 \text{ km}^2/\text{s}^2$

Chapter 5
Describing Orbits

4) $\varepsilon = -4.73 \dfrac{\text{km}^2}{\text{s}^2}$

17)

   a) $h = 64,624.02 \dfrac{\text{km}^2}{\text{s}}$

   b) $i = 96.15°$

   c) $\bar{n} = -50,036.88\hat{I} - 40,307.38\hat{J}$ (unitless)

   d) $\Omega = 218.85°$

18)

   a) $\bar{e} = 0.135\hat{I} + 0.092\hat{J} - 0.120\hat{K}$

   b) $\omega = 216.7°$

   c) $v = 319.3°$

23)

   a) $\Delta N = 240°$

   b) $\text{Lat}_{\text{max and min}} = 25°$

24)

   a) $i \approx 50°$

   b) $P = 3 \text{ hrs}$

Chapter 6
Maneuvering in Space

5)

   a) $\varepsilon_{\text{transfer}} = -29.73 \text{ km}^2/\text{s}^2$

   b) $\Delta V_1 = 0.1 \text{ km/s}$

   c) $\Delta V_2 = 0.11 \text{ km/s}$

   d) $TOF = 45.5 \text{ min}$

6) $V_{\text{circular}} = 7.78 \text{ km/s}$

11) $\Delta V_{simple} = 4.45$ km/s

12) $\Delta V_{simple} = 4.71$ km/s

13)

    a) $\varepsilon_{transfer} = -12.11$ km^2/s^2

    b) $\Delta V_1 = 2.08$ km/s

    c) $\Delta V_{combined} = 1.70$ km/s

15)

    a) TOF = 2642.6 s

    b) $\omega_{interceptor} = 0.001205$ rad/s, ω_{target}

       = 0.001173 rad/s

    c) $\alpha_{lead} = 3.10$ rad

    d) $\phi_{final} = 0.042$ rad

    e) wait time = 20.1 hrs

16)

    a) TOF = 5877 s

    b) a = 7038.8 km

    c) $\Delta V_1 = 0.23$ km/s

Chapter 7
Interplanetary Travel

8) $\Delta V = 2.32$ km/s

9)

    a) $a_{transfer} = 1.289 \times 10^8$ km

       $\varepsilon_{transfer} = -514.7$ km^2/s^2

    b) $V_{\infty\ Earth} = 2.49$ km/s

    c) $V_{\infty\ Venus} = 2.72$ km/s

    d) $\Delta V_{boost} = 3.5$ km/s

    e) $\Delta V_{retro} = 3.32$ km/s

    f) $\Delta V_{mission} = 6.82$ km/s

10) TOF = 146.1 days

11) Radius of SOI = 66,183 km

12) $\phi_{final} = 106°$

Chapter 8
Predicting Orbits

2) TOF = 1390.09 s = 23.17 min

3) TOF = 20.24 min

4) $\nu_{future} = 155.39°$

5) $\nu_{future} = 159.66°$

9) Time = 816.67 days

10) $\dot{\Omega} \cong -7.2$ deg/day, westward

11) $\nu_{future} = 216.83°$

    $a_{future} = 6876$ km

    $e_{future} = 0.11880$

    $i_{future} = 75°$

    $\Omega_{future} = 0°$

    $\omega_{future} = 0°$

Chapter 9
Getting to Orbit

8) LST = 0300 hrs

9) LST = 0320 hrs

10)

    1) One

    2) Two

12)

    a) LWST = 135°

    b) LWST = 0900 hours

    c) Wait 21 hours

15)

    e) $\alpha = 30°$

    f) $\gamma = 60.24°$

    g) $\delta = 6.91°$

    h) $LWST_{AN} = 0928$ hours

    j) $\beta_{AN} = 60.24°$

    k) $LWST_{AN} = 2032$ hours

    l) 6 hours and 2 minutes

    m) $\beta_{DN} = 119.76°$

16) Launch at 2036 $\beta_{AN} = 46.17°$

17)

    a) 13°

    b) 15 hours, 21 minutes, and 36 seconds

21) 408.7 m/s

24)

    a) $\left|\vec{V}_{burnout}\right| = 7669$ m/s

    b) $\vec{V}_{launch\ site} = 408.7$ m/s $\hat{E}$

    c) $\vec{V}_{needed} = \begin{bmatrix} 0 \\ 7260.3 \\ 0 \end{bmatrix}$ m/s

    d) $\left|\vec{V}_{needed}\right| = 7260.3$ m/s

    e) $\Delta V_{design} = 8260.3$ m/s

25) $\Delta V_{design} = 7949$ m/s

Chapter 10
Returning From Space: Entry

12) 24.813 g's, alt = 28,251 m

13) 20,348 m

19) 118,391.02 W/m^2

22) 0.34 g's

Chapter 11
Payload and Spacecraft Design

8) 6×10^{14} Hz

13) $\lambda_m = 3.574$ μm

    $E = 2.452 \times 10^4$ W/m^2

20) $\theta = 60.85°$

21) $h = 1.64 \times 10^4$ km

Chapter 12
Space Vehicle Control Systems

9) $\Delta H = 2.0$ Nm

    $\alpha = 0.002$ rad/s^2

24) $a_g = -9.6765$ m/s^2

27) $\vec{V}_{to\ go} = \begin{bmatrix} 35.11 \\ 7414.02 \\ 14.62 \end{bmatrix}_{SEZ}$ m/s

Chapter 13
Spacecraft Subsystems

6) $V_R = 300$ V

7) $i = 17.86$ A

8) $P_{OUT} = 157.41$ W/m^2

12) $TOE_{max} = 36.3$ min

13) 31.75 amp · hrs

19) $q = -4.94$ W

20) $q = -9.69$ W

21) $T = 213.43°$ K $= -59.57°$ C

22) $T = 203.37°$ K $= -69.6°$ C

33) $f_{natural} = 0.0159$ Hz

34) $K = 24.67 \times 10^6$ N/m

35) $\sigma = 127,389$ N/m^2

36) $\varepsilon = 0.0001$

37) $M = 1000$ N · m

38) $\varepsilon = 0.2, \varepsilon = 0.2$

40) $\varepsilon = 2 \times 10^{-9}$

Chapter 14
Space Transportation
Systems

8) $\Delta V = 20.10$ m/s

9) $m_{propellant} = 9.95$ kg

13) Savings of 111.2 kg

23) 1.667 years on Earth

25)
   a) 151,500 kg
   b) 101,500 kg
   c) 1178.75 m/s
   d) 91,500 kg
   e) 51,500 kg
   f) 1973.43 m/s
   g) 44,000 kg
   h) 9000 kg
   i) 6227.25 m/s
   j) 9379.43 m/s

26) $m_{i_{tot}} = 71,644.7$ kg

27) Yes, because $\Delta V_{tot} > V_{cir\ 200\ km}$

28) $F_{thrust} = 5.15$ N

Chapter 15
Space Operations and
Communication Networks

15) 7.958×10^{-4} W/m^2

16) $G = 48.85$

21) $\dfrac{S}{N} = 6.07 \times 10^{11}$

Chapter 16
Business of Space

N/A

Prepared by George E. Irvin, Jr. and Robert B. Giffen

Index

—T—